NEUROMETHODS

Series Editor
Wolfgang Walz
University of Saskatchewan
Saskatoon, Canada

For further volumes:
http://www.springer.com/series/7657

Neuromethods publishes cutting-edge methods and protocols in all areas of neuroscience as well as translational neurological and mental research. Each volume in the series offers tested laboratory protocols, step-by-step methods for reproducible lab experiments and addresses methodological controversies and pitfalls in order to aid neuroscientists in experimentation. *Neuromethods* focuses on traditional and emerging topics with wide-ranging implications to brain function, such as electrophysiology, neuroimaging, behavioral analysis, genomics, neurodegeneration, translational research and clinical trials. *Neuromethods* provides investigators and trainees with highly useful compendiums of key strategies and approaches for successful research in animal and human brain function including translational "bench to bedside" approaches to mental and neurological diseases.

Optogenetic Functional Neuroimaging

Edited by

Kevin Chuen Wing Chan

Neuroimaging and Visual Science Laboratory, Department of Ophthalmology, Stanford University School of Medicine, Palo Alto, CA, USA

Seong-Gi Kim

Center for Neuroscience Imaging Research, Sungkyunkwan University, Suwon, Korea (Republic of)

Humana Press

Editors
Kevin Chuen Wing Chan
Neuroimaging and Visual Science Laboratory
Department of Ophthalmology
Stanford University School of Medicine
Palo Alto, CA, USA

Seong-Gi Kim
Center for Neuroscience Imaging Research
Sungkyunkwan University
Suwon, Korea (Republic of)

ISSN 0893-2336 ISSN 1940-6045 (electronic)
Neuromethods
ISBN 978-1-0716-5177-3 ISBN 978-1-0716-5178-0 (eBook)
https://doi.org/10.1007/978-1-0716-5178-0

This Humana imprint is published by the registered company Springer Science+Business Media, LLC, part of Springer Nature.
The registered company address is: 1 New York Plaza, New York, NY 10004, U.S.A.

Preface to the Series

Experimental life sciences have two basic foundations: concepts and tools. The *Neuromethods* series focuses on the tools and techniques unique to the investigation of the nervous system and excitable cells. It will not, however, shortchange the concept side of things as care has been taken to integrate these tools within the context of the concepts and questions under investigation. In this way, the series is unique in that it not only collects protocols but also includes theoretical background information and critiques which led to the methods and their development. Thus it gives the reader a better understanding of the origin of the techniques and their potential future development. The *Neuromethods* publishing program strikes a balance between recent and exciting developments like those concerning new animal models of disease, imaging, in vivo methods, and more established techniques, including, for example, immunocytochemistry and electrophysiological technologies. New trainees in neurosciences still need a sound footing in these older methods in order to apply a critical approach to their results.

Under the guidance of its founders, Alan Boulton and Glen Baker, the *Neuromethods* series has been a success since its first volume published through Humana Press in 1985. The series continues to flourish through many changes over the years. It is now published under the umbrella of Springer Protocols. While methods involving brain research have changed a lot since the series started, the publishing environment and technology have changed even more radically. Neuromethods has the distinct layout and style of the Springer Protocols program, designed specifically for readability and ease of reference in a laboratory setting.

The careful application of methods is potentially the most important step in the process of scientific inquiry. In the past, new methodologies led the way in developing new disciplines in the biological and medical sciences. For example, Physiology emerged out of Anatomy in the 19th century by harnessing new methods based on the newly discovered phenomenon of electricity. Nowadays, the relationships between disciplines and methods are more complex. Methods are now widely shared between disciplines and research areas. New developments in electronic publishing make it possible for scientists that encounter new methods to quickly find sources of information electronically. The design of individual volumes and chapters in this series takes this new access technology into account. Springer Protocols makes it possible to download single protocols separately. In addition, Springer makes its print-on-demand technology available globally. A print copy can therefore be acquired quickly and for a competitive price anywhere in the world.

Saskatoon, Canada ***Wolfgang Walz***

Preface

Optogenetics has proven to be a highly useful tool to delineate the functions of distinct cellular groups or proteins. When combined with neuroimaging, it allows the investigation of neuronal dynamics upon selective manipulation of neural systems at multiple scales. Optogenetic methods for functional brain mapping have been applied in experiments ranging from in vitro investigation of microcircuits, to in vivo probing of inter-regional cortical and subcortical connections, to examination of global connections within the whole brain in both awake and resting states. This volume provides an overview of the optogenetic functional neuroimaging methodologies and their applications to the study of neural mechanisms from cells to circuits in intact and dysfunctional brains. The chapters in this book are organized into two parts detailing (1) multi-modal imaging methodologies for optogenetic dissection of brain functions and (2) applications for multi-scale optogenetic functional assessments of neural mechanisms. In the *Neuromethods* series style, chapters include the kind of detail and key advice from specialists needed to achieve successful results in researchers' laboratories. The authoritative and cutting-edge book *Optogenetic Functional Neuroimaging: Methodologies & Applications* aims to serve as a useful practical guide for researchers to further their studies in this field.

Stanford, CA, USA — ***Kevin Chuen Wing Chan***
Suwon, Korea (Republic of) — ***Seong-Gi Kim***

Contents

Contributors

NMACHI ANUMBA • *Wallace H. Coulter Department of Biomedical Engineering, Georgia Institute of Technology and Emory University, Atlanta, GA, USA*

LAUREN N. BELOATE • *Department of Biomedical Engineering, The Pennsylvania State University, University Park, PA, USA*

RUSSELL W. CHAN • *E-SENSE Innovation and Technology Limited, Hong Kong, China; School of Biomedical Engineering, The University of Hong Kong, Hong Kong, China*

TZU-HAO HARRY CHAO • *Center for Animal MRI, Biomedical Research Imaging Center, University of North Carolina at Chapel Hill, Chapel Hill, NC, USA; Department of Neurology, University of North Carolina at Chapel Hill, Chapel Hill, NC, USA*

LI MIN CHEN • *Vanderbilt University Institute of Imaging Science, Vanderbilt University Medical Center, Nashville, TN, USA; Department of Radiology and Radiological Sciences, Vanderbilt University Medical Center, Nashville, TN, USA; Department of Biomedical Engineering, Vanderbilt University, Nashville, TN, USA*

MYUNGHWAN CHOI • *School of Biological Sciences, Seoul National University, Seoul, Republic of Korea; The Institute of Molecular Biology and Genetics, Seoul, Republic of Korea*

KAI-HSIANG CHUANG • *School of Biomedical Sciences, The University of Queensland, Brisbane, QLD, Australia*

LAUREN DALEY • *Wallace H. Coulter Department of Biomedical Engineering, Georgia Institute of Technology and Emory University, Atlanta, GA, USA*

BRADLEY J. EDELMAN • *Emotion Research Department, Max Planck Institute for Psychiatry, Munich, Germany; Brain-Wide Circuits for Behavior Research Group, Max Planck Institute of Biological Intelligence, Planegg, Germany*

CORNELIUS FABER • *Clinic of Radiology, University of Münster, Münster, Germany*

MITSUHIRO FUKUDA • *Department of Radiology, University of Pittsburgh, Pittsburgh, PA, USA*

SOUPARNO GHOSH • *Expedition Medicines, Cambridge, MA, USA*

JOHN C. GORE • *Vanderbilt University Institute of Imaging Science, Vanderbilt University Medical Center, Nashville, TN, USA; Department of Radiology and Radiological Sciences, Vanderbilt University Medical Center, Nashville, TN, USA; Department of Biomedical Engineering, Vanderbilt University, Nashville, TN, USA; Department of Physics and Astronomy, Vanderbilt University, Nashville, TN, USA; Department of Molecular Physiology and Biophysics, Vanderbilt University, Nashville, TN, USA*

XU HAN • *Department of Biomedical Engineering, The Pennsylvania State University, University Park, PA, USA*

YI HE • *State Key Laboratory of Biomedical Imaging Science and System, Chinese Academy of Sciences, Shenzhen, Guangdong, China; Shenzhen University of Advanced Technology, Shenzhen, Guangdong, China; Paul C. Lauterbur Research Center for Biomedical Imaging, Shenzhen Institute of Advanced Technology, Chinese Academy of Sciences, Shenzhen, Guangdong, China; National Innovation Center for Advanced Medical Devices, Shenzhen, Guangdong, China*

LI-MING HSU • *Center for Animal MRI, Biomedical Research Imaging Center, University of North Carolina at Chapel Hill, Chapel Hill, NC, USA; Department of Neurology, University of North Carolina at Chapel Hill, Chapel Hill, NC, USA*
WON BEOM JUNG • *Emotion, Cognition & Behavior Research Group, Korea Brain Research Institute (KBRI), Daegu, Republic of Korea*
NATHALIE JUST • *Laboratory for Functional and Metabolic Imaging (LIFMET), Ecole Polytechnique Fédérale de Lausanne (EPFL), Bâtiment CH, Lausanne, Switzerland*
SHELLA KEILHOLZ • *Wallace H. Coulter Department of Biomedical Engineering, Georgia Institute of Technology and Emory University, Atlanta, GA, USA*
MICHAEL A. KELBERMAN • *Department of Human Genetics, Emory University, Atlanta, GA, USA; Department of Molecular, Cellular & Developmental Biology, University of Colorado Boulder, Boulder, CO, USA*
SEONG-GI KIM • *Center for Neuroscience Imaging Research, Institute for Basic Science, Suwon, Republic of Korea; Department of Biomedical Engineering, Sungkyunkwan University, Suwon, Republic of Korea*
SEONGHOON KIM • *Tsinghua University, Institute for Brain and Cognitive Sciences, Beijing, China; Hangzhou Zhuoxi Institute of Brain and Intelligence, Hangzhou, China*
NAN LI • *Advanced Imaging Research Center, UT Southwestern Medical Center, Dallas, TX, USA; Department of Neuroscience, UT Southwestern Medical Center, Dallas, TX, USA; Peter O'Donnell Jr. Brain Institute, UT Southwestern Medical Center, Dallas, TX, USA*
MINGYAO LIANG • *State Key Laboratory of Biomedical Imaging Science and System, Chinese Academy of Sciences, Shenzhen, Guangdong, China; Shenzhen University of Advanced Technology, Shenzhen, Guangdong, China; Paul C. Lauterbur Research Center for Biomedical Imaging, Shenzhen Institute of Advanced Technology, Chinese Academy of Sciences, Shenzhen, Guangdong, China; National Innovation Center for Advanced Medical Devices, Shenzhen, Guangdong, China*
ZHIFENG LIANG • *Institute of Neuroscience, Center for Excellence in Brain Science and Intelligence Technology, Chinese Academy of Sciences, Shanghai, China*
ALEXIA MARRIOTT • *Department of Human Genetics, Emory University, Atlanta, GA, USA*
HYUN SEOK MOON • *Center for Neuroscience Imaging Research, Institute for Basic Science, Suwon, Republic of Korea; Department of Biomedical Engineering, Sungkyunkwan University, Suwon, Republic of Korea*
RANDY NONNEMAN • *Center for Animal MRI, Biomedical Research Imaging Center, University of North Carolina at Chapel Hill, Chapel Hill, NC, USA; Department of Neurology, University of North Carolina at Chapel Hill, Chapel Hill, NC, USA*
WEN-JU PAN • *Wallace H. Coulter Department of Biomedical Engineering, Georgia Institute of Technology and Emory University, Atlanta, GA, USA*
BRUNO PRADIER • *Clinic of Radiology, University of Münster, Münster, Germany*
ADIYA RAKYMZHAN • *Department of Bioengineering, University of Pittsburgh, Pittsburgh, PA, USA; Center for the Neural Basis of Cognition, University of Pittsburgh and Carnegie Mellon University, Pittsburgh, PA, USA*
YEN-YU IAN SHIH • *Center for Animal MRI, Biomedical Research Imaging Center, University of North Carolina at Chapel Hill, Chapel Hill, NC, USA; Department of Neurology, University of North Carolina at Chapel Hill, Chapel Hill, NC, USA*

SHENG SONG • *Center for Animal MRI, Biomedical Research Imaging Center, University of North Carolina at Chapel Hill, Chapel Hill, NC, USA; Department of Neurology, University of North Carolina at Chapel Hill, Chapel Hill, NC, USA*
ANDREW E. TOADER • *College of Medicine, Lake Erie College of Osteopathic Medicine, Pittsburgh, PA, USA; Department of Radiology, University of Pittsburgh, Pittsburgh, PA, USA*
CHUANJUN TONG • *Institute of Neuroscience, Center for Excellence in Brain Science and Intelligence Technology, Chinese Academy of Sciences, Shanghai, China*
ALBERTO L. VAZQUEZ • *Department of Bioengineering, University of Pittsburgh, Pittsburgh, PA, USA; Center for the Neural Basis of Cognition, University of Pittsburgh and Carnegie Mellon University, Pittsburgh, PA, USA; Department of Radiology, University of Pittsburgh, Pittsburgh, PA, USA; Center for Neuroscience, University of Pittsburgh, Pittsburgh, PA, USA*
LYDIA WACHSMUTH • *Clinic of Radiology, University of Münster, Münster, Germany*
FENG WANG • *Vanderbilt University Institute of Imaging Science, Vanderbilt University Medical Center, Nashville, TN, USA; Department of Radiology and Radiological Sciences, Vanderbilt University Medical Center, Nashville, TN, USA*
TZU-WEN WINNIE WANG • *Center for Animal MRI, Biomedical Research Imaging Center, University of North Carolina at Chapel Hill, Chapel Hill, NC, USA; Department of Neurology, University of North Carolina at Chapel Hill, Chapel Hill, NC, USA*
XUNDA WANG • *Department of Neurosurgery, Stanford University School of Medicine, Stanford, CA, USA*
DAVID WEINSHENKER • *Department of Human Genetics, Emory University, Atlanta, GA, USA*
ZEPING XIE • *State Key Laboratory of Biomedical Imaging Science and System, Chinese Academy of Sciences, Shenzhen, Guangdong, China; Shenzhen University of Advanced Technology, Shenzhen, Guangdong, China; Paul C. Lauterbur Research Center for Biomedical Imaging, Shenzhen Institute of Advanced Technology, Chinese Academy of Sciences, Shenzhen, Guangdong, China; National Innovation Center for Advanced Medical Devices, Shenzhen, Guangdong, China*
PAI-FENG YANG • *Vanderbilt University Institute of Imaging Science, Vanderbilt University Medical Center, Nashville, TN, USA; Department of Radiology and Radiological Sciences, Vanderbilt University Medical Center, Nashville, TN, USA*
RUI YANG • *Advanced Imaging Research Center, UT Southwestern Medical Center, Dallas, TX, USA*
ZHANGYAN YANG • *Vanderbilt University Institute of Imaging Science, Vanderbilt University Medical Center, Nashville, TN, USA; Department of Biomedical Engineering, Vanderbilt University, Nashville, TN, USA*
XIN YU • *Athinoula A. Martinos Center for Biomedical Imaging, Massachusetts General Hospital and Harvard Medical School, Charlestown, MA, USA*
JIANYU YUAN • *State Key Laboratory of Biomedical Imaging Science and System, Chinese Academy of Sciences, Shenzhen, Guangdong, China; Shenzhen University of Advanced Technology, Shenzhen, Guangdong, China; Paul C. Lauterbur Research Center for Biomedical Imaging, Shenzhen Institute of Advanced Technology, Chinese Academy of Sciences, Shenzhen, Guangdong, China; National Innovation Center for Advanced Medical Devices, Shenzhen, Guangdong, China*

Kaiwei Zhang • *Institute of Neuroscience, Center for Excellence in Brain Science and Intelligence Technology, Chinese Academy of Sciences, Shanghai, China*
Nanyin Zhang • *Department of Biomedical Engineering, The Pennsylvania State University, University Park, PA, USA; The Huck Institutes of the Life Sciences, The Pennsylvania State University, University Park, PA, USA*
Ning Zheng • *Vanderbilt University Institute of Imaging Science, Vanderbilt University Medical Center, Nashville, TN, USA; Department of Radiology and Radiological Sciences, Vanderbilt University Medical Center, Nashville, TN, USA*
Yijuan Zou • *Changping Laboratory, Beijing, China*

Part I

Multi-modal Imaging Methodologies for Optogenetic Dissection of Brain Functions

Chapter 1

High-Resolution and Multimodal Optogenetic fMRI of Brain Dynamics

Yi He, Jianyu Yuan, Mingyao Liang, Zeping Xie and Xin Yu

Abstract

Functional magnetic resonance imaging (fMRI) has emerged as an essential and highly valuable tool within the neuroscience toolkit, offering noninvasive and high-resolution insights into the functional architecture of the brain. We introduce an advanced line-scanning fMRI protocol that achieves exceptional spatiotemporal fidelity, enabling layer-specific mapping of neuronal circuitry with high precision. Furthermore, single-vessel fMRI is highlighted as a powerful approach for resolving hemodynamic signals at the level of individual microvessels, providing valuable insights into the complex neurovascular coupling mechanisms. These imaging strategies, when combined with optogenetic control and fiber photometry readouts, allow for precise circuit-level manipulation and simultaneous recording, establishing a causal framework for understanding brain function. In conclusion, these cutting-edge fMRI techniques represent a major methodological milestone in functional neuroimaging, opening new avenues for high-resolution and multimodal exploration of the living brain.

Keywords Single-vessel fMRI, Line-scanning fMRI, Fiber photometry, Optogenetics

1 Introduction

Functional Magnetic Resonance Imaging (fMRI) has transformed our understanding of brain function by enabling noninvasive, high-resolution mapping of neural activity in both space and time. Fundamental to this technique is the blood oxygen level-dependent (BOLD) contrast [1], which relies on the magnetic properties of deoxygenated hemoglobin to identify subtle changes in blood oxygenation ensuing from neuronal activity. This physiological manifestation emerges from the intricate interaction among neuronal activation, cerebral blood flow, and oxygen metabolism, constituting a potent indicator for brain function. Consequently, BOLD fMRI captures localized neuronal processes related to the intricate neurovascular coupling mechanisms, which are essential to both normal and pathological brain states [2]. Through detecting these dynamic changes, BOLD fMRI has evolved into an essential means in both clinical and research

Kevin Chuen Wing Chan and Seong-Gi Kim (eds.), *Optogenetic Functional Neuroimaging*, Neuromethods, vol. 232,
https://doi.org/10.1007/978-1-0716-5178-0_1,

arenas, enhancing our understanding of healthy brain function and the pathophysiology of neurological disorders.

Recent advancements in high-field fMRI technology have substantially improved spatial resolution, now enabling it to reach the submillimeter range [3]. Remarkably, the signal-to-noise ratio (SNR) of fMRI signals has been enhanced through the development of implanted radiofrequency (RF) coils or inductively coupled detectors [5, 7]. High-field fMRI has detected layer-specific responses to both bottom-up and top-down processing, indicating laminar specificity aligned with neuronal circuit architecture in both humans and animals. However, conventional fMRI, limited by its spatiotemporal resolution, has been inadequate for investigating the laminar-specific signal correlations and coherence patterns that could shed light on the neuronal circuits involved in functional connectivity (FC).

To address this, we have developed a line-scanning fMRI method that substantially improves both spatial and temporal resolution, enabling detailed profiling across diverse cortical layers in rat brains [8]. This advanced approach has been effectively utilized to explore circuit-specific neuronal activity by capturing hemodynamic responses with high spatiotemporal precision and providing a potent tool for probing the intricate functions and interactions within distinct cortical circuits. Nonetheless, precisely attributing fMRI signals to specific vascular activities remains a major challenge.

Single-vessel fMRI represents a substantial advancement in neuroimaging, affording the capacity to dissect the contributions of individual blood vessels to the overall hemodynamic response. This approach, exhibiting a spatial resolution ranging from 50 to 100 μm in the rodent brain, enables the mapping of cerebral blood flow, volume, and oxygenation alterations in individual penetrating cortical arterioles and surface venules [3, 4]. To achieve this, single-vessel fMRI employs fast low-angle shot (FLASH) with reshuffled k-space or balanced steady-state free precession (bSSFP) methods [11]. These approaches possess a narrowly targeted field-of-view (FOV) engineered to optimize SNR while concurrently maintaining the high spatiotemporal resolution necessary for detailed vascular imaging. Furthermore, a radial encoding-based single-vessel fMRI approach furnishes 50×50 μm resolution and effectively discriminates between intravascular and extravascular effects in microvenules. These advances are redefining our comprehension of brain function and pathology, possessing the potential to augment insights into neural activity and vascular dynamics.

Recently, a 2D fast line-scan DIANA (Direct Imaging of Neuronal Activity) fMRI has been reported to possess the capability of directly mapping spike activity with high temporal (5 ms) and spatial (0.22 mm) resolution, thereby permitting detailed visualization of sequential neuronal activity propagation and temporal dynamics across functionally connected brain regions [12]. This method diminished the repetition time (TR) to 5 ms and implemented a repetitive 5 Hz stimulation paradigm in alignment with the k-space acquisition scheme. Nevertheless, it continues to encounter challenges attributable to the deficiency in repeatability and reliability in capturing direct neuronal events, thereby emphasizing the intricacy of directly correlating fMRI signals with actual neuronal activity [13, 14]. Despite these obstacles, fMRI proffers promising opportunities for advancing our understanding of the neural basis of cognition and behavior.

Consequently, it would be beneficial for future studies to augment our understanding of neurovascular coupling and the precise vascular contributions to fMRI signals. This will entail leveraging high-resolution imaging techniques, integrating multimodal approaches, and employing advanced analytical methods to dissect the complex interplay between neuronal activity and hemodynamic responses. The integration of fMRI with optogenetics and fiber photometry enables the precise manipulation and monitoring of specific neuronal populations while observing the consequent changes in brain activity. Optogenetics provides the ability to activate or inhibit neurons with light, affording precise control over neural circuits, while fiber photometry enables real-time measurement of neuronal activity through fluorescent indicators [15, 16]. Collectively, these methods enhance the understanding of neural circuits' roles in diverse brain states by correlating specific neuronal dynamics with broader brain activity captured by fMRI.

2 Materials

2.1 Animals

All experiments were conducted using Sprague-Dawley rats weighing 250–400 g and aged between 8 and 12 weeks. Animals were housed in standard ventilated cages under a 12-h light/dark cycle with ad libitum access to food and water. All procedures adhered to institutional animal care guidelines and were approved by the appropriate ethics committee. Prior to surgery, rats were acclimatized to handling and the experimental environment to reduce stress and ensure physiological stability during imaging sessions. To enable cell-type -specific calcium imaging, the rats aged 3–4 weeks were injected with adeno-associated viruses (AAVs) carrying GCaMP6f under either neuron-specific (hSyn, CaMKII) or astrocyte-specific (GFAP) promoters. For optogenetic activation, AAV vectors encoding Channelrhodopsin-2 (ChR2; e.g., AAV5-CaMKIIa-hChR2(H134R)-EYFP) were delivered to the somatosensory cortex via stereotaxic surgery.

2.2 Anesthesia and Physiological Monitoring

Anesthesia was induced with 3% isoflurane and maintained at 1%–1.5% in oxygen during surgical procedures. For imaging, α-chloralose was administered as a bolus (40 mg/kg, intravenous), followed by continuous infusion at 25 mg/kg/h to maintain stable anesthesia with minimal interference on neurovascular coupling (*see* **Note 1**). Throughout surgery and imaging, physiological parameters including body temperature, heart rate, and oxygen saturation were continuously monitored using a pulse oximeter and maintained within physiological ranges. A feedback-controlled heating pad was used to maintain core temperature at 37°C ± 0.5°C.

2.3 Surgical Tools and Consumables

A stereotaxic frame (Kopf Instruments) with ear and tooth bars was used to secure the animal's head for precise craniotomy and implantation. For cranial access, a high-speed dental drill fitted with microburrs (~0.5 mm diameter) was used to create burr holes without damaging the underlying dura. A nanoliter injector system (e.g., WPI Nanoject III) paired with 35-gauge beveled stainless steel needles or pulled glass micropipettes allowed for precise viral delivery at defined depths. To secure optical fibers, dental acrylic or light-cured dental cement (e.g., Metabond or 3MESPE) was applied around the ferrule or baseplate. Wound closure was achieved using surgical sutures (e.g., 5-0 silk or absorbable Vicryl) or tissue adhesive. Postoperative care included administration of analgesics (e.g., buprenorphine or meloxicam) and antibiotics when necessary, in accordance with institutional animal care guidelines.

2.4 Optical and Electrical Stimulation Systems

Optogenetic stimulation was delivered through a 473 nm blue laser connected to the implanted optical fiber. Stimulation parameters (e.g., 10 ms pulses at 10–40 Hz) were controlled by a Master-9 AMPI pulse generator, which received TTL triggers from the MRI system to ensure precise timing synchronization. For somatosensory activation via electrical stimulation, platinum electrodes were inserted subcutaneously into the forepaw or whisker pad, delivering square wave pulses (1–3 mA, 3–5 Hz) in block-design paradigms. All stimulation protocols were carefully timed with respect to the fMRI acquisition to minimize motion and physiological artifacts.

2.5 MRI Equipment and Fiber Photometry System

High-field MRI was performed using a 9.4 T or 14.1T small-animal scanner (Bruker Biospec), equipped with a 6 cm gradient insert. A surface transceive coil or a custom-built phased-array coil was used to maximize signal sensitivity. Functional imaging employed two primary sequences: a FLASH-based 2D line-scanning and balanced steady-state free precession (bSSFP) protocol. For line-scanning, k-space lines were reshuffled across repetitions to reconstruct images with a temporal resolution of 6 ms. Imaging data were acquired with high spatial resolution ($100 \times 100\,\mu m^2$ in-plane) and were synchronized with stimulation events through TTL triggering.

Calcium signals were recorded using a fiber photometry system consisting of a 470 nm LED or laser source, dichroic mirror, emission filter (500–550 nm), and a low-noise photodiode or photomultiplier tube (PMT). Signals were amplified by a voltage amplifier (DHPVA-100; Femto) and acquired by the analog input module of a Biopac MP 150 system (5-K sampling rate). A TTL signal from the scanner or stimulation controller was used to synchronize optical and MR recordings. Data were preprocessed to remove baseline drift and normalized for intertrial comparisons.

2.6 Data Acquisition and Processing

Raw MRI and photometry data were preprocessed using a combination of AFNI (Analysis of Functional NeuroImages) and MATLAB. Preprocessing included motion correction, slice timing correction, spatial smoothing, and temporal filtering. Region-of-interest (ROI) signals were extracted, and functional connectivity was assessed using correlation matrices, seed-based analysis, and spectral coherence. Coregistration of photometry and fMRI signals was achieved using timestamp alignment and cross-correlation. Visualizations were generated using custom scripts or tools such as ImageJ, AFNI, and FSL.

3 Methods

3.1 Line-Scanning fMRI

Gradient-Echo-Based Line-Scanning fMRI (GELINE)

The line-scan imaging methodology has been widely recognized in the field of magnetic resonance (MR) and has been utilized for several decades [17, 18]. By employing a high-field (11.7 T) magnetic resonance imaging (MRI) scanner, Silva et al. explored the spatial and temporal dynamics of BOLD fMRI response across different cortical layers during somatosensory stimulation [19]. This study not only demonstrates the capability of high-field fMRI to provide detailed insights into the temporal and spatial dynamics of brain activity but also highlights the importance of layer-specific analysis in understanding the complex interactions between neuronal activity and hemodynamic responses. We have further refined the line-scanning fMRI methodology to achieve exceptional spatial (50 μ m) and temporal resolution (50 ms) [8]. Previous studies have shown remarkable variability in onset times and amplitudes across cortical layers, where the superficial and deeper layers exhibited delayed onset and different signal amplitudes [19]. Based on these findings, this line-scanning fMRI approach noninvasively captures fMRI signals with high spatial and temporal resolution, enabling detailed analysis of early onset times. To alleviate potential aliasing effects along the phase-encoding axis, we adopted two saturation slices to eliminate signals originating from outside the field-of-view (FOV).

Initially, the line-scanning approach was incorporated into the Fast Low Angle Shot (FLASH) fMRI technique, thereby achieving an exceptionally high temporal resolution (Fig. 1A) [3]. The line-scanning method disclosed distinct fMRI onset profiles within the two circuits subsequent to stimulation, specifically an early fMRI onset peak at L4 in the FP-S1 and two separate onset peaks corresponding

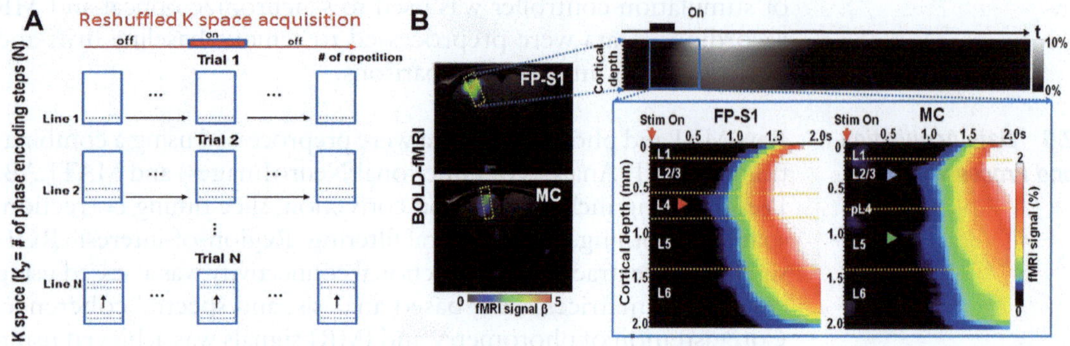

Fig. 1 The line-scanning fMRI method. (**A**) An overview of the k-space acquisition within the FLASH-fMRI method is presented. During each trial of the block-design experiment (depicted by the red line), a single k-space line was obtained for each image. Subsequently, the k-space for each image was populated with one line per trial. The experimental trials were repeated based on the number of phase-encoding steps. (**B**) The BOLD functional maps of FP-S1 and MC are depicted in the Echo-Planar-Imaging (EPI) images (left panel). The fMRI line profile was retrieved from the activated FP-S1 and MC (within the dotted rectangular area). The stimulation paradigm was configured as 1 s off, 1 s on, followed by 13 s off. An fMRI percentage change map corresponding to the initial events, both during and after the stimulus is presented. The y-axis denotes the cortical depth, while the x-axis represents time. The arrowheads indicate the earliest fMRI onset(s) in FP-S1 (marked by the red arrowhead) and in MC (indicated by the blue and green arrowheads). The red arrow symbolizes the commencement of the stimulus

to somatomotor inputs into L2/3 and L5 of the MC (Fig. 1B) [8]. This method constitutes a potent tool for the detailed mapping of brain functions and the investigation of cortical plasticity with high resolution and specificity, furnishing substantial insights into the relationship between neural inputs and fMRI signals. The FLASH-fMRI method was employed to augment the SNR for a more precise characterization of the fMRI onset. This breakthrough enables us to identify laminar-specific neural inputs along the cortical thickness, thus affording a novel perspective on neural circuitry (*see* **Note 2**).

Subsequently, Choi et al. have extended the line-scanning method to multislice acquisition, thereby enabling the concurrent collection of line profiles from diverse cortical regions [20]. Multislice line-scanning fMRI (MS-LS) attains a high spatial (50 μm) and temporal (100 ms) resolution, permitting the detection of distinct BOLD responses within diverse cortical layers. These findings disclose that the positive BOLD responses are strongly correlated with layer 4 (L4) owing to thalamocortical projections, whereas negative BOLD responses are more pronounced in layers 2/3 (L2/3), indicating a brain state-dependent laminar-fMRI response pattern influenced by corticocortical interactions. In addition to evoked conditions, the study also investigated resting-state (rs-) fMRI, uncovering robust lag time differences among layers 2/3, 4, and 5 in multiple cortices, which are indicative of low-frequency signal propagation from caudal to rostral slices. The MS-LS technique permitted the characterization of trial-specific intralaminar and interlaminar functional connectivity, affording a distinct insight into the laminar-specific information flow within and between cortical regions. This innovative mapping scheme not only augments the understanding of cortical microcircuits under various brain states but also possesses substantial potential for future research in functional connectivity and brain state dynamics.

Furthermore, bilateral line-scanning fMRI (BiLS) was carried out by expanding the slice dimension from 1 to 2 for the purpose of capturing fMRI signals in both the left and right FP-S1 cortices [21]. This was accomplished by modifying the phase and slice encoding directions in contrast to the conventional line-scanning method and by integrating two saturation slices to preclude aliasing. Choi et al. recently investigated laminar-specific interhemispheric connectivity, focusing on the forepaw somatosensory cortices in rat brains, enabling the detection of distinct spectral features in resting-state (rs-) fMRI signals. The study discerned two principal types of bilateral fluctuation spectral features: ultraslow oscillations (<0.04 Hz) traversing all cortical layers and Layer 2/3-specific BOLD signals at 0.05 Hz and 0.08–0.1 Hz. These findings imply that L2/3-specific signals are presumably associated with neuronal circuit-specific activity driven by callosal projections, accentuating the ability of the BiLS method to dissect frequency-specific interhemispheric low-frequency fluctuations with laminar specificity. The capacity of this technique to attain such high resolution and specificity represents a substantial advancement in fMRI technology, furnishing more profound insights into the underlying neuronal circuits of interhemispheric connectivity. Collectively, the line-scanning fMRI technique achieves ultrafast temporal resolution by restructuring single k-space line acquisitions over multiple repetitions in accordance with a block design stimulation paradigm [22], rendering it a potent tool for probing neuronal activity and responses in animal models.

Spin-Echo-Based Line-Scanning fMRI (SELINE)

Another key innovation is the adoption of an alpha-180 spin-echo-based line-scanning fMRI (SELINE) method [23]. In contrast to the gradient-echo-based line-scanning fMRI (GELINE), which depends on two saturation RF pulses to suppress signals outside the region of interest (ROI), the SELINE method utilizes a refocusing 180° RF pulse perpendicular to the excitation slice. This alteration eliminates the necessity for additional saturation RF pulses, thus enhancing the boundary definition of the reduced ROI and resolving aliasing problems. The SELINE method also exhibits high spatiotemporal resolution (with a repetition time of 200 ms) while effectively eliminating the large draining vein effect, which has been a substantial limitation in GELINE. Moreover, the SELINE method augments microvascular sensitivity across diverse cortical layers, affording sharper 2D slice profiles, and reduced background signals in comparison to GELINE. This advancement is of crucial importance for mapping laminar-specific BOLD responses, since, it substantially mitigates magnetic susceptibility effects at superficial layers and enhances spatial specificity regarding deeper cortical layers. Through demonstrating that the peak BOLD signals in SELINE are positioned at layer 4 (L4) instead of the superficial layer (L1), as observed in GELINE, the SELINE method offers a more precise portrayal of laminar-specific neural activities. Furthermore, the capacity to reduce the repetition time to 200 ms without compromising spatial resolution underscores the SELINE method's ability to capture fast hemodynamic changes with high precision.

3.2 Single-Vessel fMRI

Line-Scanning Based Single-Vessel fMRI

Building on the high-resolution line-scanning fMRI, the application of single-vessel fMRI permits a more detailed analysis, facilitating the precise dissection of hemodynamic signals from individual arterioles and venules [3, 9]. Single-vessel fMRI surmounts the limitations of traditional fMRI, which entails large voxel sizes incorporating multiple vascular and neural elements, thus attenuating the specificity of the signals. A high-field MRI system with a multigradient-echo (MGE) sequence was employed to map individual arterioles and venules. Through the utilization of both sensory (electrical stimulation) and optogenetic stimulation, the study distinguished the BOLD signal, mainly originating from venules, from the CBV signal, which was predominantly attributable to arterioles. This differentiation was enabled by the distinct T2* relaxation times and the consequent MRI signal differences between oxygenated and deoxygenated blood. The study employed a combination of BOLD and CBV fMRI in conjunction with a line-scanning approach to map the hemodynamic responses at the individual vessel level. The findings revealed that the peak BOLD signal coincided with venule regions, while the peak CBV signal overlapped with arteriole regions. Similar results have been observed with in-flow based time-of-flight microangiography for vessel mapping [24]. Furthermore, the temporal dynamics of the hemodynamic responses were delineated, revealing that the CBV signals from arterioles exhibited an earlier onset and time to peak in comparison to the BOLD signals from venules, which is consistent with our previous finding of an earlier onset in the arterioles and capillaries voxels than in the venule voxels [25]. Deoxygenated blood in venules possesses a shorter T2*, resulting in a more rapid signal loss compared to surrounding tissues or arterioles. At longer echo times (TE), this leads to diminished venule signals, whereas arterioles and surrounding tissues retain their brightness, thereby creating a distinct contrast (Fig. 2A). Multigradient echo (MGE) images at varying echo times (TEs) enabled the anatomical mapping of penetrating arterioles and venules, giving rise to an A-V map (Fig. 2B).

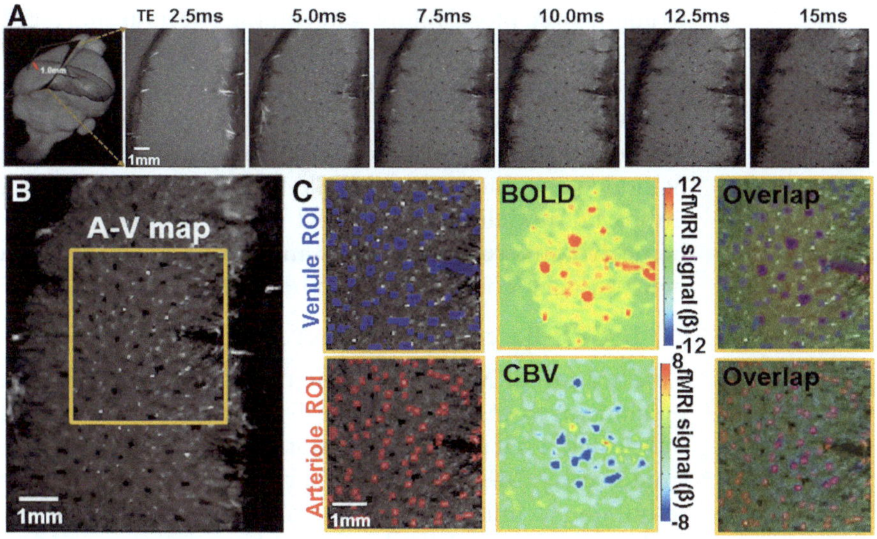

Fig. 2 The single-vessel fMRI method [3]. (**A**) A 2D MGE slice extracted from the deep layer cortex of the forepaw S1 cortex (ranging from 1.0 to 1.5 mm, left) at varying echo times (TEs), which is representative of data from five rats. (**B**) The A-V map derived from (description). Arterioles and venules manifest as bright and dark voxels, respectively. (**C**) A zoomed area (the box indicated in B) with venule (colored blue) and arteriole (colored red) regions of interest (ROIs) highlighted (left panel), along with the BOLD (upper) and CBV (lower) fMRI maps on the same 2D slice (middle panel), and the overlays (where active voxels are depicted in purple)

The FLASH-fMRI method was utilized to successfully detect both BOLD and CBV signals from the same 2D slice for the construction of the A-V map. The peak BOLD signal was preponderantly associated with penetrating venule regions of interest (ROIs), whereas the peak CBV signal primarily corresponded to penetrating arteriole ROIs (Fig. 2C).

Multi-Echo Single-Vessel fMRI

Multi-echo single-vessel (MESV) fMRI, which builds upon the foundation laid by line-scanning-based approaches, exploits multiple echo acquisitions to further improve temporal resolution and augment the sensitivity of single-vessel imaging with enhanced contrast-to-noise ratios (CNR) [4]. By employing a 2D multi-echo acquisition technique, the MESV-fMRI attains a high sampling rate of 100 ms across multiple echo times (ranging from 3 to 30 ms), thereby significantly improving the temporal and spatial resolution. This capability enables the mapping of vessel-specific T2* changes, affording higher contrast-to-noise ratios (CNRs) compared to traditional T2*-weighted fMRI at a specific echo time. The principal strength of the MESV-fMRI method resides in its capacity to distinguish between the hemodynamic responses of arterioles and venules. The study illustrates that, whereas venule voxels display enhanced T2*-weighted signals on account of extravascular effects at longer echo times, arteriole voxels manifest substantial fMRI signal increases due to augmented blood inflow and volume effects at shorter echo times. This differentiation is of crucial importance for the semiquantitative estimation of oxygen saturation levels (Y) and their changes (ΔY) following neuronal activation, relying on vessel-specific T_2* mapping. Through furnishing detailed temporal dynamics and vessel-specific hemodynamic parameters, MESV-fMRI provides a solid platform for improved modeling of the fMRI signal with respect to underlying hemodynamic responses, thus augmenting our understanding of neurovascular coupling in both health and disease.

Real-Time Single-Vessel fMRI

(1) **bSSFP-Based Single-Vessel fMRI**

Line-scanning fMRI entails the reconstruction of k-space, which precludes it from performing real-time imaging. Therefore, we have developed a balanced steady-state free precession (bSSFP) based single-vessel fMRI method to surmount the limitations of k-space reshuffling techniques, providing real-time acquisition while retaining high spatiotemporal resolution (with a resolution of 100 μm). This approach enables the precise mapping of neurovascular oscillatory dynamics, as well as the spatial distribution of fMRI signal fluctuations within both arteriolar and venous networks [26]. This method augments the spatial specificity of fMRI signals, thus enhancing the capacity to distinguish between signals originating from arterioles and venules. A prominent feature of the single-vessel fMRI approach elaborated in this study lies in its capacity to map the dynamic correlation patterns of hemodynamic signals within an ultraslow frequency range (0.01–0.04 Hz). We have demonstrated that in anesthetized rats (*see* **Note 1**), the BOLD signals from venules and the CBV signals from arterioles exhibit strong correlations with the calcium signals simultaneously recorded in neighboring neurons, thereby highlighting the neurovascular coupling. Moreover, in awake human subjects, this technique discloses that BOLD signals in veins are highly correlated, presenting vessel-specific spatial patterns that are consistent with conventional functional connectivity maps. This capacity to map vessel-specific hemodynamic fluctuations furnishes valuable insights into the vascular contributions to resting-state fMRI signals and the underlying neurovascular mechanisms.

Moreover, by employing phase contrast MRI (PC-MRI) at high spatial resolution, we managed to measure vessel-specific cerebral blood flow-related velocity (CBFv) [10]. In vivo experiments conducted on anesthetized rats demonstrated that the technique was capable of effectively mapping the blood velocity in individual arterioles and venules, in accordance with optical measurement benchmarks. The study also investigated the complementary nature of PC-MRI in relation to existing fMRI techniques like BOLD and CBV, accentuating the method's capacity to furnish distinct hemodynamic information regarding different vessel types. This advancement bears significant implications for the study of neurovascular coupling and brain disorders, providing a potent instrument for detailed hemodynamic assessment in both research and clinical settings.

(2) **Radial Encoding Single-Vessel fMRI**

Although the balanced steady-state free precession (bSSFP) technique enables real-time hemodynamic monitoring with a high signal-to-noise ratio (SNR), it is prone to banding artifacts arising from field inhomogeneities. Consequently, frequent shimming adjustments are necessary to uphold image quality. These artifacts can undermine the accuracy of hemodynamic data, especially when high spatial resolution is demanded. To surmount these limitations, radial encoding single-vessel fMRI has been developed, affording substantial improvements in both spatial and temporal resolution. This technique utilizes a radial acquisition scheme, which continuously updates the center of the k-space, thereby attaining an outstanding in-plane resolution of $50 \times 50\,\mu m$ along with a sampling rate of 1–2 Hz [6].

Radial single-vessel fMRI presents several crucial advantages, such as enhanced spatiotemporal resolution, decreased vulnerability to artifacts, and the capacity to precisely map hemodynamic responses within individual vessels. Its flexibility in adjusting the field-of-view (FOV) and projection numbers augments imaging efficiency. Meanwhile, the continuous updating of k-space centers guarantees uniform contrast and precise detection of vascular dynamics. These characteristics render radial single-vessel fMRI especially efficacious for detailed and artifact-free imaging of the microvasculature within the brain.

3.3 High-Resolution fMRI with Optical Methods

fMRI with Optogenetics

Optogenetic fMRI (ofMRI) furnishes insights into this issue by integrating optogenetic control with high-field fMRI, thereby facilitating the visualization of causal effects arising from specific cell types in accordance with their genetic identity, location, and axonal targets [27]. The integration of optogenetics with high-resolution fMRI offered distinctive insights into the direct impacts of neuronal activity on vascular responses, thus augmenting the understanding of neurovascular coupling at the microvascular level. In our prior study, through genetically modifying rats to express channelrhodopsin-2 (ChR2) in specific neurons, neuronal activity was precisely controlled by means of light stimulation transmitted via a fiber optic cable directly to the targeted brain region [3]. Subsequently, the activation of these neurons by light pulses was meticulously monitored using single-vessel fMRI, which enabled the observation and mapping of the resulting hemodynamic changes within individual microvessels, thereby furnishing insights into the localized vascular responses to neural activity.

Chen et al. integrate optogenetics with line-scanning fMRI to explore layer-specific BOLD responses in rodent brains [7]. Optogenetic stimulation was applied to the right forepaw somatosensory cortex (FP-S1) of the rodents. By using an implanted Inductively Coupled Detector (ICD), the SNR of fMRI images was significantly enhanced. This enhancement enables more precise mapping of brain activity, especially in focal brain regions, and it is of crucial importance for understanding complex neuronal circuits.

fMRI with Fiber Photometry

The integration of fMRI with fiber photometry facilitates the simultaneous measurement of brain-wide hemodynamic responses and cell-type-specific neuronal activity. The utilization of genetically encoded calcium indicators such as GCaMP6 permits real-time monitoring of intracellular calcium levels ([Ca^{2+}]) in neurons. Given calcium's function as a second messenger in neurotransmission, these intracellular calcium levels serve as an indicator of neuronal activity. The integration of fMRI and fiber photometry enabled a comprehensive analysis of the correlation between cellular-level dynamics and broader neurovascular processes, thereby furnishing valuable insights into the mechanisms of brain function and recovery subsequent to injury.

Combining Echo Planar Imaging (EPI) fMRI with neuronal calcium recordings offers a solid framework for analyzing the intricate relationships between neuronal activity and hemodynamic responses within the brain. EPI fMRI and neuronal calcium recordings were employed to correlate pupil-linked brain state changes with neural activity under anesthesia, thereby demonstrating how fluctuations in neuronal calcium levels correspond with global brain activity patterns obtained through fMRI [28]. Likewise, we employed these techniques to investigate the recovery processes from coma induced by brainstem injuries in rats [29]. Through mapping the recovery of connectivity between subcortical arousal nuclei and cortical regions, the study illustrated how neuronal calcium dynamics were capable of predicting and correlating with the recovery of functional connectivity across the brain, as observed through EPI fMRI. This furnished crucial insights into the network-based mechanisms that are vital for the reemergence of consciousness after a coma. These studies emphasize the significance of integrating EPI fMRI with neuronal calcium recordings for attaining a comprehensive understanding of the spatial and temporal dynamics of brain function, thereby augmenting our capacity to interpret complex neurophysiological processes in both health and disease states.

In our study, we have demonstrated the significant role of correlating calcium signals with fMRI data. Specifically, we have shown that the temporal alignment of neuronal calcium spikes with BOLD and CBV fluctuations establishes a direct and quantifiable link between neuronal activity and vascular responses [26]. This correlation not only validates the neurovascular coupling but also augments the precision of fMRI. By mapping these dynamics down to the individual vessel level, it significantly reduces the disparity between neuronal events and their hemodynamic signatures as witnessed in imaging studies (*see* **Note 3**). This detailed approach enables us to acquire a more refined understanding of the underlying mechanisms of brain function that are captured by functional imaging.

Another study introduces an innovative integration of optogenetics, fiber photometry, and fMRI. It employs an MRI-guided robotic arm (MgRA) to augment the precision and flexibility of optical fiber positioning for brain interventions [30]. It illustrates the capacity to map the functional connectivity of brain regions, including the basal forebrain and thalamic nuclei, as well as their responses to specific neural manipulations. This is of crucial importance for understanding the neural substrates of behaviors and identifying potential pathways for treating neurological disorders. Likewise, we explored the brain-wide network effects via optogenetic activation of the corpus callosum, in conjunction with simultaneous fMRI and calcium recording [9]. This method permitted precise manipulation and real-time observation of neural dynamics across extensive brain networks, thereby emphasizing how specific neural interventions are capable of modulating interconnected brain regions and influencing overall brain function. The approach furnishes profound insights into the neurophysiological impacts of targeted neural stimulation and its implications for brain-wide activity and connectivity. A comprehensive study conducted by Drew et al. employed a multimodal fMRI platform that integrated optogenetic stimulation with concurrent calcium imaging and fMRI [31]. This approach enabled the global mapping of brain network dynamics, thereby furnishing a detailed view of the functional connectivity and interactions within the brain.

The integration of fMRI with optogenetic stimulation is substantially enhanced by the utilization of an Inductively Coupled Detector (ICD), an innovative detection device designed specifically to augment fMRI capabilities [7]. By augmenting the SNR of fMRI signals, ICDs enable the precise detection of subtle neural fluctuations elicited by optogenetic techniques. This enhancement is of crucial importance for enhancing the spatial resolution of BOLD signal mapping in deep brain structures, where attaining high-resolution imaging poses the greatest challenge.

While fMRI has predominantly focused on capturing the effects of neuronal activity, the incorporation of astrocyte fiber photometry into fMRI offers a complementary and essential perspective. It reveals the significant role that glial cells, specifically astrocytes, play in neurovascular coupling and the broader modulation of brain function. Astrocytes have emerged as critical regulators of neurovascular coupling, influencing cerebral blood flow (CBF) through their Ca^{2+} signaling mechanisms. We investigated the intricate dynamics of astrocytic calcium signaling in connection with fMRI signals, thereby furnishing crucial insights into the neurovascular coupling mediated by astrocytes [32]. We captured real-time astrocytic calcium activities that were directly correlated with both positive and negative BOLD signals in fMRI. Through this dual measurement approach, we were able to distinguish between evoked and intrinsic astrocytic calcium signals, each of which was associated with distinct fMRI responses.

4 Notes

Note 1: It is important to consider that the use of anesthesia can significantly influence the frequency and amplitude of resting-state hemodynamic fluctuations. For example, α-chloralose anesthesia maintains ultraslow (<0.1 Hz) vasomotor oscillations more effectively than isoflurane, making it preferable for studying intrinsic neurovascular dynamics [11, 33]. However, anesthesia itself suppresses spontaneous neuronal activity, and its effects on neural-glial coupling must be interpreted with caution. When planning optogenetic activation during fMRI, the laser stimulation parameters, such as light intensity, pulse duration, and frequency, should be carefully calibrated to avoid photothermal artifacts or nonspecific vascular responses. This is especially important when targeting cortical layers near large vessels, as over-illumination may elicit unintended hemodynamic changes unrelated to neuronal excitation.

Note 2: Line-scanning methods, while powerful in terms of temporal resolution, require post hoc image reconstruction due to the reshuffling of k-space lines across time. This limits real-time visualization but enables unique capabilities such as laminar resolution of fMRI onset [8] and layer-specific vascular dynamics. Investigators should ensure proper synchronization of stimulation paradigms with the scan timing and carefully consider potential signal aliasing. Taken together, these methodological notes underscore the complexity and precision required for interpreting high-resolution optogenetic fMRI datasets.

Note 3: Single-vessel fMRI demands extremely high spatial resolution and stability. The bSSFP sequence provides a high signal-to-noise ratio (SNR) with minimal distortion, but care must be taken to identify and isolate vessel-specific signals using preacquired A-V maps and appropriate seed-based analyses or independent component analyses (ICA). CBV and BOLD signals exhibit distinct spatial distributions: CBV signals primarily originate from arterioles, whereas BOLD fluctuations localize to venules. Misinterpretation of these signals without anatomical validation can lead to erroneous functional connectivity mapping [33, 34]. Additionally, the ultraslow oscillations observed in both BOLD and CBV signals can arise from coordinated vascular mechanisms, such as vasomotion, but also reflect neuronal or astrocytic calcium activity. Simultaneous photometry and fMRI recordings help disambiguate these sources, although precise temporal alignment is critical.

This chapter presents an in-depth exploration of advanced high-resolution and multimodal fMRI approaches. Line-scanning fMRI enables precise, layer-specific analysis of cortical circuits and provides enhanced spatiotemporal resolution for capturing the dynamics of neurovascular coupling with submillimeter precision. Single-vessel fMRI facilitates the isolation and characterization of hemodynamic responses within individual arterioles and venules, thereby furnishing detailed insights into vascular dynamics at the level of these specific blood vessels. The integration of optogenetics with these fMRI methods enables targeted neuronal activation and real-time observation of the resulting hemodynamic changes. Moreover, the combination of fMRI with fiber photometry enables the simultaneous recording of calcium dynamics and BOLD signals, thereby offering a comprehensive framework for analyzing neurovascular interactions with high precision. These methodological advancements have substantially augmented the ability to investigate brain function with unprecedented resolution and specificity, thereby expanding the boundaries of what can be achieved in neuroimaging. In summary, the chapter delineates the methodological advancements in fMRI, which have remarkably enlarged the ability to investigate brain function with unprecedented resolution and specificity.

Acknowledgments

We acknowledge support from the National Natural Science Foundation of China (No. 82201447, and No. RLZY20231001-01), the Strategic Priority Research Program of the Chinese Academy of Sciences (XDB0930000), Shenzhen Science and Technology Program (JSGGKQTD20210831174329010), National Key Research and Development Program of China (2023YFF0714200, and 2024YFC3406602), the Chinese Academy of Sciences Project for Young Scientists in Basic Research (YSBR-114), Shenzhen Outstanding Talents Training Fund, and Key Laboratory for Magnetic Resonance and Multimodality Imaging of Guangdong Province (2023B1212060052).

Conflict of Interest Xin Yu is a cofounder of MRIBOT LLC.

References

1. Ogawa S, Lee TM, Kay AR, Tank DW (1990) Brain magnetic resonance imaging with contrast dependent on blood oxygenation. Proc Natl Acad Sci U S A 87:9868–9872
2. Hillman EMC (2014) Coupling mechanism and significance of the BOLD signal: a status report. Annu Rev Neurosci 37:161–181
3. Yu X et al (2016) Sensory and optogenetically driven single-vessel fMRI. Nat Methods 13:337–340
4. He Y, Wang M, Yu X (2020) High spatiotemporal vessel-specific hemodynamic mapping with multi-echo single-vessel fMRI. J Cereb Blood Flow Metab 40:2098–2114
5. Hike D et al (2025) High-resolution awake mouse fMRI at 14 Tesla. eLife 13:RP95528
6. Jiang Y, Pais-Roldán P, Pohmann R, Yu X (2024) High spatiotemporal resolution radial encoding single-vessel fMRI. Adv Sci 11:2309218

7. Chen Y et al (2022) Focal fMRI signal enhancement with implantable inductively coupled detectors. Neuroimage 247:118793
8. Yu X, Qian C, Chen D, Dodd SJ, Koretsky AP (2014) Deciphering laminar-specific neural inputs with line-scanning fMRI. Nat Methods 11:55–58
9. Chen X et al (2019) Mapping optogenetically-driven single-vessel fMRI with concurrent neuronal calcium recordings in the rat hippocampus. Nat Commun 10:5239
10. Chen X et al (2021) Assessment of single-vessel cerebral blood velocity by phase contrast fMRI. PLoS Biol 19:e3000923
11. He Y et al (2018) Ultra-slow single-vessel BOLD and CBV-based fMRI spatiotemporal dynamics and their correlation with neuronal intracellular calcium signals. Neuron 97:925–939.e5
12. Toi PT et al (2022) In vivo direct imaging of neuronal activity at high temporospatial resolution. Science 378:160–168
13. Choi S-H et al (2024) No replication of direct neuronal activity–related (DIANA) fMRI in anesthetized mice. Sci Adv 10:eadl0999
14. Phi Van VD, Sen S, Jasanoff A (2024) A different interpretation of the DIANA fMRI signal. Sci Adv 10:eadl2034
15. Sych Y, Chernysheva M, Sumanovski LT, Helmchen F (2019) High-density multi-fiber photometry for studying large-scale brain circuit dynamics. Nat Methods 16:553–560
16. Bamberg E, Gärtner W, Trauner D (2018) Introduction: optogenetics and photopharmacology. Chem Rev 118:10627–10628
17. Raimondo L et al (2021) A line through the brain: implementation of human line-scanning at 7T for ultra-high spatiotemporal resolution fMRI. J Cereb Blood Flow Metab 41:2831–2843. https://doi.org/10.1177/0271678X211037266
18. Mansfield P, Maudsley AA (1976) Line scan proton spin imaging in biological structures by NMR. Phys Med Biol 21:847–852
19. Silva AC, Koretsky AP (2002) Laminar specificity of functional MRI onset times during somatosensory stimulation in rat. Proc Natl Acad Sci U S A 99:15182–15187
20. Choi S et al (2022) Laminar-specific functional connectivity mapping with multi-slice line-scanning fMRI. Cereb Cortex 32:4492–4501
21. Choi S, Chen Y, Zeng H, Biswal B, Yu X (2023) Identifying the distinct spectral dynamics of laminar-specific interhemispheric connectivity with bilateral line-scanning fMRI. J Cereb Blood Flow Metab 43:1115–1129
22. Hike D et al (2024) Implementation of 2D line-scanning method. Apert Neuro 4:92792
23. Choi S et al (2024) Alpha-180 spin-echo-based line-scanning method for high-resolution laminar-specific fMRI in animals. Imaging Neurosci 2:1–14
24. Bolan PJ, Yacoub E, Garwood M, Ugurbil K, Harel N (2006) In vivo micro-MRI of intracortical neurovasculature. Neuroimage 32:62–69
25. Yu X et al (2012) Direct imaging of macrovascular and microvascular contributions to BOLD fMRI in layers IV–V of the rat whisker–barrel cortex. Neuroimage 59:1451–1460
26. He Y et al (2018) Ultra-slow single-vessel BOLD and CBV-based fMRI spatiotemporal dynamics and their correlation with neuronal intracellular calcium signals. Neuron 97:925–939.e5
27. Lee JH et al (2010) Global and local fMRI signals driven by neurons defined optogenetically by type and wiring. Nature 465:788–792
28. Pais-Roldán P et al (2020) Indexing brain state-dependent pupil dynamics with simultaneous fMRI and optical fiber calcium recording. Proc Natl Acad Sci U S A 117:6875–6882
29. Pais-Roldán P et al (2019) Multimodal assessment of recovery from coma in a rat model of diffuse brainstem tegmentum injury. Neuroimage 189:615–630
30. Chen Y, Pais-Roldan P, Chen X, Frosz MH, Yu X (2019) MRI-guided robotic arm drives optogenetic fMRI with concurrent Ca^{2+} recording. Nat Commun 10, 2536
31. Drew PJ, Mateo C, Turner KL, Yu X, Kleinfeld D (2020) Ultra-slow oscillations in fMRI and resting-state connectivity: neuronal and vascular contributions and technical confounds. Neuron 107:782–804
32. Wang M, He Y, Sejnowski TJ, Yu X (2018) Brain-state dependent astrocytic Ca^{2+} signals are coupled to both positive and negative BOLD-fMRI signals. Proc Natl Acad Sci U S A 115:E1647–E1656
33. Yu X et al (2016) Sensory and optogenetically driven single-vessel fMRI. Nat Methods 13:337–340
34. He Y, Wang M, Yu X (2020) High spatiotemporal vessel-specific hemodynamic mapping with multi-echo single-vessel fMRI. J Cereb Blood Flow Metab 40:2098–2114

Chapter 2

Laminar fMRI Using Layer-Specific Optogenetic Stimulations

Russell W. Chan

Abstract

Laminar optogenetic functional magnetic resonance imaging (ofMRI), which combines high-resolution fMRI with layer-specific optogenetic stimulation, provides unprecedented insights into the functional organization of the neocortex and brain-wide networks. This technique allows for precise manipulation and observation of neural activity within specific cortical layers, offering valuable data on the distinct roles and interactions of these layers and remote brain regions. By elucidating the neural origins of laminar fMRI signals, we can enhance our understanding of brain-wide networks and the specific contributions of each cortical layer to overall brain function. This knowledge is crucial for developing targeted therapies for neurological disorders, potentially leading to more effective treatments for conditions, such as epilepsy, schizophrenia, and Alzheimer's disease. In this chapter, the laminar ofMRI methods and examples are described. Transgenic mice expressing Cre-recombinase in specific neocortical layers and viral vectors to express channelrhodopsin-2 (ChR2) and enhanced yellow fluorescent protein (EYFP) (ChR2–EYFP) in these neurons were employed. Stereotaxic surgeries were performed to deliver the viral vectors and implant fiber-optic cannulas for optogenetic stimulation. Histological analyses validated the expression of ChR2. High-resolution fMRI was conducted to measure layer-specific cortical responses. It is foreseen that in the next decade, laminar ofMRI will contribute to unraveling several key aspects of cortical processing and brain-wide networks, transforming both basic neuroscience and clinical neurology research.

Keywords Optogenetics, fMRI, Cortical layers, Brain networks, Whole brain imaging

1 Introduction

The neocortex consists of six interconnected but distinct layers that have different projections and functions. Studying the six layers of the neocortex is crucial for understanding the brain's complex functions, as each layer has distinct roles in processing sensory information, motor control, and cognitive functions. By examining these layers, we can uncover the intricate neural circuits and mechanisms underlying various brain activities. This research holds significant scientific value, offering potential breakthroughs in treating neurological conditions, such as epilepsy, schizophrenia, and Alzheimer's disease, ultimately advancing clinical neurology and improving patient

Kevin Chuen Wing Chan and Seong-Gi Kim (eds.), *Optogenetic Functional Neuroimaging*, Neuromethods, vol. 232,
https://doi.org/10.1007/978-1-0716-5178-0_2,

outcomes. Historically, it has been difficult to selectively modulate each layer since they are anatomically intermingled.

Recent advances in molecular genetics have made it possible to selectively express transgenes in specific neocortical neurons [1–4], including Cre-recombinase driver lines [5] and optogenetic tools [6]. This is achieved as some neurons have layer-specific pathways that share distinct neurochemical identities, suggesting that the activation of such neurons may result in unique brain-wide responses. In particular, some neurons in L2/3 express the dopamine receptor D3 (Drd3); some neurons of L4 express sodium channel non-voltage-gated 1 alpha (Scnn1a); some in L5 express retinol-binding protein 4 (Rbp4); and some neurons in L6 express neurotensin receptor type 1 (Ntsr1). Advances in molecular genetics have made it possible to selectively express transgenes. Several studies have exploited this capability to selectively excite specific layers and measure the downstream effects using in vivo electrophysiology. For example, in the somatosensory cortex, L2/3 was shown to suppress L5. The effect of this suppression is to enhance the selectivity and output range of downstream activity for finely tuned cortical coding [7]. L4 activity of the somatosensory cortex was also shown to directly inhibit L5, which sharpened the spatial representation of L5 neurons [3]. In addition, L6 corticothalamic neurons in the visual cortex have been shown to activate L5 cortical outputs [8] and to regulate the strength of cortical responses [9]. Besides characterizing the downstream activities within the cortex, long-range pathways can also be investigated, e.g., the corticotectal projections in the visual system [10] and corticofugal projections in the auditory system [11]. Functional magnetic resonance imaging (fMRI) plays a pivotal role in neuroscience research by providing noninvasive, high-resolution images of brain activity. It allows researchers to map functional areas, study neural connectivity, and observe brain responses to various stimuli. In clinical neurology, fMRI aids in diagnosing and monitoring neurological disorders, planning surgical interventions, and assessing treatment efficacy, thereby enhancing our understanding and management of brain diseases. Conventional fMRI is not used to study cortical layers, as it is performed at a macroscopic scale of spatial resolution. Recent developments of high spatial resolution fMRI provide new opportunities for in vivo measurements of mesoscale layer-specific cortical responses [12–14]. This could be used to tackle key cognitive neuroscience questions by distinguishing bottom-up from top-down cortical responses and investigating the interactions between the two [15]. High-resolution fMRI could also be used to examine influential theories of brain-wide circuit function by assigning specific computational roles to different cortical layers [16]. Despite advancements, many aspects of the laminar neocortex and laminar fMRI remain unknown. The precise functions of individual layers and their interactions, as well as how individual layers contribute to the brain-wide networks, are not fully understood. Furthermore, the layer-evoked brain-wide fMRI networks and mesoscale layer-specific fMRI representation of neuronal activity have yet to be established. Optogenetics stimulation, which enables targeted control of specific neurons, can help elucidate these mysteries by allowing precise manipulation and observation of neural activity. By applying optogenetics and fMRI [17–23], we can precisely manipulate and visualize neural activity, revealing insights into neural circuitry and brain disorders. This approach promises to reveal new insights into the functional organization of the neocortex and its role in health and disease.

The primary motor cortex (M1) plays a prominent role in motor behavior and cortical control of movements through large-scale networks. Long-range inputs converge onto M1 through cortico-cortical, thalamocortical, and neuromodulatory projections, while M1 outputs project to spinal/bulbar motor centers, striatum, thalamus, subthalamus, red nucleus, and pons [24, 25]. Like the sensory cortices, M1 consists of a stack of interconnected but distinct layers. Traditionally, layers 2/3 (L2/3) are considered to be a prominent source of intracortical excitation of L5 [26, 27]. Additionally, L5 has output connections to the ventral anterior thalamus (VA), striatum, and spinal cord, while L6 has output connections to the ventral lateral thalamus (VL) [26]. Recently, L4 has been proposed to have prototypical synaptic circuit connectivity [28].

Here, M1 is used as an example to illustrate the methods used to study laminar fMRI using layer-specific optogenetic stimulations [19]. Targeted optogenetic stimulation of Drd3, Scnn1a, Rbp4, and Ntsr1 neurons in M1 (cells which are specific to L2/3, L4, L5, and L6, respectively) and high-resolution fMRI were combined to study how activity originating in these layers influences brain-wide regions, including thalamus, caudate putamen (Cpu), and the M1. Histology and immunohistochemistry were applied to validate optogenetic expression. Subsequently, those regions, both remotely and locally, were characterized using in vivo electrophysiology to delineate the neuronal underpinnings of the brain-wide fMRI responses and the mesoscale laminar fMRI responses, respectively.

2 Materials

2.1 Animal Preparation: Transgenic Mice

Mice (CD-1, male, 35–50 g) used for experiments in this study were heterozygous for Drd3-Cre (within L2/3), Scnn1-Cre (within L4), Rbp4-Cre (within L5), or Ntsr1-Cre (within L6). These transgenic mouse lines were obtained from the Adesnik Lab, UC Berkeley [1–4]. Mice were housed in cohorts of five or fewer with a light:dark cycle of 12:12 h with access to food and water ad libitum. All experimental procedures and animal husbandry should be performed in strict accordance with the NIH and Institutional Animal Care and Use Committee guidelines. All stereotaxic surgeries were performed on mice at 8–10 weeks of age.

2.2 Viral Vectors for ofMRI

A double-floxed inverted open reading frame (DIO) recombinant AAV5 virus (titer of 4×10^{12} particles/mL) was used to express ChR2–EYFP in Cre-expressing neurons. The recombinant AAV vector was obtained from the University of North Carolina viral vector core.

2.3 ofMRI Experiments

All MRI experiments were carried out on a 7 Tesla Bruker-converted Biospec small animal MRI system using a transmit-only birdcage coil in combination with a custom-designed actively decoupled receive-only surface coil.

3 Methods

3.1 Viral Expression and Stereotaxic Surgery for ofMRI

Animals were anesthetized with isoflurane (induction 5%, maintenance 1.5%–2%; Sigma-Aldrich) and secured in a stereotactic frame with nonrupturing ear bars (Kopf Instruments). A heating pad was used to maintain body temperature, and sterile ocular lubricant was applied to the eyes to prevent desiccation during surgery. Buprenorphine (1 mg/kg) was injected subcutaneously for analgesia. After a midline incision along the scalp, a small craniotomy and viral injection/cannula implantation were performed at the primary motor cortex (+0.75 mm AP [anterior–posterior], −1.5 mm ML [medial–lateral], injection and fiber implantation at +0.15 mm, + 0.35 mm, +0.6 mm, or +0.8 mm DV [dorsal–ventral] for L2/3, L4, L5, and L6, respectively). A total of 1.5 μL of the AAV5/DIO-ChR2–EYFP virus was delivered using a 10 μL syringe with a 34 G metal needle (World Precision Instruments Inc.) at a 75 nL/min flow rate driven by a microsyringe pump controller. Following the injection, the injection needle was held in place for 10 min before slowly retracting it from the brain. A custom-designed fiber-optic cannula was mounted and secured on the skull using light-cured dental cement (Kuraray Inc.), with the optical fiber extending from the cannula's base to the desired depth (0.1 mm above the injection site). Following surgery, mice were kept on a heating pad until recovery from anesthesia and were given carprofen (5 mg/kg, subcutaneously [s.c.]) daily for 2 days to minimize postoperative discomfort. All experiments were conducted at least 4 weeks following virus injection to ensure optimal ChR2 expression. Probe locations were validated in all animals used for ofMRI with T2-weighed structural MRI images (Fig. 1).

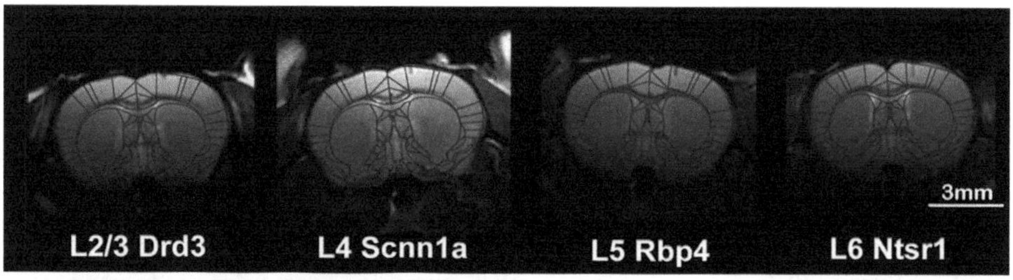

Fig. 1 Representative T2-weighted structural MRI images illustrate fiber locations. Representative T2-weighted images from each of the four neuron-specific Cre-lines illustrate fiber locations.Reproduced from Chan et al. Neuroimage, 2022 with permission [19]

3.2 Histology and Immunohistochemistry to Validate Optogenetic Expression

To confirm the precise targeting of ChR2 to L2/3, L4, L5, or L6, a cohort of mice injected with the DIO-recombinant virus was deeply anesthetized with pentobarbital and transcardially perfused with 0.1 M PBS followed by ice-cold 4% paraformaldehyde (PFA) in PBS. Brains were extracted and fixed in 4% PFA overnight at 4°C. The brains were equilibrated in 20% sucrose in PBS at 4°C overnight. Axial sections of 50 μm thickness were prepared on a freezing microtome (Micron HM550 Microtome, Thermo Scientific Inc.). For immunohistochemistry, free-floating sections were washed with 0.1 M PBS for 20 min at room temperature. Sections were then exposed to 200 ng/mL of DAPI in PBS at room temperature for 20 min. Slices were then washed and mounted using Fluoromount-G (SouthernBiotech, Birmingham, AL). Immunofluorescence was assessed with a laser confocal microscope (LSM 880 inverted confocal with Airyscan, Zeiss Inc.). The images were then analyzed to validate the expression of ChR2. Since the primary motor cortex is 1000 μm thick, and since the AAV5 spread is roughly 500 μm, the region of interest (Fig. 2) was defined as a 500 μm × 1000 μm (width × height) rectangular box centered at the injection site. Anatomical landmarks, including the midline, corpus callosum, cortex, and striatum, were identified, and the ROI was defined based on the mouse brain atlas and injection site location to ensure consistency. The number of cells with expressions of ChR2 divided by the total number of cells in the injection site was defined as sensitivity, and the number of cells with expression of ChR2 in that specific layer divided by the total number of cells with expression was defined as specificity. For sensitivity, 12%, 15%, 23%, and 15% of cells identified within the injection area were ChR2–EYFP-positive for Drd3-Cre (within L2/3), Scnn1-Cre (within L4), Rbp4-Cre (within L5), or Ntsr1-Cre (within L6) mouse lines, respectively. All mouse lines are at 100% specificity. Furthermore, ChR2 expression was observed in L2/3 M1 neurons and their L5 projections for the Drd3 Cre-line, in L4 M1 neurons and their L2/3 projections for the Scnn1a Cre-line, in L5 M1 neurons and their L2/3 projections for the Rbp4

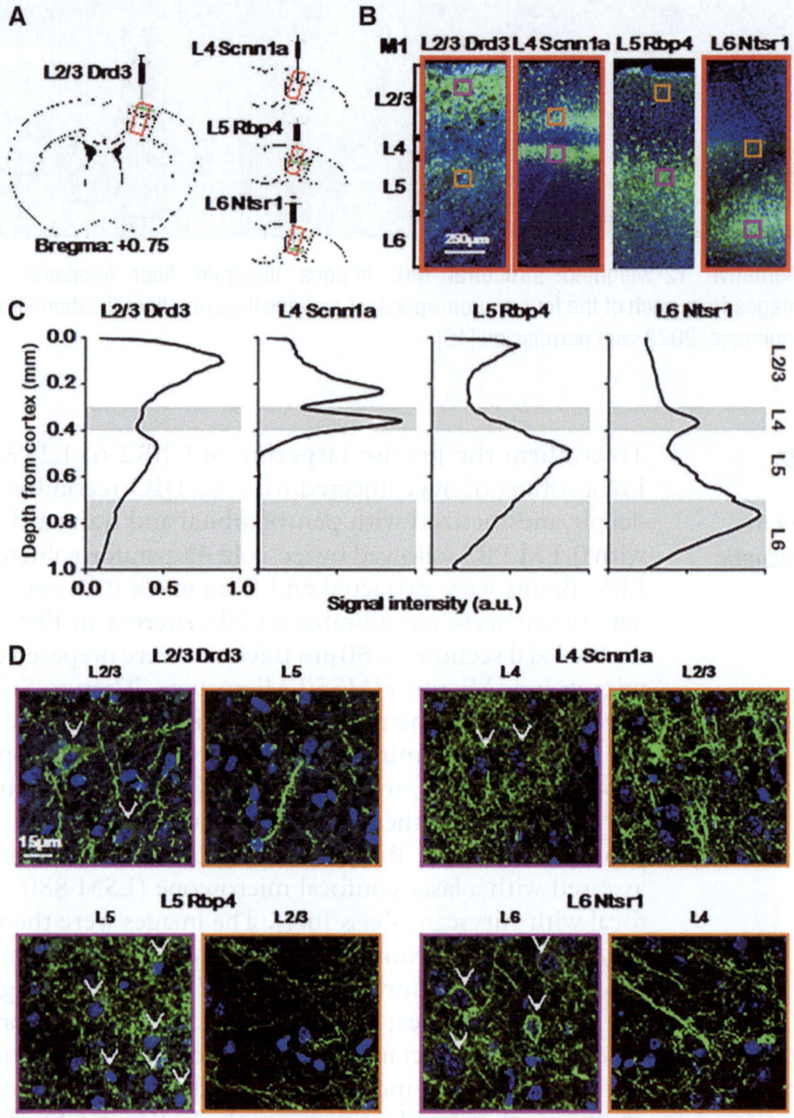

Fig. 2 Histology reveals ChR2 expression of Drd3 L2/3, Scnn1a L4, Rbp4 L5, and Ntsr1 L6 Cre-line mice in their respective M1 layers and their projections. (**A**) Illustration presents the viral injection sites in the respective layers of M1. (**B**) Representative fluorescence images from the four layer-specific Cre-lines show robust ChR2–EYFP expression throughout their respective layers and intracortical projections in the M1. (**C**) Normalized signal intensity plots along the M1 cortical depth show a local peak throughout their respective layers and intracortical projections. (**D**) Merged representative confocal images of ChR2–EYFP and stained nuclear marker DAPI confirmed colocalization of ChR2–EYFP and M1 neurons throughout their respective layers (white arrows), but not their intracortical projections. Together, ChR2–EYFP expression is observed in L2/3 M1 neurons and L5 projections for Drd3 L2/3 Cre-line, in L4 M1 neurons and L2/3 projections for Scnn1a L4 Cre-line, in L5 M1 neurons and L2/3 projections for Rbp4 L5 Cre-line, and in L6 M1 neurons and L4 projections for Ntsr1 L6 Cre-line.Reproduced from Chan et al. Neuroimage, 2022 with permission [19]

Cre-line, and in L6 M1 neurons and their L4 projections for the Ntsr1 Cre-line (Fig. 2).

3.3 ofMRI Experiments

Animals were initially anesthetized in an induction chamber with 5% isoflurane before placement into the restraining apparatus. The restrained animal was then placed onto the MRI-compatible cradle with ears, teeth, and head secured. To maximize signal-to-noise ratio, the receiver coil was first placed on top of the head and centered over the brain, and then placed into the isocenter of the magnet. The animals were lightly anesthetized and sedated using a combination of isoflurane and dexmedetomidine (0.25% isoflurane mixed with O_2 and medical air; an initial bolus of 0.1 mg/kg dexmedetomidine; continuous infusion at 0.1 mg/kg/h during scanning). During all fMRI experiments, continuous physiological monitoring was performed using an MRI-compatible system (SA Instruments). Vital signs were within normal physiological ranges (rectal temperature: 36.5–37.5°C, heart rate: 260–420 beats/min, breathing: 80–120 breathes/min, oxygen saturation: >90%) throughout the duration of the experiments.

Fourteen contiguous 1.0-mm slices were positioned in the transverse orientation according to the mouse brain atlas to cover the whole brain. T2-weighted high-resolution anatomical images were acquired prior to fMRI to check for brain damage and confirm accurate probe location. These anatomical images were acquired using a rapid acquisition with relaxation enhancement (RARE) sequence with a field of view (FOV) = 20 × 20 mm^2, matrix = 256 × 256, RARE factor = 8, echo time (TE) = 33 ms, repetition time (TR) = 4000 ms. ofMRI measurements were obtained for the same slices using a single-shot Gradient-Echo Echo-Planar-Imaging (GE–EPI) sequence with FOV = 20 × 20 mm^2, matrix = 75 × 75, flip angle = 40°, TE = 14 ms, TR = 1000 ms.

The MRI scanner and laser for optogenetic stimulation were synchronized using a Master pulse stimulator system (A.M.P.I.). The light delivery system was kept outside the magnet, and long optical cables (within 5 m) were used to deliver light into the scanner bore. Blue light was delivered using a 473-nm DPSS laser, measured before scanning as 3 mW at the fiber tip (250 μm), corresponding to a light intensity of 60 mW/mm^2 for optogenetic stimulation. To determine the frequency-dependent spatiotemporal characteristics of evoked M1 layer-specific responses, four frequencies were used (5, 10, 20, and 40 Hz with 30% duty cycle). M1 layer-specific neurons were stimulated with a standard block design paradigm that consisted of 10-s light-on and 30-s light-off periods. Three to four trials were recorded for each frequency in a pseudorandom manner in each animal.

3.4 fMRI Data Analysis

For each animal, all fMRI images were corrected for slice timing differences, realigned to the first image of the first fMRI session, and spatially smoothed with a Gaussian kernel (FWHM = 2 pixels) using SPM12 (Wellcome Department of Imaging Neuroscience, University College, London). The fMRI data were then temporally linear detrended and temporally band-pass filtered (0.01–0.25 Hz) to eliminate baseline drift caused by system instability and physiological noise. For detrending, the linear trend was first calculated using linear regression of the global temporal signal obtained from the whole brain. Then, detrending was applied to the temporal signal of each voxel using the obtained linear trend. T2-weighted images from each animal were registered to a custom-made brain template acquired with the same settings. Registration was performed by affine transformation and Gaussian smoothing to maximize normalized mutual information (SPM12). Only then were the fMRI images resliced correspondingly.

A general linear model (GLM) was applied. Student's *t* test was performed to identify activated voxels using the threshold $p<0.001$. The mapped responses were then compared between each animal to assess the data quality before group-level analysis. At the group level, realigned, registered, and resliced images corresponding to the same fMRI session were averaged across all animals. To quantify the activations from the activation maps, *t* values within the ROI were extracted and averaged, while fractions of ROI positively and negatively modulated were calculated with a voxel-wise threshold of $p<0.05$ as positively or negatively modulated. BOLD temporal profiles for each stimulation frequency were extracted from identical ROIs delineated from the mouse brain atlas by regional averaging across the whole ROI. Area under the curve of the BOLD profiles was calculated to quantify the strength of the BOLD responses for each animal, and then averaged across animals within the same group.

3.5 In Vivo Electrophysiology

In vivo electrophysiology was performed to directly measure the neuronal activity of the ipsilateral M1, CPu, and VL (Fig. 3). Similar to ofMRI experiments, the animals were lightly anesthetized and sedated using a combination of isoflurane and dexmedetomidine (0.25% isoflurane mixed with O_2 and medical air; an initial bolus of 0.1 mg/kg dexmedetomidine; continuous infusion at 0.1 mg/kg/h during experiment). Throughout the procedure, body temperature was maintained at 37°C using a thermoresistive heating pad (FHC, Inc., ME, USA). After securing the animal within a stereotaxic frame, an optrode composed of an optical fiber glued to the 16-channel linear-array electrode (NeuroNexus Technologies, MI, USA; A1x16-5mm-150-703-A16 electrode array) was inserted at the stimulation site (+0.75 mm AP [anterior–posterior], −1.5 mm ML [medial–lateral], covering between +0.00 mm and + 2.40 mm DV [dorsal–ventral]) and a 16-channel microelectrode array (NeuroNexus Technologies, MI, USA; A1x16-5mm-50-703-A16 electrode array) was inserted at the ipsilateral VL (−1.35 mm AP, +1.00 mm ML, covering between +3.30 mm and +3.70 mm DV). Identical optogenetic

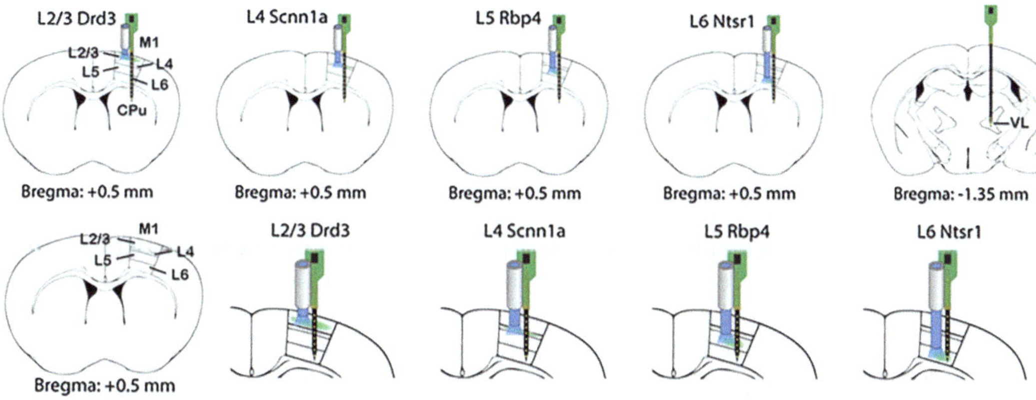

Fig. 3 LFP and spike recordings in M1 for all stimulations of neurons specific to different M1 layers and in CPu and VL, and laminar LFP and spike recordings along M1 cortical depth. (Top) The schematic shows recording optrode and electrode locations in the ipsilateral M1, CPu, and VL. (Bottom) The schematic shows the laminar recording optrode location in the ipsilateral M1 along cortical depth.Reproduced from Chan et al. Neuroimage, 2022 with permission [19]

stimulation paradigms were applied. Recordings were acquired at a 40 kHz sampling rate using the Plexon OmniPlex system with PlexControl software (Plexon Inc., TX, USA). The Plexon multichannel acquisition processor was used to amplify and low-pass filter (cutoff frequency: 200 Hz) the acquired signal for local field potential (LFP) recordings or high-pass filter (cutoff frequency: 300 Hz) for spike recordings.

3.6 Electrophysiology Data Analysis

For LFP recordings, raw data were initially imported into custom-written Matlab software and a notch filter centered at 60 Hz was applied to remove power-line noise. Individual animal LFPs were averaged across stimulation blocks to generate a single LFP trace for each stimulation frequency, which was then averaged across animals. For spike recordings, two-tailed paired *t* tests were used to identify significant changes in firing rate within each unit between prestimulation and stimulation periods (10 s each). Only units with significant change (either increase or decrease) were included in the following analysis. Peri-event time histograms were averaged across stimulation blocks and units to generate an average peri-event time histogram for each stimulation frequency. One-way ANOVA followed by Bonferroni's post hoc test was applied to compare prestimulation, stimulation, and poststimulation periods.

Although BOLD-fMRI and LFP responses reflect the overall signal within a region, when all recorded units are considered for exploring the association between spiking, BOLD-fMRI and LFP, different spiking parameters should be examined as well, namely percentage of positively modulated neurons, percentage of negatively modulated neurons, percentage of modulated neurons, average percentage spike rate change of positively modulated neurons, average percentage spike rate change of negatively modulated neurons, average percentage spike rate change of all modulated neurons, average absolute value of percentage spike rate change of all modulated neurons, average percentage spike rate change of all neurons, average absolute value of percentage spike rate change of all neurons, total percentage spike rate change of positively modulated neurons, total percentage spike rate change of negatively modulated neurons, total percentage spike rate change of all modulated neuron, total absolute value of percentage spike rate change of all modulated neurons, total percentage spike rate change of all neurons, and total absolute value of percentage spike rate change of all neurons.

4 Notes

Laminar ofMRI, combining high-resolution fMRI with layer-specific optogenetic stimulation, offers unprecedented insights into the functional organization of the neocortex and brain-wide networks. This technique allows for precise manipulation and observation of neural activity within specific cortical layers, providing valuable data on the distinct roles and interactions of these layers as well as different remote brain regions. Elucidating the neural origins of laminar fMRI signals will enhance our understanding of brain-wide networks and the specific contributions of each cortical layer to overall brain function. This knowledge is crucial for developing targeted therapies for neurological disorders, potentially leading to more effective treatments for conditions, such as epilepsy, schizophrenia, and Alzheimer's disease.

It is foreseen that in the next decade, laminar ofMRI will contribute to unraveling several key aspects of cortical processing, including the differentiation between bottom-up and top-down cortical responses and the specific computational roles of different cortical layers. Future directions for laminar ofMRI research include refining the spatial and temporal resolution of fMRI, developing additional fMRI sequences resulting in new imaging contrast, developing more sophisticated optogenetic tools, and expanding the application of this technique to other brain regions and animal models. By advancing our understanding of the neocortex's laminar structure and function, as well as brain-wide networks, laminar ofMRI holds the promise of transforming both basic neuroscience and clinical neurology research.

Conflict of Interest

The author has no conflicts of interest to declare that are relevant to the content of this chapter.

References

1. Gong S, Doughty M, Harbaugh CR, Cummins A, Hatten ME, Heintz N, Gerfen CR (2007) Targeting Cre recombinase to specific neuron populations with bacterial artificial chromosome constructs. J Neurosci 27(37):9817–9823. https://doi.org/10.1523/JNEUROSCI.2707-07.2007
2. Madisen L, Zwingman TA, Sunkin SM, Oh SW, Zariwala HA, Gu H, Ng LL, Palmiter RD, Hawrylycz MJ, Jones AR, Lein ES, Zeng H (2010) A robust and high-throughput Cre reporting and characterization system for the whole mouse brain. Nat Neurosci 13(1):133–140. https://doi.org/10.1038/nn.2467
3. Pluta S, Naka A, Veit J, Telian G, Yao L, Hakim R, Taylor D, Adesnik H (2015) A direct translaminar inhibitory circuit tunes cortical output. Nat Neurosci 18(11):1631–1640. https://doi.org/10.1038/nn.4123
4. Adesnik H, Scanziani M (2010) Lateral competition for cortical space by layer-specific horizontal circuits. Nature 464(7292):1155–1160. https://doi.org/10.1038/nature08935
5. Daigle TL, Madisen L, Hage TA, Valley MT, Knoblich U, Larsen RS, Takeno MM, Huang L, Gu H, Larsen R, Mills M, Bosma-Moody A, Siverts LA, Walker M, Graybuck LT, Yao Z, Fong O, Nguyen TN, Garren E, Lenz GH, Chavarha M, Pendergraft J, Harrington J, Hirokawa KE, Harris JA, Nicovich PR, McGraw MJ, Ollerenshaw DR, Smith KA, Baker CA, Ting JT, Sunkin SM, Lecoq J, Lin MZ, Boyden ES, Murphy GJ, da Costa NM, Waters J, Li L, Tasic B, Zeng H (2018) A suite of transgenic driver and reporter mouse lines with enhanced brain-cell-type targeting and functionality. Cell 174(2):465–480. https://doi.org/10.1016/j.cell.2018.06.035
6. Boyden ES, Zhang F, Bamberg E, Nagel G, Deisseroth K (2005) Millisecond-timescale, genetically targeted optical control of neural activity. Nat Neurosci 8(9):1263–1268. https://doi.org/10.1038/nn1525
7. Pluta SR, Telian GI, Naka A, Adesnik H (2019). Superficial layers suppress the deep layers to fine tune cortical coding. J Neurosci. https://doi.org/10.1523/JNEUROSCI.1459-18.2018.
8. Bortone DS, Olsen SR, Scanziani M (2014) Translaminar inhibitory cells recruited by layer 6 corticothalamic neurons suppress visual cortex. Neuron 82(2):474–485. https://doi.org/10.1016/j.neuron.2014.02.021
9. Kim J, Matney CJ, Blankenship A, Hestrin S, Brown SP (2014) Layer 6 corticothalamic neurons activate a cortical output layer, layer 5a. J Neurosci 34(29):9656–9664. https://doi.org/10.1523/JNEUROSCI.1325-14.2014
10. Xiong XR, Liang F, Zingg B, Ji XY, Ibrahim LA, Tao HW, Zhang LI (2015) Auditory cortex controls sound-driven innate defense behaviour through corticofugal projections to inferior colliculus. Nat Commun 6:7224. https://doi.org/10.1038/ncomms8224
11. Liang F, Xiong XR, Zingg B, Ji XY, Zhang LI, Tao HW (2015) Sensory cortical control of a visually induced arrest behavior via corticotectal projections. Neuron 86(3):755–767. https://doi.org/10.1016/j.neuron.2015.03.048
12. Huber L, Handwerker DA, Jangraw DC, Chen G, Hall A, Stuber C, Gonzalez-Castillo J, Ivanov D., Marrett S, Guidi M, Goense J, Poser BA, Bandettini PA (2017) High-resolution CBV-fMRI allows mapping of laminar activity and connectivity of cortical input and output in human M1. Neuron 96(6):1253–1263. https://doi.org/10.1016/j.neuron.2017.11.005
13. Muckli L, De Martino F, Vizioli L, Petro LS, Smith FW, Ugurbil K, Goebel R, Yacoub E (2015) Contextual feedback to superficial layers of V1. Curr Biol 25(20):2690–2695. https://doi.org/10.1016/j.cub.2015.08.057
14. De Martino F, Moerel M, Ugurbil K, Goebel R, Yacoub E, Formisano E (2015) Frequency preference and attention effects across cortical depths in the human primary auditory cortex. Proc Natl Acad Sci USA 112(52):16036–16041. https://doi.org/10.1073/pnas.1507552112
15. Lawrence SJD, Formisano E, Muckli L, de Lange FP (2019) Laminar fMRI: applications for cognitive neuroscience. Neuroimage 197:785–791. https://doi.org/10.1016/j.neuroimage.2017.07.004
16. Stephan KE, Petzschner FH, Kasper L, Bayer J, Wellstein KV, Stefanics G, Pruessmann KP, Heinzle J (2019) Laminar fMRI and computational theories of brain function. Neuroimage 197:699–706. https://doi.org/10.1016/j.neuroimage.2017.11.001
17. Chan RW, Leong ATL, Ho LC, Gao PP, Wong EC, Dong CM, Wang X, He J, Chan YS, Lim LW, Wu EX (2017) Low-frequency hippocampal-cortical activity drives brain-wide resting-state functional MRI connectivity. Proc Natl Acad Sci USA 114(33):E6972–E6981. https://doi.org/10.1073/pnas.1703309114
18. Lee JH, Durand R, Gradinaru V, Zhang F, Goshen I, Kim DS, Fenno LE, Ramakrishnan C, Deisseroth K (2010) Global and local fMRI signals driven by neurons defined optogenetically by type and wiring. Nature 465(7299):788–792. https://doi.org/10.1038/nature09108
19. Chan RW, Cron GO, Asaad M, Edelman BJ, Lee HJ, Adesnik H, Feinberg D, Lee JH (2022) Distinct local and brain-wide networks

are activated by optogenetic stimulation of neurons specific to each layer of motor cortex. Neuroimage 263:119640. https://doi.org/10.1016/j.neuroimage.2022.119640
20. Chan RW, Hamilton-Fletcher G, Edelman BJ, Faiq MA, Sajitha TA, Moeller S, Chan KC (2024) NOise Reduction with DIstribution Corrected (NORDIC) principal component analysis improves brain activity detection across rodent and human functional MRI contexts. Imaging Neurosci (Camb) 2:1–18. https://doi.org/10.1162/imag_a_00325
21. Choy M, Dadgar-Kiani E, Cron GO, Duffy BA, Schmid F, Edelman BJ, Asaad M, Chan RW, Vahdat S, Lee JH (2022) Repeated hippocampal seizures lead to brain-wide reorganization of circuits and seizure propagation pathways. Neuron 110(2):221–236. https://doi.org/10.1016/j.neuron.2021.10.010
22. Leong AT, Chan RW, Gao PP, Chan YS, Tsia KK, Yung WH, Wu EX (2016) Long-range projections coordinate distributed brain-wide neural activity with a specific spatiotemporal profile. Proc Natl Acad Sci USA 113(51):E8306–E8315. https://doi.org/10.1073/pnas.1616361113
23. Leong ATL, Gu Y, Chan YS, Zheng H, Dong CM, Chan RW, Wang X, Liu Y, Tan LH, Wu EX (2019) Optogenetic fMRI interrogation of brain-wide central vestibular pathways. Proc Natl Acad Sci USA 116(20):10122–10129. https://doi.org/10.1073/pnas.1812453116
24. Ebbesen CL, Brecht M (2017) Motor cortex - to act or not to act? Nat Rev Neurosci 18(11):694–705. https://doi.org/10.1038/nrn.2017.119
25. Weiler N, Wood L, Yu J, Solla SA, Shepherd GM (2008) Top-down laminar organization of the excitatory network in motor cortex. Nat Neurosci 11(3):360–366. https://doi.org/10.1038/nn2049
26. Gerfen CR, Economo MN, Chandrashekar J (2018) Long distance projections of cortical pyramidal neurons. J Neurosci Res 96(9):1467–1475. https://doi.org/10.1002/jnr.23978
27. Anderson CT, Sheets PL, Kiritani T, Shepherd GM (2010) Sublayer-specific microcircuits of corticospinal and corticostriatal neurons in motor cortex. Nat Neurosci 13(6):739–744. https://doi.org/10.1038/nn.2538
28. Yamawaki N, Borges K, Suter BA, Harris KD, Shepherd GM (2014) A genuine layer 4 in motor cortex with prototypical synaptic circuit connectivity. eLife 3:e05422. https://doi.org/10.7554/eLife.05422

Chapter 3

High-Throughput, Brain-Wide Effective Connectivity Mapping by fMRI with Patterned Optogenetics

Hyun Seok Moon, Seonghoon Kim, Myunghwan Choi and Seong-Gi Kim

Abstract

Functional magnetic resonance imaging (fMRI) combined with optogenetic neural manipulation provides a powerful framework for mapping brain-wide effective connectivity (EC). In this study, we developed a high-throughput approach integrating spatiotemporally programmable optogenetic stimulation with fMRI, enabling precise and flexible neural manipulation across the mouse cortex. Using a digital micromirror device (DMD) coupled with an optical fiber bundle inside an MR scanner, we delivered patterned photostimulation while acquiring fMRI data. By sequentially stimulating nine atlas-defined cortical regions within a single fMRI run, we achieved high-throughput mapping of cortical EC. This approach also offers versatility in experimental design, such as functional localizer-based stimulation planning or viral tracing-based circuit-specific targeting, facilitating a more refined investigation of functional networks. The integration of mouse fMRI with flexible stimulation paradigms provides a precise method for studying large-scale neural interactions, paving the way for high-throughput exploration of circuit-level mechanisms underlying cognition and behavior.

Keywords Mouse fMRI, Effective connectivity, Patterned optogenetics, Viral tracing, Cranial window, DMD

1 Introduction

Functional magnetic resonance imaging (fMRI) is a powerful tool for investigating neural activity across the entire brain, providing valuable insights into behavior and neurological or psychiatric disorders. Leveraging its noninvasive mapping capability, fMRI has been widely used to study brain-wide connectivity in live subjects. However, its low temporal resolution, typically in the order of seconds, limits its ability to determine the directionality of neural influence. This poses a challenge for understanding causal interactions in brain networks.

To address this limitation, fMRI has been combined with direct stimulation techniques targeting specific brain regions, such as electrical microstimulation [1-3], infrared stimulation [4, 5], or optogenetics [6-25]. Optogenetics is a genetic tool that enables the

Kevin Chuen Wing Chan and Seong-Gi Kim (eds.), *Optogenetic Functional Neuroimaging*, Neuromethods, vol. 232, https://doi.org/10.1007/978-1-0716-5178-0_3,

modulation of genetically defined neuronal populations through light illumination [26] and offers distinct advantages over other methods, particularly in its ability to selectively target specific cell types. Integration with fMRI enables precise neural manipulation while monitoring whole-brain responses, facilitating the investigation of causal interactions exerted from one to others—referred to as effective connectivity (EC) [27]—by systematically assessing how localized neural manipulation influences distant brain areas. With its high spatiotemporal precision and cell-type specificity, optogenetic fMRI has contributed to dissecting complex brain networks at the circuit level.

In conventional optogenetic fMRI, light delivery for activating optogenetic actuators is mediated by an optical fiber surgically implanted in the target area [6, 10]. This approach can be easily implemented by connecting a fiber cannula to the light source and is particularly useful for targeting deep brain structures. However, this method significantly limits experimental flexibility, as the location of stimulation is fixed before experiments, where typically only one or a few anatomically predefined regions can be studied per subject. Additionally, the fiber diameter is predetermined, restricting the activation area to a circular-shaped region around the fiber tip. To overcome these constraints, alternative optogenetic stimulation approaches have been developed to enable more flexible spatial control of light delivery, such as focal beam scanning [28, 29] or patterned illumination [30, 31]. However, integrating these approaches with fMRI remains challenging due to technical limitations. The confined space within an MRI bore restricts the use of adjustable optics, and the presence of metal or electromagnetic components in optical systems can interfere with MRI signal acquisition. As a result, implementing an advanced optogenetic system with fMRI has yet to be successfully achieved.

Recent advances in fiber bundle-relay applications offer a promising solution for integrating optical systems with MRI, preventing magnetic interference between them [32]. This approach employs a fiber bundle consisting of thousands of individual fiber cores, preserving the spatial information of the input light field and delivering it to the other end. Adopting this system, we developed a hybrid system integrating cortex-wide optical imaging and patterned optogenetics with mouse fMRI for efficient mapping of EC in cell-type-specific channelrhodopsin-2 (ChR2) expressing mice. Additionally, our method allows for in-situ stimulation targeting based on functional localizers or circuit-specific viral tracing, which has been unfeasible in optogenetic fMRI. In this chapter, we will describe the detailed methodology for high-throughput mapping of EC from cortical regions, focusing on three approaches for defining stimulation targets: (1) mouse brain atlas, (2) functional localizer, and (3) retrograde viral tracing.

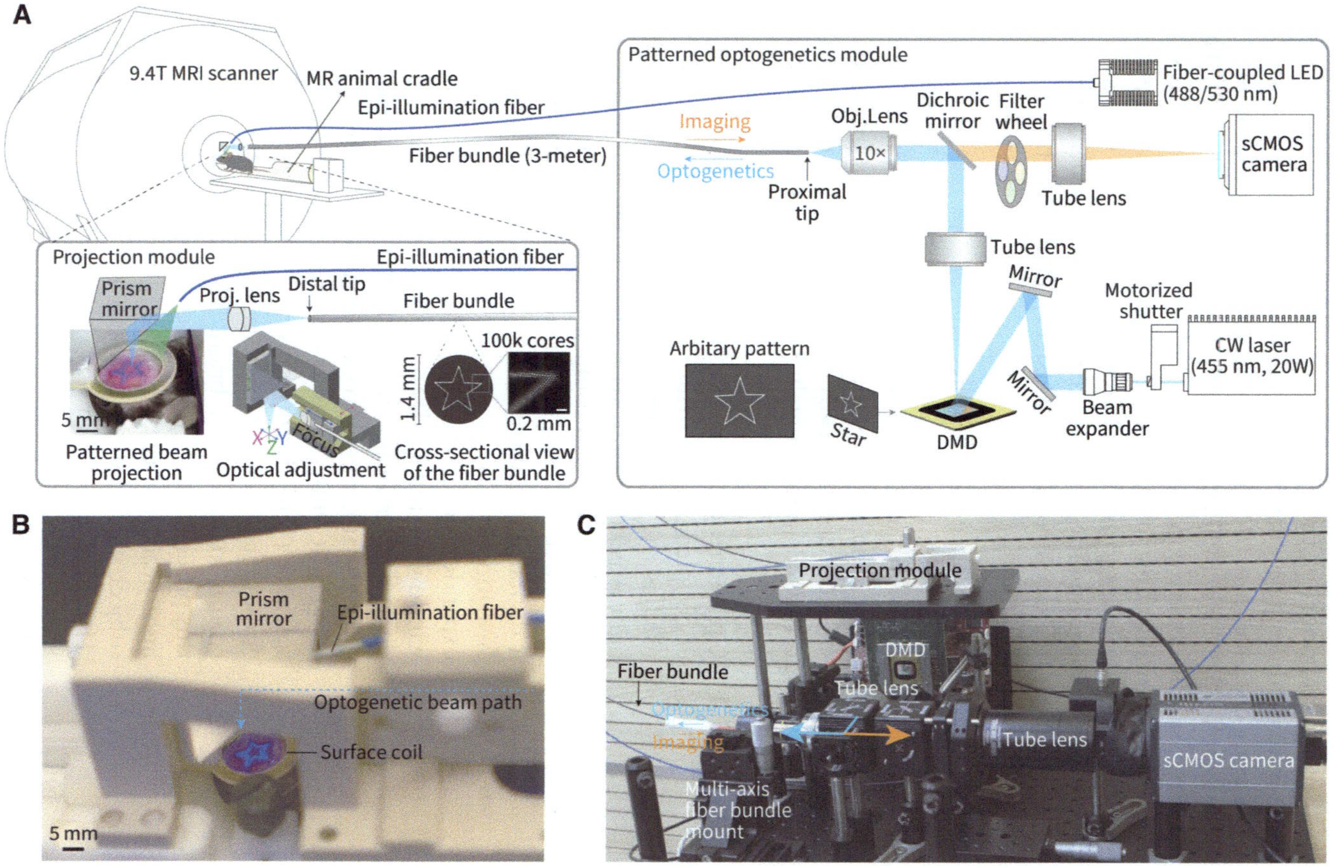

Fig. 1 Patterned optogenetic fMRI setup. (**A**) To prevent MRI interference, the optical system was divided into two modules—a patterned optogenetics module located 3 m away from an MRI scanner and a projection module mounted on the animal cradle inside the MRI bore. Those modules are connected by a fiber bundle for bidirectional optogenetic stimulation and optical imaging. The patterned optogenetics module consists of a fiber-coupled 450 nm laser and a digital micromirror device (DMD) to generate and shape light patterns. Additionally, an optical imaging system captures fluorescence and reflectance signals using an sCMOS camera, which receives signals from the mouse brain. The projection module, made of MRI-compatible plastic, precisely projected the patterned light onto the mouse cortex, with a three-axis translation stage allowing fine adjustments to the field of view and focus. For optical imaging, an additional epi-illumination fiber was installed, connected to a 488 or 530 nm LED. (**B** and **C**) Photographs of the projection module (**B**) and the patterned optogenetics module (**C**). Reprinted from [39] with permission from Elsevier

2 Materials

2.1 Optics System

To prevent interference between the MRI scanner and optoelectronic components, the optical system was designed with two separate modules: (1) a patterned optogenetics module, positioned at a safe distance from the MRI scanner, and (2) a projection module, attached to the animal cradle inside the MRI bore (Fig. 1). These two modules were connected via a high-density fiber bundle, which bidirectionally transmits optogenetic stimulation to the mouse brain and optical signals to an sCMOS camera for imaging. Optical components, including the laser, shutter, DMD, and camera, were controlled by a TTL control device (Master9, AMPI) synchronized with MRI data collection.

(1) Patterned optogenetics module.
The optogenetic module combines light beam pattern generation for optogenetic stimulation with an optical readout component. The whole setup was securely mounted on an optical breadboard (60×90 cm; MB6090, Thorlabs), which was tightly fixed approximately 3 m away from the magnet (see Fig. 1C).

(1.1) Optogenetic patterns were generated as follows: a fiber-coupled laser (450 nm; GP-450-100-20, RPMC) was used as the light source for optogenetics. The emitted light was collimated to a 4 mm in diameter beam and expanded threefold using a beam expander (GBE03-A, Thorlabs). The expanded beam was then directed to a digital micromirror device (DMD; DLP9000, Texas Instruments) via a set of mirrors. The size of the expanded beam was adjusted to fully cover the active window of the DMD (2560×1600 pixels). The incident circular beam on the DMD was sculpted into a desired pattern, controlled by a DMD digital controller (DLPC900, Texas Instruments) via DMD control software (DLP-LightCrafter, Texas Instruments). This computer-generated pattern will define the activation region in the dorsal brain. The key requirements for this process are: (1) precise co-registration between the light beam and the brain, and (2) in situ generation of subject-specific beam patterns for fMRI studies.
The patterned beam was then reflected perpendicularly off the DMD window, passed through a tube lens (TTL200A, Thorlabs), and further reflected by a long-pass dichroic mirror (470 nm; T470lpxr, Chroma). It was then focused onto the proximal tip of the fiber bundle (100,000 cores, 1.6 mm outer diameter, 3 m long;

FIGH 100–1500 N, Fujikura) by adjusting its three-axis position relative to an objective lens (10×, NA0.25, PLN10×, Olympus).

(1.2) An optical readout setup can capture fluorescence and reflectance images of the mouse cortex. Optical signals from the fiber bundle were transmitted back through the objective lens and dichroic mirror, then passed through a filter wheel equipped with a long-pass filter (FF01-496/LP, Thorlabs) for switching between fluorescence and reflectance imaging modes. The optical signals were then captured by an sCMOS camera (2048 × 2048 pixels, Zyla 4.2, Andor) through a tube lens. To deliver excitation light across the cortex, a multimode fiber-optic patch cord (MFP 600/660/2000-0.22 10 m SMA-ZF1.25, Doric Lenses) was attached to the projection module and coupled with a 530 nm LED (M530L4, Thorlabs), which was modulated by an external LED driver (DC4104, Thorlabs). The optical imaging system is necessary for co-registering the brain with the beam coordinates and adjusting the beam focus onto the cortex. Additionally, simultaneous fMRI and optical imaging measurements enable researchers to investigate the neural correlates of fMRI signals.

(2) Projection module.

The movable projection module, which houses an achromatic projection lens and a prism mirror, is MRI-compatible as it is made of polyether ether ketone (PEEK). It was mounted onto the standard mouse MRI cradle. The fiber bundle was inserted into the guiding hole of the projection module, with its distal tip positioned in front of the achromatic projection lens (ACH 6.25 × 7.5 VIS 0 TS, Edmund Optics) and precisely adjusted for optimal focus. To align the field-of-view (FOV) for both the camera and the optogenetic beam patterns onto the cortex, the positions of both the fiber tip and the projection lens were fine-tuned using a 2-axis translation stage, and the prism mirror was adjusted accordingly. This process was assisted by real-time optical imaging monitoring using an sCMOS camera. The epi-illumination fiber, secured to the projection module, provided uniform illumination across the cortex. The projection module was designed to fit within the 8.6-cm diameter radiofrequency (RF) volume coil.

(3) Fiber bundle.

An optical fiber bundle was used to transmit light between the patterned optogenetic module and the projection module. The fiber bundle was selected based on high light transmission capability, a large number of cores for high spatial resolution, flexibility for easier integration into the setup, and commercial availability. For this purpose, we chose a 3-m-long fiber bundle with 100,000 silica-based cores and a 1.5 mm fiber diameter (model: FIGH 100–1500 N, Fujikura; https://www.fujikura.co.jp/eng/products/optical/optical-fibers/07/2050110_12902.html), which has an imageable diameter of 1.4 mm. The outer diameter, including the protective coating, is 1.7 mm, and the minimum bending diameter (i.e., how tightly the fiber can be bent without causing damage or significant signal loss) is 40 cm. This allows the fiber bundle to be easily integrated into mouse MRI research setups. To cover the entire cortex of the mouse, the FOV was set to 14 mm in diameter, achieved by magnifying the image from the original fiber bundle tenfold using a 10× projection lens. Each core's diameter is approximately 4 μm, resulting in a spatial resolution of approximately 40 μm.

2.2 Animals

Cortex-wide optogenetic stimulation requires transgenic mouse models with brain-wide opsin expression. In this study, we used Thy1–ChR2–EYFP mice (B6. Cg-Tg(Thy1–COP4/EYFP)9Gfng/J; stock #007615; Jackson Laboratory, Bar Harbor, ME) [33], which have light-sensitive ChR2 expression enriched in layer 5 pyramidal neurons throughout the cortex. Additionally, for fluorescence image-based optogenetic stimulation experiments, we used one C57BL/6 mouse (Orient Bio, South Korea). All procedures were approved by the Institutional Animal Care and Use Committee of Sungkyunkwan University in accordance with standards for humane animal care from the Animal Welfare Act and the National Institutes of Health (NIH) Guide for the Care and Use of Laboratory Animals.

2.3 MRI Data Acquisition

Anatomical and functional MRI datasets were collected using a 9.4 T/30 cm MRI scanner (Bruker Biospec, Billerica, MA), equipped with an actively shielded 12-cm gradient system operating at a maximum strength of 66 G/cm and a rise time of 141 ms. An 86-mm volume coil was used for RF transmission, and a single-loop surface coil with an inner diameter of 15 mm and an outer diameter of 20 mm (Bruker Biospec, Billerica, MA) was used for signal reception. An important adoption of our approach was the utilization of existing Bruker hardware, including the RF coil and mouse cradle. To minimize field inhomogeneity, local shimming was performed based on the B_0 map within the elliptical shim volume inside the brain.

High-resolution anatomical imaging: a high-resolution 3D T2-weighted anatomical image was obtained using rapid acquisition with refocused echoes (RARE) with the following parameters: FOV = 16 mm (readout, left–right) × 10 mm (ventral–dorsal) × 10 mm (rostral–caudal), encompassing the position of fluorescence tubes; spatial resolution = 0.1 × 0.1 × 0.1 mm^3; repetition time (TR)/echo time (TE) = 1500/36 ms; RARE factor = 16; and total acquisition time = 16 min and 30 s.

Functional echo-planar imaging (EPI): we acquired cerebral blood volume (CBV)-weighted fMRI to enhance the sensitivity and specificity of fMRI responses. Functional images were acquired using 2D or 3D single-shot gradient-echo EPI sequence with the following parameters: for 2D, FOV = 16 mm (readout, left–right) 8 mm (ventral–dorsal); 18 coronal slices with a slice thickness of 0.5 mm; spatial resolution = $0.167 \times 0.167 \times 0.5$ mm^3; TR/TE = 1000/8.35 ms (temporal resolution of 1 s); flip angle (FA) = 47°; and scan time per fMRI run ≤440 s. For 3D, FOV = 16 mm (readout, left–right) × 12 mm (rostral–caudal) × 8 mm (ventral–dorsal); spatial resolution = 0.2 mm isotropic; TR/TE = 50/8 ms (temporal resolution of 2 s); FA = 12°; and scan time per fMRI run ≤580 s. To mitigate aliasing effects along the phase-encoding direction, saturation slabs were applied outside the FOV.

3 Methods

3.1 Animal Preparation

Optical transparency is essential for cortex-wide optogenetic stimulation and wide-field optical imaging. To achieve this, a thinned-skull cranial window was prepared over the dorsal cortex (Fig. 2). Mice were initially anesthetized with 4% isoflurane and secured in a stereotaxic frame (Narishige, Tokyo, Japan). Isoflurane was then reduced to 2% and maintained throughout the procedure. Body temperature was maintained at 37°C using a heating pad placed under the body. We first shaved the scalp and administered meloxicam (1 mg/kg) under the scalp to reduce pain and inflammation. The scalp was then incised, and the periosteum was removed using a surgical blade

Fig. 2 Thinned-skull cranial window. A dental resin wall was made around the cranial window to retain agarose gel. The wall fits within the 15-mm inner diameter of the MRI receiver coil. The wall was then filled with agarose solution in D_2O. For MRI-optical image co-registration, three to four fiducial markers—small PE-10 tube segments filled with rhodamine B solution in H_2O—were placed at the circumference of the wall. These markers were visible in both MRI and fluorescence imaging, while the D_2O agarose gel is invisible. Finally, a coverslip was placed over the agarose well and sealed with biocompatible silicone. Reprinted from [39] with permission from Elsevier

to expose the dorsal skull. The entire dorsal skull surface was then carefully thinned using a handheld drill (*see* **Note 1**). After clearing debris and moisture on the skull, a cyanoacrylate glue (Loctite 401, Henkel, Dusseldorf, Germany) was thinly and uniformly applied to enhance optical clarity. Last, we constructed a 1 mm-high light-curing dental resin wall (~15 mm in diameter) around the skull. This wall allowed the 15-mm diameter receiver coil to be positioned around it (Fig. 2A). During these surgical procedures, careful removal of blood clots and minimization of air bubbles within the resin wall helped reduce image artifacts. Following surgery, the mice were returned to their cages and monitored for 2 days to ensure full recovery before experiments.

On the day of fMRI experiments, anesthesia was induced by placing the mice in a chamber with 4% isoflurane for 4 min. Following induction, isoflurane was discontinued, and a bolus of ketamine (100 mg/kg) and xylazine (10 mg/kg) mixture was initially administered intraperitoneally (IP), followed by continuous intravenous (IV) infusion of ketamine (35–45 mg/kg/h) and xylazine (1.75–2.25 mg/kg/h) through the tail vein for anesthesia maintenance [34]. For the continuous administration of anesthetics and contrast agents, a 31 G needle was inserted into the tail vein and connected to a 25 G Y-connector (SCY25, Instech Laboratories, Plymouth Meeting, PA) for the delivery of a ketamine/xylazine mixture and monocrystalline iron oxide nanoparticles to provide CBV-weighted imaging contrast.

To minimize susceptibility artifacts, maintain optical clarity, and enable co-registration between optical images and MRI, three additional steps were implemented. (1) Initially, the skull surface was thoroughly cleaned with a wet cotton swab to remove dust. The dental resin wall was then filled with 0.5% agarose solution in deuterium oxide (D_2O), prepared by heating in a microwave. Dissipated air bubbles were removed from the agarose solution to minimize magnetic susceptibility artifacts [35]. (2) For MRI-optical image co-registration, we placed three or four fiducial markers immersed in agarose solution at the circumference of the resin wall (Fig. 2B). These markers are PE-10 tube segments filled with rhodamine B solution in H_2O, made by melt-cutting of the tube at 2 mm intervals (*see* **Note 2**). Note that the fiducial markers are visible in both MRI and fluorescence optical imaging, while the D_2O agarose gel is invisible in both imaging modalities. (3) To prevent evaporation of the agarose solution, a cover glass was placed over the agarose well and securely sealed with biocompatible silicone adhesive (Kwik-Sil, World Precision Instrument, Sarasota, FL). All steps were completed within a few minutes before the agarose solution cooled and solidified.

3.2 Designing Patterned Optogenetic Patterns

To ensure precise stimulation targeting, a multiple-step registration process was performed as follows: (1) registration between the DMD and sCMOS images, (2) registration between 2D optical images and

3D MRI, and (3) registration of MRI data with the Allen Mouse Brain Atlas (for atlas-based stimulation patterns). This process relied on high-resolution anatomical MRI and fluorescent optical images to achieve accurate spatial correspondence across modalities.

Step 1. The DMD and sCMOS images were aligned and co-registered using a point-based registration algorithm (Fig. 3A). This process involved three steps: (1) an arbitrary pattern (a star shape in this case) was input into the DMD, (2) the reflected pattern at the proximal tip of the fiber bundle was captured by the sCMOS camera, and (3) multiple corners of the star-shaped pattern were selected as

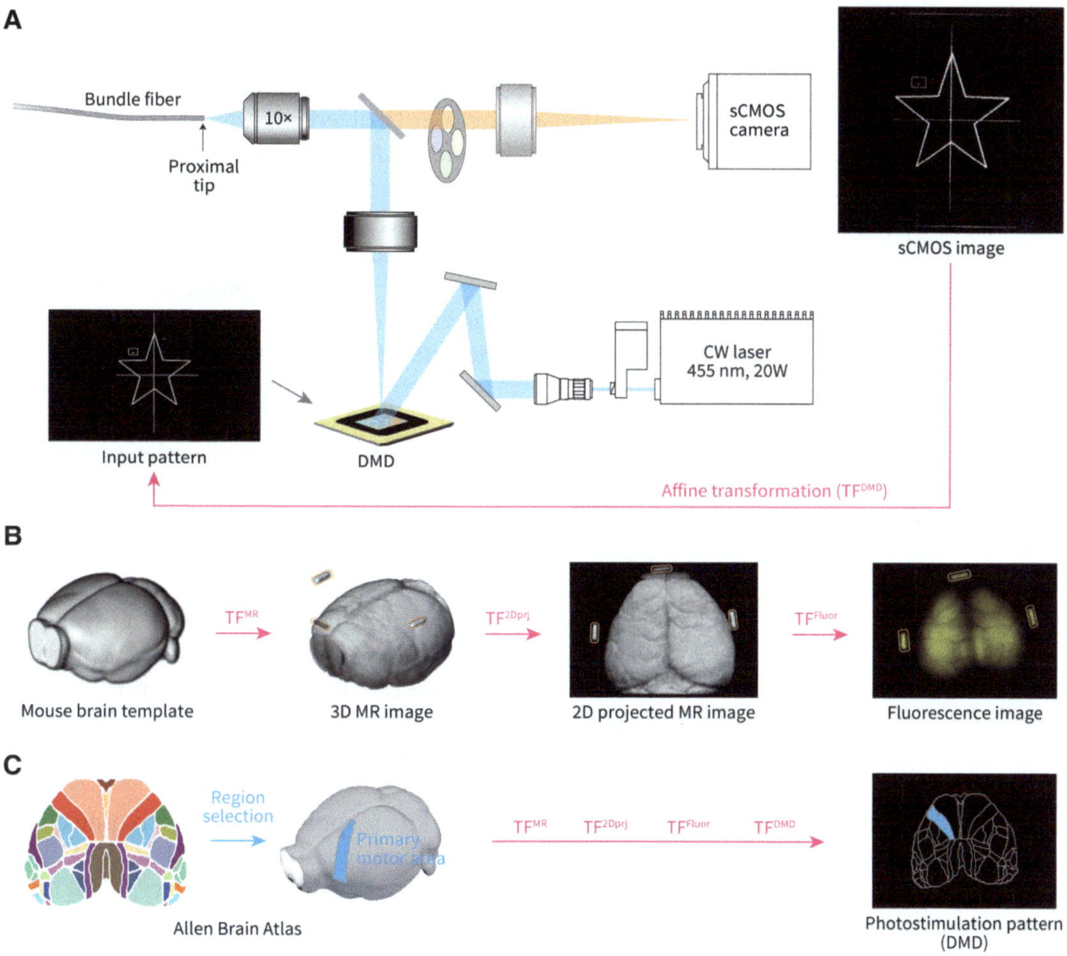

Fig. 3 Co-registration between MRI and the optics system. (**A**) For DMD–sCMOS co-registration, a polygon-shaped pattern was projected from the DMD, reflected at the proximal tip of the fiber bundle, and captured by the sCMOS. The sCMOS image was then co-registered to the DMD input pattern, generating a transformation matrix TF^{DMD}. (**B**) To define atlas-based stimulus patterns, a standard brain template was first co-registered to individual anatomical images (TF^{MR}), 2D-projected onto a dorsal plane (TF^{2Dprj}), and co-registered to a fluorescence image (TF^{Fluor}). (**C**) Finally, Allen mouse brain atlas underwent these transformations, producing individualized stimulation patterns. Reprinted from [39] with permission from Elsevier

reference points in both the sCMOS image and DMD input pattern. The sCMOS image was then affine-transformed to align with the original input pattern based on the reference points (MATLAB fitgeotrans). This procedure produced the transformation matrix (TF^{DMD}) between the DMD and sCMOS.

Step 2. To generate subject-specific, atlas-based stimulus patterns, the 3D MRI was co-registered with the 2D optical image of the cortical surface (Fig. 3B). We used a T2-weighted MRI mouse brain template in the same spatial coordinates as the Allen Mouse Brain Atlas [36], previously generated in separate sessions. For each subject, the mouse brain template was co-registered to the individual's anatomical MRI image (containing the fiducial markers) through a nonlinear transformation (ANTs antsRegistrationSyN), accounting for individual variations. This process generated the transformation matrix and warp field (TF^{MR}), enabling precise alignment between the subject's brain and the standardized atlas. Next, the most superficial voxels from the individual MR image were projected onto a dorsal plane (TF^{2Dprj}). The resulting 2D projection image was then affine-transformed to align with the fluorescence optical image (MATLAB fitgeotrans), using fiducial markers as reference points, yielding a transformation matrix TF^{fluor}. If optogenetic stimulation patterns were designed based on intrinsic optical imaging or fluorescent images instead of the Allen Brain Atlas, the mouse brain MRI template (and the TF^{MR} transformation matrix) would not be necessary. Instead, co-registration between optical images and individual anatomical MRI scans would be sufficient for stimulus targeting.

Step 3. Individualized photostimulus patterns based on the mouse brain atlas were generated through sequential transformations described above (Fig. 3C). The resulting binary stimulus patterns further underwent Gaussian filtering (MATLAB imgaussfilt; sigma = 10) to smooth pixelated edges. Next, the filtered patterns were binarized using a threshold of 0.7 to define stimulation regions. Additionally, the borders were eroded (MATLAB imerode) by up to 40 μm to introduce a spatial buffer between adjacent regions, minimizing the potential overlap between those regions.

3.3 High-Throughput EC Mapping, Based on Atlas-Based Patterned Optogenetic Stimulation

For high-throughput cortical EC mapping, we selected nine cortical atlas-based regions for optogenetic stimulation (Fig. 4A). These included: unilateral motor (primary, MOp; secondary, MOs) areas, primary somatosensory (barrel field, SSp-bfd; upper and lower limb, SSp-ul/ll) areas, visual (primary, VISp; anterior and rostrolateral, VISa/rl) areas, auditory area (AUD), medial (antero- and posteromedial visual, VISam/pm) area, and bilateral retrosplenial areas (dorsal and ventral, RSPd/v).

Each region was stimulated sequentially within a single fMRI run (a scan time of 580 s). Stimulation parameters were set at an irradiance of 3.6 mW/mm^2 (*see* **Note 3**), a frequency of 20 Hz, and a 20% duty cycle, with 10-s stimulation blocks followed by 50-s recovery periods. Each fMRI run consisted of a 40-s baseline period, followed by nine stimulation periods (each containing 10 s of stimulation and 50 s of rest). We collected and averaged three fMRI runs per animal ($N=8$).

For fMRI experiments, we employed CBV-weighted fMRI instead of conventional blood-oxygenation-level-dependent (BOLD) fMRI to enhance the sensitivity and specificity of fMRI responses [37], with the injection of 20 mg/kg long-lasting iron oxide nanoparticles (Feraheme, AMAG Pharmaceuticals, Waltham, MA). Functional images were acquired using a 3D gradient-echo EPI sequence. Increased neural activity leads to a decrease in CBV-weighted fMRI signals, reflecting an increase in actual CBV. fMRI data from each animal underwent motion correction to the first volume of the first run (AFNI 3dvolreg). The dataset was co-registered to the individual T2-weighted anatomical images using a rigid-body linear transformation (FSL flirt), and then normalized to the mouse brain template [38] using the inverse of the transformation matrix and warp field (TF^{MR}) (ANTs antsApplyTransforms). To improve the signal-to-noise ratio (SNR), spatial smoothing was applied using a Gaussian kernel with a full-width half maximum (FWHM) of one voxel size (0.2 mm). All datasets were then converted to percent signal change relative to the mean signal intensity during the initial 40-s baseline period.

Patterned optogenetic stimulation elicited robust responses in the stimulated regions and their downstream networks. Example time courses from a single subject are shown in Fig. 4B. Note that decreases in CBV-weighted fMRI signals indicate increases in CBV. To quantify EC from the preprocessed fMRI data, we first defined a hemodynamic response function in a data-driven way [39]. For this, time courses from only those exhibiting significant responses ($p < 0.001$; one-tailed z-test), compared to stimulus-free fMRI runs, were selected. The chosen time courses were then averaged and fitted by a double gamma-variate function. The resulting HRF was then used as a regressor in a general linear model to quantify voxel-wise responses (Fig. 4B). Finally, EC strength was calculated as the sum of voxel-wise fMRI responses within each anatomically defined region-of-interest (ROI) across all animals, normalized by the sum of responses in the stimulated ROIs across all animals (Fig. 4C).

3.4 EC Mapping of a Functionally Localized Region

The experiments described so far have focused on mapping EC from atlas-defined cortical regions, which are identified by the co-registration of individual brain images to a standardized template based on anatomical features. Although this approach offers a structured framework for investigating large-scale connectivity, it assumes that functional regions are anatomically consistent across subjects. However, the precise location of a region functionally relevant to a specific brain function can vary across subjects [40]. In such cases, it will be more beneficial to define ROIs using a functional localizer.

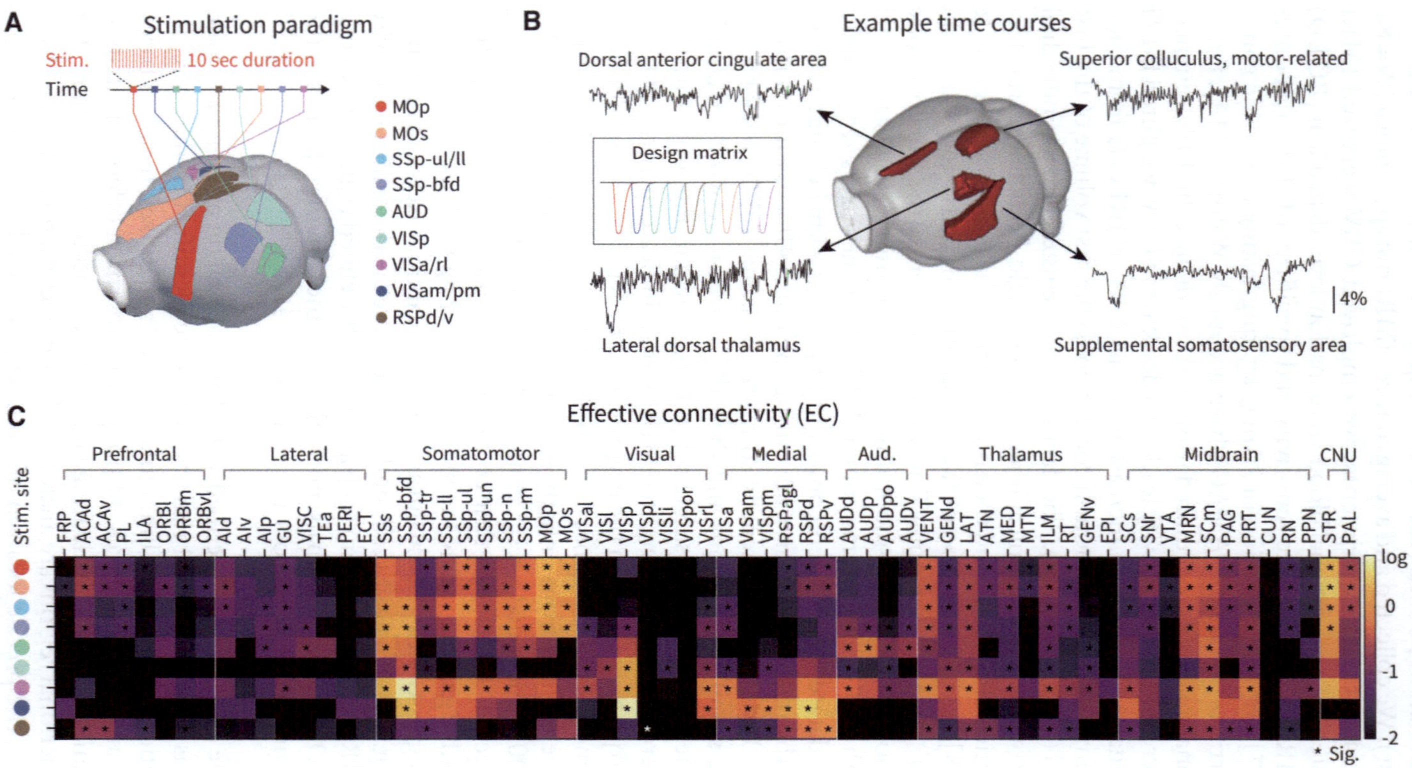

Fig. 4 EC mapping of atlas-based cortical regions. (**A**) Nine cortical regions were sequentially stimulated within a single fMRI run. (**B**) Example time courses from one subject, with the design matrix shown in the inset. (**C**) EC matrix ($N=8$) from nine cortical regions to 65 ROIs ipsilateral across nine network modules, where the percentage response in each ROI is normalized to the stimulation site. Statistical significance of each connection is determined by a one-sample t test with a false discovery rate (FDR) correction (corrected $p<0.05$). FRP frontal pole, cerebral cortex; ACAd anterior cingulate area, dorsal part; ACAv anterior cingulate area, ventral part; PL prelimbic area; ILA infralimbic area; ORBl orbital cortex, lateral part; ORBm orbital cortex, medial part; ORBvl orbital cortex, ventrolateral part; ALd agranular insular area, dorsal part; ALv agranular insular area, ventral part; ALp agranular insular area, posterior part; GU gustatory area; VISC visceral area; TEa temporal association areas; PERI perirhinal area; ECT ectorhinal area; SSs supplemental somatosensory area; SSp-bfd primary somatosensory area, barrel field; SSp-tr primary somatosensory area, trunk; SSp-ll primary somatosensory area, lower limb; SSp-ul primary somatosensory area, upper limb; SSp-un primary somatosensory area, unassigned; SSp-n primary somatosensory area, nose; SSp-m primary somatosensory area, mouth; Mop primary motor area;

This approach is challenging with conventional optical fiber-based optogenetics, as it does not allow for flexible modification of the stimulation shape or location. In contrast, our method offers a great advantage by enabling in-situ stimulation targeting, allowing for data-driven stimulation targeting.

To demonstrate this capability, we generated optogenetic stimulation patterns based on a functional ROI of sensory stimulation (Fig. 5A). For this experiment, CBV-weighted fMRI (2D EPI) and intrinsic optical imaging under 530 nm illumination were simultaneously acquired with 20-s electrical forepaw stimulation (4 Hz frequency, 500 μs pulse duration, 0.5 mA). Optical imaging was performed with a temporal resolution of 1 s and a 512 × 512 image size (with 4 × 4 pixel binning).

The fMRI results showed highly significant activation in the primary somatosensory cortex contralateral to the stimulated forepaw, as well as in the thalamic areas involved in sensory processing. Concurrently, forepaw stimulation elicited a significant decrease in CBV-weighted optical signals in the localized forelimb region (blue area in Fig. 5A), indicating an increase in CBV. Both imaging modalities can be used for stimulation planning. However, in this specific application, optical imaging was chosen to generate an optogenetic stimulation pattern due to its superior spatial resolution. Optogenetic stimulation of the functionally defined ROI revealed downstream connections from the somatosensory cortex, with activity patterns largely consistent with those observed during forepaw stimulation. This approach enables the dissection of upstream and downstream neural pathways involved in sensory processing.

3.5 Circuit-Specific EC Mapping

Understanding how distinct neuronal pathways contribute to brain function requires circuit-specific mapping of EC. Unlike conventional approaches that define regions based on anatomy or functional responses, circuit-specific targeting enables precise control of circuit activity by selectively stimulating neurons labeled with anterograde

Fig. 4 (continued): Mos secondary motor area; VISal anterolateral visual area; VISl lateral visual area; VISp primary visual area; VISpl posterolateral visual area; VISli laterointermediate area; VISpos postrhinal area; VISrl rostrolateral visual area; VISa anterior visual area; VISam anteromedial visual area; VISpm posteromedial visual area; RSPagl retrosplenial area, lateral agranular part; RSPd retrosplenial area, dorsal part; RSPv retrosplenial area, ventral part; AUDd dorsal auditory area; AUDp primary auditory area; AUDpo posterior auditory area; AUDv ventral auditory area; VENT ventral group of the dorsal thalamus; GENd geniculate group, dorsal thalamus; LAT lateral group of the dorsal thalamus; ATN anterior group of the dorsal thalamus; MED medial group of the dorsal thalamus; MTN medial group of the dorsal thalamus; ILM intralaminar group of the dorsal thalamus; RT reticular nucleus group of the thalamus; GENv geniculate group, ventral thalamus; EPI epithalamus; SCs superior colliculus, sensory related; SNr substantia nigra, reticular part; VTA ventral tegmental area; MRN midbrain reticular nucleus; SCm superior colliculus, motor related; PAG periaqueductal gray; PRT pretectal region; CUN cuneiform nucleus; RN red nucleus; PPN pedunculopontine nucleus; STR striatum; PAL pallidum. Reprinted from [39] with permission from Elsevier

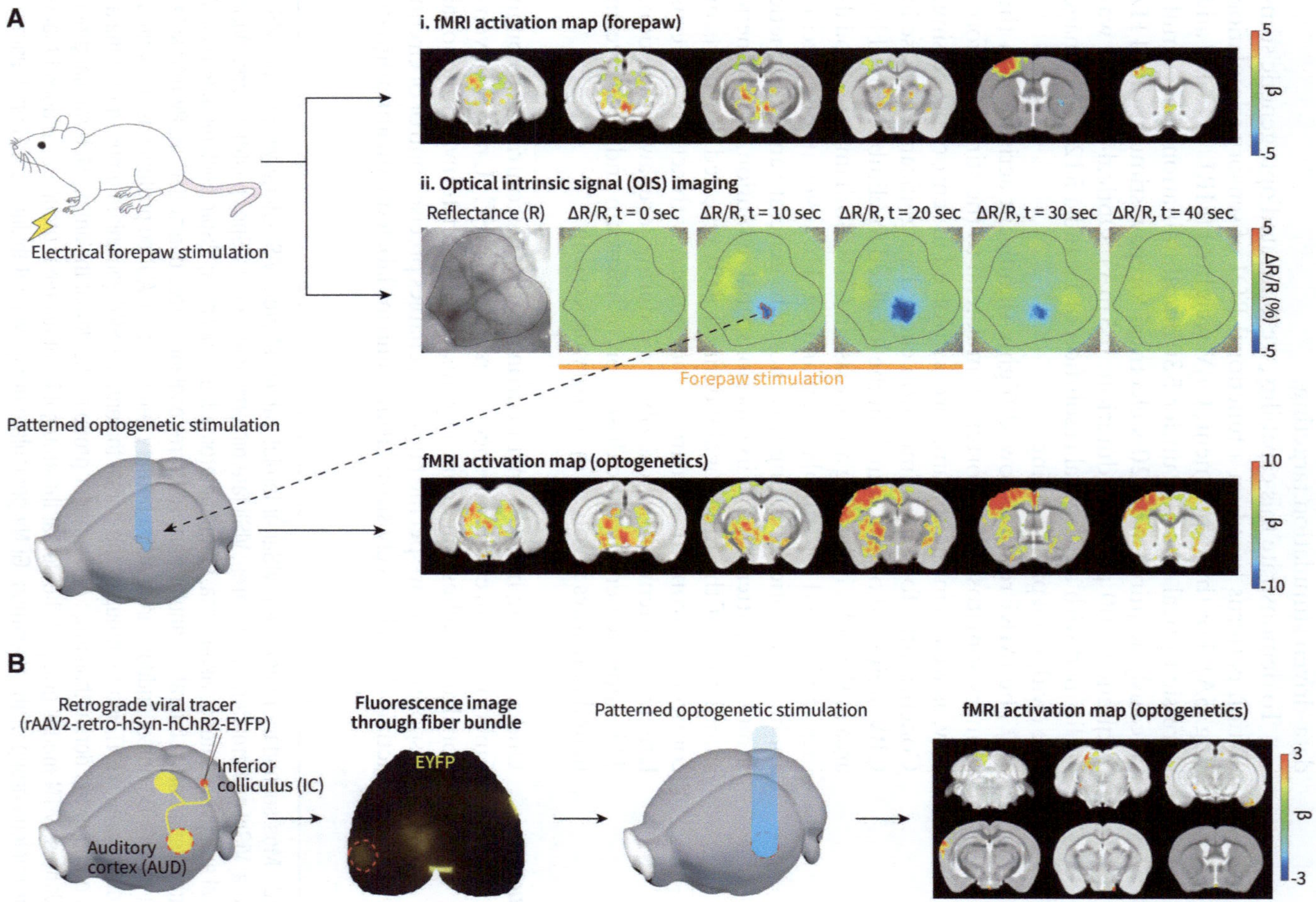

Fig. 5 In-situ stimulation planning strategies. (**A**) The somatosensory cortex related to the right forepaw was identified by simultaneous fMRI and optical intrinsic signal (OIS) imaging with electrical forepaw stimulation. In OIS imaging, a decrease in reflectance (R) indicates an increase in CBV. The responsive area, outlined in red, was extracted from OIS imaging and used as the optogenetic stimulation target. (**B**) A retrograde viral tracer encoding ChR2 and yellow fluorescence was injected into the left inferior colliculus (IC) to identify cortical regions projecting to IC, enabling targeted circuit-specific investigation. Reprinted from [39] with permission from Elsevier

or retrograde tracers [41]. This method ensures that optogenetic activation is restricted to neurons belonging to a specific projection pathway, allowing for a more detailed investigation of connectivity at the circuit level.

As a proof-of-concept, we employed retrograde viral tracing to selectively stimulate neurons projecting from the auditory cortex to the inferior colliculus (Fig. 5B). A retrograde viral vector encoding channelrhodopsin-2 (ChR2) and yellow fluorescent protein (EYFP) (rAAV2-retro-hSyn-hChR2-EYFP, a gift from Karl Deisseroth; Addgene viral prep #26973-AAVrg) was injected into the left inferior colliculus (AP −5.3 mm, ML +1.0 mm from the bregma). To ensure robust retrograde labeling, 200 nL of the virus was delivered at two depths (0.7 mm and 1.4 mm).

After a 4-week postinjection period to allow stable viral expression, optogenetic fMRI experiments were conducted using stimulation patterns guided by yellow fluorescence imaging, in which ChR2 expression was localized to cells in the auditory cortex that project to the inferior colliculus. Optogenetic stimulation resulted in significant activation in both the auditory cortex and the inferior colliculus. This method extends beyond traditional anatomical or functional ROI-based mapping by directly targeting defined neural projections, allowing for the investigation of top-down versus bottom-up processing, hierarchical connectivity, and circuit-level dysfunction in neurological and psychiatric disorders.

4 Notes

1. Since the skull thinning is applied over a large area, it is important to avoid making the skull excessively thin. If the skull becomes too thin—similar to cases of small-area thinning (<50 μm thickness) [42]—the risk of cracks and potential damage increases. Avoid thinning the skull until it gets softened. To prevent overheating, which may cause tissue damage, it is recommended to intermittently immerse the skull in saline. Before resuming drilling, ensure the skull is completely dry.
2. We recommend using titanium forceps without teeth. Heat the forceps with a gas torch for a few seconds until they reach the appropriate temperature. Be cautious not to overheat, as excessive temperature can cause the tube to melt out. After heating, pinch the tube with the forceps for approximately 10 s, allowing it to cool while maintaining pressure. Repeat this process at ~2 mm intervals along the tube. After cutting the tube into pieces using microsurgical scissors, ensure that each segment is completely sealed by gently pressing on it and visually inspecting for any leakage or air bubbles inside the tube.

3. The irradiance used in this study was determined based on Monte Carlo simulations and experiments in a brain slice [39]. We aimed to achieve sufficient penetration depth to cover the entire cortical layers, thereby maximizing the fMRI signals. However, penetration depth varies depending on the size of the stimulation pattern—larger stimulation areas result in deeper penetration. An irradiance of 3.6 mW/mm^2 is generally effective for relatively large areas (e.g., a circular shape with a diameter >2 mm), whereas smaller stimulation patterns may require higher irradiance levels for effective activation.

Conflict of Interest

The author has no conflicts of interest to declare that are relevant to the content of this chapter.

References

1. Tolias AS, Sultan F, Augath M, Oeltermann A, Tehovnik EJ, Schiller PH, Logothetis NK (2005) Mapping cortical activity elicited with electrical microstimulation using FMRI in the macaque. Neuron 48(6):901–911
2. Matsui T, Tamura K, Koyano KW, Takeuchi D, Adachi Y, Osada T, Miyashita Y (2011) Direct comparison of spontaneous functional connectivity and effective connectivity measured by intracortical microstimulation: an fMRI study in macaque monkeys. Cereb Cortex 21(10):2348–2356
3. Han M-J, Park C-U, Kang S, Kim B, Nikolaidis A, Milham MP, Hong SJ, Kim S-G, Baeg E (2021) Mapping functional gradients of the striatal circuit using simultaneous microelectric stimulation and ultrahigh-field fMRI in non-human primates. NeuroImage 236:118077
4. Xu AG, Qian M, Tian F, Xu B, Friedman RM, Wang J, Song X, Sun Y, Chernov MM, Cayce JM (2019) Focal infrared neural stimulation with high-field functional MRI: a rapid way to map mesoscale brain connectomes. Sci Adv 5(4):eaau7046
5. Shi S, Xu AG, Rui Y-Y, Zhang X, Romanski LM, Gothard KM, Roe AW (2021) Infrared neural stimulation with 7T fMRI: a rapid in vivo method for mapping cortical connections of primate amygdala. NeuroImage 231:117818
6. Lee JH, Durand R, Gradinaru V, Zhang F, Goshen I, Kim D-S, Fenno LE, Ramakrishnan C, Deisseroth K (2010) Global and local fMRI signals driven by neurons defined optogenetically by type and wiring. Nature 465(7299):788–792
7. Moon HS, Jiang H, Vo TT, Jung WB, Vazquez AL, Kim S-G (2021) Contribution of excitatory and inhibitory neuronal activity to BOLD fMRI. Cereb Cortex 31(9):4053–4067
8. Jung WB, Im GH, Jiang H, Kim S-G (2021) Early fMRI responses to somatosensory and optogenetic stimulation reflect neural information flow. Proc Natl Acad Sci USA 118(11):e2023265118
9. Jung WB, Jiang H, Lee S, Kim S-G (2022) Dissection of brain-wide resting-state and functional somatosensory circuits by fMRI with optogenetic silencing. Proc Natl Acad Sci USA 119(4):e2113313119
10. Desai M, Kahn I, Knoblich U, Bernstein J, Atallah H, Yang A, Kopell N, Buckner RL, Graybiel AM, Moore CI (2011) Mapping brain networks in awake mice using combined optical neural control and fMRI. J Neurophysiol 105(3):1393–1405
11. Weitz AJ, Fang Z, Lee HJ, Fisher RS, Smith WC, Choy M, Liu J, Lin P, Rosenberg M, Lee JH (2015) Optogenetic fMRI reveals distinct, frequency-dependent networks recruited by dorsal and intermediate hippocampus stimulations. NeuroImage 107:229–241
12. Weitz AJ, Lee HJ, Choy M, Lee JH (2019) Thalamic input to orbitofrontal cortex drives brain-wide, frequency-dependent inhibition mediated by GABA and zona incerta. Neuron 104(6):1153–1167.e1154
13. Bernal-Casas D, Lee HJ, Weitz AJ, Lee JH (2017) Studying brain circuit function with

dynamic causal modeling for optogenetic fMRI. Neuron 93(3):522–532. https://doi.org/10.1016/j.neuron.2016.12.035
14. Lee HJ, Weitz AJ, Bernal-Casas D, Duffy BA, Choy M, Kravitz AV, Kreitzer AC, Lee JH (2016) Activation of direct and indirect pathway medium spiny neurons drives distinct brain-wide responses. Neuron 91(2):412–424
15. Grandjean J, Corcoba A, Kahn MC, Upton AL, Deneris ES, Seifritz E, Helmchen F, Mansuy IM, Mann EO, Rudin M (2019) A brain-wide functional map of the serotonergic responses to acute stress and fluoxetine. Nat Commun 10(1):1–10
16. Leong AT, Gu Y, Chan Y-S, Zheng H, Dong CM, Chan RW, Wang X, Liu Y, Tan LH, Wu EX (2019) Optogenetic fMRI interrogation of brain-wide central vestibular pathways. Proc Natl Acad Sci USA 116(20):10122–10129
17. Liu J, Lee HJ, Weitz AJ, Fang Z, Lin P, Choy M, Fisher R, Pinskiy V, Tolpygo A, Mitra P (2015) Frequency-selective control of cortical and subcortical networks by central thalamus. Elife 4:e09215
18. Lim DH, Mohajerani MH, LeDue J, Boyd J, Chen S, Murphy TH (2012) In vivo large-scale cortical mapping using channelrhodopsin-2 stimulation in transgenic mice reveals asymmetric and reciprocal relationships between cortical areas. Front Neural Circuits 6:11
19. Ferenczi EA, Zalocusky KA, Liston C, Grosenick L, Warden MR, Amatya D, Katovich K., Mehta H, Patenaude B, Ramakrishnan C (2016) Prefrontal cortical regulation of brainwide circuit dynamics and reward-related behavior. Science 351(6268):aac9698
20. Ryali S, Shih Y-YI, Chen T, Kochalka J, Albaugh D, Fang Z, Supekar K, Lee JH, Menon V (2016) Combining optogenetic stimulation and fMRI to validate a multivariate dynamical systems model for estimating causal brain interactions. NeuroImage 132:398–405
21. Liang Z, Watson GD, Alloway KD, Lee G, Neuberger T, Zhang N (2015) Mapping the functional network of medial prefrontal cortex by combining optogenetics and fMRI in awake rats. NeuroImage 117:114–123
22. Cover CG, Kesner AJ, Ukani S, Stein EA, Ikemoto S, Yang Y, Lu H (2021) Whole brain dynamics during optogenetic self-stimulation of the medial prefrontal cortex in mice. Commun Biol 4(1):1–8
23. Dinh TNA, Moon HS, Kim S-G (2024) Separation of bimodal fMRI responses in mouse somatosensory areas into V1 and non-V1 contributions. Sci Rep 14(1):6302
24. Vo Thanh T, Im Geun H, Han K, Suh M, Drew Patrick J, Kim S-G (2023) Parvalbumin interneuron activity drives fast inhibition-induced vasoconstriction followed by slow substance P-mediated vasodilation. Proc Natl Acad Sci USA 120(18):e2220777120
25. Lee J-Y, You T, Lee C-H, Im GH, Seo H, Woo C-W, Kim S-G (2022) Role of anterior cingulate cortex inputs to periaqueductal gray for pain avoidance. Curr Biol 32(13):2834–2847.e2835
26. Boyden ES, Zhang F, Bamberg E, Nagel G, Deisseroth K (2005) Millisecond-timescale, genetically targeted optical control of neural activity. Nat Neurosci 8(9):1263
27. Friston KJ (1994) Functional and effective connectivity in neuroimaging: a synthesis. Hum Brain Mapp 2(1–2):56–78
28. Bauer AQ, Kraft AW, Baxter GA, Wright PW, Reisman MD, Bice AR, Park JJ, Bruchas MR, Snyder AZ, Lee JM, Culver JP (2018) Effective connectivity measured using optogenetically evoked hemodynamic signals exhibits topography distinct from resting state functional connectivity in the mouse. Cereb Cortex 28(1):370–386
29. Pinto L, Rajan K, DePasquale B, Thiberge SY, Tank DW, Brody CD (2019) Task-dependent changes in the large-scale dynamics and necessity of cortical regions. Neuron 104(4):810–824.e819
30. Leifer AM, Fang-Yen C, Gershow M, Alkema MJ, Samuel AD (2011) Optogenetic manipulation of neural activity in freely moving *Caenorhabditis elegans*. Nat Methods 8(2):147–152
31. Allen, J, 2017. Application of patterned illumination using a DMD for optogenetic control of signaling. Nature Publishing Group, New York
32. Lake EM, Ge X, Shen X, Herman P, Hyder F, Cardin JA, Higley MJ, Scheinost D, Papademetris X, Crair MC (2020) Simultaneous cortex-wide fluorescence Ca2+ imaging and whole-brain fMRI. Nat Methods 17(12):1262–1271
33. Arenkiel BR, Peca J, Davison IG, Feliciano C, Deisseroth K, Augustine GJ, Ehlers MD, Feng G (2007) In vivo light-induced activation of neural circuitry in transgenic mice expressing channelrhodopsin-2. Neuron 54(2):205–218
34. You T, Im GH, Kim S-G (2021) Characterization of brain-wide somatosensory BOLD fMRI in mice under dexmedetomidine/isoflurane and ketamine/xylazine. Sci Rep 11(1):13110
35. Rosendal F, Frandsen J, Chakravarty MM, Bjarkam CR, Pedersen M, Sangill R, Sørensen JC (2009) New surgical technique reduces the susceptibility artefact at air–tissue interfaces on in vivo cerebral MRI in the Göttingen minipig. Brain Res Bull 80(6):403–407
36. Wang Q, Ding S-L, Li Y, Royall J, Feng D, Lesnar P, Graddis N, Naeemi M, Facer B, Ho A (2020) The Allen mouse brain common coordinate framework: a 3D reference atlas. Cell 181(4):936–953.e920

37. Kim SG, Harel N, Jin T, Kim T, Lee P, Zhao F (2013) Cerebral blood volume MRI with intravascular superparamagnetic iron oxide nanoparticles. NMR Biomed 26(8): 949–962
38. Jung WB, Shim H-J, Kim S-G (2019) Mouse BOLD fMRI at ultrahigh field detects somatosensory networks including thalamic nuclei. NeuroImage 195:203–214
39. Kim S, Moon S, Vo TT, Kim C-H, Im GH, Lee S, Choi M, Kim S-G (2023) Whole-brain mapping of effective connectivity by fMRI with cortex-wide patterned optogenetics. Neuron 111(11):1732–1747.e1736
40. Saxe R, Brett M, Kanwisher N (2006) Divide and conquer: a defense of functional localizers. NeuroImage 30(4):1088–1096
41. Yizhar O, Fenno LE, Davidson TJ, Mogri M, Deisseroth K (2011) Optogenetics in neural systems. Neuron 71(1):9–34
42. Yang G, Pan F, Parkhurst CN, Grutzendler J, Gan W-B (2010) Thinned-skull cranial window technique for long-term imaging of the cortex in live mice. Nat Protoc 5(2):201–208

Check for updates

Chapter 4

Optogenetic fMRI in Awake Mice

Yijuan Zou, Chuanjun Tong, Kaiwei Zhang and Zhifeng Liang

Abstract

Optogenetic functional magnetic resonance imaging (ofMRI) is a powerful technique for investigating brain-wide dynamic responses to optogenetic modulation of specific cell populations. This protocol outlines a method for performing ofMRI in awake mice, integrating optogenetics with awake fMRI at 9.4 T using a high signal-to-noise ratio cryogenic coil. The detailed procedures include the preparation of flexible fiber optic implants, surgeries for opsin-encoding viral vector injection, fiber implantation, and head holder fixation, as well as habituation training for awake imaging, ofMRI scanning, and data analysis. The entire pipeline has been optimized to ensure high-quality data by enabling imaging within the limited space of the cryogenic coil and minimizing image distortion and head motion. In summary, ofMRI in awake mice enables precise local neural perturbation while providing whole-brain access, offering significant potential for advancing neuroscience research and preclinical imaging.

Keywords fMRI, Awake mice, Optogenetics, Neural perturbation, Brain

1 Introduction

Functional magnetic imaging (fMRI) has been widely used to examine the large-scale functional neural networks, primarily through measurement of the blood-oxygen-level-dependent (BOLD) signal [1, 2]. Meanwhile, optogenetics has emerged as a powerful tool for dissecting circuitry-level neuroscience [3–5]. Various types of light-sensitive opsins can be expressed in specific cell populations transgenically or virally, allowing precise modulation of cellular activities through light [3]. The combination of these two techniques, termed optogenetic functional magnetic resonance imaging (ofMRI), provides a method to investigate the brain-wide response to local, precise neural modulation [6–10].

So far, most rodent fMRI studies, including of MRI, have been performed under anesthesia. However, anesthesia in rodent fMRI,

The authors "Yijuan Zou" and "Chuanjun Tong" contributed equally.

The author "Zhifeng Liang" is the lead contact.

Kevin Chuen Wing Chan and Seong-Gi Kim (eds.), *Optogenetic Functional Neuroimaging*, Neuromethods, vol. 232, https://doi.org/10.1007/978-1-0716-5178-0_4,

particularly in mouse studies, presents several disadvantages. First, the mouse's small body size (~25 g) makes it challenging to maintain stable physiological conditions under anesthesia. Without mechanical ventilation, the blood gas parameters (e.g., pCO_2) in spontaneously breathing mice are often in abnormal ranges, and thus, lead to unstable or reduced BOLD response [11], although ketamine anesthetized mice still exhibited relatively robust fMRI responses [12]. Furthermore, anesthesia by definition impacts neural activities, the extent of that is dependent on specific anesthetics. Meanwhile, most human fMRI studies are conducted in the awake condition, making direct comparison between human and animal studies difficult. It is not surprising that recent studies reported that anesthesia resulted in different brain metabolism [13], functional connection [14-16], and optogenetic activation [9, 10] from the awake state. Consequently, there is a growing need for awake animal imaging [7, 8, 17-20]. Despite the progress in the past, performing awake animal fMRI, particularly in combination with optogenetics, remains challenging due to issues, such as head motion and stress.

Performing ofMRI at high field also brings other technical challenges related to the choice of implanted optical fiber. Image artifacts (signal dropout and distortion) caused by brain implants and surgery can be profound if not carefully optimized, particularly in functional images. The primary cause for such image artefacts lies in the B_0 distortion due to the mismatch of magnetic susceptibility between the implanted optical fiber and the surrounding brain tissues. In addition, optic fiber implants need to be mechanically stable and biocompatible, allowing for longitudinal scanning over more than 1 month. Furthermore, the limited space when using cryogenic coils (or other phased array coils with no opening) necessitates the use of flexible optical fibers or fibers with a large angle. Overall, it is critical to choose the proper materials of optic fiber implants for acquiring high-quality ofMRI images.

Conducting ofMRI in awake mice provides access to a huge amount of mouse neuroscience resources and tools to the animal neuroimaging field, such as numerous transgenic strains for disease modeling or targeting specific cell types in the brain. Therefore, awake mouse ofMRI holds great potential for addressing numerous neuroscience and neuroimaging questions [8]. In this chapter, we described an optimized experimental protocol and data preprocessing and analysis pipeline for awake mouse ofMRI. As a demonstration, we present findings from our recent work [8], which used ofMRI to stimulate basal forebrain (BF) cholinergic (ChAT), glutamatergic (VGLUT2), parvalbumin-positive (PV), and somatostatin-positive (SOM) GABAergic neurons in awake mice.

2 Materials

In this protocol, ChAT-IRES-Cre, VGLUT2-IRES-Cre, PV-IRES-Cre, and SOM-IRES-Cre mice (Jackson stock #: 006410, 016963, 008069, and 013044, respectively) were obtained from Jackson Laboratory. Wild-type mice were purchased from local vendors. Mice were group-housed (5–6/cage) under a 12-h light/dark cycle (light on from 7 a.m. to 7 p.m.) with food and water ad libitum. The viral vector AAV2/9-hSyn-DIO-ChrimsonR-mCherry-WPRE-hGH pA (Titer: 4.76×10^{12} v.g./mL; BrainVTA Co., China) was used.

To reduce artifacts in MRI and accommodate the limited space of RF coil, polymethylmethacrylate (PMMA) optical fiber was used. Figure 1 shows the artifacts induced by PMMA and silica fiber implants in T2-weighted and EPI images. Compared to the commonly used silica optical fiber, the artifacts of PMMA optical fiber with the same diameter (200 μm) are much smaller in both T2-weighted and echoplanar imaging (EPI) images. Such an effect can be attributed to the susceptibility index of PMMA ($\chi = -9.06$ ppm) [21], which is very close to that of water ($\chi = -9.05$ ppm), resulting in greatly reduced perturbation of the B0 field. Furthermore, the plasticity of PMMA allows bending of the optical fiber, which is essential for conducting ofMRI in radiofrequency (RF) coils with no opening. The PMMA optical fiber used in this study was purchased from Inper Ltd (https://www.inper.com/).

In addition to the aforementioned materials, the following materials were used in stereotaxic surgery: isoflurane, saline, iodophor, 75% alcohol, ophthalmic ointment, 3% hydrogen peroxide, absorbable

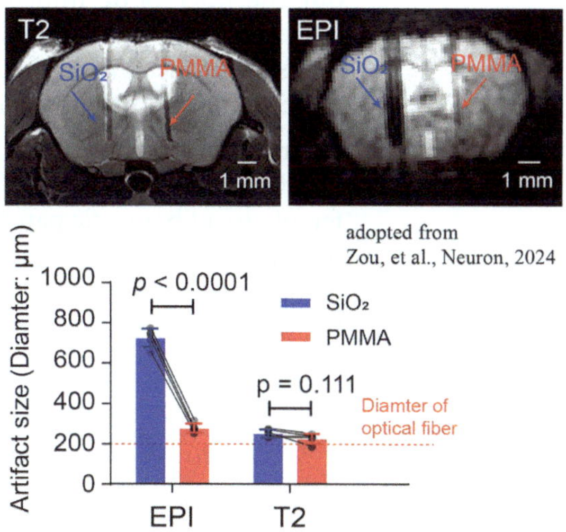

Fig. 1 The artifacts caused by PMMA or silica optofiber implants with the same diameter (200 μm) in T2-weighted and EPI images, respectively. This graph was adopted from Zou et al. [8] Neuron 2024

gelatin sponge, tissue adhesive, blue light-curing self-etch adhesive (3 M ESPE Adper Easy One), nanoliter injector (Nanoject III, Drummond Scientific Company), dust-free paper, light-curing flowable dental resin, dental drill, dental cement, polyethylene glycol 2000 (PEG2000), custom-made head holder, and fiber-straightening device.

The following materials were used in habituation and ofMRI scanning: ceramic screws, light source, fiber optic patch cable, power meter, ferrule sleeve, Kwik-cast silicone casting compound (WPI), custom-made animal bed, and black silicone eye patches.

The following materials were used in histology: 0.1 M phosphate-buffered saline (PBS), 4% paraformaldehyde (PFA), 30% sucrose solution, OCT compound (NEG-50, Thermo Scientific), mounting media containing DAPI (#F6057, Sigma), and an epifluorescence microscope (VS120, Olympus).

All custom 3D-printed model drawings in this protocol are available online at https://doi.org/10.12412/BSDC.1735609439.20001

3 Methods

3.1 Preparation of Fiber Optic Implant

1. Place the fiber optic implant into the custom-designed and 3D-printed fiber straightening device, as shown in Fig. 1. Immerse the device with a fiber optic implant in 100°C water for 20 s, and dry it completely. This process uses the plasticity of PMMA material to make the fiber straighter (*see* **Note 1**).
2. Melt PEG2000 sealed in the syringe over hot water and avoid mixing water into PEG. Apply the melted PEG into the designated slot as shown in Fig. 1. After the PEG solidifies, take out the fiber optic implant from the fiber-straighten device. Covering the fiber optic implant with PEG largely overcomes the issue of the PMMA fiber being too flexible to remain completely straight during the implantation process.
3. Insert a fiber optic implant into a hole in the skull model, then bend, and fix it to the surface of the skull model. Couple the ferrule to a fiber optic patch cable with a ferrule sleeve and connect the patch cable to a laser or LED light source. Measure the attenuation rate of at least five fiber implants and calculate the mean attenuation rate of these fiber optic implants.

$$\text{Output Efficiency} = \frac{\text{Power at the tip of the fiber optic implant}}{\text{Power at the end of the patch cable}}$$

3.2 Stereotaxic Surgery of Virus Injection, Fiber Optic, and Head Holder Implantation

1. Use a programmable nanoliter injector to load enough AAV with a fill rate of 10 nL/s.
2. Induce anesthesia in the mouse under 5% isoflurane anesthesia. Place the animal into a stereotaxic frame with a heating

pad and use 1%–1.5% isoflurane to maintain anesthesia. Shave the head and disinfect the scalp with betadine and 70% ethanol three times, respectively. Apply ophthalmic ointment to the eyes to prevent dryness.

3. Make a midline incision approximately 15 mm long. Cutoff excess skin on both sides of the incision to make the incision triangular. Remove the muscle covering the skull posterior to the lambda point. If bleeding occurs, use an absorbable gelatin sponge to stop bleeding. Use a cotton swab dipped in 3% hydrogen peroxide to clean the skull, and then, use a spatula to remove the periosteum and other associated soft tissues. After that, use cotton swabs dipped in saline to repeatedly clean the skull, and dry the skull surface with dust-free paper. Avoid pressing the skull forcibly, as it may cause subdural hematoma (*see* **Note 2**).
4. Apply a layer of light-curing self-etch adhesive to cover all exposed skull. Use lambda and bregma to position the drill over the coordinate for the region of interest (ROI). In this protocol, the ROI is the basal forebrain (AP 0 mm, ML ± 1.3 mm, DV −5.0 mm from dura). Make a small craniotomy (400–500 μm) on the skull over the ROI with a dental drill carefully without puncturing the brain.
5. Use a sharp hook to break the dura, then slowly insert the needle of the nanoliter injector into the ROI in the brain, and keep still for 5 min. Inject 300–500 nL of the vector solution at a rate of 1 nL/s. After the injection is completed, wait 10 min before withdrawing the needle slowly.
6. Dry the surface of the skull. Insert the fiber optic implant to the target depth of 0.2 mm above the ROI coordinates.
7. Drain the overflow of liquid from the craniotomy with dust-free paper. Apply the light-curing flowable dental resin to the fiber optic implant, and cure the resin with blue light for 20 s. Gently melt the PEG surrounding, the fiber optic implant with two cotton swabs that have been dipped in 100°C water.
8. Dry the skull surface with dust-free paper (*see* **Note 3**). Adhere the skin surrounding the skull to the edge of the skull with tissue adhesive. Apply the dental resin on the occipital bone, and place a custom-made head holder closely on the interparietal bone. Cure the resin completely with blue light. Additional dental resin is applied around the holder and cured.
9. Use tweezers to clamp the ferrule to bend the fiber close to the head holder, and quickly fix the fiber optic implant to the skull and head holder with the dental resin. Apply more

dental resin to cover the whole fiber optic implant to protect the fiber optic implant because the subsequent dental cement to be applied will melt the PMMA material (*see* **Note 4**). Cover other areas of the exposed skull with a thick, smooth layer of dental cement.

10. After surgery, place the mouse in the postoperative incubator for anesthesia recovery before returning to its home cage (*see* **Note 5**). Allow the mouse 2 weeks for recovery before habituation training.

3.3 Habituation

To reduce the stress level and head motion during imaging, mice are habituated to the imaging environment of head fixation and loud acoustic noise for 1 week [20].

1. Anesthetize the animal and place its head in the head holder, fixate the head holder in the grooves of the training bed, and use a piece of tape to secure the head holder to the bed (Fig. 2A). If needed, use more tapes to secure the body to the bed (*see* **Note 6**). Depending on the purpose of the specific project, slightly different setups of head and body restraining can be used.
2. The length of the habituation and the intensity of the acoustic noise increase gradually each day. The maximum training time depends on the time of the actual scan. The acoustic noise should be recorded from the scanner running the same pulse sequence for actual imaging. In this protocol, the detailed habituation schedule is listed in Table 1. Again, modifications of the habituation schedule can be made depending on the specific project.

3.4 Optogenetic Functional MRI

1. Use the Arduino Uno board (https://www.arduino.cc/) with custom-written scripts (https://doi.org/10.12412/BSDC.1735609439.20001) to synchronize the light delivery with fMRI scanning and control the stimulation parameters. Scanner triggers can be sent out at the beginning of the scan or each frame (or even slice), depending on the requirement of

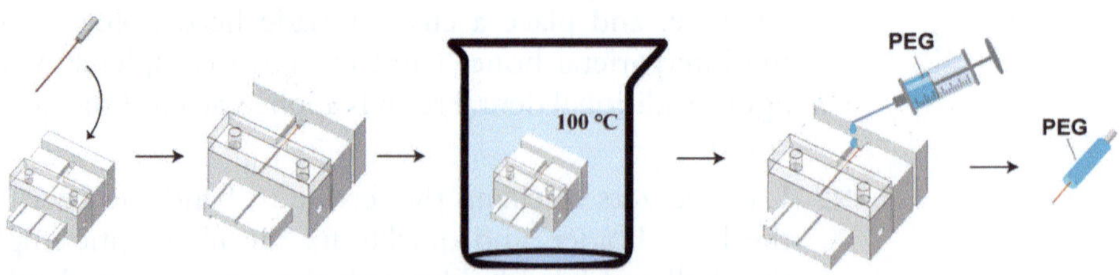

Fig. 2 Schematic diagram of wrapping an optical fiber implant with PEG in a custom-designed fiber-straightening device

the stimulation protocol (*see* **Note 7**). The light source is typically placed outside the magnet room, and a long fiber optic patch cable (6 m) is used to deliver light into the bore center of the scanner.

2. Lightly anesthetize the animal and fix the head on the animal bed with two ceramic screws (Fig. 3). Apply a thick, smooth layer of Kwik-cast smoothly over the skull to reduce distortion caused by air. Mouse's eyes are covered with two black silicone eye patches, which prevent confounding visual stimulation by light leakage between the interface of the ferrule implant and the patch cable during optogenetic stimulation. Secure the animal bed in the scanner cradle. Use two pieces of tape to secure the body to the cradle (Fig. 2B). Place the respiratory pillow under the mouse's chest to monitor its respiratory rate.

3. Connect the fiber optic patch cable to a light source and measure the output power at the tip of the patch cable's ferrule with a power meter. Adjust to the appropriate power level to produce the desired output (5 mW in this protocol) at the tip

Table 1
Habituation schedule for awake mouse ofMRI

	Day 1	Day 2	Day 3	Day 4	Day 5	Day 6	Day 7
Duration of habituation	60 min	60 min	60 min	60 min	60 min	60 min	60 min
Acoustic noise	–	60 dB	60 dB	90 dB	90 dB	110 dB	110 dB

"–" denotes absence.

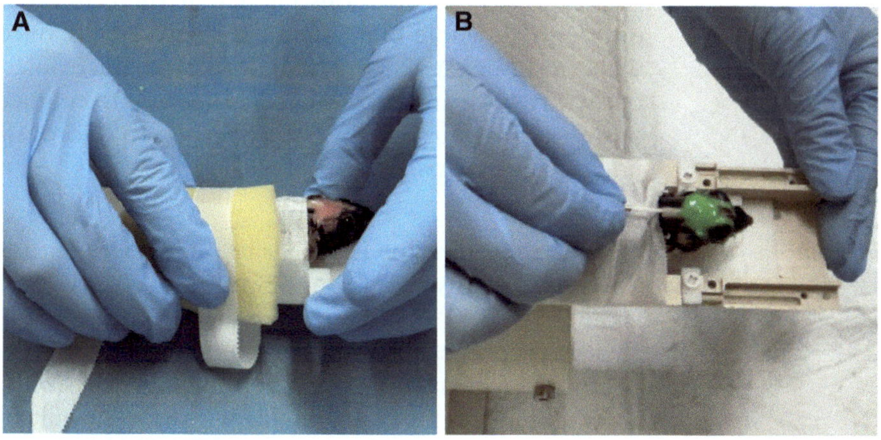

Fig. 3 Schematic diagram of wrapping an optical fiber implant with PEG in the fiber-straightening device. (**A**) The setup of daily training. Using tapes to secure the head holder to the bed during daily training. (**B**) The setup of actual imaging. Using ceramic screws to secure the head holder to the bed during actual imaging

of the fiber optic implanted inside the brain, based on the calculated mean output efficiency in Step 3.1.3. Couple the fiber optic cable to the fiber optic implant with a ferrule sleeve.

4. Insert the cradle with the mouse into the model of the cryogenic coil to test if it can move smoothly into the space underneath the coil (*see* **Note 8**). If successful, insert the cradle into the bore center.
5. Start the fMRI scan protocol, typically including a localizer scan (adjust the animal's position if not in the isocenter), T2-weighted anatomical imaging (check for any structural damage to the brain and confirm the optical fiber location), and an EPI scan for functional imaging. Before starting EPI scans, acquire a field map and perform field map-based local shimming within the mouse brain. Biband-EPI was acquired using the multiband EPI sequence [20] with the following parameters: repetition time (TR) = 500 ms, echo time (TE) = 15 ms, flip angle = 38.8°, bandwidth = 300 kHz, field of view = 15 × 12 mm^2, matrix size = 100 × 80, nominal slice thickness = 0.48 mm (slice thickness 0.4 mm with a gap of 0.08 mm), 898 or 982 volumes (scan 1–3: 898 volumes and scan 4–6: 982 volumes), and 38 axial slices per scan. During fMRI acquisition, verify that the light delivery is functioning correctly (*see* **Note 9**).
6. Additionally, the photothermal effect may cause nonspecific activation. To ensure that activation resulting from manipulation of specific cell types during ofMRI is not due to any nonspecific effects, a control group should follow the same stimulation protocol. In the current work [8], due to the use of four Cre mouse lines, we employed a group of wild-type mice injected with the same opsin virus as a control group. Although, we also demonstrated that there is no significant difference in whole-brain activation between Cre mice injected with opsin-negative virus and wild-type mice injected with opsin virus, we still recommend using Cre mice injected with opsin-negative virus as the control group to rigorously exclude nonspecific effects.

3.5 ofMRI Data Analysis

The data analysis pipeline is similar to our previous study [8]. All data are processed using custom scripts in MATLAB (MathWorks, Natick, MA) and SPM12 (http://www.fil.ion.ucl.ac.uk/spm/). Briefly, the preprocessing steps include the following: converting raw data to.nii format (note that the nominal voxel size is enlarged ten times to improve the compatibility with SPM default parameters optimized for human brain size); use ITK-SNAP (http://www.itksnap.org/) to extract the mouse brain; realign the images of each scan to the first volume of each scan, which produces six realignment parameters that can be used for head motion correction; for group analysis, EPI images are registered to a mouse brain template (http://www.imaging.org.au/AMBMC/Model); spatial smoothing with a 0.4 mm isotropic Gaussian kernel.

For conventional general linear model (GLM) analysis, first-level analysis is conducted for individual EPI scans with 22 regressors (six head motion parameters and their first order derivatives and time courses of ten nonbrain tissue-based principal components [PCs]). As the mouse hemodynamic response function (HRF) differs considerably from the human HRF, it is important to modify the default HRF of SPM12 based on our empirically estimated HRF of awake mice in the previous study [17]. Replace the default \spm12\spm_hrf.m file in SPM12 with our spm_hrf.m file (https://doi.org/10.12412/BSDC.1735609439.20001), which has been modified based on the empirically estimated HRF from mice [17]. For the second-level analysis, one sample t-test is conducted to generate the group activation maps with FDR-corrected $p < 0.05$ and cluster size > 10 voxels. To evaluate the time series of ROIs, the Allen Institute's mouse brain atlas (http://mouse.brain-map.org/static/atlas) is normalized to the template images.

Figure 4 shows representative data resulting from optogenetic stimulation of four neuron populations in BF. Results showed that activation of VGLUT2 neurons in the BF region led to the strongest and most widespread BOLD signal increases across the brain compared to the other three cell types (Fig. 4A). ChAT neuron activation resulted in positive BOLD signals across extensive cortical and subcortical areas, including the prefrontal cortex (PFC), ACA, RSP, orbital (ORB), visual (VIS), auditory (AUD), and entorhinal (ENT) regions, as well as a significant portion of the striatum and thalamus (Fig. 4B). In the PV-Cre group, significant positive activation was observed in the ACA (Fig. 4C). In the SOM-Cre group, significant positive activations were noted in the ACA, striatum, and anterior thalamus (Fig. 4D). In summary, ofMRI in awake mice is a useful tool for studying system-level neuroscience.

3.6 Confirmation of Opsin Expression

After the scan, expression of opsin at the intended ROI needs to be confirmed through histology. In this example, histological verification of virus expression in BF neurons is shown in Fig. 5.

1. Mouse is deeply anesthetized and immediately perfused using 0.1 M PBS, followed by 4% PFA. The brain tissues are removed and postfixed overnight in 4% PFA.
2. Transfer the brain to a 30% sucrose solution at 4°C for 3 days until the brain sinks to the bottom of the solution. This step is to remove water from the brain and prevent it from freezing later. Brain samples are embedded with OCT compound and cut into 50-μm sections using a cryostat. Brain sections were stored at −20°C before further processing.

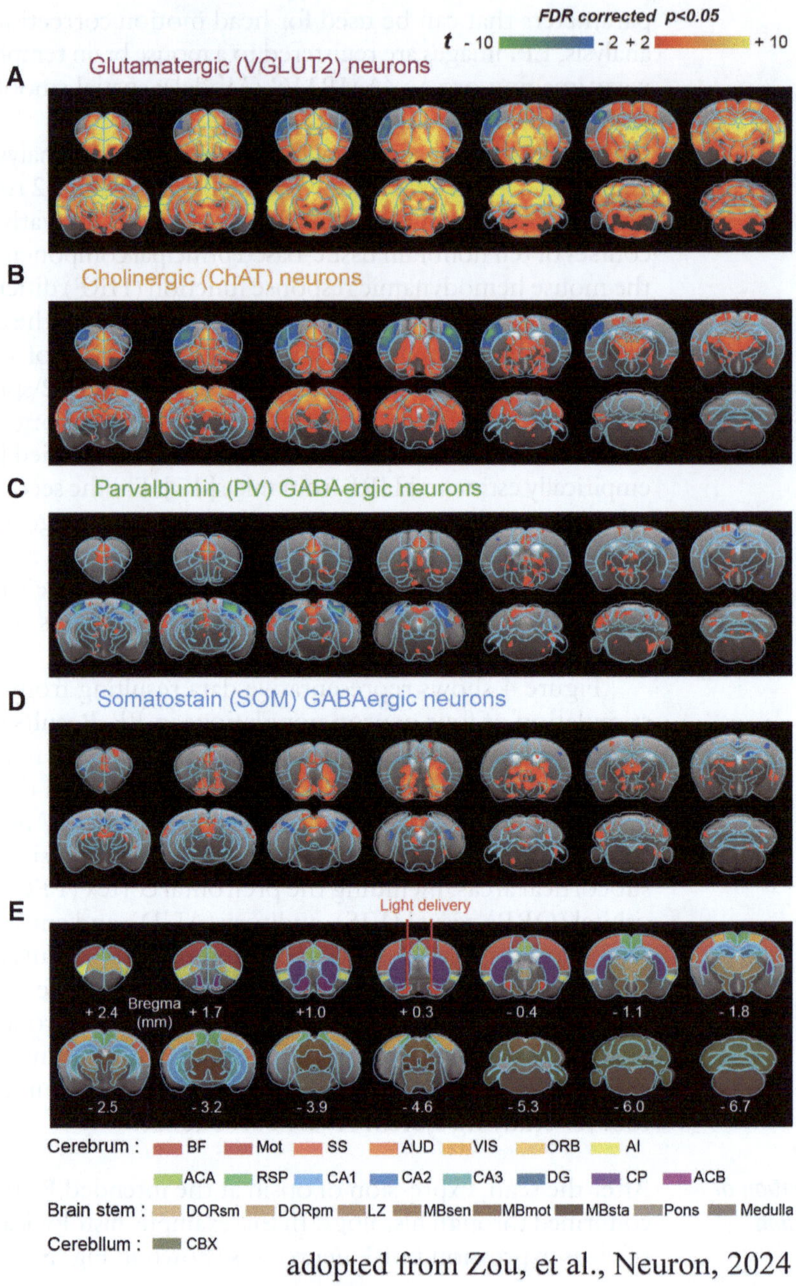

Fig. 4 Brain-wide activation map under optogenetic stimulations of specific BF neuron populations. FDR corrected p <0.05 and cluster size >10 voxels. This graph was adopted from Zou et al. [8] Neuron 2024

3. For histology examination, brain sections are washed in PBS and coverslipped with mounting media containing DAPI. The fluorescence images are captured using an epifluorescence microscope. Confirm whether fluorescence is expressed in the targeted brain region.

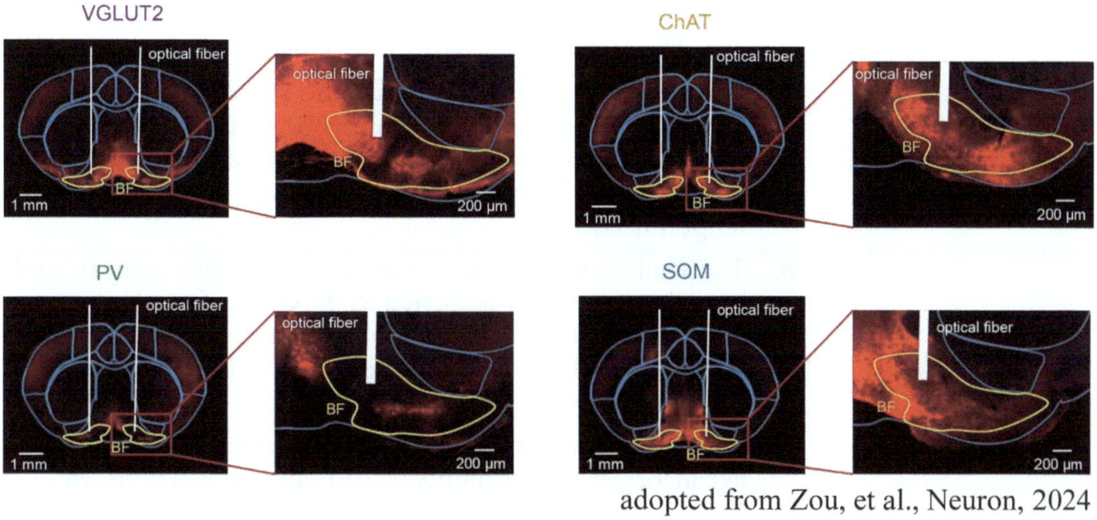

Fig. 5 Histological verification of virus expression in BF neurons. This graph was adopted from Zou et al. [8] Neuron 2024

4 Notes

1. Immersing the PMMA optical fibers, which are fixed in holes that are perfectly horizontal, in boiling water can conveniently and quickly reshape the optical fibers without significantly affecting their light transmission.
2. As the mouse skull is thin, avoid applying excessive pressure on the skull throughout the surgery to avoid subdural hemorrhage. Signal dropout and distortion may occur around the bleeding site.
3. Ensure the skull is dry before applying the dental resin and dental cement every time, otherwise the ceramic ferrule implant and head holder may not be firmly fixed on the head and could fall off during the study.
4. Before the final step of the stereotactic surgery, which is covering with dental cement, ensure that all PMMA materials of the optical fibers have been completely encapsulated by the dental resin. Uncured dental cement can melt PMMA in our experience.
5. Keeping multiple mice in one cage is acceptable, but ensure the trough is removed to prevent it from jamming the fiber implants.
6. Do not use screws to pin the head holder to the bed during daily training. This prevents the threads in the head holder

from being eroded, which could affect the head fixation during actual imaging.

7. For Bruker scanners, triggers are also sent out during the pre-scan calibration period, which can be variable depending on the sequence parameters. Thus, it is necessary to measure the actual number of triggers for a given sequence and modify the script if needed.
8. The vacuum seal of the cryogenic coil is fragile and can easily break if a large force is applied to the coil surface. This is more likely to happen when the headpost protrudes from the animal's head. Therefore, make sure each animal is fitted in the mock device before moving it to the actual cryogenic coil.
9. It is not common that triggering does not work due to issues like incorrect trigger numbers, bad cable connections, etc.

5 Conclusion

The method described in this chapter provides a detailed protocol for conducting ofMRI in awake mice, combining the advantages of ofMRI, and awake imaging. Compared with fMRI under anesthesia, awake mouse fMRI overcomes the confounding effects of anesthesia on the brain [22-24]. Naturally, the significant benefits of awake imaging also come with certain challenges. In addition to a significantly increased animal preparation time, the stress and head motion also need to be carefully handled. However, the need to reduce stress levels and the need to reduce head motion are often contradictory to each other. For example, more strict body restraining reduces the head motion level, but may increase the stress level. The pipeline described in this chapter aims to achieve a balance, and the exact procedure and setup can be adjusted depending on the needs of specific projects.

In summary, this protocol describes a comprehensive pipeline of ofMRI paradigms in awake mice, with optimized experimental setups and a data preprocessing and analysis pipeline. This method takes full advantage of the abundance of transgenic mouse strains to investigate whole-brain activity upon precise manipulation, which will be helpful for studying system-level neuroscience questions.

Acknowledgments

This work was supported by the National Science and Technology Innovation 2030 Major Program (2021ZD0202200 and 2021ZD0200100 to Z.L.), National Natural Science Foundation of China (82402245 to C.T., 82171899 to Z.L., and Excellent

Young Scientists Fund 32322036 to Z.L.), China National Postdoctoral Program for Innovative Talents (BX20230383 to C.T.), China Postdoctoral Science Foundation (2023M743616 to C.T.), Shanghai Post-doctoral Excellence Program (C.T.), and Special Research Assistant Project of the Chinese Academy of Sciences (C.T.).

Conflict of Interest The author has no conflicts of interest to declare that are relevant to the content of this chapter.

References

1. Dolan RJ (2008) Neuroimaging of cognition: past, present, and future. Neuron 60: 496–502. https://doi.org/10.1016/j.neuron.2008.10.038
2. Friston KJ (2009) Modalities, modes, and models in functional neuroimaging. Science 326:399–403. https://doi.org/10.1126/science.1174521
3. Wentz CT, Oettl L-L, Kelsch W (2013) Optogenetics in psychiatric animal models. Cell Tissue Res 354:61–68. https://doi.org/10.1007/s00441-013-1651-5
4. Duebel J, Marazova K, Sahel J-A (2015) Optogenetics. Curr Opin Ophthalmol 26:226–232. https://doi.org/10.1097/ICU.0000000000000140
5. Kim CK, Adhikari A, Deisseroth K (2017) Integration of optogenetics with complementary methodologies in systems neuroscience. Nat Rev Neurosci 18:222–235. https://doi.org/10.1038/nrn.2017.15
6. Lee JH, Durand R, Gradinaru V, Zhang F, Goshen I, Kim D-S, Fenno LE, Ramakrishnan C, Deisseroth K (2010) Global and local fMRI signals driven by neurons defined optogenetically by type and wiring. Nature 465:788–792. https://doi.org/10.1038/nature09108
7. Liang Z, Watson GDR, Alloway KD, Lee G, Neuberger T, Zhang N (2015) Mapping the functional network of medial prefrontal cortex by combining optogenetics and fMRI in awake rats. NeuroImage 117:114–123. https://doi.org/10.1016/j.neuroimage.2015.05.036
8. Zou Y, Tong C, Peng W, Qiu Y, Li J, Xia Y, Pei M, Zhang K, Li W, Xu M, Liang Z (2024) Cell-type-specific optogenetic fMRI on basal forebrain reveals functional network basis of behavioral preference. Neuron 112:1342-1357.e6. https://doi.org/10.1016/j.neuron.2024.01.017
9. Ferenczi EA, Zalocusky KA, Liston C, Grosenick L, Warden MR, Amatya D, Katovich K, Mehta H, Patenaude B, Ramakrishnan C, Kalanithi P, Etkin A, Knutson B, Glover GH, Deisseroth K (2016) Prefrontal cortical regulation of brainwide circuit dynamics and reward-related behavior. Science 351:aac9698. https://doi.org/10.1126/science.aac9698
10. Hamada HT, Abe Y, Takata N, Taira M, Tanaka KF, Doya K (2024) Optogenetic activation of dorsal raphe serotonin neurons induces brain-wide activation. Nat Commun 15:4152. https://doi.org/10.1038/s41467-024-48489-6
11. Kim S, Moon HS, Vo TT, Kim C-H, Im GH, Lee S, Choi M, Kim S-G (2023) Whole-brain mapping of effective connectivity by fMRI with cortex-wide patterned optogenetics. Neuron 111:1732-1747.e6. https://doi.org/10.1016/j.neuron.2023.03.002
12. Shim H-J, Jung WB, Schlegel F, Lee J, Kim S, Lee J, Kim S-G (2018) Mouse fMRI under ketamine and xylazine anesthesia: robust contralateral somatosensory cortex activation in response to forepaw stimulation. Neuroimage 177:30–44. https://doi.org/10.1016/j.neuroimage.2018.04.062
13. Xu M, Bo B, Pei M, Chen Y, Shu CY, Qin Q, Hirschler L, Warnking JM, Barbier EL, Wei Z, Lu H, Herman P, Hyder F, Liu Z-J, Liang Z, Thompson GJ (2022) High-resolution relaxometry-based calibrated fMRI in murine brain: metabolic differences between awake and anesthetized states. J Cereb Blood Flow Metab 42:811–825. https://doi.org/10.1177/0271678X211062279
14. Paasonen J, Stenroos P, Salo RA, Kiviniemi V, Gröhn O (2018) Functional connectivity under six anesthesia protocols and the awake condition in rat brain. NeuroImage 172:9–20. https://doi.org/10.1016/j.neuroimage.2018.01.014
15. Gutierrez-Barragan D, Singh NA, Alvino FG, Coletta L, Rocchi F, De Guzman E, Galbusera A, Uboldi M, Panzeri S, Gozzi A (2022) Unique spatiotemporal fMRI dynamics in the awake mouse brain. Curr Biol 32:631–644.e6. https://doi.org/10.1016/j.cub.2021.12.015
16. Becker K, Eder M, Ranft A, von Meyer L, Zieglgänsberger W, Kochs E, Dodt H-U (2012) Low dose isoflurane exerts opposing effects on

neuronal network excitability in neocortex and hippocampus. PLoS One 7:e39346. https://doi.org/10.1371/journal.pone.0039346

17. Chen X, Tong C, Han Z, Zhang K, Bo B, Feng Y, Liang Z (2020) Sensory evoked fMRI paradigms in awake mice. Neuroimage 204:116242. https://doi.org/10.1016/j.neuroimage.2019.116242
18. Han Z, Chen W, Chen X, Zhang K, Tong C, Zhang X, Li CT, Liang Z (2019) Awake and behaving mouse fMRI during Go/No-Go task. Neuroimage 188:733–742. https://doi.org/10.1016/j.neuroimage.2019.01.002
19. Yu Y, Qiu Y, Li G, Zhang K, Bo B, Pei M, Ye J, Thompson GJ, Cang J, Fang F, Feng Y, Duan X, Tong C, Liang Z (2023) Sleep fMRI with simultaneous electrophysiology at 9.4 T in male mice. Nat Commun 14, 1651. https://doi.org/10.1038/s41467-023-37352-9
20. Xu W, Pei M, Zhang K, Tong C, Bo B, Feng J, Zhang X-Y, Liang Z (2022) A systematically optimized awake mouse fMRI paradigm. bioRxiv 2022.11.16.516376. https://doi.org/10.1101/2022.11.16.516376
21. Wapler MC, Leupold J, Dragonu I, von Elverfeld D, Zaitsev M, Wallrabe U (2014) Magnetic properties of materials for MR engineering, micro-MR and beyond. J Magn Reson 242:233–242. https://doi.org/10.1016/j.jmr.2014.02.005
22. Nishikawa K, MacIver MB (2001) Agent-selective effects of volatile anesthetics on GABAAReceptor–mediated synaptic inhibition in hippocampal interneurons. Anesthesiology 94:340–347. https://doi.org/10.1097/00000542-200102000-00025
23. Schlegel F, Schroeter A, Rudin M (2015) The hemodynamic response to somatosensory stimulation in mice depends on the anesthetic used: implications on analysis of mouse fMRI data. NeuroImage 116:40–49. https://doi.org/10.1016/j.neuroimage.2015.05.013
24. Shim H-J, Jung WB, Schlegel F, Lee J, Kim S, Lee J, Kim S-G (2018) Mouse fMRI under ketamine and xylazine anesthesia: robust contralateral somatosensory cortex activation in response to forepaw stimulation. Neuroimage 177:30–44. https://doi.org/10.1016/j.neuroimage.2018.04.062

Check for updates

Chapter 5

Shedding Light on Brain Activity: How to Fuse Optical Signals with BOLD fMRI

Bruno Pradier, Lydia Wachsmuth and Cornelius Faber

Abstract

Functional magnetic resonance imaging (fMRI) is a crucial tool for studying brain activity. It mostly uses the blood oxygen level dependent (BOLD) contrast to identify activated brain regions through neurovascular coupling. Therefore, fMRI alone does not provide direct information on neuronal activity or metabolic turnover. This limitation can be addressed by optical sensors, especially genetically encoded sensor proteins, which allow for the cell-specific detection of various biological signals, such as calcium, neurotransmitters, or metabolites. Combining fiber-based detection of signals from optical sensors with fMRI (ofMRI) can yield valuable insights into brain function and advance the understanding of neurovascular coupling, neurotransmitter regulation, or metabolic turnover. ofMRI may further be used to assess brain states or to clarify how anesthesia or drugs impact brain physiology. However, the integration of fiber-based optical recordings into fMRI poses challenges and creates pitfalls that must be avoided.

This chapter describes the experimental procedures for acute simultaneous fMRI and optical recordings using genetically encoded sensor proteins, focusing on protocols for mice and rats. The complete experiment is divided into two parts: after the initial stereotaxic surgery for vector injection, several weeks for sensor protein expression have to be allowed. The actual experiment starts with a surgery for fiber implantation, immediately followed by ofMRI measurement. Protocols for resting-state and task fMRI are provided. The chapter aims to provide a practical guide for ofMRI, lists many potential pitfalls and details how these can be avoided. However, the chapter neither covers hardware details for optical recordings nor data analysis procedures.

Keywords BOLD, ofMRI, Neurovascular coupling, Optogenetics, Calcium recordings, Fiber photometry, Genetically encoded sensors, Anesthesia, Brain state, Stereotaxic surgery

1 Introduction

Functional magnetic resonance imaging (fMRI) is one of the most important tools to investigate how the brains of humans and other mammals work. Despite more than three decades of intense research, the exact mechanism of neurovascular coupling is still elusive. Most fMRI studies exploit the BOLD contrast [1] to detect activated regions of the brain indirectly via neurovascular coupling. Therefore, direct information on neuronal activity or metabolic turnover is

Kevin Chuen Wing Chan and Seong-Gi Kim (eds.), *Optogenetic Functional Neuroimaging*, Neuromethods, vol. 232, https://doi.org/10.1007/978-1-0716-5178-0_5,

usually not available from fMRI measurements alone. This limitation can be addressed by optical sensors [2], which can be either small molecules (chemical sensors) or genetically encoded proteins. The latter offers broad functionality, enabling the detection of calcium [3, 4], neurotransmitters [5–7], metabolites [8–10], and membrane voltage [11]. GCaMP variants are widely used calcium indicators, with versions optimized for either faster kinetics (e.g., GCaMP6f) or larger signal amplitudes (e.g., GCaMP6s) [3]. GRAB [5, 7] and SnFR sensors [6] detect neurotransmitters and modulators, such as glutamate, GABA, dopamine, serotonin, and acetylcholine, while metabolite sensors are available to detect, for example, glucose, lactate, and pyruvate. These optical readouts can be more easily combined with fMRI than electrical recording methods [12–14], since light does not directly interfere with electromagnetic fields in the MR frequency range. Fiber-based recordings detect population signals from hundreds to thousands of neurons in a spatially confined region [13]. When this fiber photometry signal is combined with simultaneous fMRI (ofMRI), it provides valuable insights into key neuroscience questions, including how neurovascular coupling and neurotransmitters regulate brain function [15–17]. It addresses metabolic turnover [18], providing insights into energy use in cognition and disease. By identifying distinct brain states [19, 20] and individual signatures [21], it helps define what makes each brain unique. It also clarifies how anesthesia and drugs alter brain physiology and network dynamics [22, 23], advancing our understanding of consciousness and pharmacological effects.

Independent of the desired application, fiber-based recordings during an MRI scan create challenges that must be considered [24]. This chapter will describe all experimental procedures of simultaneous fMRI and optical recordings using genetically encoded sensors. We will not address the use of microscopes inside the MR scanner [25], nor will we discuss chemical [2] or FRET-based sensors [4, 8, 9, 18]. We will outline procedures for mice and rats and comment where differences are relevant. To keep the chapter concise, we will not detail the hardware for optical detection. In our lab, we use a custom-built optical stand, but commercial systems for the detection of fluorescent reporters are also available from several vendors.

Therefore, the Methods section will describe a two-part procedure, which is separated by a waiting period of several weeks to allow for expression of the genetically encoded sensors. In the first part, stereotaxic surgery (Sect. 3.1) is performed, followed by injection of the reporter system into the brain (Sect. 3.2). Typically, a few weeks later, the second part begins with another stereotaxic surgery (Sect. 3.1) for fiber implantation (Sect. 3.3) immediately followed by the terminal ofMRI experiment (Sects. 3.4 and 3.5). Protocols are provided for (1) resting-state fMRI (Sect. 3.5.1), which can be used to study brain networks in different brain states under varying anesthetic or physiological conditions, and (2) task fMRI (Sect. 3.5.2), which can be used to study hemodynamic coupling or brain networks in response to any applied stimulation paradigm. Finally, we briefly address general aspects of data analysis (Sect. 3.6). However, due to the wide variety of different software packages and custom procedures, a detailed description is beyond the scope of this chapter.

2 Materials

2.1 Animals and Animal Handling

Protocols described herein are optimized for rats (<200 g) and mice (20–30 g). Ensure permissions and compliance with local animal welfare regulations.

Anesthesia

1. Isoflurane or ketamine/xylazine (*see* **Notes 1 and 2**).
2. Compressed air: medical O_2 (80:20).
3. Vaporizer suitable for mice and rats (make sure the device is calibrated at regular intervals).
4. Anesthesia induction chamber.
5. Medetomidine, and in case of survival experiments, antisedan to antagonize medetomidine.

Chemicals and Equipment for Surgery

1. Analgesia for systemic peri- and post-surgical pain relief, as defined in your approved animal use protocol (e.g., buprenorphine, carprofen, meloxicam, metamizole).
2. Analgesia for local peri-surgical pain relief (skin/periosteum: lidocain spray (10 mg/shot) and ropivacain-HCl (2 mg/mL), ear canal: 2.5% lidocain and prilocain ointment (25 mg/g+ 25 mg/g).
3. Sterile physiological saline.
4. Eye ointment (5% dexpanthenol).
5. Povidone-iodine (100 mg/mL).
6. Electric shaver.
7. Stereotaxic frame.
8. Stereomicroscope (magnification 45×) with an articulating arm stand (providing greater degrees of freedom so the working area remains unobstructed by stereotaxic arms) and a table clamp. An optional Barlow lens can be used to increase the working distance between the objective and the area of interest.
9. Feedback thermometer with rectal temperature probe.
10. Optional: respiration monitor.
11. Electrocautery.

12. Gelatin sponge (*see* **Note 3**).
13. Hamilton syringe 5–10 μL attached to glass capillary (*see* **Note 4**).
14. Syringe pump for continuous substance administration (e.g., anesthesia or viral vector).
15. Glass bead sterilizer.
16. Dental drill with ceramic or diamond round drill bits.
17. Glue for fiber fixation on the cranium (*see* **Note 5**).
18. Low-melting agarose.
19. Paper tape.
20. Sterile surgical consumables (e.g., gloves, syringes, drapes, scalpel blades, needles, cotton tips).
21. Optional: MR-compatible ventilator (CWE, Ardmore, PA, USA, MRI-1; *see* **Note 6**).
22. Optional: Intubation stand.

2.2 Scanner and Coils

1. Small animal MRI scanner with a magnetic field gradient system of preferentially 0.5 T/m or higher. For further specifications, *see* **Note 7**.
2. Radio-frequency coil setup, preferentially a volume resonator for transmission and a separate receive coil: surface loop coil with 1 or 2 cm diameter or phased-array surface coil with fiber lead-through (*see* **Note 8**).
3. The MR animal cradle is equipped with a three-point head fixation with a bite bar and blunt ear plugs to minimize motion-related artifacts.
4. Modified functional MR sequence, which waits for a trigger signal from the external optical recording setup to ensure synchronization of all recordings.
5. Small animal physiology monitoring systems, at least with a respiration pillow and a rectal temperature probe, are used to continuously monitor respiratory rate and core body temperature. Optional: CO_2 analyzer, blood pressure transducer, ECG.
6. Warming device for an animal cradle. Warm water circulation is preferred over warm air flow, as the limited space within the bore of the optical setup may not provide sufficient airflow. Airflow also increases the risk of drying out the open skull during acute ofMRI experiments.

2.3 Equipment for Optical Setup and Stimulation

1. Required is an optical device capable of sensor excitation and fluorescence emission detection. Such systems can be acquired from commercial vendors but can also be custom-built (see Fig. 1). Relevant components are listed below. For more information, *see* **Note 9**.

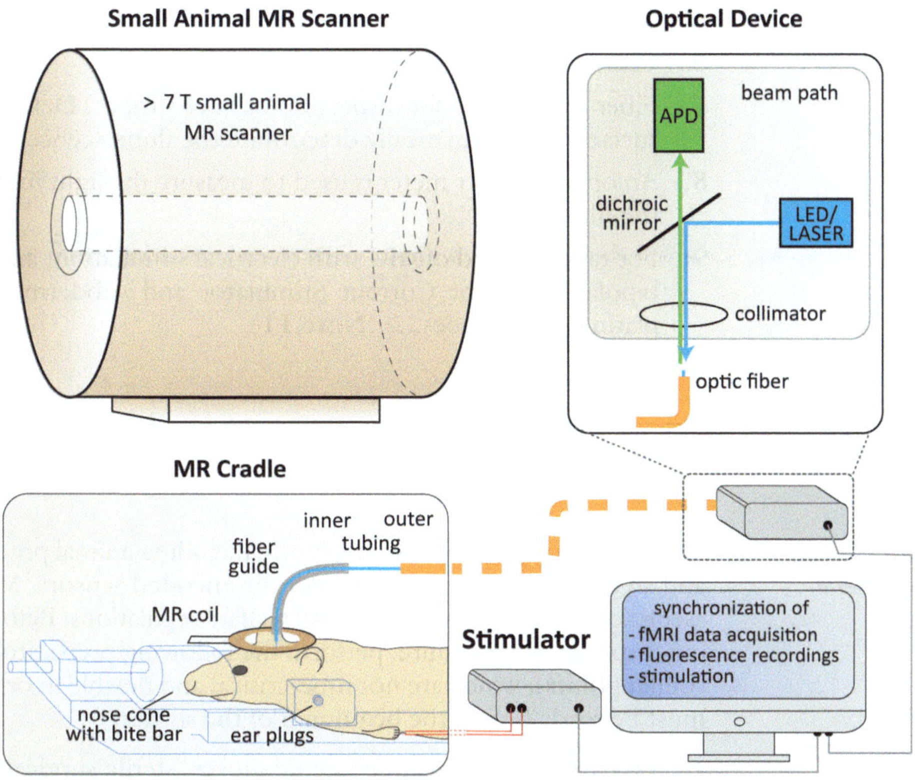

Fig. 1 Setup for simultaneous fMRI and fiber-optic recordings. Use a small animal MR scanner for functional MR image acquisition. For ofMRI experiments, animals are positioned with a bite bar and earplugs on an MR cradle. The implanted optic fiber is passed through the MR coil. Note that only the fully dismantled fiber tip (light blue) is inserted into the brain, while the inner tubing (dark blue) is fixed with glue to the cranium and passed through the fiber guide (gray). Thereby, the bend fiber can run parallel to the animal bed inside the magnet bore without tension. The rigid outer fiber tubing (orange) is taped to the cradle to promote stability. The fiber is connected to the optical device outside the scanner room. The optical stimulation and detection unit comprises a light source (LED/LASER) for excitation and a fluorescence detector unit. The excitation light is reflected by a dichroic mirror into the fiber and guided to the brain. The fiber collects the population signal. The emitted light passes through the dichroic mirror and is detected by an avalanche photodiode (APD). The voltage signal from the detector is filtered by an antialiasing filter and then sampled at 2 kHz. A computer with a custom-written program (e.g., LabVIEW) controls fMRI data acquisition, recording of fluorescence light, and stimulation

2. Optical multimode fibers (diameter: 200–400 μm, numerical aperture: 0.39–0.48). *See* **Note 10**.
3. Software for signal acquisition and synchronization of sensor recordings, fMRI, and stimulation. Such software may be included in commercial systems. For custom-built optical systems, software such as LABVIEW (National Instruments, Austin, TX, USA) can be used.
4. Angled fiber guide (*see* **Note 10**).

5. Safety goggles.
6. Glass cutter.
7. Fiber microscope for inspection of fiber surface. Defects at the surface may dramatically deteriorate the fluorescence signal.
8. An optical power meter is used to measure the light intensity at the fiber tip.
9. For task-induced ofMRI with electrical stimulation, an isolated bipolar Constant Current Stimulator and subdermal needle platinum electrodes (*see* **Note 11**).

3 Methods

3.1 Stereotaxic Surgery, General

Surgery duration is approximately 1 h, including animal preparation and stereotaxic injection of genetically encoded sensors. Maintain aseptic conditions, adhere to animal welfare regulations. Before starting the surgical procedure, perform the following preparatory steps (bullet points), which are not time-critical and flexible in order, but must be ready before the beginning of the surgery:

- Prepare surgical location: sterile gloves, sterile surgical drapes, sterilized instruments, a clear, clean, and disinfected area, and a stereotaxic frame.
- Prepare optical sensor: get an aliquot with the viral vector from the −80°C freezer, keep the box at −20°C as long as needed to avoid thawing (*see* **Note 12**).
- Check the animal's health status and general appearance (*see* **Note 13**).
- Apply analgesia 30 min before the start of surgery (*see* **Note 1**).

1. Place the animal in the induction chamber and initiate anesthesia with 5% isoflurane in 1 L/min O_2. Continuous application of 1%–2% isoflurane, depending on the respiration rate. Apply eye ointment to prevent dehydration of the cornea.
2. Shave the animal's head and wipe away loose hair with a wet tissue, remote from the surgery location to avoid loose hair in the aseptic environment.
3. Place the animal in the stereotaxic frame with the tooth bar and blunt ear plugs (*see* **Note 14**). Apply local anesthetic (lidocaine ointment) to the plugs (*see* **Note 1**).
4. Monitor body temperature continuously via rectal temperature probe (target temperature 36.5–37.5°C), ideally combined with a feedback thermometer. Monitor respiration rate

(target range 60–80 bpm), adjust isoflurane dose accordingly (*see* **Note 2**).

5. Cover the animal with a surgical drape to maintain an aseptic environment and to keep the animal warm. Note that visual monitoring of breathing is no longer possible this way. We, therefore, implemented an auditory respiration signal monitoring via a pressure sensor on the animal's abdomen.
6. Disinfect skin with povidone-iodine, wipe the first shot off, and keep the second shot on the skin.
7. Switch to sterile surgical gloves, avoid leaving the surgical site from then on until the incision has been closed.
8. Apply one shot of lidocaine spray to the disinfected skin.
9. Check reflexes to ensure surgical anesthetic depth: absence of toe pinch and eyelid reflexes; monitor respiration rate to ensure persistent surgical depth during surgery.
10. Cut skin with a scalpel (hold the scalpel slightly oblique to avoid damaging the cranial bone, 1–2 cm long midline incision between the level of eyes and ears (depending on the target area(s)), retract the scalp with surgical hemostats.
11. Stop potential bleeding at the skin incision and on the cranial bone with a cotton tip, gelatin sponge, or cautiously with electrocautery (*see* **Note 3**).
12. Apply ropivacaine to the cranial bone.
13. Remove connective tissue and periosteum from the bone surface by gently scraping the bone with the scalpel blade. Caution: Remove only as much connective tissue from the bone surface as necessary to have a clean area for drilling.
14. From here on, we recommend using a stereomicroscope to identify Bregma.
15. Move pointer to target region (*see* **Note 15**).
16. Higher pointer (keep in medial-lateral/anterior-posterior position to confirm centrality of drill hole), start drilling (approximately 1 mm diameter) in short intervals to prevent tissue heating and damaging the brain. Drill until only an alabaster-like thin layer is remaining (*see* **Note 3**).
17. Use a sterile needle tip (27 G) to cautiously flip off remaining bone and open the dura.

3.2 Reporter Injection

Experiments with genetically encoded sensors require two stereotaxic surgeries conducted several weeks apart. The initial survival surgery introduces a suitable vector system (e.g., an adeno-associated virus [AAV]) into the brain target area.

- Fill the Hamilton glass syringe with viral vector, take care to avoid air bubbles, and insert it into the syringe pump (*see* **Note 4**).

1. Position the needle/silica capillary in the center of the drill hole with the tip touching the surface of the brain, record Dorso–Ventral (D–V) position on the stereotactic frame; slowly lower the needle/silica capillary to the target position, make sure not to bend the capillary (bending indicates remaining bone or dura, *see* **Note 4**).
2. Inject viral construct (volume depending on the size of the target region in the range of 0.1–2.0 μL) with syringe pump at 0.25 μL/min. When targeting larger regions (e.g., cortex), the injection volume can be delivered stepwise (starting at the deepest position) and distributed along the vertical axis.
3. Following injection, leave the capillary in place for at least 2–5 min. Then, remove capillary slowly to prevent loss of injected fluid.
4. Moisten retracted skin and open skull with saline.
5. Close incision with 3–4 single knots (Nylon, 4 USP).
6. Put the animal in a warm cage on tissue to prevent the drowsy animal from choking on bedding. Closely monitor the animal until it has fully recovered. Transfer the animal to the home cage. We have good experience with group housing of animals after injections.
7. Score animal well-being and assess incision for potential inflammation (reddish appearance, weeping sores, swelling, suture opening) for at least 2 days, reapply s.c. analgesia. According to our experience, animals tolerate stereotaxic surgery well and recover without any complications other than a transient weight loss of less than 2%–3% in a small number of animals (*see* **Note 1**).

3.3 Fiber Implant Fixation and Preparation for Combined fMRI and Calcium-Recordings

For the optical recordings, the optical fiber needs to be implanted above the sensor expressing area—a procedure taking approximately 1 h and starting as outlined in Sect. 3.1. Be well prepared and carefully follow all surgery preparation steps—there are many details to consider. Again, the preparatory steps listed in bullet points are not time-critical and flexible in order, but must be ready before the beginning of surgery.

For mouse surgery, we position the mouse in the MR cradle, which is then integrated into a rat stereotaxic frame. This minimizes the movement of the animal once the fiber is fixated. For rat surgery, we position the animal in the rat stereotaxic frame and only transfer it to the animal cradle after fiber fixation (Step 14 below).

- Turn on the LED or LASER and the optical setup 1 h in advance.

- Prepare surgical location: sterile gloves, sterile surgical drapes, sterilized instruments, and a clear, clean, and disinfected area.
- Check the animal's health status and general appearance (*see* **Note 13**).
- Prepare the glue curing lamp (if needed) and the tools for gluing, according to the manufacturer's instructions (*see* **Note 5**).
- Prepare 1% of low-melting agar. Bring to a boil in the microwave. Afterward, keep it at 50°C on a hot plate with a stirring magnet.
- Prepare a piece of Parafilm™ (~4×4 cm^2) with a 3 mm cut reaching to the center on one side.
- Prepare sufficient strips of thin paper tape, some longer (~3 cm) and some shorter (~1 cm).
- Place the MR animal cradle next to the stereotaxic apparatus and prewarm it with a heating lamp.
- Guide the fiber through the MRI coil.
- Prepare fibers and angled fiber guide (*see* **Note 16**). Make sure to adjust light intensity to working conditions after preparation.
- Fixate the fiber on the lab bench with the fiber front end fixed on the stereotaxic frame to minimize unintended fiber movement during implantation.
- Turn on the warm water circulation of the MR animal cradle.

When all preparations are finished, follow the steps below. Make sure to apply analgesia consistently 30 min before the start of surgery and log time stamps to ensure consistent timing across experiments with different animals (*see* **Notes 1 and 13**).

1. Perform the general stereotaxic surgery following Steps 1–12 in Sect. 3.1.
2. Follow Step 13 in Sect. 3.1. Consider removing sufficient periosteum to provide adequate adherence at the glue-bone contact area (*see* **Note 5**).
3. Follow Step 14 in Sect. 3.1.
4. Follow Step 15 in Sect. 3.1, while adjusting the coordinates for the fiber implant (*see* **Note 15**).
5. Follow Steps 16 and 17 in Sect. 3.1.
6. Soak any fluid from within the drill hole.
7. Fixate the fiber tip on the stereotaxic arm and align it vertically above the drill hole. Make sure to have a clear view of the fiber tip and the drill hole through the stereomicroscope.
8. Lower the fiber tip onto the brain surface. Record D–V dura position.
9. Start optical recording and slowly move the fiber in 100 μm steps toward the target position inside the brain (*see* **Note 17**).

10. Once the final target position is reached, adjust the excitation light intensity to balance signal-to-noise and photo bleaching. Stop optical recording.
11. Prepare glue according to the manufacturer's instructions (*see* **Note 5**). Make sure that the glue is air bubble-free.
12. Apply a small drop of glue between the fiber and the bone skull. Cover the fiber with glue by pulling the drop up to 2 mm along the fiber. Cure glue according to the manufacturer's instructions.
13. Mold the glue to create a stable, cone-shaped base for the fiber (*see* **Note 5**).
14. If the animal is not in the MR cradle already, place it there now.
15. Aspirate warm 1% agar into the pipette and test the agar's temperature on the back of your hand to ensure it is at body temperature. Carefully apply bubble-free drops over the exposed skull, gradually building a uniform layer of approximately 3 mm thickness.
16. Once the agar layer solidifies, place the piece of Parafilm over the head to protect the surface coil from wet agar and to prevent drying of agar.
17. Position the MR coil onto the animal's head and fix it on the cradle.
18. Carefully move the angled guide along the fiber using tweezers to position the guide at the base of the fiber. The fiber should run approximately parallel to the animal's body without putting any tension on the implantation site.
19. Secure the fiber with tape in the MR cradle at multiple positions. Make sure to avoid contact of optical fiber with the animals' bodies to reduce motion artifacts in sensor recordings.

3.4 Preparation of ofMRI Experiments

The preparation of the MRI examination, including positioning and waiting time, takes approximately 20–40 min. The duration may vary with the choice of anesthetics (*see* **Note 2**).

1. Connect the animal to the MR scanner monitoring devices (at least a respiration pillow and a rectal temperature probe), if applicable, insert s.c. catheter for injection anesthesia or sensory stimulation devices for stimulus/task-induced ofMRI studies (*see* **Note 11**).
2. Move the animal with the cradle to the center of the MR scanner.
3. Run tri-pilot (localizer, GE TR/TE 100/2.3 ms, 1 average, scan time 12 s, 128^2 matrix, slice thickness 1 mm, field of view

[FOV] 4 cm^2) and position the target region in the isocenter of the scanner.

4. Confirm that the fiber-cranium assembly is still functioning by observing the optical sensor signal.
5. When using medetomidine sedation, this is a good time point to start s.c. bolus injection (*see* **Note 2**).
6. Run a 2D multislice scan in sagittal view (mouse: GE TR/TE 15/3 ms, 3 averages, scan time 12 s, 192^2 matrix, slice thickness 0.5 mm, FOV 2 cm^2; rat: RARE TR/TE 2000/12.5 ms, 2 averages, scan time 2 min, 256^2 matrix, slice thickness 1 mm, FOV 3 cm^2) as additional reference scan for reproducible positioning (Fig. 2A).
7. Acquire axial anatomical reference images T2w 2D RARE (mouse: TR/TE 6000/25.8 ms, RARE factor 8, 2 averages, FOV 18 × 12 mm^2, 180 × 120 matrix, 0.5 mm slice thickness, 18 slices; rat: TR/TE 2000/12.7, RARE factor 8, FOV 28 × 26 mm^2, 256^2 matrix, 1.2 mm slice thickness, 12 slices). The slice package should be placed according to Fig. 2A–D.
8. Acquire B_0 map and run shim adjustment routines (Fig. 2E, *see* **Note 7**).
9. Run a single repetition BOLD fMRI scan (for both mouse and rat experiments, use single-shot GE-EPI imaging with fat suppression, a left-to-right read-out, and a TR/TE of 1000/18 ms, flip angle of 60°. For mouse imaging, use a 90 × 60 matrix, 18x12 mm^2 FOV, and 18 axial slices with 0.5 mm thickness. For rat imaging, use an 80 × 80 matrix, 28 × 26 mm^2 FOV, and 12 axial slices with 1.2 mm thickness.
10. Setup optical signal recording parameters, such as light intensity, sampling frequency, and recording duration (*see* **Note 9**).

3.5 ofMRI Experiment

- The duration of an ofMRI session may vary considerably with the number of fMRI scans. These usually have a duration of 5–10 min each, depending on the number of repetitions.
- A stable physiological state is essential for successful ofMRI experiments, which requires maintaining a body temperature of 36.5–37.5°C, a breathing rate of 80–160 bpm in mice or 60–120 bpm in rats (depending on anesthesia), and a defined brain state as evidenced by stable optical recordings (*see* **Note 2**).
- For studies involving multiple animals, it is critical to maintain consistent timing across surgeries, animal preparation, and waiting periods between different runs or conditions (e.g., anesthesia transitions, stimulation paradigms, pharmacological challenges, etc., *see* **Note 13**).

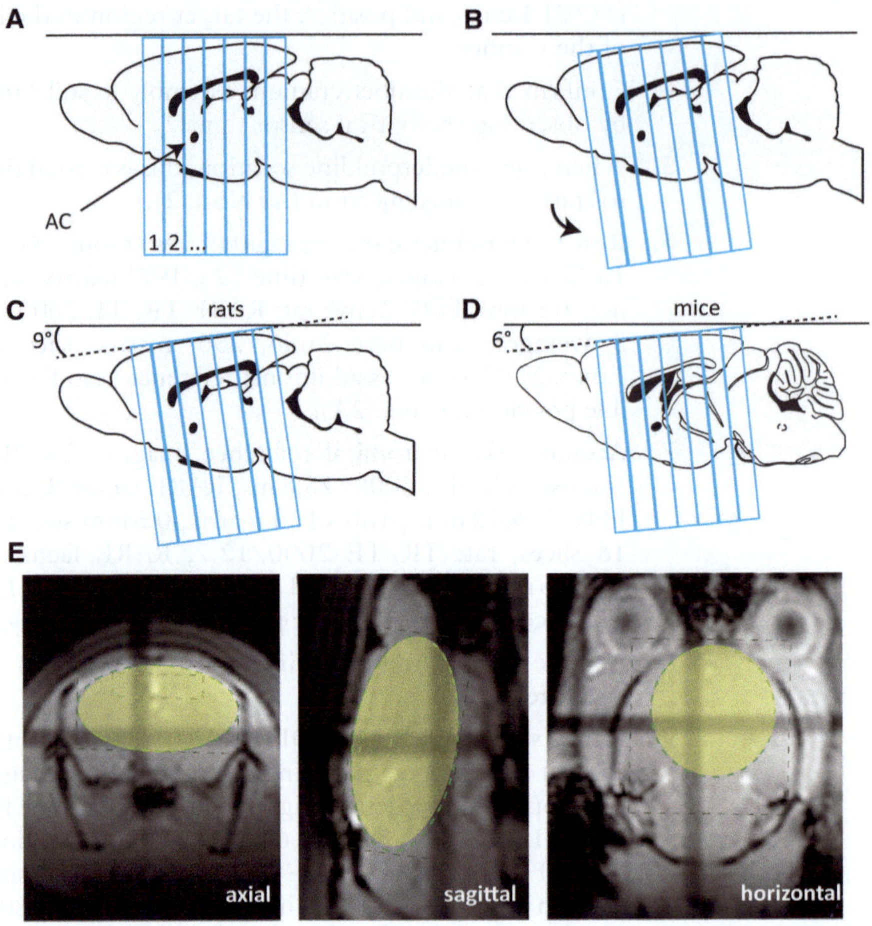

Fig. 2 Positioning the slice package and shim volume. (**A**) Reproducible positioning is facilitated by using the anterior commissure (AC) as a landmark and aligning the slice package in head-foot (**H–F**) direction with the cortical skull in the sagittal view. Depending on the area of interest, the AC should be placed in the same position in each slice package when measuring multiple animals. (**B**) When Bregma and Lambda are leveled (straight black line), the brain typically angles downward (dotted black line), and the slice package needs to be rotated to match slice orientation in the Paxinos atlas. (**C**) and (**D**) The rotation angle depends on the species and is approximately 9° in rats (**C**) and 6° in mice (**D**). (**E**) Ellipsoid shim volume (yellow) overlaid onto three localizer slices

Resting-State ofMRI Experiment

1. Duplicate single repetition scan from Step 9 in Sect. 3.4 and set number of repetitions to 600 (i.e., 10 min, can be adjusted to shorter or longer acquisition times).
2. Start BOLD fMRI scan. After automatic adjustments, the pulse program should wait for an external trigger from the optical setup.
3. Start optical recording via software. This will send a trigger signal to the MR scanner to begin synchronized ofMRI recordings.

4. Multiple fMRI scans are possible by repeating Steps 1–3, for example, when investigating changes in brain states over time or under different anesthetic conditions. Alternatively, different sequence parameters (e.g., shorter TR for ultrafast EPI acquisition) or imaging sequences (e.g., functional DWI, spin-echo EPI) can be applied.
5. At the end of the terminal experiment, remove the cradle with the animal from the scanner.
6. Increase isoflurane anesthesia and remove the fiber.
7. Proceed as detailed in your animal use protocol (e.g., cardiac perfusion for brain extraction for later histological validation).

Stimulus/Task-induced ofMRI Experiment

1. Duplicate single repetition scan from Step 9 in Sect. 3.4 and set number of repetitions to 600 (i.e., 10 min, can be adjusted to shorter or longer acquisition times).
2. Define the stimulation parameters. For non-noxious electrical stimulation, use a 1 ms pulse at 1–1.5 mA, delivered at 9 Hz for 10 s followed by 20 s of rest, repeated for 10 min to yield 20 blocks (*see* **Note 11**).
3. Start BOLD fMRI scan. After automatic adjustments, the pulse program should wait for an external trigger from the optical setup.
4. Start optical recording and stimulation paradigm via software. This will send a trigger signal to the MR scanner and the stimulation device to begin the synchronized task ofMRI recordings.
5. For further recordings, repeat Steps 1–4. Multiple runs are possible, for example, when various stimulation paradigms are applied.
6. At the end of the experiment, remove the cradle with the animal from the scanner.
7. Increase isoflurane anesthesia and remove the fiber.
8. Proceed as detailed in your animal use protocol (e.g., cardiac perfusion for brain extraction and later histological validation).

3.6 Analysis of Calcium and BOLD Data

A common approach to analyze simultaneously recorded optical time-series (e.g., calcium signals) alongside BOLD data is to compute correlation metrics, often incorporating time-shifted cross-correlation. Regression-based methods, such as general linear models (GLMs), can also use the calcium signal as a predictor for the BOLD time series. While more advanced techniques—like Granger causality or dynamic causal modeling—hold promise for exploring directional or causal relationships, their application to combined BOLD-calcium datasets remains an emerging area of research and requires careful interpretation. Regardless of the method, thorough preprocessing, such as precise time-alignment, filtering, and normalization of both datasets, is essential for obtaining robust and meaningful results.

4 Notes

1. **Consideration for Pain Management during Stereotaxic Surgery**
   Craniotomy inflicts a moderate burden on the animals and requires approval by local animal welfare authorities. Key considerations include potential side effects of analgesics, such as respiratory depression when using opioids like buprenorphine. Minimizing pain during and, if applicable, after surgery (e.g., survival procedures following viral vector injections) is essential to reduce discomfort and stress while promoting animal well-being. Maintaining a suitable depth of anesthesia during surgery, combined with systemic analgesics, is essential. Survival surgeries additionally require a 3-day post-operative pain care period with regular scoring of animal well-being and analgesic treatment. Commonly used regimens include buprenorphine (note that this is a controlled substance), carprofen, meloxicam, and metamizole. For best practices, consult with your local veterinarians or relevant literature [26, 27]. In addition to systemic analgesics, local analgesics, such as lidocaine and ropivacaine, mitigate acute pain from earplugs, skin incision, scraping of connective tissue from the skull, and penetration of bone and dura mater during surgery, as detailed in the Method section. Analgesic doses and timing of application may need to be adjusted to balance effective pain relief with experimental requirements, as some analgesics can interfere with physiological responses (e.g., suppression of seizures).
2. Anesthesia **for ofMRI**
   Anesthetics minimize animal stress and motion artifacts, but affect basic physiology (e.g., respiration parameters), as well as neural activity and hemodynamic parameters (i.e., BOLD, cerebral blood volume [CBV], or cerebral blood flow [CBF]). They thereby impact the sensitivity and specificity of fMRI signals and alter the response to stimuli or resting state networks. Depending on the choice of anesthetics, their mode of action and half-life may vary considerably. Side effects such as hypercapnia may be prevented by mechanical ventilation (*see* **Note 6**). While we use a sedation with combined isoflurane and medetomidine for functional studies in rats and mice

[22, 23], other anesthetic regimens (e.g., ketamine/xylazine, propofol, alpha chloralose) [28, 29] have also been shown to produce reliable results. Specifically, medetomidine is administered as a bolus followed by continuous subcutaneous infusion (for rat: bolus 40 μg/kg and continuous infusion 50 μg/kg/h; for mouse: bolus 50 μg/kg and continuous infusion 100 μg/kg/h). Make sure to note the time when starting the bolus injection. While isoflurane is gradually reduced over 20 min (to ~0.4% in rats, 0.2% in mice), closely monitor the respiratory rate (target ~90 bpm in rats, 120–140 bpm in mice). If respiratory rates suddenly drop, lower the isoflurane immediately. Consider a waiting period of approximately 40 min following bolus injection to allow formation of stable brain states before starting functional recordings. These waiting periods vary considerably depending on the anesthesia used. When performing cohort studies, consistent timing of surgery, fiber implantation, and start of ofMRI experiments across different animals is crucial for comparable optical and BOLD fMRI data.

3. **Bleeding**
   Bleeding at the skin incision site can create severe signal voids and geometric distortions in MR images. This bleeding can be controlled using electrocautery. Additionally, drilling may cause bleeding because blood vessels traverse both the skull and the dura. If bleeding occurs inside the drill hole, a gelatin sponge should be used to stem the flow, allowing several minutes for hemostasis. Bleeding during the viral vector injection is gradually resorbed from the parenchyma, yet, it may compromise fMRI and fluorescence readings weeks later. Blood that seeps beneath the skull or into the fiber canal during fiber implantation may trigger cortical spreading depression and additionally compromise fluorescent readings. Note that bleeding tends to be more common under isoflurane anesthesia; for surgeries in areas prone to bleeding (e.g., the olfactory bulb), consider alternatives such as ketamine/xylazine.

4. **Reporter Injection**
   Reporter injection should be as little invasive as possible. Several options are available. Either manual injection is possible or syringe pumps can be employed. When injecting manually, obstruction is easily recognized. On the other hand, volumes can be administered more slowly with a syringe pump. Stainless steel needles (30 G) are relatively large and may cause tissue damage associated with blood–brain barrier leakage and astrocyte activation. Cores of brain tissue may enter the needle and obstruct the flow of the solution. Glass capillaries customized with a puller can be used alternatively. However,

there is a risk of breaking the glass tip when the remaining bone is obstructing the injection canal.

An alternative method is the use of custom-made silica capillaries made from fused silica capillary tubing with polyimide coating (Polymicro Technologies, www.molex.com, cat. No. TSP075150 for injection tubing and TSP250350 for structural stabilization). These are more robust and can be cleaned and reused: cut a 2–3 cm segment of Tygon™ tubing (inner/outer diameter 0.01/0.03, www.coleparmer.com, cat. No. YO-06420-01). Thicken both ends by briefly heating them near a flame. Insert a piece of thick silica tubing (comparable length to the Tygon™ segment) into one end of the Tygon™ tubing. Secure the connection with a drop of glue (e.g., UV, cyanoacrylate, or epoxy). Slide a piece of thin silica tubing (long enough to reach from the syringe pump to the animal's head) at least 5 mm into the thick silica tubing in the glued assembly. Fix it in place with another drop of glue. Thread another piece of thick silica tubing onto the free end of the thin tubing. Ensure the thin injection tubing protrudes sufficiently to reach the target brain region, as only this thin portion will be inserted into the parenchyma. Secure with glue. Attach the free end of the Tygon™ tubing to a Hamilton syringe. Test for leaks with distilled water. Store the assembled system in a dry, dust-free environment.

5. **Cranium Preparation and Fiber Fixation**

Fiber implantation into the brain introduces susceptibility artifacts in T2*-weighted images. These artifacts arise primarily at the interface between materials differing in susceptibility (e.g., air vs. adhesive edge vs. bone), creating magnetic field inhomogeneities that become more pronounced at higher field strengths. Choosing the right adhesive and application strategy for fiber fixation onto the skull is essential for minimizing susceptibility artifact formation [30]. In general, during the preparation of adhesives, great care should be taken to prevent the inclusion of air bubbles. Furthermore, strictly avoid the application of glue on the dismantled glass fiber, as dispersion artifacts will interfere with optical recordings.

Preparing the cranium involves removing the periosteum and eventually drying the surface. For virus injection alone, only a smaller area of the cranium has to be prepared, whereas for fiber implant fixation, a larger area is recommended to promote secure glue adherence. Roughening the skull surface can further improve bond strength by increasing the contact area for the adhesive. Although chemical etching is commonly used for this purpose, it was originally developed for harder tissues such as dental enamel and may be too aggressive for the thinner and more elastic mouse cranium. Here, etching gel can potentially diffuse across the cranial sutures, causing damage to the underlying brain tissue. This concern is potentially relevant for rat crania as well. Therefore, we recommend considering mechanical methods, such as dental ceramic grinders, to achieve sufficient roughening without risking excessive tissue damage.

Adhesives that can be thinly spread or shaped into small patches generally produce fewer and shallower artifacts, while those that can only be applied in thicker, spherical patches tend to yield larger artifacts. Such factors must be weighed alongside mechanical demands—some adhesives offer superior bonding but can be difficult to apply precisely or cause extensive artifacts when used in larger quantities. Additionally, controlling viscosity via partial curing or layering can help mold the adhesive to the cranium's surface, while balancing tradeoffs between artifact minimization and implant stability [30].

In practice, apply multiple small droplets of adhesive ($<$ 2 mm) sequentially in approximately 120° steps around the fiber onto the dry cranium surface, continuing until the fiber is fully encased and the drill hole is completely covered with adhesive for maximum stability. Curing between each application minimizes inadvertent spread, and refining the techniques can further enhance both adhesion and overall patch shape. Finally, adhesives should be spread thinly and flatly to keep susceptibility artifacts as superficial and distant from target regions as possible, since artifacts form most pronouncedly at the adhesive's edges.

6. **Ventilation**

Anesthesia-induced hypercapnia lowers the BOLD signal. When ventilating animals, blood gases can be maintained in the physiological range by adjusting respiratory minute volume. We systematically performed BOLD fMRI and calcium-recordings in ventilated and spontaneously breathing rats and detected robust BOLD responses even under deep isoflurane anesthesia [22]. In non-ventilated animals, BOLD responses were not detectable under these conditions. This was most likely due to hypercapnia caused by respiratory depression, as administered CO_2 had the same effect in ventilated animals. Ventilation of animals boosts BOLD contrast; on the other hand, it is an invasive and technically challenging procedure. A recent paper assists decision-making whether to ventilate mice or rats for your specific research goal [31].

Note that for acute experiments, intubation should be performed before fiber implantation. During surgery, animals can breathe through the tracheostomy. The tubus will be connected to the ventilator when positioning the animal in the cradle. Continuous monitoring of exhaled air can easily be performed by connecting the expiratory valve of the ventilator with a CO_2 analyzer (Micro CapStar End-Tidal CO2 Analyzer, CWE, Inc., Ardmore, PA, USA). Administration of a muscle relaxant (e.g., pancuronium bromide, 2 mg/kg bolus followed by continuous infusion of 1.5 mg/kg/h, for rats) facilitates ventilation (no breathing against the machine) and minimizes motion artifacts. Be aware: when muscle relaxants are used, animals cannot breathe on their own! For survival experiments, antagonize pancuronium with a cholinesterase inhibitor, e.g., neostigmine, at the end of the experiment.

7. **Preparation for MRI Examination**
Robust rodent fMRI data can be acquired at field strengths of 7 T or higher, with higher field strengths providing improved SNR. However, a signal void at the fiber implantation side may be more pronounced due to increased susceptibility artifacts at higher field strengths. Recent standardization efforts of rodent fMRI, drawing on the collective experience of the neuroimaging community, guide MR protocol setup and parameter selection [32]. Notably, the field dependency of TE and FA should be considered [33].
Since susceptibility differences are not only the source of BOLD contrast but also a cause of artifacts at bone-brain and tissue-air interfaces, it is crucial to achieve high field homogeneity in the area of interest. Ideally, sophisticated shim adjustment routines should be based on B_0 maps. The shim volume should fully encompass the brain while excluding the skull and air-filled inner ear (see Fig. 2). Shimming performance can be evaluated using localized MR spectroscopy (full width at half maximum of a local water spectrum, e.g., STEAM) or a single-repetition GE-EPI sequence. B_0 homogeneity within the brain will minimize geometric distortions.
For functional BOLD fMRI, single-shot GE-EPI sequences are most commonly used with axial slices (with slice geometry copied from anatomical images) and a TR of 1000 ms. The duration of functional scans is usually between 5 and 30 min, finding a compromise between averaging and maintaining the stable animal's physiology.

8. **Coil Selection**
Depending on how the optic fiber is attached to the animal's head, the MR detector, a radiofrequency coil, may need a fiber pass-through. We work with single-loop surface coils or array coils: whatever best accommodates our needs regarding sensitivity, suitability for the respective target region, head fixation, and mechanical ventilation (see Fig. 3). Smaller coils offer a better signal-to-noise ratio but smaller brain coverage and penetration depth. The use of cryogenic probes is also

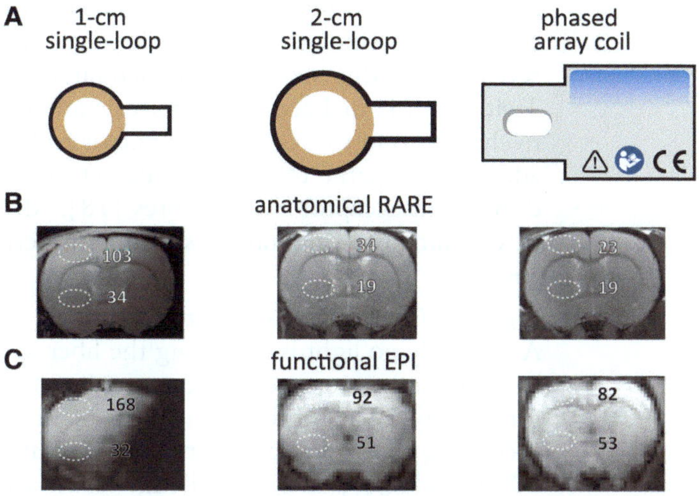

Fig. 3 Coils for ofMRI. (**A**) 1-cm and 2-cm single loop surface coil and rat array coil. (**B**) Representative slices of rat brain anatomical RARE images acquired with the respective coil. The 1-cm coil offers the best SNR (numbers for upper and lower brain areas), but no full brain coverage. The 2-cm loop and the array coil provide lower SNR but better brain coverage. The 2-cm single loop coil offers the highest degree of freedom for positioning of optical fibers on the cranium. (**C**) Corresponding slices of functional GE-EPI images. The 1-cm single loop coil provides the highest tSNR (numbers) in cortical areas, while the 2-cm and phased array coil offer higher tSNR in subcortical areas

possible, but it requires a different fiber implantation strategy—namely, specialized implants with angled, mirrored fiber tips and connectors that lie close and horizontally along the head [14].

9. **Optical Setup**

We use a custom-built setup for optic fiber-based calcium recordings that is located outside the scanner room. The light for excitation of fluorescent calcium indicators is delivered by a 20 mW solid-state laser (Sapphire, Coherent, Dieburg, Germany) with a wavelength of 488 nm (for GCaMP). An acousto-optic modulator (AOM 3080–125, Crystal Technology, Palo Alto, CA, USA) controls laser power. Alternatively, light-emitting diodes (LEDs) of the appropriate wavelength may be used. After deflection by a dichroic mirror, transmitted light is focused by a collimator into the multimode fiber (Thorlabs, Newton, NJ, USA).

Fluorescent light emitted from the brain is guided back through the same fiber and focused on a photodetector containing an avalanche photodiode (LCSA500-01, Laser components GmbH, Olching, Germany) with an aperture of 0.5 mm. The recorded fluorescence signals are digitized at a sampling frequency of 2 kHz using a multifunction I/O device (PCI 6259, National Instruments, Austin, TX, USA) and custom-written LabVIEW-based software (LabVIEW, National Instruments). To enable FRET-based fluorescence recordings with genetically encoded fluorescent reporters like Twitch2B [4] or Laconic [8] for calcium and lactate, respectively, a second pair of dichroic mirror and light detector was added to our setup to separately detect the light emitted from donor and acceptor fluorophores [18]. Multiple setups can be combined to simultaneously target different brain regions and/or cell types.

10. **Optical Fibers and Connectors**

As virtually no light is lost along the fiber itself, its length can be extended as needed without significant attenuation. We, for example, have integrated 15 m of fiber into our setup. However, incorporating connectors typically results in up to 50% signal loss, posing a challenge for measurement sensitivity. Therefore, for our experiments, we rely on a single fiber with a freshly prepared tip (*see* **Note 16**). Nevertheless, it is possible to employ optical implants with ceramic ferrules in combination with modified fiber tips (e.g., from Doric Lenses, Quebec, QC, Canada).

When using a single-fiber approach, a fiber guide is necessary to bend the fiber at a right angle without introducing mechanical stress. We typically use 200–400 μm diameter multimode fibers with a numerical aperture of around 0.48. Note that both fiber diameter and numerical aperture determine the illuminated volume and thus the total number of cells contributing to the bulk fluorescent signal. While larger diameters increase the signal, they also heighten invasiveness and can exacerbate susceptibility artifacts.

11. **Sensory Stimulation**

Multiple stimulation modalities can be used in sensory-evoked fMRI or ofMRI experiments, including mechanical skin [34] or whisker stimulation [35], thermal (heat, warmth, or cold) [36], visual [37], olfactory [38], optogenetic [39], and electrical stimuli [22]. Regardless of the modality, establishing the appropriate stimulation paradigm is crucial and depends heavily on the stimulus type, anesthetic protocol, and overall brain state.

For any stimulation modality, key parameters include:

- Interval: Rest periods between stimulation blocks.
- Stimulation duration: Length of each stimulus block.
- Intensity: Current amplitude, temperature, or other relevant measure.
- Stimulation frequency: Pulse repetition rate.
- Stimulus pulse length: duration of a single stimulus event.

Electrical stimulation typically involves placing needle platinum electrodes subcutaneously in the animal's hind or forepaw (between digits 2 and 4) and connecting them to a constant-current stimulator [22]. One common block design with electrical stimulation consists of a 10 s pulse train at 9 Hz, followed by 20 s of baseline, repeated over 10 min (i.e., 20 blocks). Higher frequencies may yield a stronger BOLD response, although calcium signals can exhibit adaptation with repeated stimulation. Randomizing the stimulus order may help prevent predictable patterns and confounds.

12. **Viral Vector Systems**

Viral vector systems—including, but not limited to, AAV-based and lentivirus-based approaches—offer several advantages for neuroscience research [40, 41]. Lentiviruses integrate into the host genome, enabling stable transduction in dividing cells, whereas AAV typically remains episomal, minimizing insertional mutagenesis in non-dividing cells, and hence are considered safer for laboratory use. When selecting a viral vector, consider cell type, expression duration, and potential immune responses. Thanks to a rapidly expanding library of genetically encoded biosensors, researchers can monitor calcium, neurotransmitters, metabolites, and even membrane voltage in defined cellular populations. Promoters such as CamKII or Synapsin narrow expression to excitatory or general neuronal populations, while GFAP targets astrocytes. However, AAV vectors also have limitations: their ~4.8 kb packaging capacity can make it challenging to include large promoters or sensor constructs and may reduce cell-type specificity (e.g., GFAP promotor). One way to circumvent this problem is by using genetically modified animals that express Cre-recombinase in specific cell types and inject Cre-dependent actuators. Additionally, retrograde AAV serotypes enable projection-specific targeting, providing an even finer level of control over subpopulations within a given cell type.

Achieving optimum expression also relies on careful dose titration and histological validation. While the tropism of different serotypes has been thoroughly documented, in our hands, AAV1, AAV2, and AAV9 demonstrated robust neuronal expression. In addition, experimental protocols involving AAV may require extra bureaucratic steps depending on regulatory guidelines regarding the work with genetically modified organisms. To maintain viral integrity, it is crucial to prepare aliquots, store them at −80°C, and minimize freeze-thaw cycles. Despite these challenges, AAV-based methods remain a powerful and versatile platform for unraveling complex brain functions and informing the development of new therapeutic approaches.

13. **Lab Records**

Lab records should include detailed information for each animal (e.g., source, date of delivery, strain, age, weight, sex), all administered drugs and substances (including batch numbers and dosing details), and specific procedures applied. When performing survival experiments, use ear clipping or other suitable identification strategy. An animal's weight is essential for accurate drug dosing and serves as a reference for post-surgical well-being assessment. Any noteworthy occurrences during procedures—such as excessive bleeding after drilling, instability under anesthesia—should be documented. Because variables like type, depth, and duration of anesthesia can significantly affect both the animal's condition and functional readouts, it is advisable to record time stamps at relevant experimental steps. Maintaining a detailed datasheet for each animal throughout the entire experiment has proven highly beneficial and can help identify potential confounders later on.

Details to be included on this data sheet should include, but are not limited to:

- vector type (e.g., AAV or retrovirus): source, serotype, promoter, batch number, titer.
- injection volume, target region, stereotactic coordinates.
- light intensity at fiber tip, AOM settings, baseline level, and signal amplitude of calcium-transients.
- kind of anesthesia, dose, time of dose, or drug changes.
- stimulation: modality, intensity, frequency, duration.
- file name of calcium-recordings data with time stamp.
- file name/scan number of MRI data with time stamp.
- respiration rate and body temperature (continuous or at a minimum for each recording).

14. **Adaptations for the Stereotaxic Frame**

Proper three-point head fixation is of utmost importance to ensure accurate results for stereotaxic surgery. Apply local analgesic (lidocaine ointment) to the blunt earplugs. Make sure earplugs are placed at the tympanic membrane in front of the inner ear (only then will the animal's head be leveled in L–R). The incisor bar should be adjusted until the height of Lambda and Bregma skull points are in the same horizontal plane (only then will the animal's head be levelled in A–P). We use a rat stereotaxic frame for both rats and mice. For mice, we modified the anesthetic delivery system to maintain good access to the skull by replacing the standard nose cone with a silicone tube placed perpendicular to the animal's axis. A small hole in the tubing's side allows the mouse's nose to extend into the anesthetic gas flow, ensuring stable inhalation anesthesia without obstructing the surgical field.

15. **Stereotaxic Targeting**
Selection of target brain regions may be based on BOLD activation maps or brain atlases (e.g., Allen brain atlas [42] or Paxinos atlas [43, 44]). Anatomical multislice or 3D MR images can help to plan access points for drilling and paths for injection into deep brain structures, thereby limiting invasiveness or destruction of relevant brain structures. Brain coordinates are then determined with the aid of an atlas and with Bregma as the reference point [45]. It may be helpful to use a glass capillary fixed at the stereotaxic arm as a pointer for accurate targeting of Bregma. Following injection of the viral vector, a waiting period is required to allow for expression of the viral vector. While the optimal time range needs to be determined experimentally, for cohort studies, it is important to keep timing consistent.
When performing fiber implantations weeks after the injections of genetically encoded reporters, separate drill holes should be used. In the cortex (see Fig. 4), it is advantageous to perform an angled injection and later implant the optic fiber vertically above the reporter-expressing volume. This strategy prevents excessive bleeding, which may be caused by touching the callous tissue in the previous drill hole. Planning of injection and fiber implantation paths needs to be adjusted for the desired target region. Additionally, for optic fiber approaches, the target area must be free of potential gliosis or clotted blood in the injection canal, which would obstruct illumination and collection of emitted fluorescent light. To maintain a stable bone structure for optimal fiber fixation, each drill hole should be bordered by intact bone, leaving at least a small bony bridge between adjacent openings. This ensures a stable anchor point for the fiber and enhances the overall durability of the adhesive bond (*see* also **Note 5**). In contrast, for acute experiments with chemical sensors, the optic fiber can be positioned via the same drill hole as used for sensor injection. Consequently, accurate targeting of the stained regions is less of an issue than employing genetically encoded sensors.

16. **Fiber Preparation**
First, consider how many fibers you will need for your experiment. Are you interested in multisite recordings? Do you plan to measure multiple animals in a row? To prevent signal loss, we avoid using optical connectors or mating sleeves and

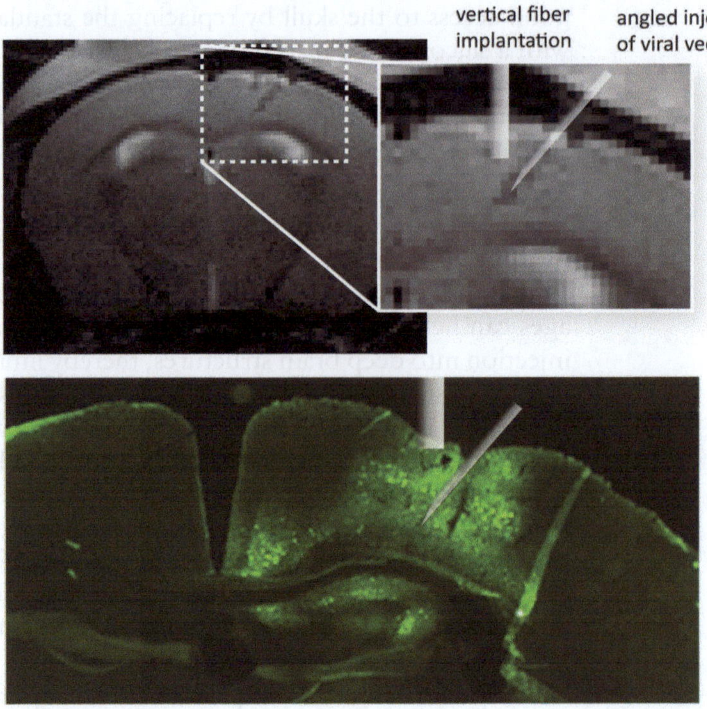

Fig. 4 Stereotaxic targeting in the mouse brain. Representative slice of the 2D anatomical image and zoom-in exemplarily illustrate the strategy for stereotaxic targeting of the sensory cortex of the hindlimb. The angled injection path from delivering the viral construct and the fiber position are both visible as a signal void in the MR image. (**B**) Schematics of the injection capillary and fiber are overlaid onto the respective histological image. Green fluorescence indicates the GCamP-expressing area

instead rely on a single fiber with a freshly prepared tip for each experiment. We recommend preparing fibers one day in advance to allow for any necessary adjustments. First, remove approximately 20 cm of the outer tubing so that the fiber ending has only its thin sheath. To create angled guides from glass or plastic cannulas, gently heat them (e.g., over a low flame or with a heat gun) until pliable, then bend them to the desired angle or rectangular shape. Before cutting the fiber, thread both the angled guides and the surface coil onto it, positioning them near the outer tubing. Next, plan the insertion depth for the recording site, allowing an extra ~1 mm for trimming the fiber. Use a fiber stripping tool to remove the inner tubing, then incise the fiber with a diamond cutter and break it as needed, inspecting the cut surface under a fiber microscope to ensure a clean, even break. Finally, confirm that the length of the fully dismantled fiber tip aligns with the distance from the brain surface to your target region. Throughout the procedure, wear safety goggles to protect your eyes, and use a power meter to measure the light intensity at the fiber tip—assessing dark current and background levels under defined light intensity settings (light working condition should be approximately 2 mW). Document settings (*see* **Note 13**). Be aware that excessive power may cause tissue heating and result in brain damage.

17. **Fiber Positioning for Optical Recording**
When targeting the cortex, the fluorescent signal from the sensor-expressing cells should become detectable once the fiber is placed on the brain surface, though maximum signal intensity may require a brief delay. Even before reaching its peak, the signal should increase as the fiber approaches the target region and exhibit some degree of modulation.
Under isoflurane anesthesia, spontaneous synchronous neuronal firing—referred to as burst suppression—can be observed across the cortex. This activity alternates between periods of synchronized neuronal firing (bursts or "up-states") and periods of relative quiescence (suppression), with the frequency of burst-suppression transitions depending on anesthetic depth. Observing these patterns indicates correct fiber placement, and cortex-wide up-states can also be triggered by simple stimuli like hand claps.
For deep-brain targets, gradually advance the fiber while continuously recording, positioning it close enough to the region of interest for robust signal detection. A sudden surge in fluorescence followed by signal loss for 10–20 min may indicate spreading depression; if this occurs, pause recordings and allow sufficient time for the tissue to recover.

Conflict of Interest

The authors haves no conflicts of interest to declare that are relevant to the content of this chapter.

References

1. Buxton RB (2013) The physics of functional magnetic resonance imaging (fMRI). Rep Prog Phys 76(9). https://doi.org/10.1088/0034-4885/76/9/096601
2. Göbel W, Helmchen F (2007) In vivo calcium imaging of neural network function. Physiology 22(6):358–365. https://doi.org/10.1152/physiol.00032.2007
3. Chen TW, Wardill TJ, Sun Y, Pulver SR, Renninger SL, Baohan A, Schreiter ER, Kerr RA, Orger MB, Jayaraman V, Looger LL, Svoboda K, Kim DS (2013) Ultrasensitive fluorescent proteins for imaging neuronal activity. Nature 499(7458):295. https://doi.org/10.1038/nature12354
4. Thestrup T, Litzlbauer J, Bartholomaus I, Mues M, Russo L, Dana H, Kovalchuk Y, Liang Y, Kalamakis G, Laukat Y, Becker S, Witte G, Geiger A, Allen T, Rome LC, Chen TW, Kim DS, Garaschuk O, Griesinger C, Griesbeck O (2014) Optimized ratiometric calcium sensors for functional in vivo imaging of neurons and T lymphocytes. Nat Methods 11(2):175–182. https://doi.org/10.1038/nmeth.2773
5. Zhuo YZ, Luo B, Yi XY, Dong H, Miao XL, Wan JX, Williams JT, Campbell MG, Cai RY, Qian TR, Li FL, Weber SJ, Wang L, Li BZ, Wei Y, Li GC, Wang H, Zheng Y, Zhao YL, Wolf ME, Zhu YJ, Watabe-Uchida M, Li YL (2024) Improved green and red GRAB sensors

for monitoring dopaminergic activity in vivo. Nat Methods 21(1): 680–691. https://doi.org/10.1038/s41592-023-02100-w
6. Aggarwal A, Liu R, Chen Y, Ralowicz AJ, Bergerson SJ, Tomaska F, Mohar B, Hanson TL, Hasseman JP, Reep D, Tsegaye G, Yao PT, Ji X, Kloos M, Walpita D, Patel R, Mohr MA, Tillberg PW, Looger LL, Marvin JS, Hoppa MB, Konnerth A, Kleinfeld D, Schreiter ER, Podgorski K, Team GP (2023) Glutamate indicators with improved activation kinetics and localization for imaging synaptic transmission. Nat Methods 20(6):925. https://doi.org/10.1038/s41592-023-01863-6
7. Xia XJ, Li YL (2025) A high-performance GRAB sensor reveals differences in the dynamics and molecular regulation between neuropeptide and neurotransmitter release. Nat Commun 16(1):819. https://doi.org/10.1038/s41467-025-56129-w
8. San Martin A, Ceballo S, Ruminot I, Lerchundi R, Frommer WB, Barros LF (2013) A genetically encoded FRET lactate sensor and its use to detect the Warburg effect in single cancer cells. PLOS One 8(2):e57712. https://doi.org/10.1371/journal.pone.0057712
9. Mächler P, Wyss MT, Elsayed M, Stobart J, Gutierrez R, von Faber-Castell A, Kaelin V, Zuend M, San Martín A, Romero-Gómez I, Baeza-Lehnert F, Lengacher S, Schneider BL, Aebischer P, Magistretti PJ, Barros LF, Weber B (2016) In vivo evidence for a lactate gradient from astrocytes to neurons. Cell Metab 23(1):94–102. https://doi.org/10.1016/j.cmet.2015.10.010
10. Keller JP, Marvin JS, Lacin H, Lemon WC, Shea J, Kim S, Lee RT, Koyama M, Keller PJ, Looger LL (2021) Glucose imaging in multiple model organisms with an engineered single-wavelength sensor. Cell Rep 35(12):109284. https://doi.org/10.1016/j.celrep.2021.109284
11. Bando Y, Grimm C, Cornejo VH, Yuste R (2019) Genetic voltage indicators. BMC Biol 17(1):71. https://doi.org/10.1186/s12915-019-0682-0
12. Schulz K, Sydekum E, Krueppel R, Engelbrecht CJ, Schlegel F, Schroter A, Rudin M, Helmchen F (2012) Simultaneous BOLD fMRI and fiber-optic calcium recording in rat neocortex. Nat Methods 9(6):597–602. https://doi.org/10.1038/nmeth.2013
13. Schmid F, Wachsmuth L, Schwalm M, Prouvot PH, Jubal ER, Fois C, Pramanik G, Zimmer C, Faber C, Stroh A (2016) Assessing sensory versus optogenetic network activation by combining (o)fMRI with optical Ca2+ recordings. J Cerebr Blood F Met 36(11):1885–1900. https://doi.org/10.1177/0271678x15619428
14. Schlegel F, Sych Y, Schroeter A, Stobart J, Weber B, Helmchen F, Rudin M (2018) Fiber-optic implant for simultaneous fluorescence-based calcium recordings and BOLD fMRI in mice. Nat Protoc 13(5):840–855. https://doi.org/10.1038/nprot.2018.003
15. Wang MS, He Y, Sejnowski TJ, Yu X (2018) Brain-state dependent astrocytic Ca signals are coupled to both positive and negative BOLD-fMRI signals. Proc Natl Acad Sci USA 115(7):E1647–E1656. https://doi.org/10.1073/pnas.1711692115
16. He Y, Wang MS, Chen XM, Pohmann R, Polimeni JR, Scheffler K, Rosen BR, Kleinfeld D, Yu X (2018) Ultra-slow single-vessel BOLD and CBV-based fMRI spatiotemporal dynamics and their correlation with neuronal intracellular calcium signals. Neuron 97(4):925. https://doi.org/10.1016/j.neuron.2018.01.025
17. Zhou XA, Jiang Y, Gomez-Cid L, Yu X (2025) Elucidating hemodynamics and neuro-glio-vascular signaling using rodent fMRI. Trends Neurosci 48(3):227–241. https://doi.org/10.1016/j.tins.2024.12.010
18. Lambers H, Wachsmuth L, Thomas D, Boumezbeur F, Hoesker V, Pradier B, Faber C (2022) Fiber-based lactate recordings with fluorescence resonance energy transfer sensors by applying an magnetic resonance-informed correction of hemodynamic artifacts. Neurophotonics 9(3):032212. https://doi.org/10.1117/1.NPh.9.3.032212
19. Schwalm M, Schmid F, Wachsmuth L, Backhaus H, Kronfeld A, Jury FA, Prouvot PH, Fois C, Albers F, Van Alst T, Faber C, Stroh A (2017) Cortex-wide BOLD fMRI activity reflects locally-recorded slow oscillation-associated calcium waves. eLife 6: e27602. https://doi.org/10.7554/eLife.27602
20. Segeroth M, Wachsmuth L, Gagel M, Albers F, Hess A, Faber C (2023) Disentangling the impact of cerebrospinal fluid formation and neuronal activity on solute clearance from the brain. Fluids Barriers CNS 20(1):43. https://doi.org/10.1186/s12987-023-00443-2
21. Mandino F, Horien C, Shen X, Desrosiers-Grégoire G, Luo W, Markicevic M, Constable RT, Papademetris X, Chakravarty MM, Betzel RF, Lake EMR (2025) Multimodal identification of the mouse brain using simultaneous Ca2+ imaging and fMRI. Commun Biol 8:665. https://doi.org/10.1038/s42003-025-08037-4
22. van Alst TM, Wachsmuth L, Datunashvili M, Albers F, Just N, Budde T, Faber C (2019) Anesthesia differentially modulates neuronal and vascular contributions to the BOLD signal. Neuroimage 195:89–103. https://doi.org/10.1016/j.neuroimage.2019.03.057
23. Pradier B, Wachsmuth L, Nagelmann N, Segelcke D, Kreitz S, Hess A, Pogatzki-Zahn EM, Faber C (2021) Combined resting state-fMRI and calcium recordings show stable brain

states for task-induced fMRI in mice under combined ISO/MED anesthesia. Neuroimage 245:118626. https://doi.org/10.1016/j.neuroimage.2021.118626
24. Albers F, Wachsmuth L, Schache D, Lambers H, Faber C (2019) Functional MRI readouts from BOLD and diffusion measurements differentially respond to optogenetic activation and tissue heating. Front Neurosci 13:01104. https://doi.org/10.3389/fnins.2019.01104
25. Lake EMR, Ge X, Shen X, Herman P, Hyder F, Cardin JA, Higley MJ, Scheinost D, Papademetris X, Crair MC, Constable RT (2020) Simultaneous cortex-wide fluorescence Ca(2+) imaging and whole-brain fMRI. Nature methods 17(12):1262–1271. https://doi.org/10.1038/s41592-020-00984-6
26. GV-SOLAS (n.d.) Dokumente. https://www.gv-solas.de/dokumente. Accessed 25 Feb 2026
27. National Research Council (US), Committee for the Update of the Guide for the Care and Use of Laboratory Animals; Institute for Laboratory Animal Research (US) (2011) Guide for the care and use of laboratory animals, 8th edn. National Academies Press, Washington, DC
28. Reimann HM, Niendorf T (2020) The (un) conscious mouse as a model for human brain functions: key principles of anesthesia and their impact on translational neuroimaging. Front Syst Neurosci 14:8. https://doi.org/10.3389/fnsys.2020.00008
29. Grandjean J, Schroeter A, Batata I, Rudin M (2014) Optimization of anesthesia protocol for resting-state fMRI in mice based on differential effects of anesthetics on functional connectivity patterns. Neuroimage 102:838–847. https://doi.org/10.1016/j.neuroimage.2014.08.043
30. Szinyei AZ, Maus B, Schmid JQ, Klimek M, Segelcke D, Pogatzki-Zahn EM, Pradier B, Faber C (2025) Systematic evaluation of adhesives for implant fixation in multimodal functional brain MRI. Magn Reson Mater Phy 38:191–205. https://doi.org/10.1007/s10334-024-01220-4
31. Just N, Hoehn M (2025) To intubate or not? Balancing anesthesia in rodent fMRI: strategies to mitigate confounding effects. Cereb Cortex 35(2):bhae499. https://doi.org/10.1093/cercor/bhae499
32. Grandjean J, Desrosiers-Gregoire G, Anckaerts C, et al (2023) A consensus protocol for functional connectivity analysis in the rat brain. Nat Neurosci 26:673-681. https://doi.org/10.1038/s41593-023-01286-8
33. Grandjean Lab (n.d.) MultiRat (GitHub repository). https://github.com/grandjeanlab/MultiRat. Accessed 25 Feb 2026
34. Chen HF, Lambers H, Nagelmann N, Sandbrink M, Segelcke D, Pogatzki-Zahn E, Faber C, Pradier B (2023) Generation of a whole-brain hemodynamic response function and sex-specific differences in cerebral processing of mechano-sensation in mice detected by BOLD fMRI. Front Neurosci 17:1187328. https://doi.org/10.3389/fnins.2023.1187328
35. Alonso BD, Sergeyeva M, Brune K, Hess A (2012) Lateralization of responses to vibrissal stimulation: connectivity and information integration in the rat sensory-motor cortex assessed with fMRI. Neuroimage 62(3): 2101–2109. https://doi.org/10.1016/j.neuroimage.2012.05.045
36. Reimann HM, Hentschel J, Marek J, Huelnhagen T, Todiras M, Kox S, Waiczies S, Hodge R, Bader M, Pohlmann A, Niendorf T (2016) Normothermic mouse functional MRI of acute focal thermostimulation for probing nociception. Sci Rep 6:17230. https://doi.org/10.1038/srep17230
37. Schmid F, Wachsmuth L, Albers F, Schwalm M, Stroh A, Faber C (2017) True and apparent optogenetic BOLD fMRI signals. Magn Reson Med 77(1):126–136. https://doi.org/10.1002/mrm.26095
38. Boido D, Rungta RL, Osmanski B, Roche M, Tsurugizawa T, Le Bihan D, Ciobanu L, Charpak S (2019) Mesoscopic and mioscopic imaging of sensory responses in the same animal. J Cerebr Blood F Met 39:61–61
39. Albers F, Wachsmuth L, Schache D, Lambers H, Faber C (2019) Functional MRI readouts from bold and diffusion measurements differentially respond to optogenetic activation and tissue heating. Front Neurosci 13:1104. https://doi.org/10.3389/fnins.2019.01104
40. Yizhar O, Fenno LE, Davidson TJ, Mogri M, Deisseroth K (2011) Optogenetics in neural systems. Neuron 71(1):9–34. https://doi.org/10.1016/j.neuron.2011.06.004
41. Naso MF, Tomkowicz B, Perry WL, Strohl WR (2017) Adeno-associated virus (AAV) as a vector for gene therapy. Biodrugs 31(4):317–334. https://doi.org/10.1007/s40259-017-0234-5
42. Allen Institute (n.d.) Brain Map: Atlas and anatomy resources. https://portal.brain-map.org/anatomy
43. Paxinos G, Watson C (2006) The rat brain in stereotaxic coordinates, Hard cover edition. Elsevier Science, Amsterdam
44. Paxinos G, Franklin KBJ (2001) The mouse brain in stereotactic coordinates. Academic, San Diego
45. Cecyn MN, Abrahao KP (2023) Where do you measure the Bregma for rodent stereotaxic surgery? Ibro Neurosci Rep 15:143–148. https://doi.org/10.1016/j.ibneur.2023.07.003

Chapter 6

Multimodal Neuroimaging with Optogenetic Functional Ultrasound and Functional Magnetic Resonance Imaging

Bradley J. Edelman

Abstract

The combination of optogenetic manipulations and neural recordings is a powerful approach to studying brain processes associated with specific neural circuits. The link between neural circuits and overall brain function further relies on comprehensive datasets that consist of both behavioral observations and large-scale neural recordings. In particular, functional ultrasound (fUS) and functional MRI (fMRI) are brain-wide imaging modalities that can be performed in rodents alongside optogenetic perturbations. These modalities record similar hemodynamic signals and, at the same time, complement one another by enabling neural recordings under different behavioral contexts. Despite the similarities between these modalities and their increasing utility with neural circuit investigations, multimodal experiments are far and few between. This chapter focuses on how optogenetics can be combined with both fUS imaging and fMRI in rodents, highlighting key experimental considerations, surgical preparation, and timelines that optimize data quality.

Keywords Optogenetics, Functional ultrasound, Functional MRI, Neural circuits, Anesthesia, Mouse behavior, Cranial window, Head-fixed

1 Introduction

Optogenetics is a powerful approach to perturb neural circuits in mice and to observe the resulting behavioral consequences. Combining the activation or inhibition of local cell populations with large-scale neural recordings further enriches gain-of-function and loss-of-function experiments by providing a comprehensive overview of network-level effects [1–5]. To date, optogenetic manipulations are often paired with local recording techniques such as electrophysiology, two-photon imaging, or fiber photometry, as these approaches also facilitate freely moving experiments. However, for these approaches to be effective, a *priori* information is necessary to identify downstream network hubs and recording sites that are relevant to the circuit of interest. To avoid uncertainty, whole-brain recording techniques can provide researchers with an overview of global activation patterns that encompass multiple and distant regions. Traditionally, whole-brain

Kevin Chuen Wing Chan and Seong-Gi Kim (eds.), *Optogenetic Functional Neuroimaging*, Neuromethods, vol. 232, https://doi.org/10.1007/978-1-0716-5178-0_6,

activation patterns associated with a particular behavior or manipulation have been extracted through the use of markers for immediate early genes (e.g., c-fos). This technique labels recently active cells in the brain (e.g., after an intervention) and provides researchers with relevant target regions for further investigation. However, this approach lacks temporal resolution and can be substituted with functional neuroimaging modalities that provide large-scale time-resolved neural information. Within the realm of functional recording techniques, there are various examples that can be readily combined with optogenetic manipulations (Fig. 1), each of which exhibits its own advantages. In particular, functional magnetic resonance imaging (fMRI) and functional ultrasound (fUS) imaging provide whole-brain information at a spatial scale of <250 μm and a temporal resolution of around 2 Hz. These two modalities share many characteristics in terms of the functional information that is extracted from the brain, but also exhibit varying degrees of portability, behavioral compatibility, and more (Table 1).

While both fUS and fMRI measure a hemodynamical readout of brain activity, the principles of neurovascular coupling suggest that these signals reflect the underlying neurophysiology [6, 7].

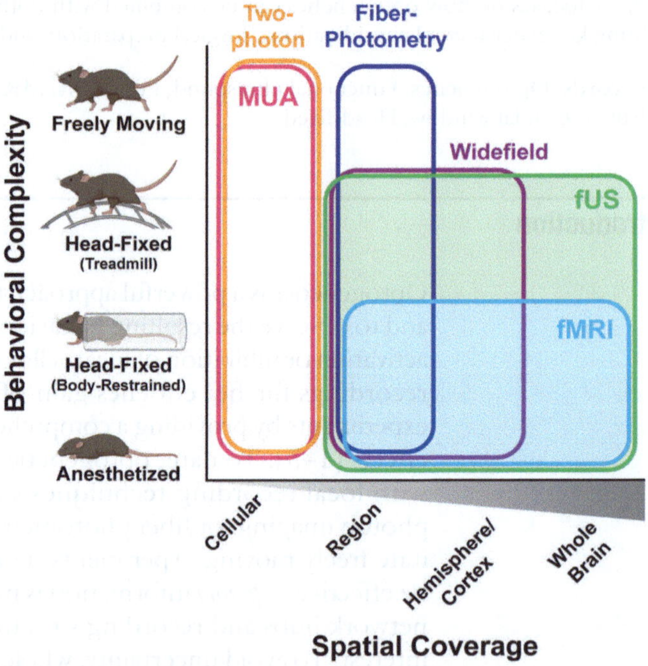

Fig. 1 Combining optogenetics, neural recordings, and mouse behavior. This schematic compares the spatial coverage and behavioral complexity that can generally be achieved with various neural recording modalities that can be combined with optogenetics. This book chapter focuses on multimodal neuroimaging using fMRI and fUS (shaded blue and green boxes). *MUA* multiunit activity, *fUS* functional ultrasound, *fMRI* functional magnetic resonance imaging

Table 1
Comparison of various fUS and fMRI parameters.

	fUS	fMRI
Contrast	CBV, ULM	CBV, BOLD, structural (T1, T2, T2*), DTI, NMR, etc.
Spatiotemporal resolution	$240 \times 240 \times 180\,\mu m^3$ @ 350 ms [5]	$230 \times 230 \times 600\,\mu m^3$ @ 1000 ms [51]
	$220 \times 280 \times 175\,\mu m^3$ @ 200 ms [50]	$200 \times 200 \times 300\,\mu m^3$ @ 350 ms [43]
Invasiveness	Cranial window	Noninvasive
Behavior	Head-fixed (3D), freely moving (2D)	Restrained
Cost	\$200–300k	>\$1.5 million
Space	Mobile cart	Shielded room/self-shielded magnet + technical room
Translation	Limited feasibility studies	Clinical and basic research

Importantly, both of these modalities are becoming increasingly compatible with awake rodent recordings and complex behaviors that facilitate the study of extrasensory neuroscientific principles [8, 9]. While performing the same experiments with both fUS and fMRI can be redundant in certain situations, doing so can also enable researchers to leverage the advantages of each modality in different contexts. For example, whole-brain anesthetized fMRI can robustly map neural circuits using well-established and standardized acquisition and analysis pipelines. Subsequent fUS imaging in head-fixed (whole-brain) or even freely moving contexts (single slice) can provide complementary information regarding the behavioral consequences of a manipulation, as well as confirmation of neural activation patterns in the awake state. Aside from this example, other combinations of these two modalities have demonstrated the value of multimodal imaging using fUS and fMRI for studying neurovascular coupling, resting-state brain networks, and functional brain mapping [10–13].

2 Materials

2.1 fUS Acquisition System

An ultrafast ultrasound controller and proper transducers are required for fUS acquisition. A standard 128-channel linear probe and Vantage 128 system are recommended for the methods described here. Stock systems can acquire 2D Doppler images at 0.5–1.0 Hz, with custom GPU additions increasing this sampling rate to nearly 10 Hz. This setup is, therefore, sufficient for comparing an individual fUS plane to that of fMRI at a similar temporal resolution. For 3D imaging, a matrix probe and a Vantage 256 (with a 4:1 multiplexer) can be used to acquire whole-brain information at ~2 Hz [5]; however, when combining fUS with optogenetics, linear probes (especially low-profile versions) are far more flexible in terms of positioning on top of the brain.

2.2 Preclinical MR Scanner

An MRI scanner and compatible transmit/receive coils are necessary for optogenetic fMRI. The MRI equipment used with the methods described here is a 7 T Bruker BioSpec 70/30 USR small animal MR system operating at 300 MHz. Surface head coils or 1–2 cm loop receive coils provide flexibility in terms of optic fiber placement, and importantly also with a cranial window to complement fUS. Alternative coils, such as CryoProbes, can also be used for increased SNR and resolution; however, proper compatibility with fUS imaging should be carefully planned.

2.3 Animals

Longitudinal studies benefit from using the same mice for both optogenetic fUS and fMRI; however, if necessary, different cohorts of animals can be imaged with each modality. While the latter scenario will weaken the statistical power of the experiment, this may be necessary if the required equipment for each technique is not available in the same location or institution. In theory, various rodent models are compatible with the multimodal imaging approaches described here, with mice offering the most flexibility in terms of optogenetic capabilities. Furthermore, the small size of mice generally prevents them from detaching a chronic cranial implant, and therefore, reduces attrition throughout experiments, especially those involving fUS. The methods described here have been validated with 8–10 week-old C57BL6 mice.

2.4 Choice of Opsin + Fiber Setup

The choice of opsin used for optogenetic fUS or fMRI can play a large role in the effectiveness of the experiment and the overall effort required by the researchers. Channelrhodopsin (ChR) and halorhodopsin constructs are commonly used to activate or inhibit cell populations, respectively, but may not be ideal when accounting for different sources of light leakage. First, to eliminate light leakage at the fiber coupling point, black putty should be placed around the implanted ferrules after patch cables from the laser are attached. This strategy is effective when the skull is left intact and also when using low light intensities alongside the installation of a large cranial window. However, opsins, cell populations, or behavioral effects that require high light intensities can be problematic when paired with a cranial window, as light can escape into the environment. Thus, in such scenarios, optogenetic stimulation of neural tissue can manifest as visible patterned light sources that can induce potentially unwanted visual circuit activation during anesthetized recordings or bias animal behavior during awake recordings (especially in dark contexts such as the bore of an MRI scanner or a behavior rig). Placing a constant nuisance light of a similar wavelength to the laser near the animal can mitigate the effects of light leakage and any unexpected behavioral effects and/or visual pathway activation.

Step-function opsins (SFOs) are another option that can reduce the effect of light leakage by inducing a stable and chronic perturbation with small bursts of light delivery rather than continuous stimulation [14]. Both excitatory and inhibitory variants (e.g., the stabilized step-function opsin—SSFO [15], the step-waveform inhibitory ChR—SwiChR [16]) of SFOs exist and can induce modulation on the order of tens of minutes with only short bursts of light. These opsins have been rarely employed by optogenetic fMRI studies [17, 18], but offer a potential solution to decouple light stimuli from neuronal manipulations and to enable increased specificity of circuit investigation. While light leakage should be accounted for, light itself has also been shown to cause local vasodilation effects, which can create artificial activation or deactivation patterns without inducing local neuronal firing [19–21]. This effect must be carefully considered, for both fUS and fMRI, and control cohorts should always be utilized to ensure that the elicited brain patterns are driven by the circuit of interest rather than other physiological effects such as tissue heating.

In addition to biology-related limitations, certain brain targets can be problematic for either fUS or fMRI due to physical interference between the implanted fiber and the recording device. For example, for both fUS and fMRI, the physical proximity between the brain and the transmit and/or receive device is of utmost importance. Therefore, efforts aimed at creating low-profile and streamlined implants greatly benefit the signal quality of both imaging modalities. In particular, prism fibers that bend the light path are an elegant solution to minimize upward protrusions that can interfere with high-resolution CryoProbes when acquiring fMRI data [2, 22]. Nevertheless, these fibers often run along the surface of the head and may lie between the brain and an ultrasound probe, potentially causing signal dropout. For this reason, the majority of optogenetic fUS studies implant fibers at an angle to accommodate the presence of an ultrasound probe, which acquires information from the non-obstructed field of view [4, 11, 23–25]. To avoid obstructed fields of view for fUS specifically, red-shifted opsins provide an alternative solution by activating specific cell populations through the lateral skull without the need for a physically implanted fiber [26]. Similarly, thinning the skull can also allow superficial cell populations to be activated by blue light sources [3]. Overall, researchers should consider many factors when deciding on which opsin and fiber setup to use for optogenetic fUS and fMRI experiments to ensure that imaging with both modalities is possible and that the data/results are comparable.

2.5 Cranial Window Implant

As to be further described in Sect. 3.2, fUS penetration depth and data quality need to be optimized when imaging through a cranial window. While some fUS approaches perform transcranial imaging with mice at young ages (<10 weeks old) [27, 28], penetration depth may still be limited, and longitudinal imaging can become problematic as mice age. fUS imaging can also be acquired from an acute cranial window after skull removal [29]; however, this approach has various limitations. Aside from animal welfare and discomfort, any local edema (visible or not) present after skull removal can negatively impact brain function and impair the ability to acquire high-quality fUS data. Therefore, it is recommended to install a chronic cranial window implant that allows the animals to fully recover while preserving the ability to perform fUS imaging (or e.g., optical imaging) for extended periods of time. There are currently various open-source options for cranial window implants (see [30] for further discussion), some of which have also demonstrated compatibility with optogenetic stimulations [24, 29]. However, these options either only allow for acute fUS recordings or provide relatively unspecific optogenetic targeting. The ChrOnic Multimodal imaging and Behavioral Observation (COMBO) window is another option that has been specifically designed for fUS imaging and optogenetic stimulations, and is recommended here [23]. While the COMBO window has not been extensively tested with fMRI, similar but more bulky implants have been shown to accommodate optogenetic fMRI data acquisition in anesthetized mice [11].

2.6 fUS and fMRI Atlas Registration

Proper alignment of fUS and fMRI data is paramount to comparing the results from these two modalities, and optimal results benefit from acquiring data from both modalities in the same animals to ensure consistent and identifiable landmarks. There are currently numerous fMRI toolboxes such as SPM, ANTS, RABIES, FSL, etc. [31–33] that provide both affine and nonlinear registration tools to align individual subject data to a common anatomical atlas or group template. From the fUS perspective, some tools originally developed for fMRI have also enabled the alignment of fUS data to brain atlases/templates. For example, MR angiographic (MRA) data acquired in the same native space as anatomical MRI data can be registered based on the transformation parameters defined by the desired anatomical warping. Registered MRA data can then be more easily aligned with vascular fUS data to bridge these modalities (Fig. 2A), either by registering these two data types together or to a common template.

Other options for fUS registration often rely on visual alignment and affine transformations [34]. Exemplary approaches, such as those described in Brunner et al. [35] and Nouhoum et al. [36], provide interactive software to manually or automatically register volumetric (3D) vascular fUS data to the Allen Brain Atlas. However, single-slice (2D) fUS imaging is still more common than volumetric imaging and can be more difficult to properly register across animals or to the correct brain location. When using a linear ultrasound probe, multiple slices can first be acquired as a reference scan (with a variable number of slices and spacing) that can be used for volumetric registration purposes. Functional scans acquired at one of these slices

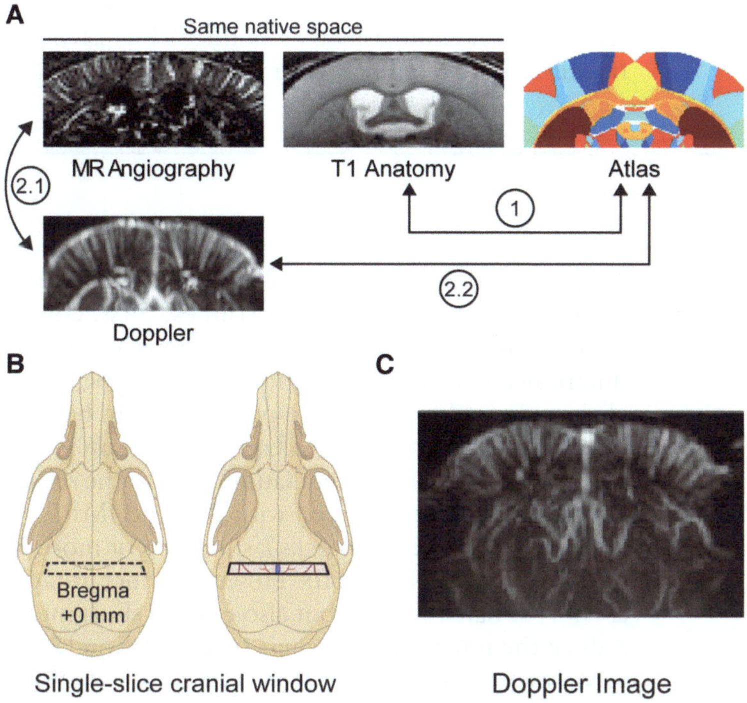

Fig. 2 Overview of multimodal data registration. (**A**) Existing tools are available to register anatomical MRI data to a common atlas or template (1). Functional and angiographic data acquired in the same native space can be registered using the same transformation parameters defined by the anatomical warping in (1). fUS data can be registered to the same atlas or template using multiple approaches. One option is to register Doppler information to MR angiographic data (2.1). As these two approaches measure similar vascular information from the brain, proper alignment can be easier to confirm in this case. The same transformation from (1) can then also be applied to the aligned Doppler data. Another option is to directly register Doppler information to an atlas/template (2.2). Recent tools allow for such an approach using either automatic or manual alignment. In either case, quality control measures should be implemented to ensure proper alignment. MR images are provided by courtesy of Dr. Giovanna Ielacqua. (**B**) For fUS, a single-slice cranial window with a known distance from Bregma can also provide a reference Doppler image that can then be used to identify other planes when performing either 2D or 3D functional acquisitions

can then be warped according to the volumetric transformation parameters. Alternatively, deep learning tools now show promise for automatically identifying any given coronal slice of rodent brain [37], and nonlinear registration algorithms can help fine-tune the alignment for group-level averaging and segmentation [38].

Despite the utility of this recent progress, creating a single-slice (or very thin) cranial window (e.g., in an example animal) exactly at the desired location, relative to bregma, provides a simple and sufficient reference for identifying the Doppler image of interest [10, 11]. In this case, it is important that the window is drilled straight across the skull, which can be outlined using standard stereotaxic devices and then completed by hand. In all future experiments/animals, the ultrasound probe can be placed above the brain and adjusted until the same vascular landmarks are identified (Fig. 2B). This slice can then be utilized as a reference for any other imaging plane that is desired.

3 Methods

3.1 Experimental Design Considerations

There are multiple parameters that should be considered when designing optogenetic fUS or fMRI studies. First, common parameters associated with optogenetic stimulations are stimulation intensity, duration, and frequency. The former should be titrated to a level that elicits a desired behavioral response; this can be performed prior to optogenetic fUS or fMRI to ensure that brain-wide effects reflect a particular behavioral manipulation. Due to the relatively slow hemodynamic response observed with fUS and fMRI, stimulation trials are typically applied in blocks lasting 10 s or more with an intertrial interval lasting less than 60 s. However, other instances utilize blocks as short as half a second, but also employ short intertrial intervals to reduce the temporal sparseness of stimulation [2]. Finally, it is also recommended to perform recording runs with different stimulation frequencies, as many previous works have observed strong frequency-dependent network effects (more so than intensity or duration) in response to optogenetic manipulations [4, 39, 40] (Fig. 3A, B). In addition to these parameters, the number of stimulations should be limited to accommodate both the adaptation of the neural response and the comfort of the animal (if performing awake recordings) or the depth of anesthesia (if performing anesthetized recordings). While a higher number of stimulations can lead to more averaging and reduced variability, in some cases, it becomes clear that the evoked response diminishes over time, which can significantly affect subsequent analyses such as those involving general linear models (Fig. 3C). Once parameters have been decided, they should be preprogrammed into a pulse generator that is synchronized with either the ultrasound controller or the MR scanner. This can be done with open-source hardware, such as an Arduino, or with a commercial solution, such as a Master 9 (A.M.P.I.) or an OTPG4/8 (Doric Lenses).

If stimulation and imaging are being performed on awake animals, researchers should also monitor the animals' physiological stability/parameters and behavioral state throughout each experimental session. Even when an animal is habituated to a behavior rig and experimental paradigm, repeated stimulations in a head-fixed context can agitate the animal and cause erratic behaviors (Fig. 3D) (*see* **Note 1**). Nevertheless, this concept should be approached with caution depending on whether researchers desire only a stimulation-evoked behavior or if they hope to increase the occurrence of a generalized behavior throughout a stimulation session, independent of stimulation timing. Overall, the internal and behavioral state of an animal can be difficult to control, and it is, therefore, recommended to also perform anesthetized recordings in the same cohort. While this condition involves the somewhat unnatural state of anesthesia, it also significantly reduces signal variability across trials and dampens any movement-related artifacts that may occur.

3.2 Surgical Preparation for Multimodal Imaging

Longitudinal optogenetic studies with fUS and fMRI should be carefully planned such that data quality in both modalities is optimized and comparable. Finding a compromise for each modality can be challenging as the ideal experimental preparation varies across the two techniques. Where fMRI data quality increases with reduced distance between the brain and the receive coil (among other factors), fUS data quality increases with the removal of the skull and the installation of a chronic prosthetic skull (Fig. 4A, B). The latter preparation can at times involve a bulky implant that limits the possible proximity of MR receive coils to the brain. Thus, researchers can choose to perform fUS and fMRI transcranially, potentially sacrificing fUS data quality, or after the installation of a cranial window, potentially sacrificing fMRI data quality. Alternatively, to optimize each approach and as is recommended here, fMRI can be performed with an intact skull, followed by the installation of a cranial window and then fUS imaging [11, 12] (Fig. 4C).

When combining either approach with optogenetics, virus injection should be performed according to standard stereotaxic surgical procedures. However, the position, orientation, and shape of the implanted optic fiber(s) should also be considered since objects protruding from the skull can physically interfere with the placement of other necessary components, such as the cranial window or MR coil (see Sect. 2.4 for details). If implanting the fiber(s) at an angle, virus injection should also be performed at the same angle to reduce the chance of location mismatch between virus expression and light delivery. Finally, during this procedure, the fibers should be secured in place, but researchers should avoid placing excess dental cement over regions of the skull that will later be removed as part of the cranial window procedure (*see* **Note 2**). By contrast, if using a red-shifted opsin or a thinned-skull option, no fiber implantation is required as external light sources can be utilized for these cases.

3.3 COMBO Window Installation

The COMBO window is an open-source, 3D printable chronic cranial window implant with all components available for download at https://zenodo.org/records/11092491 [23]. Of the eight available versions of the COMBO window, researchers should identify the one that is most compatible with their experiment. If there is no optogenetic component, the "bilateral" version is recommended for performing both fUS and fMRI in the same animals. This version is physically compatible with different MR receive coils (and ear-bar head fixation) and can also be used with a metallic head plate for head fixation on various behavior rigs compatible with fUS probes.

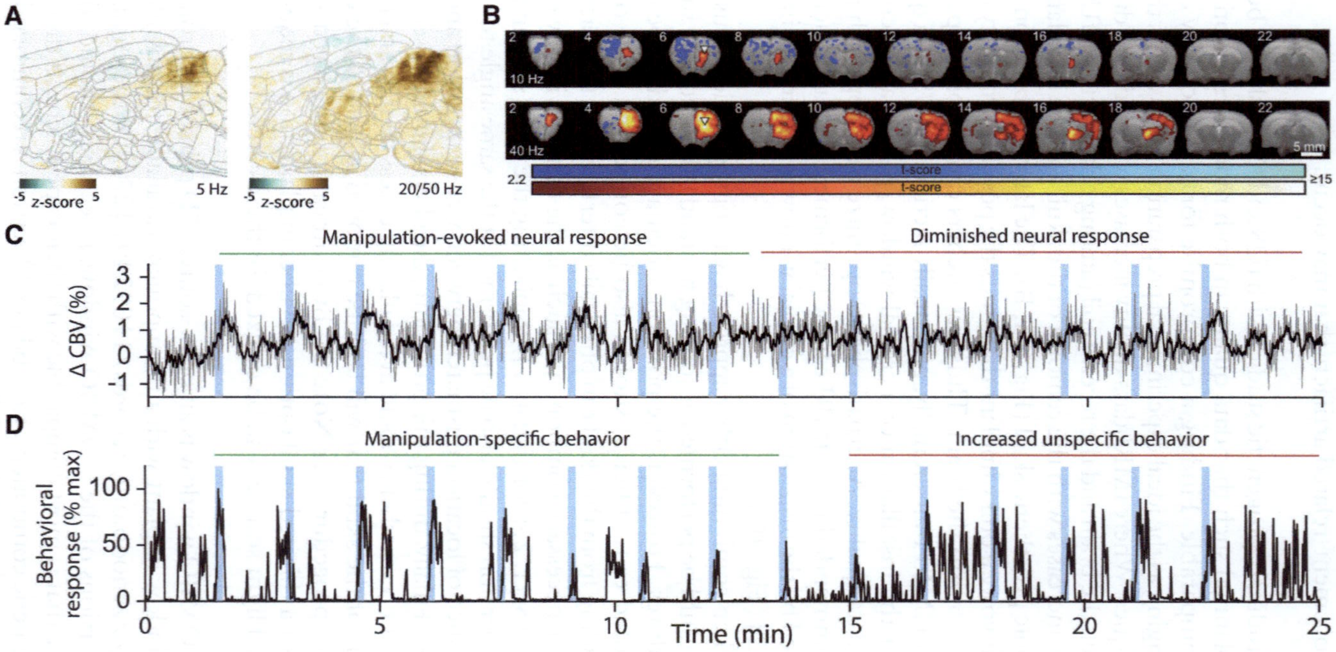

Fig. 3 Experimental design considerations. (**A**) Dissimilar fUS activation maps in response to optogenetic stimulation of the superior colliculus at different stimulation frequencies. Left—low frequency (5 Hz), right—high frequency (20/50 Hz). Adapted from [4] with permission. (**B**) Dissimilar fMRI activation maps in response to thalamocortical optogenetic stimulation of the ventrolateral orbitofrontal cortex at different frequencies. Top—low frequency (10 Hz), bottom—high frequency (40 Hz). Adapted from [40] with permission. (**C**) Example midbrain trace in response to 15 optogenetic stimulation trials (vertical blue bars, 10 s long, every 90 s) in an anesthetized mouse. A clear evoked neural response is observed in the first seven trials. This response is diminished in the latter eight trials, demonstrating an adaptation effect. The gray and black lines represent raw and smoothed data, respectively. (**D**) Example behavioral response to 15 optogenetic stimulation trials in an awake mouse. A behavioral response (black line) is specific and time-locked to the first eight trials, but becomes more erratic and un-specific in the latter seven trials, suggesting that the animal may be agitated. The fUS and behavior data presented here can be found in a publicly-available Edmond repository located at https://doi.org/10.17617/3.5KYYHK

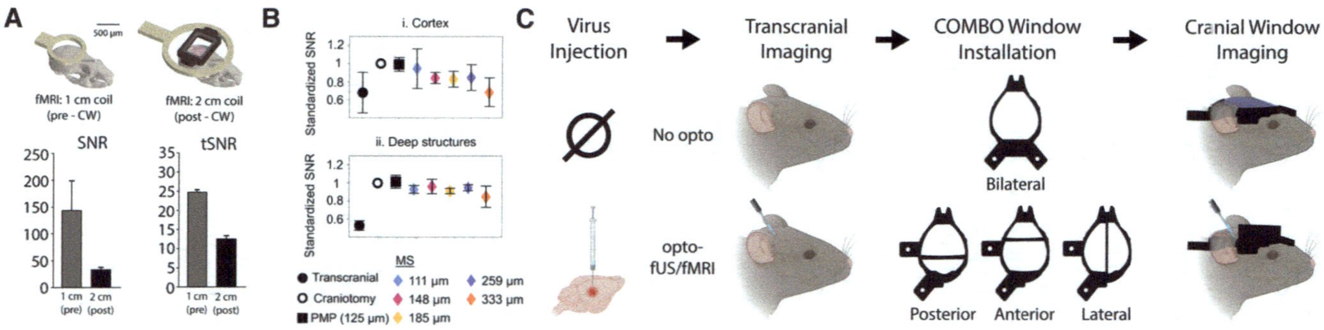

Fig. 4 Optimal preparation for optogenetic fUS and fMRI. (**A**) The signal-to-noise (SNR) and temporal SNR of gradient echo echoplanar images dramatically increase before a cranial window surgery and implant installation. Bulky implants may require the use of receive coils that exhibit an inherently lower SNR, suggesting that the cranial window itself is not fully responsible for reduced signal quality. Adapted from [11] with permission. (**B**) Transcranial Doppler imaging exhibits a notable reduction in SNR compared to an open craniotomy or a thin (125 µm thick) polymethylpentene (PMP) prosthetic film. Adapted from [52] with permission. (**C**) To optimize (t)SNR for multimodal whole-brain functional imaging, fMRI and fUS can be performed before and after cranial window implantation, respectively. If utilizing optogenetics, virus injection and fiber implantation should be performed such that the dental cement cap does not interfere with either fMRI receive coils or the future placement of the cranial window

Otherwise, one of the seven versions compatible with optic fiber implantation is recommended; however, these were not specifically designed for compatibility with MR receive coils and may require slight customization. Alternatively, other chronic cranial window implants [41] can be tested; however, it should be noted that glass windows will not be compatible with fUS.

If using the COMBO window, the implant should be prepared and examined a few days before implantation to ensure that it is both sealed and transparent (*see* **Note 3**). During the installation surgery, the animal should be secured either in a stereotaxic frame or with a bite bar that provides access to the full dorsal skull, as increased access to the skull near the ears is advantageous. After exposing the skull and gently removing the surrounding muscles (i.e., temporal muscle, trapezius), fit the preconstructed COMBO window to the skull around the implanted fiber(s) (if present). Identify areas where dried glue or dental cement from fiber implantation interferes with the COMBO window being able to sit flush against the skull. If necessary, carefully grind down these regions with a dental drill, and be careful not to damage or displace the fiber(s). Once the COMBO window fits flush against the skull, set it aside and begin drilling the cranial window.

The dorsal skull is easiest to remove when two bone islands are created that cover each hemisphere (Fig. 5A). While this involves drilling the bone on top of the superior sagittal sinus (SSS), thinning the skull in this location helps to both visualize the SSS and to remove the bone above it. The size of the bone islands in the rostral-caudal axis can be adjusted to accommodate fibers and the specific version of the COMBO window being utilized (Fig. 5B). Once both bone islands have been created and are floating on top of the skull, submerge them in saline solution and gently remove one island with a pair of fine forceps while holding the other in place on top of the brain. At this stage, assess how strongly the remaining bone island is attached to the dura by gently lifting it from the lateral edge—tissue connections will be strongest on top of a sinus (e.g., SSS, transverse sinus, etc.). Run the tip of a microprobe under the surface of the remaining bone island to detach any dura connections and subsequently remove the remaining skull. While keeping the dura moist, attach the COMBO window to the skull with super glue placing rostral and caudal to the window (*see* **Note 4**). Once secured, the full perimeter of the COMBO window should be sealed and secured to the remaining exposed skull using a combination of super glue and dental cement. Allow the animal to recover for at least 1 week (*see* **Note 5**), after which the head plate can be attached according to Edelman et al. [23].

3.4 Anesthesia and Animal Habituation

There are many publications that detail anesthetic regimens and their effects on fUS or fMRI signals [3, 24, 42]. A cocktail of fentanyl (0.05 mg/kg), midazolam (5 mg/kg), and medetomidine (0.5 mg/kg) injected intraperitoneally, or a combination of isoflurane (0.5%) and

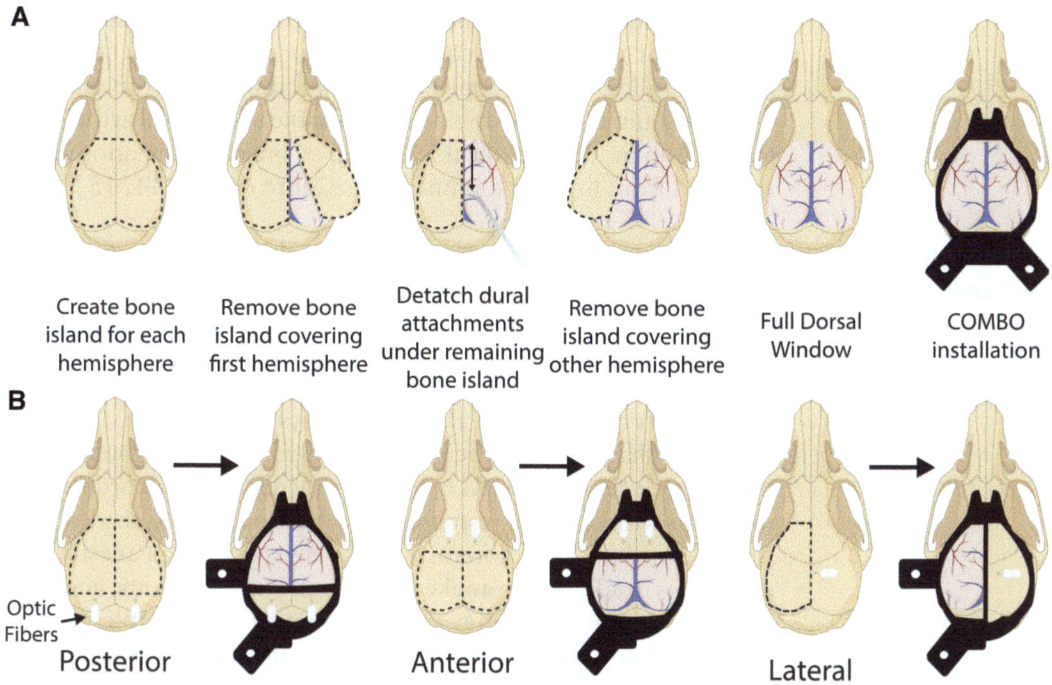

Fig. 5 COMBO window installation. (**A**) Installation of a chronic cranial window can be performed before fUS imaging to improve signal quality. The dorsal skull should be removed in different sections to minimize damage to the dura and brain tissue. Damage to the superior sagittal sinus (SSS) (or other large superficial veins) can be ameliorated by significantly thinning the skull directly above it. In the case shown here, once two separate bone islands are floating on top of the dura, one can be removed quickly. The remaining dura attachments beneath the remaining bone island can then be visualized and separated with a microprobe. This should allow the second bone island to be removed without ripping the dura and/or the SSS. (**B**) Different versions of the COMBO window can be utilized to accommodate optic fibers implanted for optogenetic manipulations. The "posterior" and "anterior" versions allow for a bilateral fiber implantation, whereas the "lateral" and quartered versions (not shown) allow for a unilateral fiber implantation

a continuous infusion of dexmedetomidine (0.2 mg/kg/h), paired with an initial bolus of dexmedetomidine (0.1 mg/kg), works well for preserving the neural response and suppressing movement [11, 23]. When performing anesthetized recordings, habituation to the recording setup is not necessary; however, a heating source should be placed under the animal to maintain constant temperature throughout the session. It is currently standard practice to record breathing rate during fMRI recordings; however, this is less common for fUS, although still highly recommended. If performing awake imaging, animals should be gradually habituated to the cradle/running wheel over the course of a week or more [23, 43]. Slowly increasing the duration of head fixation by 15 min in subsequent days is usually sufficient for proper habituation. In addition, other environmental items required for the experiment (i.e., probe, coil, tastant spout, electrodes, etc.) can be gradually introduced in later habituation sessions. For both anesthetized and awake imaging, it is recommended that a sample image from a functional sequence (e.g. one-repetition echo-planar image, single power Doppler) be acquired to check the quality of the data. Animals that exhibit uncorrectable anatomical artifacts due to air bubbles, dura mater thickening, or brain deformation can be removed from future experiments (*see* **Note 6**) [35, 44, 45].

3.5 Optogenetic Functional Neuroimaging

There are various considerations that should be accounted for when using optogenetics with fUS imaging or fMRI, especially when comparing the two modalities. Firstly, awake recordings will always exhibit a relatively large degree of variability caused by the dynamic behavioral state of the animal. While proper habituation will help calm animals during recording sessions, there will always be sources of day-to-day variability that may be difficult to control for. This is complicated by the fact that fUS and fMRI exhibit highly dissimilar recording environments with different restraining mechanisms, noises, shadows, etc. Because of this, it is recommended to always perform anesthetized recordings in addition to awake recordings. Under the same anesthetic regimens, these recordings will be the most comparable between modalities, and will ensure that movement artifacts are minimized and that circuit activation through optogenetics is not superimposed on top of behavior-driven neural activity.

3.6 Data Analysis

Standard analysis for optogenetic fUS and fMRI involves using a general linear model (GLM) to generate whole-brain activation maps associated with the circuit of interest. Prior to obtaining these outputs, all functional data should be checked for quality and artifacts (*see* **Note 6**), and preprocessed. As with registration, there are many freely available toolboxes and analysis pipelines that have been optimized for fMRI preprocessing. By contrast, there are far fewer open-source platforms and/or standard protocols that are used to analyze fUS data, and making fUS data compatible with fMRI toolboxes is non-trivial. Because of this, it is recommended to apply modality-specific preprocessing steps, followed by common steps and postprocessing. Preprocessing for fMRI typically involves slice-timing correction, head movement estimation, susceptibility distortion correction, as well as normalization to an atlas/template (e.g., the RABIES pipeline [33]). For fUS, the primary preprocessing step involves clutter filtering of compound ultrasound image stacks to form a power Doppler image at each time point [46]. After these steps are performed, general denoising techniques can be applied to both fUS and fMRI datasets. Optional steps in this regard include, but are not limited to, spatial smoothing, temporal filtering, and movement artifact rejection. In particular, the latter two can be applied directly to voxel-wise fUS and fMRI time series via digital filter design and regression, or can be added as GLM nuisance regressors.

Once preprocessed and registered to a common anatomical space, fUS and fMRI data can be analyzed using a GLM. For this, the design matrix should at a minimum include the optogenetic block-stimulus after being convolved with a hemodynamic response function (HRF). Here, it is important to ensure that the proper HRF is being employed for each modality. For example, the blood-oxygenation-level-dependent (BOLD) fMRI contrast is generally modelled using a double-gamma response that is characterized by an initial peak and a small subsequent undershoot. By contrast, fUS data is thought to be directly proportional to cerebral blood volume (CBV) and is, therefore, modeled using a single-gamma response that features only an initial peak. Furthermore, the parameters of these HRFs (e.g., delay, dispersion, length, etc.) should be set according to consensus values specifically observed in rodents rather than those found in humans (see [47, 48] for details on these parameters). Behavioral variables of interest can also be appended to the GLM; however, these regressors should be carefully assessed for collinearity as many aspects of behavior can share temporal features [49]. In general, all regressors of interest should be orthogonal to facilitate proper interpretation of the resulting beta or contrast maps. In some cases, L1- (LASSO) or L2-regularized (Ridge) regression can account for design matrix collinearity; however, manual orthogonalization is always recommended when possible. In addition to these regressors of interest, nuisance regressors can be added for the purpose of temporal filtering (i.e., cycles of sines and cosines) and movement artifact removal (i.e., time series of head movement measurements and/or nonbrain regions-of-interest).

4 Notes

1. **Behavior:** Determining the valence of a particular optogenetic manipulation can be informative for animal compliance during optogenetic fUS or fMRI recordings. Head movement is one of the largest sources of artifacts for these modalities; despite lengthy habitation protocols, animals can become agitated during aversive manipulations and exert large forces on the head fixation. The resulting micromovements can severely impact data quality and should be minimized by titrating the number and duration of stimulations that an animal can comfortably tolerate. Initial valence can be determined using a real-time place preference paradigm and then subsequently tested in a head-fixed context.
2. **Fiber implantation:** When implanting an optic fiber(s) for optogenetic stimulations, ensure that the gap between the ferrule(s) and COMBO window frame is also completely sealed. If performing transcranial fMRI followed by fUS through a cranial window, secure the ferrules to the skull with a minimal amount of dental cement, focusing on regions of the skull

not intended for the cranial window or those that will come into contact with the implant. If needed, dental cement can be drilled down later; however, this also comes with the risk of reduced ferrule stability or of the dental cement cap being fully removed.

3. **COMBO window preparation:** The installation of a COMBO window that is prepared with film being too flat can compress and deform the brain beyond the correction of various registration techniques. On the other hand, installing a COMBO window with film that is too curved can cause the cavity to include air bubbles in the long run. Aside from the curvature, it is advised to check the film before installation to ensure that there are no regional opacities due to glue/epoxy.
4. **COMBO window installation:** The initial attachment points are most crucial to ensure chronic implantation of the COMBO window or other implants. Ensure that the skull is dry before applying the initial portion of super glue (a thin layer rather than a blob is recommended) and attaching the implant. Furthermore, ensure that subsequent glue and dental cement applied around the perimeter of the implant contact the skull and not only the nearby skin and muscles. For this step, it is recommended to use a liquid dental cement that can flow through crevices to the bone to dry rather than a light-cured dental cement that requires a primer. With this being said, be careful not to let the cement run under the implant and onto the brain while still in liquid form. If properly sealed, the window should completely fill up with cerebrospinal fluid within 1 week.
5. **Cranial window surgery post-op:** Animals should be carefully monitored for 3–5 days after cranial window surgery. Damage to the dura mater or brain during the surgery can significantly affect the animals' survival and subsequent imaging quality. Placing the animals' cage on a heating pad and providing the animals with painkillers, wet food, and subcutaneous saline injections can markedly help increase their survival in the postoperative phase.
6. **Imaging artifacts:** The COMBO window is compatible with both fUS and fMRI technology; however, there are potential limitations for each. For fUS, even though the dorsal skull is removed, the curvature of the skull can hide lateral brain tissue and cause lower visibility for some brain regions, which may or may not be acceptable depending on the research questions and regions of interest. By contrast, bone fragments leftover from the cranial window surgery and located on the surface of the brain can attenuate the fUS signal and cause shadows throughout the depth of tissue underneath. While the brain surface can be cleaned by gently flushing with saline or manually with

surgical instruments, caution should be taken so as not to damage the dura or brain tissue. In particular, trauma to the dura can accelerate thickening and can lead to reduced ultrasound penetration, and tearing of the dura can lead to brain deformation. If performing fMRI after COMBO window installation and/or fiber implantation, artifacts can be expected to a similar extent as any other cranial manipulation. For example, air bubbles underneath the implant, within the structure of the dental cement cap, or surrounding the optic fiber(s) compartment can create artifacts in echoplanar images.

Conflict of Interest

The author has no conflicts of interest to declare that are relevant to the content of this chapter.

References

1. Lee JH, Durand R, Gradinaru V et al (2010) Global and local fMRI signals driven by neurons defined optogenetically by type and wiring. Nature 465:788–792. https://doi.org/10.1038/nature09108
2. Zou Y, Tong C, Peng W et al (2024) Cell-type-specific optogenetic fMRI on basal forebrain reveals functional network basis of behavioral preference. Neuron 112:1342–1357.e6. https://doi.org/10.1016/j.neuron.2024.01.017
3. Kim S, Moon HS, Vo TT et al (2023) Whole-brain mapping of effective connectivity by fMRI with cortex-wide patterned optogenetics. Neuron 111:1732–1747.e6. https://doi.org/10.1016/j.neuron.2023.03.002
4. Sans-Dublanc A, Chrzanowska A, Reinhard K et al (2021) Optogenetic fUSI for brain-wide mapping of neural activity mediating collicular-dependent behaviors. Neuron 109:1888–1905.e10. https://doi.org/10.1016/j.neuron.2021.04.008
5. Brunner C, Grillet M, Sans-Dublanc A et al (2020) A platform for brain-wide volumetric functional ultrasound imaging and analysis of circuit dynamics in awake mice. Neuron 108:861–875.e7. https://doi.org/10.1016/j.neuron.2020.09.020
6. Aydin AK, Haselden WD, Goulam Houssen Y et al (2020) Transfer functions linking neural calcium to single voxel functional ultrasound signal. Nat Commun 11:2954. https://doi.org/10.1038/s41467-020-16774-9
7. Logothetis NK, Pauls J, Augath M et al (2001) Neurophysiological investigation of the basis of the fMRI signal. Nature 412:150–157. https://doi.org/10.1038/35084005
8. Edelman BJ, Macé E (2021) Functional ultrasound brain imaging: bridging networks, neurons, and behavior. Curr Opin Biomed Eng 18:100286. https://doi.org/10.1016/J.COBME.2021.100286
9. Mandino F, Vujic S, Grandjean J, Lake EMR (2024) Where do we stand on fMRI in awake mice?. Cerebral Cortex 34:bhad478. https://doi.org/10.1093/cercor/bhad478
10. Boido D, Rungta RL, Osmanski B-F et al (2019) Mesoscopic and microscopic imaging of sensory responses in the same animal. Nat Commun 10: 1110. https://doi.org/10.1038/s41467-019-09082-4
11. Edelman BJ, Ielacqua GD, Chan RW et al (2021) High-sensitivity detection of optogenetically-induced neural activity with functional ultrasound imaging. Neuroimage 242:118434. https://doi.org/10.1016/j.neuroimage.2021.118434
12. Soloukey S, Collée E, Verhoef L et al (2023) Human brain mapping using co-registered fUS, fMRI and ESM during awake brain surgeries: a proof-of-concept study. Neuroimage 283:120435. https://doi.org/10.1016/j.neuroimage.2023.120435
13. Hikishima K, Tsurugizawa T, Kasahara K et al (2023) Functional ultrasound reveals effects of MRI acoustic noise on brain function. Neuroimage 281:120382. https://doi.org/10.1016/j.neuroimage.2023.120382
14. Berndt A, Yizhar O, Gunaydin LA et al (2009) Bi-stable neural state switches. Nat Neurosci 12:229–234. https://doi.org/10.1038/nn.2247

15. Yizhar O, Fenno LE, Prigge M et al (2011) Neocortical excitation/inhibition balance in information processing and social dysfunction. Nature 477:171–178. https://doi.org/10.1038/nature10360
16. Berndt A, Lee SY, Ramakrishnan C, Deisseroth K (2014) Structure-guided transformation of channelrhodopsin into a light-activated chloride channel. Science 344:420–424. https://doi.org/10.1126/science.1252367
17. Takata N, Sugiura Y, Yoshida K et al (2018) Optogenetic astrocyte activation evokes BOLD fMRI response with oxygen consumption without neuronal activity modulation. GLIA 66:2013–2023. https://doi.org/10.1002/glia.23454
18. Takata N, Yoshida K, Komaki Y et al (2015) Optogenetic activation of CA1 pyramidal neurons at the dorsal and ventral hippocampus evokes distinct brain-wide responses revealed by mouse fMRI. PLoS One 10:e0121417. https://doi.org/10.1371/journal.pone.0121417
19. Christie IN, Wells JA, Southern P et al (2013) fMRI response to blue light delivery in the naïve brain: implications for combined optogenetic fMRI studies. Neuroimage 66:634–641. https://doi.org/10.1016/j.neuroimage.2012.10.074
20. Rungta RL, Osmanski B-F, Boido D et al (2017) ARTICLE Light controls cerebral blood flow in naive animals. Nature Publishing Group. https://doi.org/10.1038/ncomms14191
21. Schmid F, Wachsmuth L, Albers F et al (2017) True and apparent optogenetic BOLD fMRI signals. Magn Reson Med 77:126–136. https://doi.org/10.1002/mrm.26095
22. Grimm C, Frässle S, Steger C et al (2021) Optogenetic activation of striatal D1R and D2R cells differentially engages downstream connected areas beyond the basal ganglia. Cell Rep 37:110161. https://doi.org/10.1016/j.celrep.2021.110161
23. Edelman BJ, Siegenthaler D, Wanken P et al (2024) The COMBO window: a chronic cranial implant for multiscale circuit interrogation in mice. PLoS Biol 22:e3002664. https://doi.org/10.1371/journal.pbio.3002664
24. Rungta RL, Osmanski B-F, Boido D et al (2017) Light controls cerebral blood flow in naive animals. Nat Commun 8:14191. https://doi.org/10.1038/ncomms14191
25. Li B, Ma C, Huang Y-A et al (2023) Circuit mechanism for suppression of frontal cortical ignition during NREM sleep. Cell 186:5739–5750.e17. https://doi.org/10.1016/j.cell.2023.11.012
26. Aurup C, Pouliopoulos AN, Kwon N et al (2023) Evaluation of non-invasive optogenetic stimulation with transcranial functional ultrasound imaging. Ultrasound Med Biol 49:908–917. https://doi.org/10.1016/j.ultrasmedbio.2022.11.002
27. Ferrier J, Tiran E, Deffieux T et al (2020) Functional imaging evidence for task-induced deactivation and disconnection of a major default mode network hub in the mouse brain. Proc Natl Acad Sci 117:15270–15280. https://doi.org/10.1073/pnas.1920475117
28. Bertolo A, Ferrier J, Cazzanelli S et al (2023) High sensitivity mapping of brain-wide functional networks in awake mice using simultaneous multi-slice fUS imaging. Imaging Neurosci 1:1–18. https://doi.org/10.1162/imag_a_00030
29. Provansal M, Labernède G, Joffrois C et al (2021) Functional ultrasound imaging of the spreading activity following optogenetic stimulation of the rat visual cortex. Sci Rep 11:12603. https://doi.org/10.1038/s41598-021-91972-z
30. Cramer SW, Carter RE, Aronson JD et al (2021) Through the looking glass: a review of cranial window technology for optical access to the brain. J Neurosci Methods 354:109100. https://doi.org/10.1016/j.jneumeth.2021.109100
31. Avants BB, Tustison NJ, Song G et al (2011) A reproducible evaluation of ANTs similarity metric performance in brain image registration. Neuroimage 54:2033–2044. https://doi.org/10.1016/j.neuroimage.2010.09.025
32. Jenkinson M, Beckmann CF, Behrens TEJ et al (2012) FSL. Neuroimage 62:782–790. https://doi.org/10.1016/j.neuroimage.2011.09.015
33. Desrosiers-Gregoire G, Devenyi GA, Grandjean J, Chakravarty MM (2022) Rodent automated bold improvement of EPI sequences (RABIES): a standardized image processing and data quality platform for rodent fMRI. bioRxiv 2022.08.20.504597. https://doi.org/10.1101/2022.08.20.504597
34. Hikishima K, Tsurugizawa T, Kasahara K et al (2023) Brain-wide mapping of resting-state networks in mice using high-frame rate functional ultrasound. Neuroimage 279:120297. https://doi.org/10.1016/j.neuroimage.2023.120297
35. Brunner C, Grillet M, Urban A et al (2021) Whole-brain functional ultrasound imaging in awake head-fixed mice. Nat Protoc 16:3547–3571. https://doi.org/10.1038/s41596-021-00548-8
36. Nouhoum M, Ferrier J, Osmanski B-F et al (2021) A functional ultrasound brain GPS for automatic vascular-based neuronavigation. Sci Rep 11:15197. https://doi.org/10.1038/s41598-021-94764-7
37. Lambert T, Brunner C, Kil D et al (2024) A deep learning classification task for brain navigation in rodents using micro-Doppler ultrasound imaging. Heliyon 10:e27432. https://doi.org/10.1016/j.heliyon.2024.e27432

38. Zhong S, Agyeman K, Syed S et al (2023) Comparative evaluation of image registration techniques in functional ultrasound imaging. Imaging Neurosci 3:IMAG.a.47. https://doi.org/10.1162/IMAG.a.47
39. King AJ, Liu J, Lee HJ et al (2015) Frequency-selective control of cortical and subcortical networks by central thalamus. https://doi.org/10.7554/eLife.09215.001
40. Weitz AJ, Lee HJ, Choy MK, Lee JH (2019) Thalamic input to orbitofrontal cortex drives brain-wide, frequency-dependent inhibition mediated by GABA and zona incerta. Neuron 104:1153–1167.e4. https://doi.org/10.1016/J.NEURON.2019.09.023
41. Ghanbari L, Carter RE, Rynes ML et al (2019) Cortex-wide neural interfacing via transparent polymer skulls. Nat Commun 10:1500. https://doi.org/10.1038/s41467-019-09488-0
42. Grandjean J, Schroeter A, Batata I, Rudin M (2014) Optimization of anesthesia protocol for resting-state fMRI in mice based on differential effects of anesthetics on functional connectivity patterns. Neuroimage 102:838–847. https://doi.org/10.1016/j.neuroimage.2014.08.043
43. Gutierrez-Barragan D, Singh NA, Alvino FG et al (2022) Unique spatiotemporal fMRI dynamics in the awake mouse brain. Curr Biol 32:631–644.e6. https://doi.org/10.1016/j.cub.2021.12.015
44. Grimm C, Wenderoth N, Zerbi V (2022) An optimized protocol for assessing changes in mouse whole-brain activity using opto-fMRI. STAR Protoc 3:101761. https://doi.org/10.1016/j.xpro.2022.101761
45. Shim H-J, Im GH, Jung WB et al (2022) Protocol for mouse optogenetic fMRI at ultrahigh magnetic fields. STAR Protoc 3:101846. https://doi.org/10.1016/j.xpro.2022.101846
46. Demené C, Deffieux T, Pernot M et al (2015) Spatiotemporal clutter filtering of ultrafast ultrasound data highly increases Doppler and fUltrasound sensitivity. IEEE TMI 34(11):2271-2285. https://ieeexplore.ieee.org/abstract/document/7098422
47. Lambers H, Segeroth M, Albers F et al (2020) A cortical rat hemodynamic response function for improved detection of BOLD activation under common experimental conditions. Neuroimage 208:116446. https://doi.org/10.1016/j.neuroimage.2019.116446
48. Pisauro MA, Dhruv NT, Carandini M, Benucci A (2013) Fast hemodynamic responses in the visual cortex of the awake mouse. J Neurosci 33:18343–18351. https://doi.org/10.1523/JNEUROSCI.2130-13.2013
49. Musall S, Kaufman MT, Juavinett AL et al (2019) Single-trial neural dynamics are dominated by richly varied movements. Nature Neurosci 22:1677–1686. https://doi.org/10.1038/s41593-019-0502-4
50. Rabut C, Correia M, Finel V et al (2019) 4D functional ultrasound imaging of whole-brain activity in rodents. Nat Methods 16:994–997. https://doi.org/10.1038/s41592-019-0572-y
51. Chen X, Tong C, Han Z et al (2020) Sensory evoked fMRI paradigms in awake mice. Neuroimage 204:116242. https://doi.org/10.1016/j.neuroimage.2019.116242
52. Kim G, Rabut C, Ling B et al (2023) Microscale acoustic metamaterials as conformal sonotransparent skull prostheses. In Review

Chapter 7

Methodological Considerations for fMRI with Optogenetics, Chemogenetics, and Fiber Photometry

Sheng Song, Tzu-Hao Harry Chao, Li-Ming Hsu, Randy Nonneman, Tzu-Wen Winnie Wang and Yen-Yu Ian Shih

Abstract

This chapter discusses essential methodological considerations for integrating functional magnetic resonance imaging (fMRI) with neuromodulation techniques such as optogenetics, chemogenetics, and fiber photometry. The synergy of these approaches with fMRI provides a robust framework to investigate neural circuit dynamics, neurovascular coupling, and behavior correlates of imaging features in preclinical models. Combining fMRI with these neuromodulatory and recording tools offers a systems-level view of neural activity and the potential to reveal the functional impact of targeted activity on complex brain networks.

Keywords fMRI, Optogenetics, Chemogenetics, Fiber photometry, Neurovascular coupling, Brain networks, Rodent models

1 Introduction

The integration of functional magnetic resonance imaging (fMRI) with advanced genetically based neuromodulation and recording technologies—including optogenetics, chemogenetics, and fiber photometry—has expanded preclinical neuroscience research capabilities, particularly within rodent models [1–5]. Each of these technologies allows precise "reading" over specific neural circuits, offering cell-type specificity across a range of temporal precision, with optogenetics allowing millisecond-scale control over neural firing through light-activated channels, and chemogenetics enabling prolonged neuronal modulation via receptor-mediated drug activation [6, 7]. These tools are essential for establishing causal links between neural activity and behavioral outcomes. fMRI, meanwhile, provides spatially extensive and temporally dynamic measures of brain activity across the entire brain, making it an ideal tool for investigating the systems-level impact of targeted neural modulation or their relevance to regional activities.

Kevin Chuen Wing Chan and Seong-Gi Kim (eds.), *Optogenetic Functional Neuroimaging*, Neuromethods, vol. 232,
https://doi.org/10.1007/978-1-0716-5178-0_7,

Fiber photometry has emerged as a valuable technique for real-time monitoring of fluorescent signals, particularly for tracking cellular activity in rodent models [8]. This approach leverages implanted optical fibers to both deliver excitation light and collect emitted fluorescence, offering a minimally invasive means of assessing neuronal or neurochemical dynamics within targeted brain regions. Its versatility is demonstrated across a range of settings, from freely moving behavioral assessments [9] to head-fixed setups used in MRI [2, 5, 10], allowing for comprehensive studies of neural processes linked to specific behaviors or tasks. With various genetically encoded sensors now available, fiber photometry is increasingly utilized to capture diverse signals, such as calcium or neurotransmitter activity. This technique's robustness and cost-effectiveness, coupled with its adaptability for multiplexing and multisite monitoring, make it a powerful tool for fMRI, providing insights into brain function and circuit dynamics.

These neuromodulation and recording methods have specific methodological requirements when integrated with fMRI to ensure data fidelity and minimize artifacts. For optogenetic and fiber photometry studies, precise fiber optic setups are necessary for accurate light delivery to specific brain regions without inducing heating artifacts that could influence the blood-oxygenation-level-dependent (BOLD) signal [5]. Chemogenetic studies require carefully controlled administration of ligands, such as clozapine-N-oxide (CNO) [4, 11, 12] or deschloroclozapine (DCZ) [13], to activate Designer Receptors Exclusively Activated by Designer Drugs (DREADDs), enabling sustained neuronal modulation and corresponding BOLD signal changes over extended periods. For all methods, MR-compatible equipment and well-established surgical protocols are essential to maintain data integrity and reduce interference.

The integration of these techniques provides a versatile framework for exploring neural circuits and understanding brain function on multiple levels. Combining fMRI with neuromodulation techniques, researchers gain insights into both rapid and sustained neuronal effects across large-scale networks. These multimodal approaches open new possibilities for identifying how specific neural populations contribute to behavior and brain-wide neurovascular dynamics, promising to yield deeper insights into brain function and dysfunction. While these techniques are becoming increasingly popular in the field, this article focuses on sharing the specific protocols utilized in our laboratory, with the hope of guiding researchers in reproducing these approaches effectively.

2 Materials

2.1 Rodent Subjects and Environment

All animal protocols should be approved by the local Institutional Animal Care and Use Committee, often in accordance with the National Institutes of Health Guide for Care and Use of Laboratory

Animals. The subjects in these fMRI experiments include commonly used laboratory animal species and strains, particularly rat strains such as Sprague-Dawley, Long-Evans, and Wistar (typically 250–600 g, postnatal day 80–300) and mice in C57BL/6J background (20–25 g, 10–12 weeks old). These species and strains are standard in rodent fMRI research due to their well-characterized physiology and suitability for various experimental paradigms. We often utilize rodent models from Charles River (Wilmington, MA, USA) and Jackson Laboratory (Bar Harbor, ME, USA), or the Rat Resource & Research Center (Columbia, MO, USA) for transgenic rats. Our animals are group housed prior to surgery at animal facilities, given food and water ad libitum, and kept on a 12 h light/dark cycle.

Rats are often chosen for their larger brain size, which allows for better-resolved anatomical structures even at limited spatial resolution, and are frequently employed in pharmacological studies due to metabolic pathways that share significant similarities with humans. Additionally, rats are generally more adept at learning complex tasks, making them advantageous for behavior-correlate studies that require task acquisition and training. Mice, on the other hand, are favored for their genetic manipulability, which enables precise control over specific neural populations and pathways and is invaluable for studies targeting specific circuits and cell types. Mice are also easier to adapt to head-fixed experimental conditions, which facilitates imaging in the awake condition. Furthermore, mouse models are commonly used in disease research due to their adaptability for genetic modifications that replicate human disease phenotypes.

In these studies, rats are typically housed individually following surgical or invasive procedures to prevent interference with recovery or experimental devices, while mice are commonly group-housed throughout the study to support their social needs, unless individual housing is necessary. Both sexes are routinely included to consider sex as a biological variable, ensuring a comprehensive approach to data interpretation.

Prior to experiments, these rodent subjects are commonly provided with food and water ad libitum, and housing conditions are carefully maintained in a vivarium with controlled temperature (e.g., ~21°C) and humidity (40%–70%) to ensure a stable environment. The vivarium is usually kept on a standard 12-h light/dark cycle, aligned with the natural circadian rhythms of the animals. All animal care and handling should follow the national and Institutional guidelines for the use of laboratory animals, with protocols approved by ethics committees.

2.2 Surgical Procedures for Viral Expression

Pre-operative care: The subject is placed on a circulating water-heated pad. Ophthalmic ointment is liberally applied to the eyes to ensure comfort during procedures. Rats are deeply anaesthetized with 5% isoflurane, placed in a rodent stereotaxic frame (Kopf, Tujunga, CA, USA), and maintained at 2% isoflurane. Mice are deeply anaesthetized with 3% isoflurane and maintained at 1.5% isoflurane.

A circulating warm water pad is used to maintain body temperature at 37.0 ± 0.5°C. Depth of anesthesia is confirmed via toe pinch and eye blink responses. Anesthesia plane is continuously monitored throughout the surgery. A local anesthetic (Bupivacaine, 0.5%, s.c.) is injected at the surgical site before surgery.

Aseptic technique: To maintain a sterile environment during surgery, we have a dedicated surgical table under a surgical light. The surgical table is routinely cleaned. Animals are prepared for surgery in a separate area. Surgical fields are cleaned by hair removal, disinfection with three alternating applications, beginning with betadine, and followed by alcohol. The animal technician always wears a clean lab coat, mask, and sterile gloves during surgery. The surgical tools are cleaned using an autoclave after each animal and are kept sterile during the surgical procedures using a Germinator 500 (Electron Microscopy Sciences Cat #6611810). Sterile disposable surgical blades or suture needles are used.

Surgery: A stereotaxic frame will be used to position the head. The scalp and connective tissue are removed to expose the skull (Fig. 1a). The head is leveled by measuring lambda and bregma. Hydrogen peroxide can be used to remove blood from the surrounding area in the surgical view. A surgical blade can be used to "roughen" up the skull surface to ensure dental acrylic or C&B-Metabond (Parkell #S380) adhesion. Small holes are drilled (using a 1 mm drill bit) through the skull above the region of interest (Fig. 1b). Coordinates are relative to dura/brain surface at bregma [14, 15]. Additional holes are drilled to implant screws (usually up to four) to anchor the headplate to the skull (Fig. 1c). Place these as far away from your virus or implant holes as possible. Plastic screws from ACCU (SIK-M1.4-4-PEEK) are typically used for our fMRI studies (Fig. 1d). We have also used Brass screws (McMaster-Carr #92451A010), but found that the MR compatibility may vary across batches.

2.3 Surgical Procedures for Intracranial Implantation

Optogenetics: An adenovirus (AAV) linked to a fluorophore reporter is infused into a targeted area in the brain. This AAV is engineered to express either channel rhodopsin (ChR2), Chronos (a ChR2 variant), or enhanced yellow fluorescent protein (EYFP) [1, 3, 10, 16–20]. Most common viral vectors for optogenetic stimulation include AAV-CaMKIIa-hChR2-eYFP and control vector AAV-CaMKIIa-eYFP for targeting excitatory neurons. AAV-hSyn-hChR2-eYFP for pan-neuronal targeting, and AAV-hSyn-DIO-hChR2-eYFP for use in Cre-dependent systems are also frequently employed. These viral vectors induce transduced cells to produce light-sensitive ion channels on their membranes. For instance, in the presence of blue light, ChR2 forms sodium-permeable channels, leading to cell depolarization and activation. A subset of animals receives control virus injections, which express the fluorophore reporter (e.g., EYFP) but lack constructs such as ChR2. These controls allow visualization of

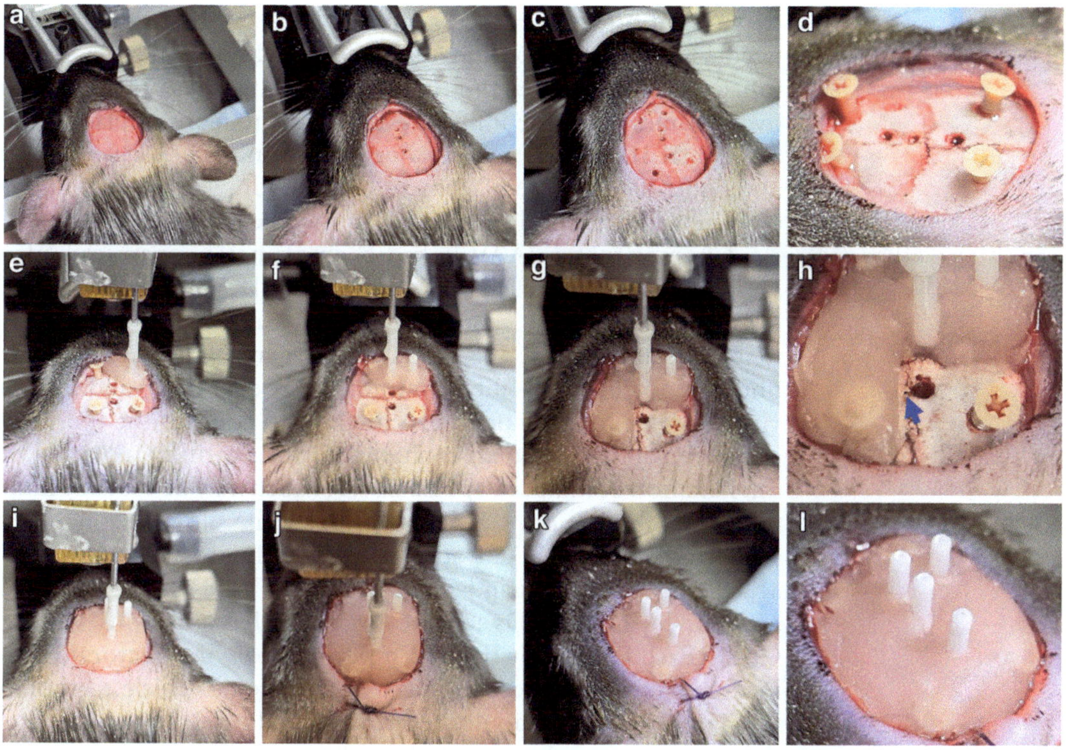

Fig. 1 Stepwise preparation of the rat skull for surgical procedures. (**a**) After skin removal, the skull is thoroughly cleaned to expose the surface for surgical preparation. (**b**) Using stereotaxic coordinates, precise drill holes are made at the target sites for virus delivery or implant positioning. (**c**) Anchor drill holes are placed at the corners of the surgical area to enhance the adhesion of dental acrylic, ensuring the stability of implants. (**d**) Plastic screws are inserted into the anchor drill holes at a slight angle, positioned away from the planned implant locations. This angled placement provides additional stability for securing the implants with dental acrylic and reduces interference with the targeted surgical sites. (**e**) Dental cement is applied to adhere the first optic fiber to one of the screws, providing a stable foundation for additional fibers. (**f**) Additional optic fibers are carefully positioned and secured sequentially using dental cement, ensuring precise placement and firm attachment. (**g**) Using India Ink, a mark is created to represent the bregma point, ensuring reference accuracy for subsequent measurements after the surgical site is covered. The final optic fiber is then implanted and secured. (**h**) Enlarged view of panel (**g**), with the blue arrow highlighting the India Ink mark, which serves as the new reference point for stereotaxic measurements in the absence of visible bregma due to the completed implantation. (**i**) The edges of the headcap are smoothed to ensure they are not attached to any skin, reducing the risk of irritation or discomfort. (**j**) One absorbable suture is placed to pull the skin tight over the headcap, ensuring proper healing and stability. (**k**) As the surgical procedure is completed, preparations for postoperative care begin, including monitoring and wound management. (**l**) Enlarged view of panel (**k**), showing the final placement of the suture and the smooth integration of the headcap, ready for postoperative care

targeted cells without light-responsive activity, serving as a baseline for comparison. The viral construct includes a specific promoter that ensures expression is restricted to neurons expressing a particular molecular marker. Using a stereotaxic apparatus, the needle is precisely positioned at the desired site and allowed to rest for 5 min. Injections may consist of 0.5–1 μL (~10^{12} viral genomes/mL), depending on the targeted region. We use a 30-gauge injection needle to the target coordinate, raised by 0.1 mm. The virus is then infused into the target brain region at a slow rate of 0.1 μL/min over 10 min. After the infusion, the needle remains in place for an additional 10 min to ensure proper diffusion of the virus. Gradual and controlled needle placement and withdrawal minimize tissue damage and prevent displacement of the injected virus solution.

Chemogenetics: AAV-linked chemogenetic constructs, a.k.a. Designer Receptors Exclusively Activated by Designer Drugs (DREADDs), enable precise, remote, and selective manipulation of specific neural circuits. These DREADD viruses are tagged with fluorophore reporters to facilitate immunohistochemical validation of viral expression. They provide a method to confirm the specificity of manipulations, stimulate multiple cell types without requiring the implantation of multiple optical fibers (which may be spatially limiting), and inhibit specific cell populations while allowing concurrent stimulation of others. Activation of DREADDs is pharmacologically achieved using inert ligands such as CNO [21–23] or DCZ [13]. In general, Gq-DREADDs activate targeted cell populations, whereas Gi-DREADDs inhibit them. Empty control viral vectors will be included to verify target specificity. An additional vehicle control for the ligand activator should be incorporated into the experimental design to ensure robust control conditions.

Fiber photometry: AAV-linked optical sensors enable cell-type-specific measurement of various aspects of neuronal activity. These include neuronal calcium influx, with various color options of sensors such as GCaMP (green [24]), jRGECO1a (red [25]), and NIR-GECO (near-infrared [26]). Some sensors also allow for monitoring neurotransmitter release, such as iGluSnFR (glutamate sensor [27]), iGABASnFR (GABA sensor [28]), $GRAB_{DA}$ (dopamine sensor [29]), $GRAB_{NE}$ (norepinephrine sensor [30]), $GRAB_{5\text{-}HT}$ (serotonin sensor [31]), GRABOT (oxytocin sensor [32]), and GACh (acetylcholine sensor [33]). These sensors not only change their emission brightness in response to neuronal activity but also remain fluorescent under baseline conditions, enabling their use as reporters for histological validation of viral expression. Motion artifacts are typically mitigated by coexpressing a control fluorescent protein, such as GFP or TdTomato [34], which provides a reference time course for motion correction. Additionally, when a spectrometer is used as the signal detector, hemodynamic artifacts can be corrected by leveraging the reference spectral time course from the control fluorescent protein, combined with the estimated photon travel distances at different wavelengths specific to the experimental setup [10].

Optical fiber implantation: Following virus injections, an optical fiber (such as an implantable optical fiber or a graded-index lens) will be implanted just above the injection site. We use Thorlabs ceramic ferrules (Thorlabs #CFMLC12U-20). Sterile screws are positioned near the drilled holes in the skull to provide stability for the implant (Fig. 1d). Using stereotaxic coordinates, the fiber is carefully placed 0.3 mm above the virus injection site (Fig. 1e).

The fiber and screws are secured to the skull with dental acrylic or C&B-Metabond (Fig. 1e–j). Once the material solidifies, the incision site is sealed with sutures around the dental cement cap to ensure the skin lies flush with the dental acrylic (Fig. 1k–l). Care is taken to preserve bregma as a reference point for future measurements, to ensure that the dried dental acrylic does not have sharp edges, and to avoid accidentally gluing the skin to the headcap during the procedure.

After a sufficient period for viral expression (e.g., 4–6 weeks), animals are anesthetized for intracranial implantation surgeries. Burr holes are created for the placement of optical fibers (for optogenetic stimulation or fiber photometry), and/or MR-compatible guide cannulas for intracranial delivery of chemogenetic actuators for targeted design. Fibers are lowered to a depth just above the intended stimulation sites, with coordinates tailored to target regions. Guide cannulas are also positioned slightly above the intended infusion area to allow the infusion cannula to reach fresh tissue during experiments. These implants are secured with a headcap made of dental cement (see video demonstration in [35]). Brass or plastic screws provide additional stability to the headcap structure by anchoring it to the skull.

Animals undergo a recovery period of 1–2 weeks post-surgery before experimental sessions, including imaging or recording, commence. Surgical implantation accuracy for virus injections and optical fiber placements is confirmed through histological analysis and anatomical MRI scans.

2.4 Post-surgical Care and Euthanasia

Neo-Poly-Dex Ophthalmic ointment is applied to the margins of the surgical site, and 5% lidocaine is used to provide local analgesia. After removal from the stereotaxic frame, the animal is placed in a clean, heated recovery cage lined with paper towels to prevent suffocation from bedding material. The animal is continuously monitored until it regains consciousness and demonstrates signs of healthy recovery. Upon regaining consciousness, the animal is administered oral meloxicam (1–1.5 mg/mL) every 24 h for up to 3 days or until full recovery. If severe pain is observed, following consultation with University Veterinary Staff, this regimen is supplemented with subcutaneous buprenorphine (0.05–0.1 mg/kg, diluted to 0.03 mg/mL with sterile 0.9% saline). Food and water are provided ad libitum.

Close monitoring is conducted to assess surgical wounds, behavior, body posture, and overall health. The body condition score and grimace scale are used to evaluate pain and discomfort throughout recovery and subsequent testing. Meloxicam is administered as needed, following recommendations from the lab or in consultation with the Division of Comparative Medicine (DCM) staff.

Postoperatively, animals are monitored daily for signs of distress, infection at the surgical site, or general discomfort. If a mild infection is detected, a topical antibacterial ointment, such as Polysporin, is applied. If systemic infection symptoms, such as fever, lethargy, or weight loss exceeding 10%, are observed, additional monitoring is initiated, and veterinary staff are consulted to determine appropriate interventions. If these measures fail to resolve the condition, humane euthanasia is performed to alleviate suffering. All procedures strictly adhere to the University of North Carolina-Chapel Hill's Institutional Animal Care and Use Committee (IACUC) guidelines.

Veterinary care is provided by a qualified animal care team within an accredited facility that adheres to high animal welfare standards. The facility typically operates under a 12-h light/dark cycle, with animals receiving food and water ad libitum. Custom feeding regimens are permitted with prior protocol approval. Animals are routinely assessed for health, and researchers are required to address any illness or unexpected mortality promptly. Medical records and experimental usage are meticulously documented to ensure comprehensive subject care.

At the conclusion of experimental procedures, animals are euthanized with an intraperitoneal injection of sodium pentobarbital (120 mg/kg) after confirmation of deep anesthesia, typically induced with approximately 4% isoflurane. Transcardial perfusion is performed using saline, followed by 4% paraformaldehyde (PFA) to preserve brain tissue for histological analysis, including verification of fiber placement and viral expression.

2.5 Histological Verification

For histological verification, rodent brains are collected after transcardial perfusion with saline, followed by fixation with 4% paraformaldehyde (PFA). The brains are stored in 4% PFA at 4°C overnight, then transferred to a 30% sucrose solution in 0.1 M phosphate buffer for 2–3 days, or until they sink to the bottom of the containers. Brain tissue is sectioned into 40 μm coronal slices using a cryostat.

Tissue slices are blocked at room temperature for 1 h in a solution containing 10% normal goat serum and 0.2% Triton X-100, followed by overnight incubation with a primary antibody at 4°C. For detecting GCaMP, chicken anti-GFP is typically used as the primary antibody at a dilution of 1:1000. After washing to remove excess primary antibody, sections are incubated with secondary antibodies, such as Alexa Fluor 647-conjugated goat anti-chicken, at room temperature for 2 h.

After further washes, the sections are mounted onto slides for imaging. Fiber placement and viral expression are verified using a confocal microscope equipped with appropriate laser lines and high-resolution objectives (Fig. 2). Images are acquired and processed with suitable software, and all evaluations are conducted in grayscale to minimize unintentional bias during analysis.

2.6 Preparation and Imaging of Rat Brain Samples for DTI-EPI MRI

The preparation of rodent brain samples for diffusion tensor imaging-echo planar imaging (DTI-EPI) follows protocols adapted from established methods commonly used for MRI studies [36–39]. The process begins with perfusing the vasculature using a heparinized saline buffer (0.1% heparin) to thoroughly flush the vasculature and remove red blood cells, improving tissue clarity. This is followed by perfusion with 10% ProHance in 10% formalin to ensure optimal fixation and enhance contrast. After perfusion, the brain is placed in 10% formalin for 16–24 h for post-fixation. Once post-fixation is complete, the tissue, including the skin, lower jaw, muscles, and nose, is carefully removed. The sample is then placed in 1X phosphate-buffered saline (PBS; pH 7.4) containing 0.5% ProHance for 14 days and stored at 4°C.

After the 14-day incubation, the brain is loaded into a 5 mL syringe. A piece of gauze is positioned at the front of the syringe to trap bubbles, and Fomblin is carefully loaded around the sample, with precautions taken to prevent the introduction of air bubbles.

To assess white matter integrity and fiber orientation, fractional anisotropy (FA) maps and FA color maps are generated from DTI-EPI data (Fig. 3). FA maps provide a quantitative measure of the anisotropy of water diffusion, helping to delineate white matter structures with high specificity. FA color maps, on the other hand, encode directional information of water diffusion using a color scheme. These maps are essential for visualizing the microstructural organization of the brain's white matter tracts and for conducting detailed analyses of neural connectivity. These high-resolution scans are also useful in validating electrode/fiber implant locations, as seen in Lai et al. [40].

2.7 Optogenetic Equipment

Optogenetic equipment encompasses a suite of tools and systems designed to deliver precise optical stimulation to genetically targeted cells or circuits, enabling researchers to control and study biological processes with high specificity. This technology integrates optics, electronics, and computational control to facilitate groundbreaking studies in neuroscience, physiology, and behavior. The core components of an optogenetic system include:

1. Optical components:
   - Light sources: High-power, stable LEDs or lasers with precise wavelengths optimized for specific opsins (e.g., 473 nm for ChR2, 590 nm for Arch, 635 nm for ChrimsonR). These sources should provide uniform illumination with minimal power fluctuations.
   - Optical couplers: Efficient fiber optic couplers for connecting light sources to patch cables while minimizing power loss.
   - Beam shaping components: Adjustable collimators and focusing lenses to align and shape the light beam for optimal transmission.

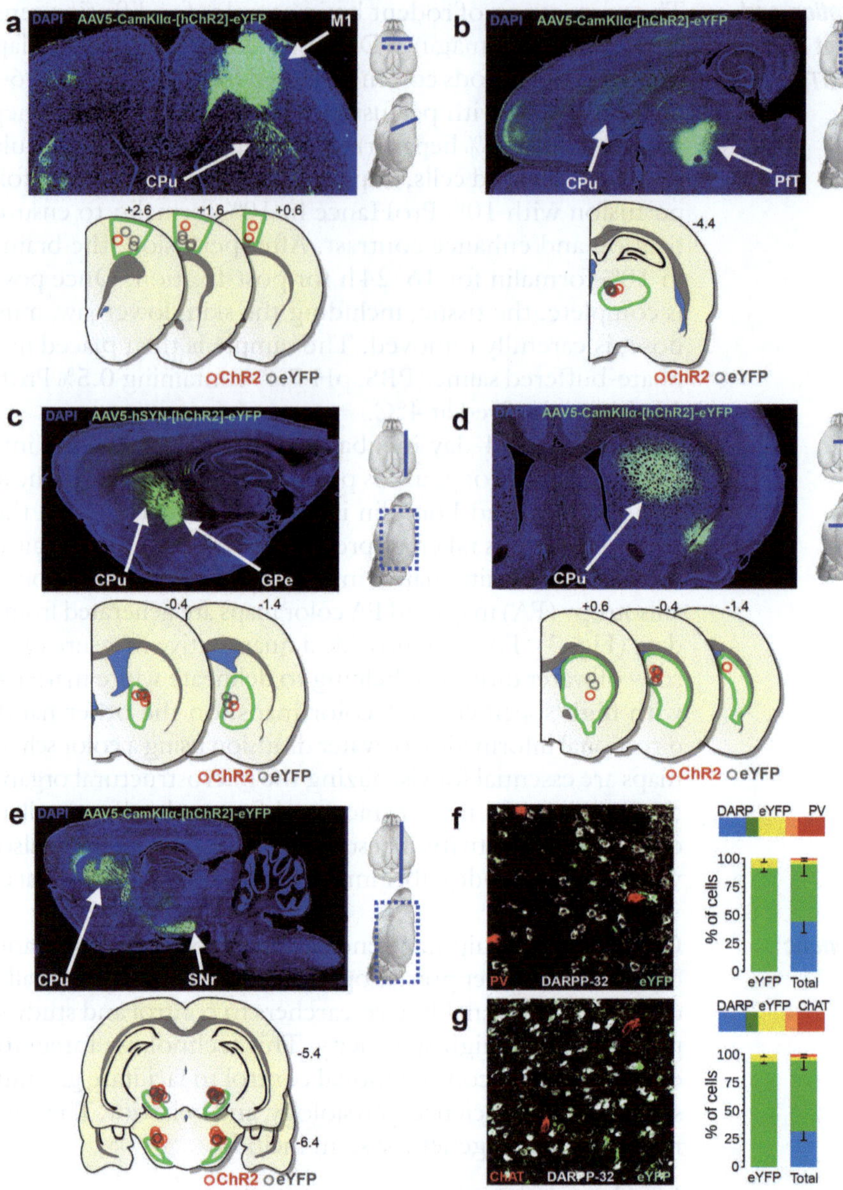

Fig. 2 Confirmation of optogenetic fMRI stimulation targets. Reproduced from Cerri et al. [1] with permission. **(a–e)** (*top*) Representative cross-section indicating spread of eYFP for each stimulation target, with 3-D models of the brain to the right, indicating the location of the histological slice. **(a–e)** (*bottom*) Locations of optical fiber tips relative to the anatomical stimulation target ROI outlined in green. Stimulation targets and areas of interest are as follows: (**a**) Viral expression in M1 injection sites and its spread to CPu. (**b**) Viral expression at the PfT injection site and its spread to the CPu. (**c**) Viral expression at the GPe injection site and its spread to CPu. (**d**) Viral expression in the CPu for direct MSN stimulation. **(e)** Viral expression at the CPu injection site and its spread to SNr stimulation sites. (**f–g**) High-resolution 20× magnification representative image and immunohistochemical verification of DARPP-32 (DARP), eYFP, and either PV (**f**) or ChAT (**g**) coexpression in striatum (PV $n = 2$ brains, 7 slices, 6195 cells; ChAT $n = 3$ brains, 11 slices, 14,181 cells). Reproduced from Cerri et al. [1] Nat Commun, 2024 with permission

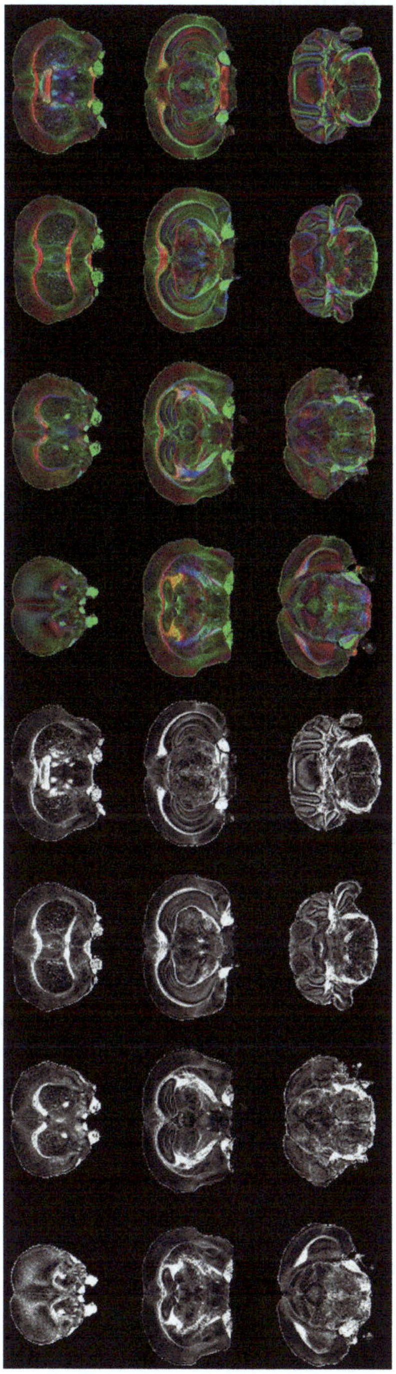

Fig. 3 Fractional anisotropy (FA) map and FA color map of the mouse brain obtained using DTI-EPI. The FA map highlights variations in diffusion anisotropy across different brain regions, while the FA color map encodes principal diffusion directions using red (left-right), green (anterior-posterior), and blue (superior-inferior)

- Wavelength-specific dichroic mirrors and filters: Dichroic mirrors and bandpass filters to ensure precise wavelength control and block unwanted light interference during experiments.

2. Light delivery system:
   - Patch cables: Flexible, low-loss optical fibers (e.g., 200 μm core, 0.37 NA) designed for efficient light transmission with minimal attenuation.
   - Implantable optical fibers and cannulas: Precision-constructed ferrule-connected optical fibers or cannulas for stable, targeted light delivery to specific brain regions.
   - Fiber rotary joints (optional): High-performance rotary joints to enable unrestricted movement of animals during stimulation.
3. Control and synchronization:
   - Light source drivers: Programmable, high-frequency LED or laser drivers for controlling intensity, pulse duration, and frequency with sub-millisecond accuracy.
   - Pulse generators and synchronization modules: Devices capable of generating stimulation patterns and synchronizing optogenetic stimulation with external systems such as fMRI, electrophysiology, or behavioral tracking setups.
   - Optical shutters: Motorized shutters for precise temporal control of light delivery.
4. Data monitoring and verification:
   - Optical power meters: Instruments for calibrating and verifying light intensity at the fiber tip to ensure consistent and safe stimulation levels.
   - Spectrometers: For real-time monitoring of wavelength and power stability during experiments.
5. Data acquisition and real-time control:
   - Control software: Integrated platforms (e.g., MATLAB, LabVIEW, or Python-based systems) for designing, executing, and synchronizing stimulation protocols with other experimental setups.
6. Computer workstation:
   - High-performance personal computers (PCs) capable of running real-time control software, managing complex stimulation protocols, and logging experimental data.

2.8 Fiber Photometry Equipment

Fiber photometry equipment is a specialized system used to measure fluorescence changes in vivo, providing real-time insights into cellular and neural activity. This technology is particularly effective for monitoring genetically encoded calcium or neurotransmitter indicators, offering high sensitivity and temporal resolution in freely behaving animals or in vivo preparations. The equipment integrates advanced optical, detection, and data acquisition components, ensuring precise and reliable experimental outcomes. The core components include:

1. Optical components:
   - Lasers or LEDs, with appropriate excitation wavelengths for targeted sensors (e.g., 488 nm for GCaMP, 561 nm for jRGECO1a, etc.).
   - An optical combiner to integrate multiple laser sources into a single output.
   - Adjustable collimators for laser beam alignment.
   - Wavelength-specific dichroic mirrors to separate excitation and emission light paths.
   - Bandpass filters specific to each excitation wavelength.
   - Bandpass filters for emitted fluorescence wavelengths.
   - Photodetectors (e.g., photomultiplier tube [PMT], avalanche photodiode detector [APD], spectrometer, etc.) for fluorescence signal detection.
   - Patch cables (appropriate core size and numerical aperture for fluorescence detection).
   - Implantable optical fibers with ferrule connectors (e.g., 200 μm core, 0.37 NA).
   - Fiber rotary joint (optional, to allow free movement of animals)

2. Data acquisition and processing:
   - Data acquisition system:
     - High-speed analog-to-digital converter (ADC).
     - Software compatible with real-time signal acquisition (e.g., LabVIEW, OceanView, etc.).
   - Synchronization hardware:
     - Pulse generators or triggers for synchronizing with other systems, such as fMRI or behavioral setups.
   - Computer workstation:
     - High-performance PC for real-time data acquisition, storage, and analysis.

2.9 Chemogenetic Actuators

Chemogenetic actuators, particularly DREADDs, enable precise and non-invasive manipulation of neuronal circuits through receptor-mediated pharmacological activation. These engineered receptors, when combined with synthetic ligands such as CNO or DCZ, offer prolonged and highly specific modulation of targeted neural pathways. Compared to optogenetics, which relies on light delivery for rapid activation or inhibition, chemogenetics provides a slower but sustained mode of neuronal modulation, making it ideal for studies examining long-term or systemic neural effects [41].

Chemogenetics integrates with fMRI to explore brain-wide network dynamics and functional connectivity. Activation of excitatory DREADDs (e.g., Gq-DREADDs) or inhibitory DREADDs (e.g., Gi-DREADDs) induces predictable changes in neuronal activity that are reflected in the BOLD signal. These changes allow researchers to delineate causal relationships between specific neural populations and their impact on large-scale brain networks. For example, the activation of locus coeruleus noradrenergic neurons using chemogenetics has been shown to modulate functional connectivity within the default mode network (DMN) and enhance synchronous low-frequency oscillations [4]. Such findings underscore the utility of chemogenetics in probing the neurovascular and network-level underpinnings of brain function.

The choice of actuator, such as CNO or DCZ, requires careful evaluation due to potential off-target effects or differences in pharmacokinetics. While CNO has been widely used due to its accessibility and established protocols [21–23], it may exhibit off-target effects, necessitating cautious protocol optimization [42]. DCZ, on the other hand, offers higher specificity and potency, making it increasingly favored in contemporary research [13]. Additionally, rigorous experimental controls, including the use of vehicle injections and empty viral vectors, are essential to isolate the effects of DREADD activation from nonspecific pharmacological influences.

By combining chemogenetics with advanced imaging techniques like fMRI, researchers can unravel the complex interplay between neuronal activity and brain-wide functional dynamics [4, 43–46], paving the way for novel therapeutic strategies and a deeper understanding of brain function.

3 Methods

Here, we summarize methods used in recent optogenetic [1], chemogenetic [4], and fiber photometry [2, 5, 10, 47, 48] studies conducted in our lab, providing practical examples for real-life projects.

3.1 Optogenetics and Chemogenetics with fMRI

1. *Optogenetic Stimulation Protocol* [1]:
   - Viral delivery and expression: An adeno-associated virus (AAV) encoding ChR2 with an excitatory CaMKIIα

promoter is injected into the CPu to selectively target excitatory medium spiny neurons (MSNs). The viral construct includes a reporter (e.g., eYFP) for site verification and is delivered at a concentration of ~10^{12} viral genomes/mL. A total volume of 0.5–1.0 μL is injected over 10 min at 0.1 μL/min, with the needle remaining in place for 10 additional minutes to ensure proper diffusion. After injection, animals recover for 4–6 weeks to allow ChR2 expression.

- Optical fiber placement: After the expression period, animals undergo a second surgery for optical fiber implantation. Fibers are precisely placed 0.3 mm above the target stimulation site using stereotaxic coordinates specific to the CPu, and they are secured to the skull with dental acrylic and anchored using plastic or brass screws to ensure experimental stability.
- Stimulation parameters: Optogenetic stimulation is conducted using a 473 nm laser or LED light source to deliver pulses through the implanted optical fibers. The protocol involves trains of 40 Hz pulses, each lasting 5 ms, chosen for their ability to evoke robust cerebral blood volume (CBV) fMRI responses. The stimulation intensity is carefully calibrated to avoid thermal artifacts while ensuring consistent activation of ChR2.
- fMRI data acquisitions: fMRI is performed using a high-field scanner (e.g., 9.4 T) with gradient-echo echo-planar imaging (EPI) sequences optimized for rodent imaging. The optogenetic stimulation protocol alternates baseline and stimulation periods, with sufficient off intervals to allow hemodynamic signals to return to baseline levels.
- Experimental controls: Control animals receive an identical viral vector encoding eYFP but lacking ChR2 to account for non-specific effects of viral expression and light delivery, while sham stimulation trials, where light is delivered without the presence of ChR2, are conducted to ensure observed responses are specific to ChR2 activation (Fig. 4).

2. *Chemogenetic Protocol* [4]:
   - DREADD expression and animal preparation: Viral vectors encoding the excitatory hM3Dq receptor under a neuron-specific promoter (e.g., hSyn), including a fluorescent reporter such as mCherry for expression confirmation, are injected into the target brain region using stereotaxic surgery. AAV serotype and titer (e.g., AAV5, 10^{12} vg/mL) are selected to optimize transduction efficiency. After surgery, animals are monitored for 7–10 days during recovery, and DREADD expression is verified through histology or fluorescence imaging after MRI scans.

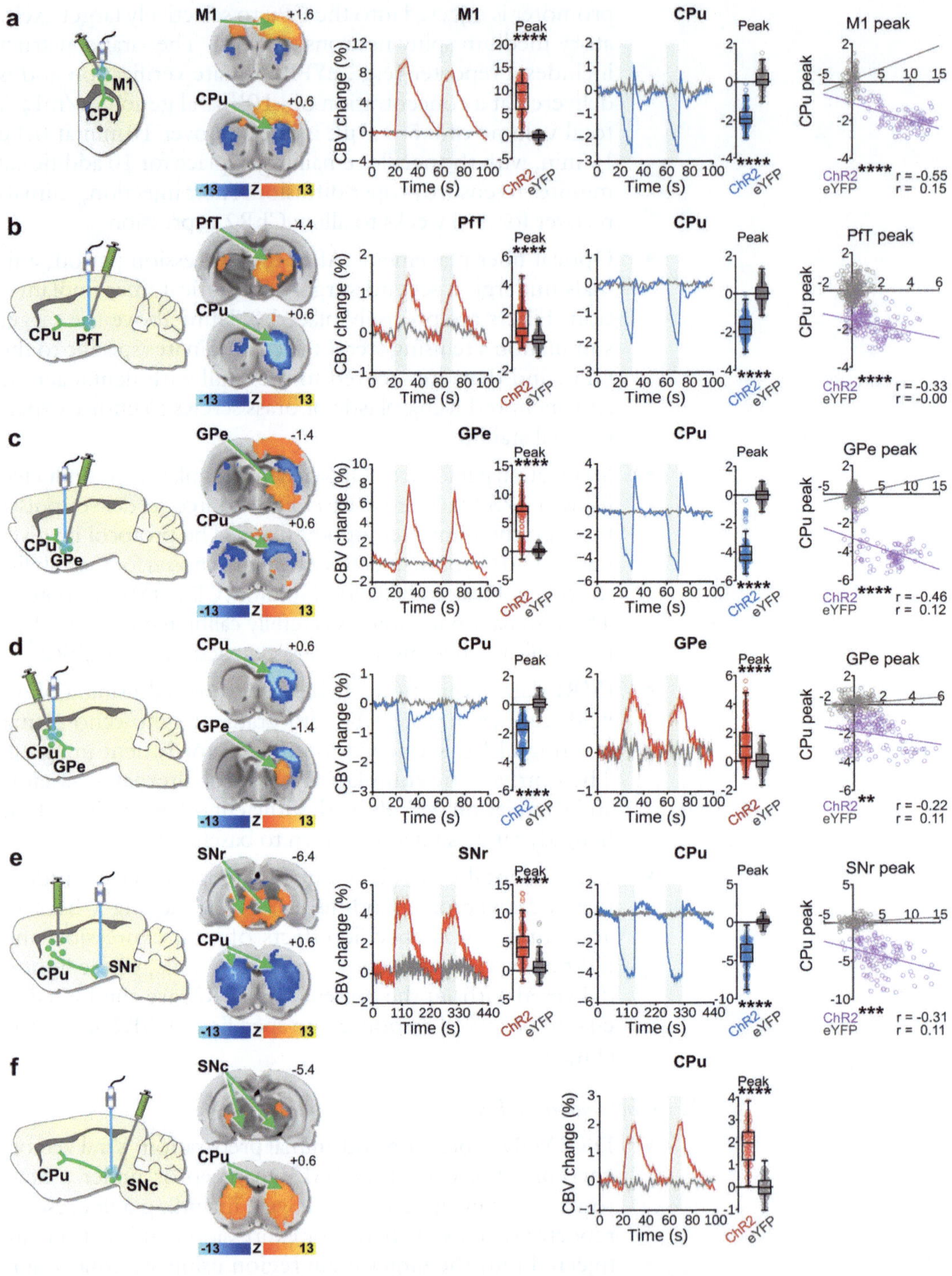

Fig. 4 fMRI responses to optogenetic stimulation of striatum-related circuits. Panels **(a–f)** show fMRI responses across various brain regions following optogenetic activation. The time-course data highlight response patterns during stimulation, while the correlation plots illustrate relationships between the stimulated targets and their projections. Control experiments using eYFP validate the specificity of responses, emphasizing the circuit-dependent nature of the observed fMRI results. Reproduced from Cerri et al. [1] Nat Commun, 2024 with permission

- Agonist administration: CNO is administered intraperitoneally at a dose of 1 mg/kg, a dosage selected based on prior studies for robust activation of DREADD-expressing neurons with minimal off-target effects. To minimize potential confounds, all injections are consistently performed at the same time of day, controlling for circadian influences on neuronal activity.
- fMRI data acquisition: Animals undergo functional magnetic resonance imaging (fMRI) using a high-field scanner (e.g., 9.4 T). Baseline scans are acquired for 10 min before CNO administration, followed by 30 min of postinjection scans to capture the temporal evolution of CNO effects.
- Experimental controls: Control animals receive identical injections of vehicle (e.g., saline with dimethyl sulfoxide) to account for nonspecific effects of the injection procedure, while additional control groups include animals injected with CNO but lacking DREADD expression to address off-target effects of CNO or its back-metabolized products (e.g., clozapine) (Fig. 5).
- Potential confounds and side effects: CNO metabolism can produce clozapine, which has known pharmacological effects; thus, low doses of CNO are used to mitigate this risk. Off-target activation of serotonergic or dopaminergic pathways is carefully considered during result interpretation, and the choice of anesthesia is critical—isoflurane at low doses is preferred to avoid suppression of neural activity.

3.2 Fiber Photometry and fMRI

1. *Concurrent fMRI and fiber-photometry recording* [2, 4, 5, 10, 47]*:*

- Animal preparation: Viral vectors encoding the desired sensor under a cell type-specific promoter (e.g., CaMKII), along with a reference fluorescent reporter such as TdTomato for motion and hemodynamic artifact correction, are injected into the target brain region using stereotaxic surgery. The choice of AAV serotype and titer (e.g., AAV5, 10^{12} vg/mL) is optimized to maximize transduction efficiency. Immediately following virus injection, optical fibers are precisely placed 0.3 mm above the target recording site and secured to the skull with dental acrylic, anchored using plastic or brass screws to ensure experimental stability. Animals are monitored during a 7–10 day recovery period after surgery. Depending on the sensor employed, 14–21 days are typically required for sufficient signal expression. Sensor and control fluorophore expression is confirmed using fiber photometry prior to formal experiments to

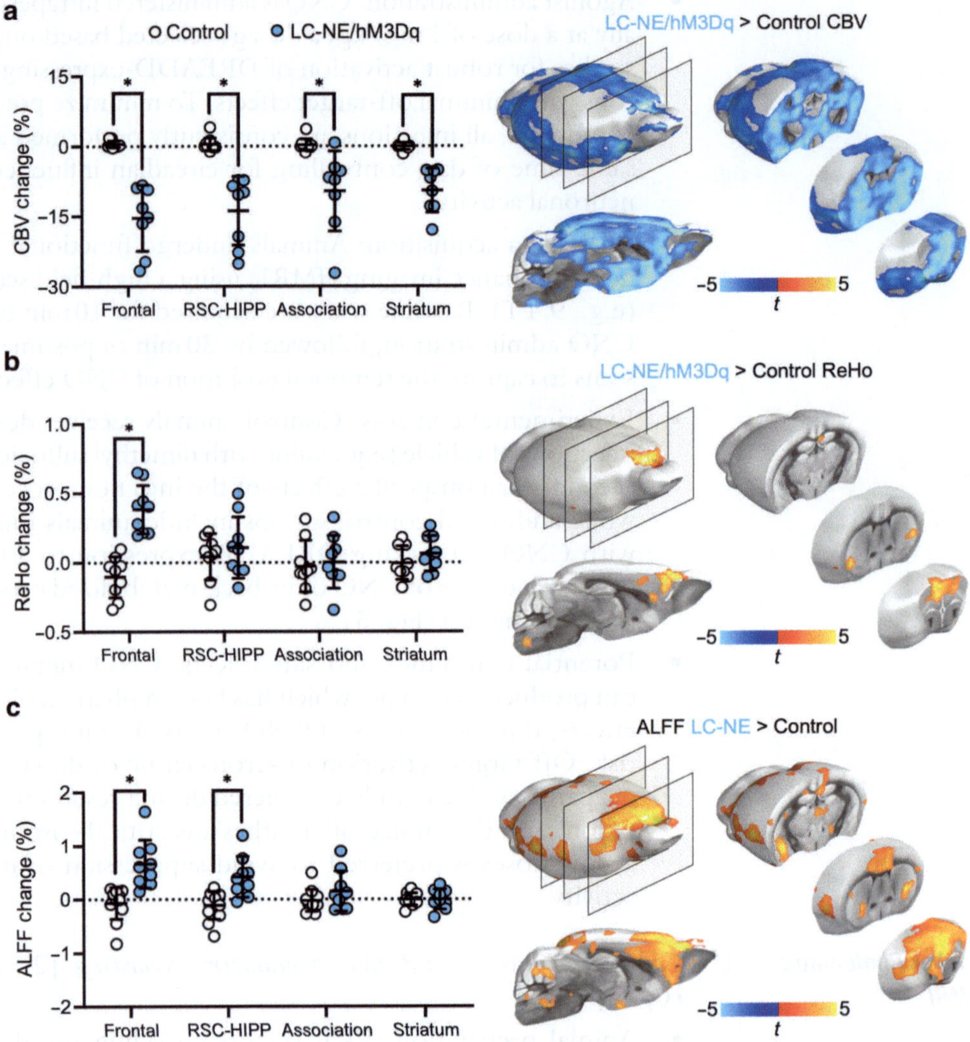

Fig. 5 Cerebral blood volume (CBV), regional homogeneity (ReHo), and amplitude of lower frequency fluctuation (ALFF) changes in default mode network (DMN) modules following activation of LC-NE neurons. **(a)** CBV decreased (post-CNO—baseline) significantly in LC-NE/hM3Dq following LC-NE activation across all DMN modules compared to controls. LC-NE activation significantly increased **(b)** ReHo change in the *Frontal* DMN module and **(c)** ALFF changes in *Frontal* and *RSC-HIPP* modules compared to controls. * $P_{\text{FDR-corrected}} < 0.005$, horizontal lines represent means, and error bars represent ± SD. The brain maps indicate the significant difference of CBV, ReHo, and ALFF changes between LC-NE/hM3Dq and controls ($P_{\mathit{3dClustSim\text{-}corrected}}$ <0.05). Reproduced from Oyarzabal et al. [4] Sci Adv, 2022 with permission

ensure adequate signal expression. After experiments are completed, histological analysis is performed to validate the recording location.

- Data acquisition: Animals undergo fMRI using a high-field scanner (e.g., 9.4 T). A custom-made single-loop surface coil, with an internal diameter of 1.6 cm and designed to allow optic fibers to pass through its center, serves as the radio-frequency transceiver. Prior to connecting the fiber-photometry patch cable, all lights in the room are turned off, and the final output power of the excitation laser (488 nm and/or 561 nm wavelength) is adjusted to balance spectral amplitudes. Once the setup processes are complete, the cradle is pushed into the MRI bore for fMRI scanning. Laser outputs and spectrometer recordings are controlled via a programmable microcontroller and synchronized with the MRI system using Transistor-Transistor Logic (TTL) pulses (Fig. 6).

2. *Correlating Fiber Photometry with BOLD* [2, 4, 5, 10, 47]:
   - Modeling the transfer function (TF): To estimate the TF between fiber photometry data and the BOLD signal, the fiber photometry data points must be resampled or binned to match the sampling rate of the BOLD data. The relationship between fiber photometry-measured neural activity and the BOLD signal can be expressed as:

$$BOLD(T) = N(T)TF(t) + c + d(T), \tag{1}$$

where the $BOLD(T)$ represents BOLD signal time-course, the $N(T)$ represents neural activity time-course, the $TF(t)$ represents an impulse hemodynamic response function with t sampling points, c is a constant for baseline offset, and $d(T)$ is for linear drift over time. Assuming $T = \{0, 1, 2, \ldots, m\}, t = \{0, 1, 2, \ldots, n\}$, this equation can be expressed as:

$$\begin{bmatrix} BOLD(0) \\ BOLD(1) \\ BOLD(2) \\ \vdots \\ BOLD(m) \end{bmatrix} = \begin{bmatrix} N(0) & 0 & 0 & \cdots & 0 & 1 & 0 \\ N(1) & N(0) & 0 & \cdots & 0 & 1 & \frac{1}{m} \\ N(2) & N(1) & N(0) & \cdots & 0 & 1 & \frac{2}{m} \\ \vdots & & & \ddots & & \vdots & \vdots \\ N(m) & N(m-1) & N(m-2) & \cdots & N(m-n) & 1 & 1 \end{bmatrix} \times \begin{bmatrix} TF(0) \\ TF(1) \\ TF(2) \\ \vdots \\ TF(n) \\ c \\ D \end{bmatrix}. \tag{2}$$

Therefore, the $TF(t)$, c and D (slope of $d(T)$) can be solved using the Ordinary Least Squares solution with known $BOLD(T)$ and $N(T)$. To avoid physiological noise contamination, $BOLD(T)$ is low-pass filtered with a cutoff frequency at 0.5 Hz before the hemodynamic response function (HRF) estimation. One unique advantage of fiber photometry over other optical techniques is its ability to probe the TF in deep brain regions, as the signals received are not limited by tissue

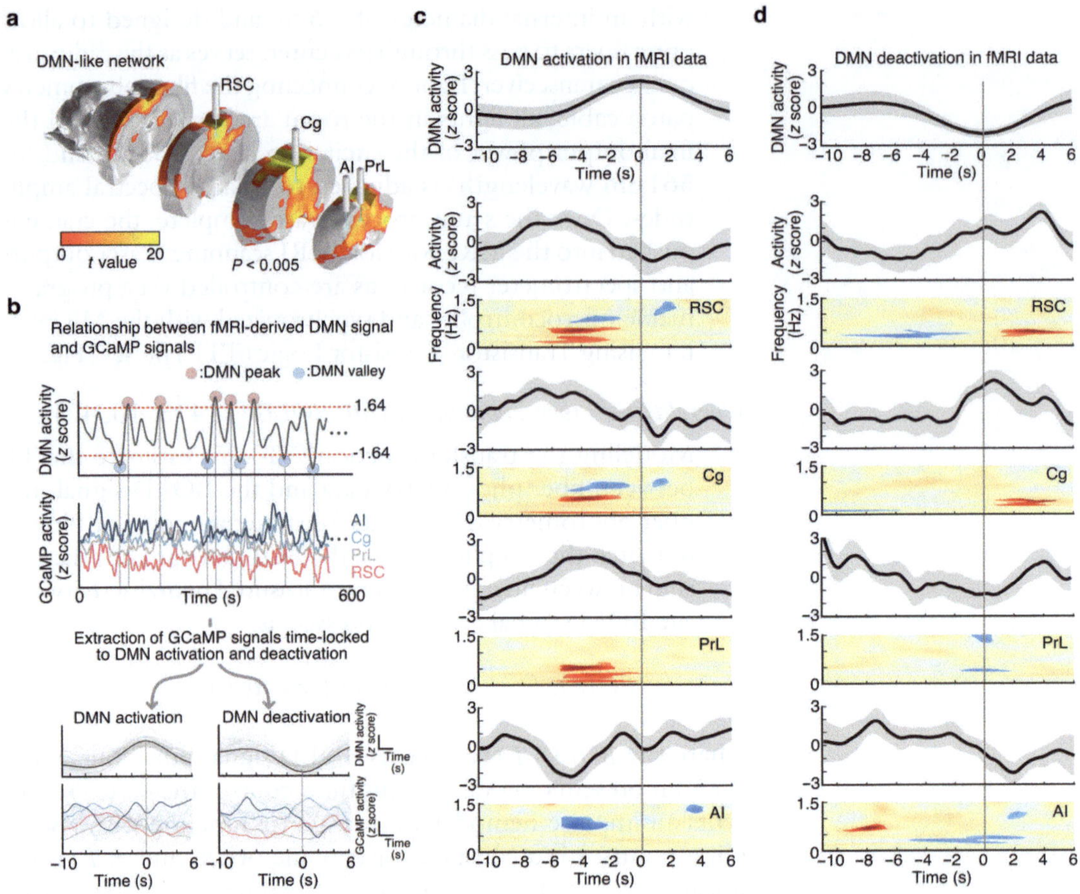

Fig. 6 Frequency-specific GCaMP power changes associated with concurrent fMRI-derived DMN activations and deactivations. (**a**) CBV-fMRI-derived DMN functional connectivity map using group-level independent component analysis (gICA; one-sample t-test, t-value >3.2, $p < 0.005$ for visualization, $n = 7$), with labels for brain regions corresponding to DMN-related nodes targeted by optical fibers for photometry recording of GCaMP. (**b**) Schematic illustrating identification of AI, Cg, PrL, and RSC neuronal activity time-locked to DMN activation and deactivation. Subject-level DMN time series were used to identify DMN activation and deactivation events ($|z| > 1.64$, $p < 0.05$). (**c**) *Top*: Averaged DMN activation peak in fMRI data. *Bottom*: Overall GCaMP power changes and statistical maps of GCaMP spectrograms from each node time-locked to DMN activation peaks ($p < 0.05$, $n_{activation\_peak} = 94$, two-tailed t-test). (**d**) *Top*: Averaged DMN deactivation peak in fMRI data. *Bottom*: Overall GCaMP power changes and statistical maps of GCaMP spectrograms from each node time-locked to DMN deactivation peaks ($p < 0.05$, $n_{deactivation\_peak} = 85$, two-tailed t-test). Reproduced from Chao et al. [47] Sci Adv, 2023 with permission

depth. Additionally, this method allows for the derivation of the TF using spontaneous activations without requiring external stimulation, making it particularly suitable for TF estimation in brain regions where determining an appropriate stimulation approach is challenging.

- Functional connectivity between neurons and large-scale brain network: With the fiber photometry-measured neural activity time course, $N(T)$, and the empirical TF between neural activity and the BOLD signal, it is possible to use $N(T) \otimes TF$ (where $\otimes$ denotes convolution) to predict the BOLD signal time course. This predicted BOLD signal can then be applied in voxel-wise General Linear Model (GLM) analysis of the fMRI data, enabling the integration of neural activity measurements with brain-wide hemodynamic responses.

4 Notes

The integration of advanced neuromodulation techniques such as optogenetics, chemogenetics, and fiber photometry with fMRI has significantly advanced our understanding of neurovascular coupling and brain network dynamics [1–5, 10, 18, 19]. Each technique offers unique advantages, providing complementary insights into the complex interplay between neuronal activity and hemodynamic responses. Optogenetics, leveraging cell-type specificity, has unraveled the intricate roles of excitatory and inhibitory neurons in shaping brain-wide activity. Chemogenetics, with its non-invasive and prolonged modulation capabilities, has elucidated the sustained effects of receptor signaling on brain networks. Fiber photometry enables real-time monitoring of cellular activity, offering a dynamic understanding of neuronal and neurochemical processes. Together, these approaches contribute to a comprehensive understanding of the causal relationships between regional brain activity and brain-wide network measures.

The combination of these neuromodulation tools with fMRI has highlighted the importance of multimodal approaches in neuroscience. For instance, optogenetics-fMRI studies have enabled the precise mapping of brain circuits with temporal and spatial specificity, revealing critical insights into the roles of neurotransmitter systems such as dopamine and serotonin in modulating brain function [1, 17–19]. Similarly, chemogenetics-fMRI studies have demonstrated the impact of cellular pathway manipulations on global network dynamics and behavior [3, 4], while fiber photometry-fMRI correlations link cellular-level signals to macroscopic brain activity [5, 10, 47].

The future of multimodal fMRI lies in several key directions. First, the development of MR-compatible neuromodulation devices, such as fiber optics, promises to enhance the integration of neuromodulation techniques with high-field fMRI systems. Second, the transition to awake animal imaging will enable more accurate modeling of naturalistic brain states, bridging the gap between animal models and human studies. Third, advancements in analytical methods, such as machine learning, dynamic network analysis, and causality modeling [3, 19, 47, 48], will uncover novel patterns and relationships in brain activity and connectivity. Finally, the exploration of combinatory neuromodulation strategies—such as simultaneous optogenetic activation and chemogenetic inhibition—will provide new avenues for investigating causal mechanisms underlying brain function.

These advancements not only hold the potential to deepen our understanding of neurovascular coupling and brain network dynamics but also pave the way for translational applications in human neuroscience. By leveraging the strengths of each technique and integrating them within the framework of fMRI, researchers can address complex questions about brain function and develop novel therapeutic strategies for neurological and psychiatric disorders.

Conflict of Interest

The author has no conflicts of interest to declare that are relevant to the content of this chapter.

References

1. Cerri DH, Albaugh DL, Walton LR, Katz B, Wang T-W, Chao T-HH, Zhang W, Nonneman RJ, Jiang J, Lee S-H, Etkin A, Hall CN, Stuber GD, Shih Y-YI (2024) Distinct neurochemical influences on fMRI response polarity in the striatum. Nat Commun 15: 1916
2. Chao T-HH, Zhang W-T, Hsu L-M, Cerri DH, Wang T-W, Shih Y-YI (2022) Computing hemodynamic response functions from concurrent spectral fiber-photometry and fMRI data. Neurophotonics 9:032205
3. Hsu LM, Shih YYI (2025) Neuromodulation in small animal fMRI. J Magn Reson Imaging 61(4):1597–1617. https://doi.org/10.1002/jmri.29575
4. Oyarzabal EA, Hsu L-M, Das M, Chao T-HH, Zhou J, Song S, Zhang W, Smith KG, Sciolino NR, Evsyukova IY, Yuan H, Lee S-H, Cui G, Jensen P, Shih Y-YI (2022) Chemogenetic stimulation of tonic locus coeruleus activity strengthens the default mode network. Sci Adv 8:eabm9898
5. Zhang W-T, Chao T-HH, Cui G, Shih Y-YI (2022) Simultaneous recording of neuronal and vascular activity in the rodent brain using fiber-photometry. STAR Protocols 3:101497
6. Deisseroth K (2015) Optogenetics: 10 years of microbial opsins in neuroscience. Nat Neurosci 18:1213–1225. https://doi.org/10.1038/nn.4091
7. Roth BL (2016) DREADDs for neuroscientists. Neuron 89:683–694
8. Cui G, Jun SB, Jin X, Luo G, Pham MD, Lovinger DM, Vogel SS, Costa RM (2014) Deep brain optical measurements of cell type-specific neural activity in behaving mice. Nat Protoc 9:1213–1228
9. Meng C, Zhou J, Papaneri A, Peddada T, Xu K, Cui G (2018) Spectrally resolved fiber photometry for multi-component analysis of brain circuits. Neuron 98:707–717.e4
10. Zhang W-T, Chao T-HH, Yang Y, Wang T-W, Lee S-H, Oyarzabal EA, Zhou J, Nonneman R, Pegard NC, Zhu H, Cui G, Shih Y-YI (2022) Spectral fiber photometry derives hemoglobin concentration changes for accurate measurement of fluorescent sensor activity. Cell Rep Methods 2:100243
11. Alexander GM, Rogan SC, Abbas AI, Armbruster BN, Pei Y, Allen JA, Nonneman RJ, Hartmann J, Moy SS, Nicolelis MA, McNamara JO, Roth BL (2009) Remote control of neuronal activity in transgenic mice expressing evolved G protein-coupled receptors. Neuron 63:27–39
12. Armbruster BN, Li X, Pausch MH, Herlitze S, Roth BL (2007) Evolving the lock to fit the key to create a family of G protein-coupled receptors potently activated by an inert ligand. Proc Natl Acad Sci USA 104:5163–5168

13. Nagai Y, Miyakawa N, Takuwa H, Hori Y, Oyama K, Ji B, Takahashi M, Huang X-P, Slocum ST, DiBerto JF, Xiong Y, Urushihata T, Hirabayashi T, Fujimoto A, Mimura K, English JG, Liu J, Inoue K-I, Kumata K, Seki C, Ono M, Shimojo M, Zhang M-R, Tomita Y, Nakahara J, Suhara T, Takada M, Higuchi M, Jin J, Roth BL, Minamimoto T (2020) Deschloroclozapine, a potent and selective chemogenetic actuator enables rapid neuronal and behavioral modulations in mice and monkeys. Nat Neurosci 23:1157–1167
14. Paxinos G, Franklin KBJ (2019) Paxinos and Franklin's the mouse brain in stereotaxic coordinates, 5th edn. Academic Press, London
15. Paxinos G, Watson C (2006) The rat brain in stereotaxic coordinates, 6th edn. Academic Press, London
16. Albaugh DL, Salzwedel A, Van Den Berge N, Gao W, Stuber GD, Shih Y-YI (2016) Functional magnetic resonance imaging of electrical and optogenetic deep brain stimulation at the rat nucleus accumbens. Sci Rep 6:31613
17. Decot HK, Namboodiri V. M. K, Gao W, McHenry JA, Jennings JH, Lee S-H, Kantak PA, Jill Kao Y-C, Das M, Witten IB, Deisseroth K, Shih Y-YI, Stuber GD (2017) Coordination of brain-wide activity dynamics by dopaminergic neurons. Neuropsychopharmacology 42:615–627
18. Li Y, Lee S-H, Yu C, Hsu L-M, Wang T-WW, Do K, Kim H-J, Shih Y-YI, Grill WM (2024) Optogenetic fMRI reveals therapeutic circuits of subthalamic nucleus deep brain stimulation. Brain Stimul 17:947–957
19. Menon V, Cerri D, Lee B, Yuan R, Lee S-H, Shih Y-YI (2023) Optogenetic stimulation of anterior insular cortex neurons in male rats reveals causal mechanisms underlying suppression of the default mode network by the salience network. Nat Commun 14: 866
20. Ryali S, Shih Y-YI, Chen T, Kochalka J, Albaugh D, Fang Z, Supekar K, Lee JH, Menon V (2016) Combining optogenetic stimulation and fMRI to validate a multivariate dynamical systems model for estimating causal brain interactions. Neuroimage 132:398–405
21. Smith KS, Bucci DJ, Luikart BW, Mahler SV (2021) Dreadds: use and application in behavioral neuroscience. Behav Neurosci 135:89–107
22. Urban DJ, Roth BL (2015) DREADDs (designer receptors exclusively activated by designer drugs): chemogenetic tools with therapeutic utility. Annu Rev Pharmacol Toxicol 55:399–417
23. Zhu H, Roth BL (2014) Silencing synapses with DREADDs. Neuron 82:723–725
24. Chen T-W, Wardill TJ, Sun Y, Pulver SR, Renninger SL, Baohan A, Schreiter ER, Kerr RA, Orger MB, Jayaraman V, Looger LL, Svoboda K, Kim DS (2013) Ultrasensitive fluorescent proteins for imaging neuronal activity. Nature 499:295–300
25. Dana H, Mohar B, Sun Y, Narayan S, Gordus A, Hasseman JP, Tsegaye G, Holt GT, Hu A, Walpita D, Patel R, Macklin JJ, Bargmann CI, Ahrens MB, Schreiter ER, Jayaraman V, Looger LL, Svoboda K, Kim DS (2016) Sensitive red protein calcium indicators for imaging neural activity. eLife 5:e12727. https://doi.org/10.7554/eLife.12727
26. Qian Y, Piatkevich KD, Mc Larney B, Abdelfattah AS, Mehta S, Murdock MH, Gottschalk S, Molina RS, Zhang W, Chen Y, Wu J, Drobizhev M, Hughes TE, Zhang J, Schreiter ER, Shoham S, Razansky D, Boyden ES, Campbell RE (2019) A genetically encoded near-infrared fluorescent calcium ion indicator. Nat Methods 16:171–174
27. Marvin JS, Borghuis BG, Tian L, Cichon J, Harnett MT, Akerboom J, Gordus A, Renninger SL, Chen T-W, Bargmann CI, Orger MB, Schreiter ER, Demb JB, Gan W-B, Hires SA, Looger LL (2013) An optimized fluorescent probe for visualizing glutamate neurotransmission. Nat Methods 10:162–170
28. Marvin JS, Shimoda Y, Magloire V, Leite M, Kawashima T, Jensen TP, Kolb I, Knott EL, Novak O, Podgorski K, Leidenheimer NJ, Rusakov DA, Ahrens MB, Kullmann DM, Looger LL (2019) A genetically encoded fluorescent sensor for in vivo imaging of GABA. Nat Methods 16:763–770
29. Sun F, Zhou J, Dai B, Qian T, Zeng J, Li X, Zhuo Y, Zhang Y, Wang Y, Qian C, Tan K, Feng J, Dong H, Lin D, Cui G, Li Y (2020) Next-generation GRAB sensors for monitoring dopaminergic activity in vivo. Nat Methods 17:1156–1166
30. Feng J, Zhang C, Lischinsky JE, Jing M, Zhou J, Wang H, Zhang Y, Dong A, Wu Z, Wu H, Chen W, Zhang P, Zou J, Hires SA, Zhu JJ, Cui G, Lin D, Du J, Li Y (2019) A genetically encoded fluorescent sensor for rapid and specific in vivo detection of norepinephrine. Neuron 102:745–761.e8
31. Wan J, Peng W, Li X, Qian T, Song K, Zeng J, Deng F, Hao S, Feng J, Zhang P, Zhang Y, Zou J, Pan S, Shin M, Venton BJ, Zhu JJ, Jing M, Xu M, Li Y (2021) A genetically encoded sensor for measuring serotonin dynamics. Nat Neurosci 24:746–752
32. Qian T, Wang H, Wang P, Geng L, Mei L, Osakada T, Wang L, Tang Y, Kania A, Grinevich V, Stoop R, Lin D, Luo M, Li Y (2023) A genetically encoded sensor measures temporal oxytocin release from different neuronal compartments. Nat Biotechnol 41:944–957

33. Jing M, Zhang P, Wang G, Feng J, Mesik L, Zeng J, Jiang H, Wang S, Looby JC, Guagliardo NA, Langma LW, Lu J, Zuo Y, Talmage DA, Role LW, Barrett PQ, Zhang LI, Luo M, Song Y, Zhu JJ, Li Y (2018) A genetically encoded fluorescent acetylcholine indicator for in vitro and in vivo studies. Nat Biotechnol 36:726–737
34. Creamer MS, Chen KS, Leifer AM, Pillow JW (2022) Correcting motion induced fluorescence artifacts in two-channel neural imaging. PLoS Comput Biol 18:e1010421
35. Younce JR, Albaugh DL, Shih YYI (2014) Deep brain stimulation with simultaneous fMRI in rodents. J Vis Exp (84):51271. https://doi.org/10.3791/51271
36. Calabrese E, Badea A, Cofer G, Qi Y, Johnson GA (2015) A diffusion MRI tractography connectome of the mouse brain and comparison with neuronal tracer data. Cereb Cortex 25:4628–4637
37. Calabrese E, Johnson GA (2013) Diffusion tensor magnetic resonance histology reveals microstructural changes in the developing rat brain. Neuroimage 79:329–339
38. Crater S, Maharjan S, Qi Y, Zhao Q, Cofer G, Cook JC, Johnson GA, Wang N (2022) Resolution and b value dependent structural connectome in ex vivo mouse brain. Neuroimage 255:119199
39. Johnson GA, Cofer GP, Gewalt SL, Hedlund LW (2002) Morphologic phenotyping with MR microscopy: the visible mouse. Radiology 222:789–793
40. Lai H-Y, Younce JR, Albaugh DL, Kao Y-CJ, Shih Y-YI (2014) Functional MRI reveals frequency-dependent responses during deep brain stimulation at the subthalamic nucleus or internal globus pallidus. Neuroimage 84:11–18
41. Kang HJ, Minamimoto T, Wess J, Roth BL (2023) Chemogenetics for cell-type-specific modulation of signalling and neuronal activity. Nat Rev Methods Primers 3:93
42. Gomez JL, Bonaventura J, Lesniak W, Mathews WB, Sysa-Shah P, Rodriguez LA, Ellis RJ, Richie CT, Harvey BK, Dannals RF, Pomper MG, Bonci A, Michaelides M (2017) Chemogenetics revealed: DREADD occupancy and activation via converted clozapine. Science 357:503–507
43. Elorette C, Fujimoto A, Stoll FM, Fujimoto SH, Bienkowska N, London L, Fleysher L, Russ BE, Rudebeck PH (2024) The neural basis of resting-state fMRI functional connectivity in fronto-limbic circuits revealed by chemogenetic manipulation. Nat Commun 15: 4669
44. Hirabayashi T, Nagai Y, Hori Y, Inoue K-I, Aoki I, Takada M, Suhara T, Higuchi M, Minamimoto T (2021) Chemogenetic sensory fMRI reveals behaviorally relevant bidirectional changes in primate somatosensory network. Neuron 109:3312–3322.e5
45. Rocchi F, Canella C, Noei S, Gutierrez-Barragan D, Coletta L, Galbusera A, Stuefer A, Vassanelli S, Pasqualetti M, Iurilli G, Panzeri S, Gozzi A (2022) Increased fMRI connectivity upon chemogenetic inhibition of the mouse prefrontal cortex. Nat Commun 13: 1056
46. Zerbi V, Floriou-Servou A, Markicevic M, Vermeiren Y, Sturman O, Privitera M, von Ziegler L, Ferrari KD, Weber B, De Deyn PP, Wenderoth N, Bohacek J (2019) Rapid reconfiguration of the functional connectome after chemogenetic locus coeruleus activation. Neuron 103:702–718.e5
47. Chao T-HH, Lee B, Hsu L-M, Cerri DH, Zhang W-T, Wang T-WW, Ryali S, Menon V, Shih Y-YI (2023) Neuronal dynamics of the default mode network and anterior insular cortex: intrinsic properties and modulation by salient stimuli. Sci Adv 9:eade5732
48. Nghiem T-AE, Lee B, Chao T-HH, Branigan NK, Mistry PK, Shih Y-YI, Menon V (2024) Space wandering in the rodent default mode network. Proc Natl Acad Sci USA 121:e2315167121

Check for updates

Chapter 8

Hemogenetic Functional MRI: The Imaging Principles and Protocols

Nan Li, Souparno Ghosh and Rui Yang

Abstract

Hemogenetic functional MRI (fMRI) is a recently developed imaging technique that enables the visualization of neural activity at a brain-wide scale by genetically encoding molecular probes that generate hemodynamic contrast in targeted populations of cells or circuits. This approach overcomes the limitations of conventional fMRI, which relies on indirect blood flow changes linked to general brain activity. The first generation of hemogenetic fMRI is based on engineered nitric oxide synthase-based reporters (NOSTICs) to produce localized hemodynamic signals in genetically targeted neuronal populations. This chapter outlines the principles behind hemogenetic fMRI, describes the molecular engineering of NOSTIC-based hemogenetic reporters, and provides a step-by-step guide to experimental protocols for in vitro evaluations of the reporters' functions and in vivo imaging procedures to apply hemogenetic fMRI to detect circuit-specific neuronal activity. The advantages, limitations, and future applications of these technologies are discussed in the context of functional neuroimaging.

Keywords Hemogenetic fMRI, NOSTIC, Neurovascular coupling, Molecular neuroimaging, Hemogenetic reporters, Genetically encodable reporters

1 Introduction: Imaging Brain Activity Using Genetically Encoded Sensors

The mammalian brain, the highly efficient computational center controlling complex behaviors and cognitive functions, consists of densely interconnected neurons organized into diverse subpopulations. These neurons relay, integrate, and process information through complex circuits. To unravel the fundamental basis of brain functions—both in healthy and diseased states—it is essential to identify the mechanisms by which information is processed within the brain. Specifically, understanding *where*, *when*, and *which* neurons are activated within a neural circuit, and determining how their activity influences downstream processes, remains a core challenge in modern neuroscience. Traditionally, techniques such as multiunit and local field potential electrophysiology recordings have been widely used to decode brain functions, providing direct measurements of neuronal

Kevin Chuen Wing Chan and Seong-Gi Kim (eds.), *Optogenetic Functional Neuroimaging*, Neuromethods, vol. 232,
https://doi.org/10.1007/978-1-0716-5178-0_8,

activity, offering high temporal resolution but really limited by the poor spatial coverage.

In the late 1990s, the advent of blood-oxygenation-level-dependent functional MRI (BOLD-fMRI) [1] transformed the landscape of functional brain imaging. By leveraging the tight coupling between neuronal activity and blood oxygenation level changes [2], BOLD-fMRI introduced a noninvasive method to map whole-brain activity in response to designed stimuli/tasks or at resting state. This breakthrough enabled neuroscientists to investigate brain-wide functional dynamics in small animals, nonhuman primates, and human subjects, ushering in a new era of functional brain mapping [3]. BOLD-fMRI catalyzed the development of computational approaches to analyze indirect hemodynamic measurements and decode brain activity. Significant efforts have also been made to dissect the pathways contributing to the hemodynamic response functions, improving our understanding of the relationship between neural events and the observed BOLD signal [1, 4]. Besides BOLD fMRI, multiple other MRI techniques have been developed to image hemodynamic-related signals, such as vascular occupancy (VASO) MRI for cerebral blood volume (CBV) mapping [5], and arterial spin labeling (ASL) -MRI for cerebral blood flow (CBF) mapping [6].

Parallel to the advancements in fMRI, another transformative innovation emerged in the field of neuroscience during the late 1990s and early 2000s: the development of genetically encoded calcium indicators (GECIs) [7]. Following the discovery and optimization of green fluorescent protein (GFP) [8], the invention of calcium-sensitive GECIs, especially GCaMPs [9, 10], marked a critical milestone in optical imaging. Combined with optical recording or imaging approaches, these genetically encodable tools enabled the visualization of intracellular calcium dynamics, a key proxy for neuronal activity. The subsequent creation of genetically encodable voltage sensors [11] and G-protein-based indicators further expanded the capability of the technology, allowing the tracking of not only calcium signals but also neuromodulatory signals, such as fluctuation levels of dopamine [12], acetylcholine [13], norepinephrine [14]. Genetically encoded indicators have transformed modern neuroscience by providing unmatched cellular and molecular specificity, allowing researchers to visualize distinct cell populations, targeted neural circuits, and specific neurochemical events using advanced techniques, such as viral transduction or transgenic strategies.

Despite their respective strengths, BOLD-fMRI and genetically encoded optical imaging techniques have notable limitations. BOLD-fMRI offers unparalleled brain-wide coverage but lacks cellular and molecular specificity. For example, studies have reported that cortical BOLD signals represent a composite of contributions from activities [15–17] of astrocytes, diverse subtypes of interneurons, and pyramidal neurons, making it difficult to disentangle their individual roles in neural computations. Conversely, genetically encoded indicators combined with optical imaging provide exceptional resolution at the cellular level but are inherently limited in their spatial coverage, as optical imaging is restricted to small fields of view and shallow depths, limiting its application to superficial cortical regions and small volumes of tissue, and posing a significant challenge in understanding how local circuits integrate and contribute to whole-brain dynamics. While multisite fiber photometry recordings [18] have attempted to address this limitation, they fall short of providing a comprehensive view of brain-wide dynamics without perturbing the brain invasively.

These limitations highlight a critical gap in neuroscience: the need for a method that can combine the cellular and molecular specificity of genetically encoded indicators with the whole-brain coverage of MRI. This is where hemogenetic fMRI emerges as a transformative approach. By integrating the specificity of genetically encoded sensors with hemodynamic signals as readouts, hemogenetic fMRI bridges the divide between local and global imaging modalities. It provides a powerful tool to simultaneously capture the fine details of neuronal activity and the broader functional context of brain-wide networks.

1.1 Principles of Hemogenetic fMRI

Hemogenetic fMRI integrates genetic targeting with hemodynamic imaging to enable the precise mapping of neuronal activity from defined sources across the entire brain. At its core, this technique leverages the brain's inherent neurovascular coupling mechanism—a fundamental process by which increased neuronal activity leads to localized elevations in cerebral blood flow (CBF) to supply nutrients and remove metabolic waste. In the murine brain, where intercapillary distances are under 30 μm, even modest neuronal events can elicit substantial local vascular responses. Under physiological conditions, an influx of calcium in neurons or astrocytes initiates intracellular cascades that culminate in the release of various vasoactive agents, including nitric oxide (NO), prostaglandins (PG), cyclooxygenase-2 (COX-2), vasoactive intestinal peptide (VIP), epoxyeicosatrienoic acids (EETs), and potassium ions, as well as vasoconstrictors, such as atrial natriuretic peptide (ANP) and 20-hydroxyeicosatetraenoic acid (20-HETE) [2, 15, 16, 19, 20]. Densely distributed astrocyte endfeet around capillary blood vessels have been reported to regulate blood vessel size and blood flow actively [21].

The concept of hemogenetic imaging utilizes these neurovascular coupling processes and combines them with engineering-specific neuronal populations with enhanced vascular responses coupled to their activity, either reflected in intracellular calcium fluctuations or action potential events, which consequently amplify the corresponding hemodynamic-based fMRI signal changes. This is achieved through the development of reporter proteins that transduce neuronal signals into localized changes in blood oxygenation level–dependent (BOLD) or cerebral blood volume (CBV) signals detectable by MRI (Fig. 1). The artificial BOLD or CBV fMRI responses generated by reporter-expressing cells are pharmacologically distinguishable from

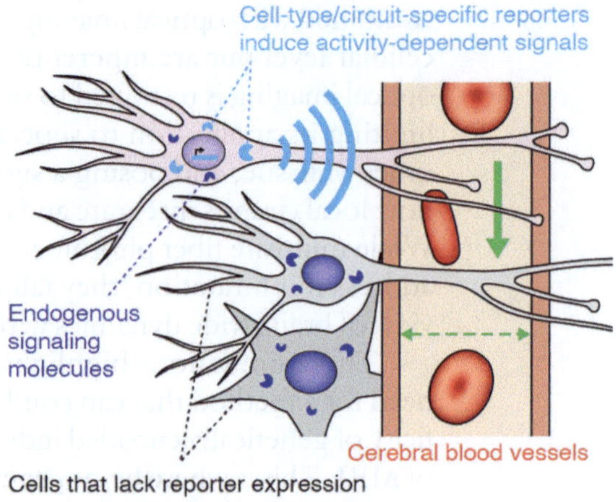

Fig. 1 Hemogenetic fMRI involves the detection of localized MRI signal changes in cerebral blood volume, blood oxygenation level, or blood flow induced by genetically encoded activity reporters (cyan). Endogenous drivers of activity-dependent signals (dark blue) are suppressed or distinguishable from the reporter signals [25]

the endogenous BOLD signal through the systematic administration of reporter-specific inhibitors, thereby allowing a direct association of hemodynamic responses with the activity of genetically defined cells in living brains.

1.2 Principles of NOSTIC-Based Hemogenetic fMRI

Among the key pathways regulating neurovascular coupling, neuronal nitric oxide synthase (nNOS) plays a particularly prominent role. Upon glutamate binding to N-methyl-D-aspartate (NMDA) receptors on postsynaptic neurons, a calcium influx is triggered, leading to elevated intracellular calcium levels. The rise in intracellular Ca^{2+} activates nNOS through calcium binding to calmodulin (CaM), which then interacts with the calmodulin-binding domain of nNOS, initiating its reversible enzymatic activity. Once activated, nNOS catalyzes the oxidation of L-arginine to produce nitric oxide (NO) and L-citrulline [22]. NO is a diffusible gas that acts as a potent vasodilator, primarily exerting its effects through the NO/cGMP signaling pathway in vascular smooth muscle cells. It has an EC_{50} in the nanomolar range [23]. Although nNOS expression is limited to only 1%–2% of cortical neurons, its localized activity is sufficient to drive robust vasodilation and produce detectable hemodynamic signals. Recent optogenetic studies have shown that brief activation (1 s duration, 1 ms pulses) of nNOS-expressing cortical neurons can significantly elevate local cerebral blood flow (CBF) [24]. Furthermore, pharmacological inhibition of NOS enzymes has been shown to reduce stimulus-evoked changes in CBF and attenuate BOLD fMRI signals by up to 60% in somatosensory paradigms [25, 26]. These findings underscore the utility of nNOS as a biological platform for engineering artificial hemodynamic responses in defined, genetically targeted cell populations.

These observations support the use of nNOS as the foundation for engineering artificial hemodynamic responses, even in relatively sparse, genetically targeted cell populations. For a defined population of cells engineered to express nNOS, one could detect the calcium activity of those cells, if the hemodynamic responses caused by NO release from calcium influx induction are detectable in BOLD fMRI and distinguishable from the background fMRI response. Leveraging this concept, under the mentorship of Dr. Alan Jasanoff at MIT, we developed the first generation of hemogenetic fMRI reporter, NOSTICs (Nitric Oxide Synthases for Targeting Image Contrast), which was introduced in *Nature Neuroscience* in 2022 [25]. NOSTIC-1 is a chimeric construct in which the catalytic domain of nNOS is replaced with that of inducible nitric oxide synthase (iNOS) [27], an isoform with a distinct inhibitor sensitivity profile. This design retains calcium-dependent activation via the nNOS CaM-binding domain [28] while incorporating the selective pharmacological properties of iNOS. As a result, NOSTIC-1 can transduce intracellular calcium elevations into spatially localized, pharmacologically isolatable hemodynamic signals (Fig. 2). These artificial BOLD responses are distinguishable from endogenous activity, enabling selective detection of calcium dynamics in targeted neuronal populations.

It is important to note that the term *hemogenetic fMRI* is not limited to the NOSTIC probes described here. Hemogenetic fMRI encompasses all techniques that utilize *hemodynamic MRI signals* as a noninvasive brain-wide readout method while integrating *genetically encoded reporters* for cell-type-specific or molecular-level targeting of brain activities. By employing genetic strategies, such as viral transfection or transgenic animal models, hemogenetic fMRI enables precise tracking of neural events across specific cell populations or circuits. In this chapter, we provide detailed experimental protocols for using NOSTIC-based hemogenetic fMRI.

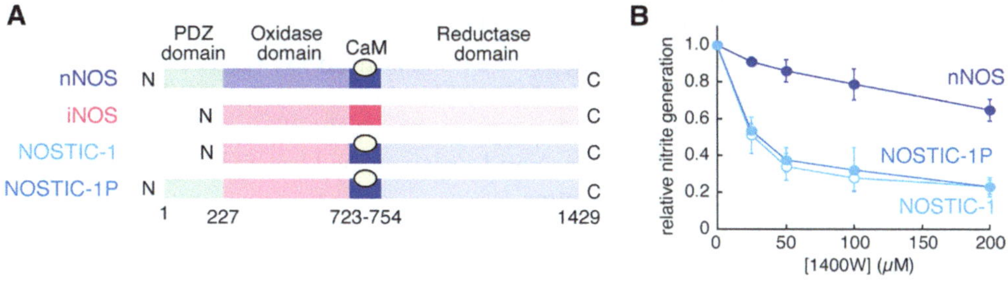

Fig. 2 NOSTIC-1 and NOSTIC-1P reporters are calcium-dependent chimeric enzymes that combine the oxidase (catalytic) domain of human iNOS with calmodulin (CaM)-binding site and reductase domains of rat nNOS at the C-terminus. NOSTIC-1P includes an additional PDZ domain from nNOS at its N-terminus, which facilitates anchoring to the postsynaptic membrane. When calcium influx increased, calcium binds to CaM, which then activates the enzyme, triggering nitric oxide (NO) production

2 Materials

Constructs Design and Selection

2.1 DNA Assembly

Currently, there are three construct options available for implementing NOSTIC-based hemogenetic fMRI. **NOSTIC-1 was** designed for general cytosolic expression in the targeted cell population. It exhibits calcium sensitivity and produces transient nitric oxide (NO) in response to elevated intracellular calcium levels, comparable to endogenous nNOS. NOSTIC-1 also features a unique pharmacological inhibition profile that selectively regulates its catalytic activity. For users aiming to target specific subcellular compartments, modified versions of the sensor are available. **NOSTIC-1P**, for instance, includes a PDZ-binding domain that anchors it to the postsynaptic density via interaction with PSD95 [29], a scaffolding protein at excitatory synapses. This configuration enables localized NO production following glutamate-induced NMDA receptor activation in postsynaptic neurons. Additionally, because endogenous **nNOS**-expressing neurons are relatively sparse in the central nervous system of rodent brains [30], users may choose to co-express nNOS selectively to amplify fMRI signals arising from defined neural activity within a desired expression time period.

DNA Sequence and Plasmids

The plasmid pCWori containing human iNOS and rat nNOS constructs was a generous gift from the laboratory of Dr. Thomas Poulos at the University of California, Irvine. The nNOS and iNOS were individually cloned into the pIRESpuro3 mammalian vector (Takara Bio USA). An inverse polymerase chain reaction (PCR) was performed to create the backbone. Inserts were amplified from the pCWori backbone using overhangs ranging from 24 to 30 nucleotides on each end. A single-step, isothermal, ligation-free procedure was used to integrate the inserts. The entire length of the coding strand was verified using Sanger sequencing (Quintara Biosciences). A similar DNA assembly strategy was used to generate chimeric genes encoding NOSTIC-1 and NOSTIC-1P. Coding sequences of NOSTIC-1 and NOSTIC-1P are shown in Table 1. All plasmids for expressing these genes in mammalian cells are available upon request from Dr. Nan Li's lab at the University of Texas Southwestern Medical Center.

2.2 Cell Culture Assays for Testing the Expression and Activity of NOSTIC

To test for expression and calcium activity of NOSTIC constructs expressed in cell cultures, we used FreeStyle 293-F cells from ThermoFisher. The cells were maintained according to the supplier's protocol. High-purity DNA suitable for transfection was obtained using the Qiagen Plasmid Maxi Kit. DNA was introduced into the cells using 293Fectin transfection reagent (ThermoFisher) following the manufacturer's protocols. At 48 h post-transfection, cells were pelleted and resuspended in Freestyle 293 Expression Medium (Thermo Fisher) containing 1 mM L-arginine and 1 mM $CaCl_2$ and seeded in six-well plates at two million cells per well. Cells were stimulated with 5 μM calcimycin A23187 (MilliporeSigma). Catalytic activity of NOSTIC was determined by quantifying nitrite from the supernatant using the Griess test (Promega).

For experiments to test for inhibitor sensitivity of NOSTIC, the following inhibitors (where applicable) were added to resuspension medium prior to stimulation of the cells: (i) NG-Nitro-L-arginine methyl ester hydrochloride (L-NAME, Tocris) and (ii) N-[[3-(Aminomethyl)phenyl]methyl]-22 ethanimidamide dihydrochloride (1400 W, Tocris).

Griess test: Nitrite standards were prepared in triplicate, starting from a 0.1 M nitrite stock solution diluted serially in Opti-MEM to yield a 100 μM standard. For the assay, 50 μL of sulfanilamide reagent was added to each standard and sample well in a 24/96-well plate, mixed gently, and after a 5-min incubation in the dark, 50 μL of *N*-(1-naphthyl)ethylenediamine (NED) reagent was added to each well. Plates were incubated for an additional 5 min in the dark, and the color of the solution was expected to change to purple or pink. The absorbance was measured at 550 nm using a plate reader.

2.3 Stable Cell Line Generation

A lentiviral backbone containing the cytomegalovirus (CMV) promoter driving enhanced GFP (eGFP) with a Blasticidin resistance cassette driven by a phosphoglycerate kinase (PGK) promoter (pLV6-eGFP-BLA) was used to generate pLV6-NOSTIC-1-BLA.

Table 1
Sequences of NOSTIC-1 and NOSTIC-1PNOSTIC-1

MHHHHHHMMACPWKFLFKTKFHQYAMNGEKDINNNVEKAPCATSSPVTQD
DLQYHNLSKQQNESPQPLVETGKKSPESLVKLDATPLSSPRHVRIKNWGSGM
TFQDTLHHKAKGILTCRSKSCLGSIMTPKSLTRGPRDKPTPPDELLPQAIEFVN
QYYGSFKEAKIEEHLARVEAVTKEIETTGTYQLTGDELIFATKQAWRNAPRCI
GRIQWSNLQVFDARSCSTAREMFEHICRHVRYSTNNGNIRSAITVFPQRSDGK
HDFRVWNAQLIRYAGYQMPDGSIRGDPANVEFTQLCIDLGWKPKYGRFDVV
PLVLQANGRDPELFEIPPDLVLEVAMEHPKYEWFRELELKWYALPAVANML
LEVGGLEFPGCPFNGWYMGTEIGVRDFCDVQRYNILEEVGRRMGLETHKLAS
LWKDQAVVEINIAVLHSFQKQNVTIMDHHSAAESFMKYMQNEYRSRGGCPA
DWIWLVPPMSGSITPVFHQEMLNYVLSPFYYYQVEAWKTHVWQDEPTKRRA
IGFKKLAEAVKFSAKLMGQAMAKRVKATILYATETGKSQAYAKTLCEIFKHA
FDAKAMSMEEYDIVHLEHEALVLVVTSTFGNGDPPENGEKFGCALMEMRHP
NSVQEERKSYKVRFNSVSSYSDSRKSSGDGPDLRDNFESTGPLANVRFSVFGL

GSRAYPHFCAFGHAVDTLLEELGGERILKMREGDELCGQEEAFRTWAKKVF
KAACDVFCVGDDVNIEKANNSLISNDRSWKRNKFRLTYVAEAPDLTQGLSN
VHKKRVSAARLLSRQNLQSPKSSRSTIFVRLHTNGNQELQYQPGDHLGVFPG
NHEDLVNALIERLEDAPPANHVVKVEMLEERNTALGVISNWKDESRLPPCTIF
QAFKYYLDITTPPTPLQLQQFASLATNEKEKQRLLVLSKGLQEYEEWKWGKN
PTMVEVLEEFPSIQMPATLLLTQLSLLQPRYYSISSSPDMYPDEVHLTVAIVSY
HTRDGEGPVHHGVCSSWLNRIQADDVVPCFVRGAPSFHLPRNPQVPCILVGP
GTGIAPFRSFWQQRQFDIQHKGMNPCPMVLVFGCRQSKIDHIYREETLQAKN
KGVFRELYTAYSREPDRPKKYVQDVLQEQLAESVYRALKEQGGHIYVCGDV
TMAADVLKAIQRIMTQQGKLSEEDAGVFISRLRDDNRYHEDIFGVTLRTYEV
TNRLRSESIAFIEESKKDADEVFSSGSDYKDDDDK

NOSTIC-1P

MHHHHHHMMEENTFGVQQIQPNVISVRLFKRKVGGLGFLVKERVSKPPVIIS
DLIRGGAAEQSGLIQAGDIILAVNDRPLVDLSYDSALEVLRGIASETHVVLILR
GPEGFTTHLETTFTGDGTPKTIRVTQPLGPPTKAVDLSHQPSASKDQSLAVDR
VTGLGNGPQHAQGHGQGAGSVSQANGVAIDPTMKSTKANLQDIGEHDELLK
EIEPVLSILNSGSKATNRGGPAKAACPWKFLFKTKFHQYAMNGEKDINNNVE
KAPCATSSPVTQDDLQYHNLSKQQNESPQPLVETGKKSPESLVKLDATPLSSP
RHVRIKNWGSGMTFQDTLHHKAKGILTCRSKSCLGSIMTPKSLTRGPRDKPTP
PDELLPQAIEFVNQYYGSFKEAKIEEHLARVEAVTKEIETTGTYQLTGDELIFA
TKQAWRNAPRCIGRIQWSNLQVFDARSCSTAREMFEHICRHVRYSTNNGNIR
SAITVFPQRSDGKHDFRVWNAQLIRYAGYQMPDGSIRGDPANVEFTQLCIDL
GWKPKYGRFDVVPLVLQANGRDPELFEIPPDLVLEVAMEHPKYEWFRELELK
WYALPAVANMLLEVGGLEFPGCPFNGWYMGTEIGVRDFCDVQRYNILEEVG
RRMGLETHKLASLWKDQAVVEINIAVLHSFQKQNVTIMDHHSAAESFMKYM
QNEYRSRGGCPADWIWLVPPMSGSITPVFHQEMLNYVLSPFYYYQVEAWKT
HVWQDEPTKRRAIGFKKLAEAVKFSAKLMGQAMAKRVKATILYATETGKSQ
AYAKTLCEIFKHAFDAKAMSMEEYDIVHLEHEALVLVVTSTFGNGDPPENGE
KFGCALMEMRHPNSVQEERKSYKVRFNSVSSYSDSRKSSGDGPDLRDNFEST
GPLANVRFSVFGLGSRAYPHFCAFGHAVDTLLEELGGERILKMREGDELCGQ
EEAFRTWAKKVFKAACDVFCVGDDVNIEKANNSLISNDRSWKRNKFRLTYV
AEAPDLTQGLSNVHKKRVSAARLLSRQNLQSPKSSRSTIFVRLHTNGNQELQ
YQPGDHLGVFPGNHEDLVNALIERLEDAPPANHVVKVEMLEERNTALGVISN
WKDESRLPPCTIFQAFKYYLDITTPPTPLQLQQFASLATNEKEKQRLLVLSKGL

QEYEEWKWGKNPTMVEVLEEFPSIQMPATLLLTQLSLLQPRYYSISSSPDMYP
DEVHLTVAIVSYHTRDGEGPVHHGVCSSWLNRIQADDVVPCFVRGAPSFHLP
RNPQVPCILVGPGTGIAPFRSFWQQRQFDIQHKGMNPCPMVLVFGCRQSKIDH
IYREETLQAKNKGVFRELYTAYSREPDRPKKYVQDVLQEQLAESVYRALKEQ
GGHIYVCGDVTMAADVLKAIQRIMTQQGKLSEEDAGVFISRLRDDNRYHEDI
FGVTLRTYEVTNRLRSESIAFIEESKKDADEVFSSGSDYKDDDDK

Key segements: 6H tag; PDZ domain; iNOS oxygenase; CaM-binding; nNOS reductase; FLAG tag

The parent plasmid pLV6-eGFP was treated with BamHI (NEB) and EcoRI (NEB) to make a linearized vector. NOSTIC-1 was amplified using PCR with primers having 20–25-base-pair overhang. The purified PCR product and the linearized backbone were assembled using Gibson assembly (NEB). DNA was transformed into stable competent cells (NEB). The entire coding strand was verified by Sanger sequencing (Quintara Biosciences). HEK 293FT cells (ThermoFisher) were maintained in accordance with the manufacturer's guidelines. Cells were co-transfected with pLV6-NOSTIC-BLA, psPAX2, and pMD2.G (Addgene) using Lipofectamine 3000 (ThermoFisher). Forty-eight hours post-transfection, the supernatant containing lentiviral particles was added to freshly seeded cells on a 10-cm tissue culture dish. Twenty-four hours following the addition of the lentiviral supernatant, the media were aspirated and changed with fresh DMEM medium (ThermoFisher) supplemented with 10% foetal bovine serum (FBS) containing 10 μg/mL blasticidin. The media was changed every 48 h till the dish was about 80% confluent. Colony selection was performed using both antibiotic resistance and fluorescent markers. Beginning on day 3 after initial infection, 10 μg/mL blasticidin and 1 μg/mL puromycin (Life Technologies) were added to the medium for selection, and selection was continued until all cells expressed the appropriate fluorescent markers.

2.4 Viral Packaging Strategies

Given the genetically encodable nature of NOSTICs, they can be used to target specific cell subtypes or defined neuronal circuits. Here, we describe two distinct viral packaging strategies tailored for different experimental objectives.

To trace and detect activity from monosynaptic inputs, we utilized long-term Herpes simplex virus (LT-HSV) type 1 vectors, produced by Dr. Rachael Neve at the Massachusetts General Hospital Gene Technology Core. These vectors are replication-deficient and capable of retrograde transport from the injection site in the rodent brain. NOSTIC-1 expression was driven by a construct containing the NOSTIC-1 gene in tandem with mCherry, separated by an internal ribosome entry site (IRES), under the control of the human EF1α promoter (LT-HSV-hEF1α-NOSTIC-1-IRES-mCherry). A control

HSV vector expressing only mCherry under the same promoter (LT-HSV-hEF1α-mCherry) was used for comparison.

To detect calcium activity in a targeted subpopulation of cells by expressing NOSTIC in the targeted cell types at the defined location, we used AAV-DJ vectors produced by Dr. Wei Xu at the University of Texas Southwestern Medical Center. For general neuronal targeting, NOSTIC was fused to a C-terminal 3xHA tag and expressed under the synapsin promoter (AAV-DJ-Syn-NOSTIC-3xHA). Figure 3A shows an example of mouse cortical neurons expressing NOSTIC-HA. For cell-type-specific and doxycycline-inducible expression, we used a Cre-dependent vector driven by the tetracycline response element (TRE) promoter (AAV-DJ-TRE-NOSTIC-3xHA in combination with AAV5-DIO-tTA), enabling precise control of expression in transgenic animals expressing Cre recombinase. Figure 3B shows astrocytes expressing NOSTIC-HA in the cortex of an Aldh1l1-Cre mouse (Jax Strain#023748) 3 weeks after viral injection.

2.5 Functional Test of Virally Delivered NOSTIC in Cell Culture

For in vitro validation of NOSTIC expression and calcium-sensitive nitric oxide (NO) production, HEK293 cells were plated in a six-well plate at a density of 0.5×10^6 cells per well (approximately 50% confluency). Within 12 h, viral vectors were added (1.5 μL of the virus with an original titer of 1 × 109 infectious units per mL (ifu/mL)) to the cell culture media (0.5 mL). After 24 h of incubation, expression was visually confirmed by checking mCherry signals. Two Opti-MEM-based solutions were prepared subsequently: solution #1 (baseline control) consisted of Opti-MEM with 1.8 mM

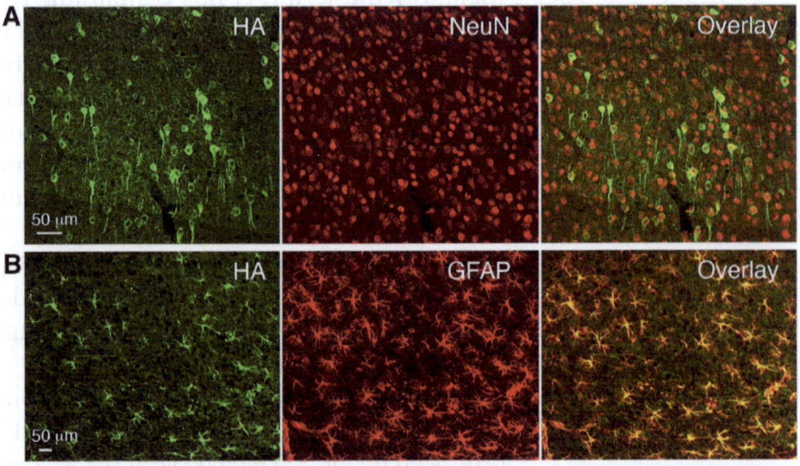

Fig. 3 **(A)** Immunostaining of mouse brain sections reveals strong expression of NOSTIC-HA (anti-HA) in cortical neurons, as indicated by colocalization with the neuronal marker NeuN (anti-NeuN), following AAV-DJ–mediated intracranial viral delivery. **(B)** Immunostaining shows robust expression of NOSTIC-HA (anti-HA) in cortical astrocytes, confirmed by colocalization with the astrocytic marker GFAP (anti-GFAP), following AAV-DJ–mediated intracranial viral delivery

$CaCl_2$ solution #2 (stimulation condition) included Opti-MEM with 1.8 mM $CaCl_2$ and 5 μM calcimycin A23187 (from a 5 mM stock in DMSO). Culture media were aspirated, and each well was washed with Opti-MEM. Top wells received 1.5 mL of solution #1 (no stimulation), and bottom wells received 1.5 mL of solution #2 (ionomycin stimulation). Following overnight incubation (about 18 h), we assessed NO production from supernatants using the Griess test assay described above, comparing the results in response to calcium stimulation, with and without NOSTIC expression in the cultured cells (Fig. 4).

2.6 Animal Preparation and Viral Injection

We used adult Sprague–Dawley rats (male, 300–350 g, Charles River Laboratory) for imaging presynaptic afferent activity to a targeted brain area via utilizing retrograde HSV viral vector to deliver NOSTIC. The rats were housed and maintained on a 12 h light/dark cycle with ad libitum access to food and water. Each animal was anesthetized with isoflurane (4% for induction, 2% for maintenance), and head-mounted on a rodent stereotaxic instrument (David Kopf Instruments), with a heating pad to keep the body temperature at 37°C. The scalp was retracted, and small holes were drilled into the skull above the target sites. 35 G needle connected with Hamilton syringes preloaded with LT-HSV-hEF1α-NOSTIC-1-IRES-mCherry (1 × 109 ifu/mL) were lowered into the brain. The viral vector was infused at a rate of 0.1 μL/min at the targeted brain region. We have conducted two studies using this circuit-specific NOSTIC hemogenetic fMRI strategy. The Hamilton syringe was slowly removed 10 min after the viral injection was completed.

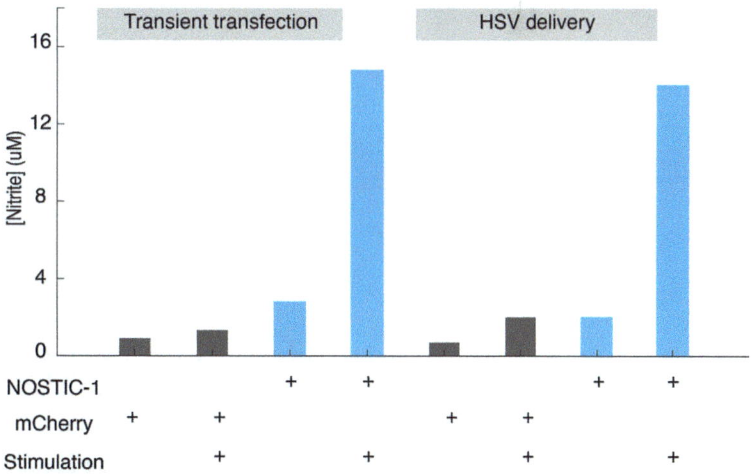

Fig. 4 Griess assay measuring nitrite levels in HEK 293FT cells either transfected with GFP or NOSTIC-1, or virally transduced with mCherry or NOSTIC-1, with and without stimulation by calcimycin A23187 [25]. Significant nitrite production—indicative of nitric oxide (NO) generation—is observed only in NOSTIC-1–expressing cells upon stimulation, confirming calcium-dependent NO production by the reporter

Bone wax was applied to each injection site using a sterile cotton applicator tip. Skin incisions were closed by sterile wound clips, and lidocaine gel (2%) was applied over the wound areas. Isoflurane was then discontinued, and each rat was removed from the stereotaxic frame and placed in a warmed cage on a heating pad to recover for 45 min. Slow-release buprenorphine (0.3 mg/kg) was administered subcutaneously to minimize pain and discomfort. Wound clips were removed 7–10 days after surgery, and when the wounds had healed.

3 Methods

Here, we demonstrate two experimental applications on the use of NOSTIC-based hemogenetic fMRI to resolve circuit-level functional connectivity in the living rodent brain.

3.1 Image Brain-Wide Inputs to the Posterior Medial Thalamic Nucleus in Somatosensory Processing

The posterior medial thalamic nucleus (POm) plays critical roles in modulating somatosensory perception, motor behavior, and well-learned sensorimotor habits [31–33]. The circuit mechanisms underlying such modulatory functions involve pathways from both cortical and subcortical brain regions, but the process whereby these disparate inputs combine to influence POm function in sensory perception is poorly understood. Here, we show the method details on using NOSTIC-1-based hemogenetic fMRI to investigate the integration of brain-wide inputs to POm during somatosensory processing. We delivered NOSTIC-1 to the posteromedial thalamus (POm, AP 3.5 mm, ML 2 mm, DV 5.8 mm) with LT-HSV viruses encoding

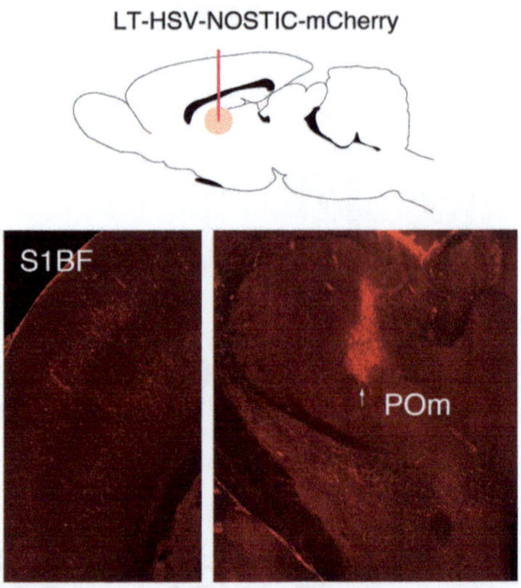

Fig. 5 (**A**) Schematic of viral delivery of NOSTIC-1 to the rat's posterior medial thalamic nucleus. (**B**) Visualization of NOSTIC-expressing cells near the injection site (***right***), and in ipsilateral S1 (***left***)

NOSTIC-1 or mCherry/eGFP-expressing control constructs, with an injection rate of 0.1 μL/min and volume of 1 μL (Fig. 5).

Forepaw sensory stimulation—Electrical stimulation (World Precision Instruments, Sarasota, FL, USA) of the left forepaw of the rats was induced by two short electrodes that were inserted between the digits. Stimulation was applied using a constant current stimulus isolator set to deliver six blocks of 3 mA 1 ms pulses repeated at 9 Hz for 10 s alternating with 40 s of rest. This procedure was applied first in the absence and then in the presence of systemically delivered 1400W (25 mg/kg i.v.). The stimulation frequency 9 Hz has been reported to induce the maximum BOLD response compared to other stimulation frequencies [34, 35].

MRI imaging protocols—Animals were intubated and mechanically ventilated using a small animal ventilator (Harvard Apparatus) set to deliver 62 breaths/min with a 6-mL stroke volume, using a 1:5 oxygen-to-air ratio. Animals were placed in a custom-built rat cradle for imaging, and a 4-channel surface receive-only radiofrequency coil (Bruker) was positioned over the head. Isoflurane anesthesia was discontinued prior to imaging, and sedation was maintained using dexmedetomidine, administered intraperitoneally as a 0.1 mg/kg bolus followed by continuous infusion at 0.1 mg/kg/h. To suppress motion, animals were paralyzed with pancuronium (1 mg/kg i.p. bolus followed by 1 mg/kg/h infusion). Body temperature was maintained using a water-circulating heating pad (Braintree Scientific). Physiological monitoring included heart rate and oxygen saturation via an MRI-compatible pulse oximeter (Nonin Medical), and respiratory rate and end-tidal CO_2 via a SurgiVet MRI-compatible capnograph (Smiths Medical).

MRI data were acquired on a 9.4 T Bruker scanner using a 70-cm-inner-diameter transmit-only linear volume coil (Bruker) for excitation and a 4-channel receive-only surface coil (Bruker) for signal detection. Scanner operation was managed using ParaVision software (Bruker), and fieldmap shimming was conducted to optimize B_0 homogeneity. High-resolution anatomical images were then acquired T2-weighted rapid acquisition with refocused echoes (RARE) pulse sequence with RARE factor = 8, bandwidth = 200 kHz, effective echo time (TE) = 30 ms, repetition time (TR) = 5 s, in-plane field of view (FOV) = 2.56 cm × 1.28 cm, in-plane resolution of 100 μm × 100 μm, and slice thickness = 1 mm. Functional imaging scan series were acquired under the same FOV, using a T2*-weighted GE-EPI sequence for detection of stimulus-induced BOLD contrast, with bandwidth = 200 kHz, a single TE of 16 ms, flip angle = 90°, TR = 2 s, in-plane resolution of 400 μm × 400 μm with a slice thickness of 1 mm. To discern NOSTIC inhibitor dependence of stimulus-induced fMRI signals, image series from each animal were acquired first in the absence and then in the presence of systemically delivered 1400W (25 mg/kg i.v.) under the same somatosensory stimulation.

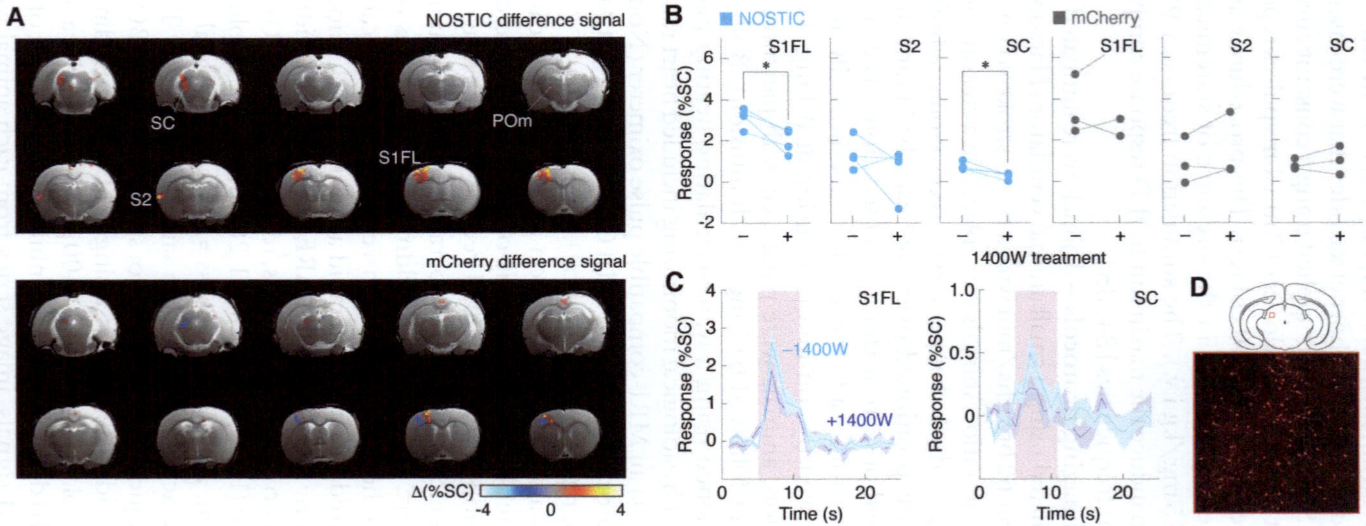

Fig. 6 Hemogenetic fMRI in the high-order somatosensory system. (**A**) Maps of average 1400W-dependent fMRI difference signal change observed in LT-HSV-NOSTIC-mCherry injected (top, $n = 4$) or LT-HSV-mCherry control-treated (bottom, $n = 3$) animals. In each case, maps show amplitude differences for voxels significantly activated (F test $p \leq 0.01$) by forepaw stimulation in the predrug condition. (**B**) Mean fMRI response amplitudes observed in S1FL, S2, and SC in rats treated with NOSTIC and mCherry-encoding viruses in the absence vs. presence of 1400W. Differences in S1FL and SC for NOSTIC animals are significant with a t test $p \leq 0.027$ ($n = 4$). (**C**) Mean time courses of fMRI signal relative to forepaw stimulation blocks (pink shading) in the absence (cyan) and presence (blue) of 1400W in S1FL (left) and SC (right). Shading denotes SEM over four animals. (**D**) Visualization of SC cells labeled by POm injection of LT-HSV-mCherry. The top schematic indicates the location of the field of view (FOV, small red square) for the histology results presented at the bottom. Scale bar = 100 μm

MRI image analysis—MRI images were reconstructed using ParaVision6 (Bruker) and imported into the AFNI software package for processing [36]. High-resolution anatomical images were registered to the Waxholm rat brain atlas, and functional time series underwent standard preprocessing in AFNI, including slice timing and motion correction, voxel-wise intensity normalization, 0.5-mm Gaussian spatial smoothing, and resampling. Brain image segmentation was performed in MATLAB, and time series were coregistered to atlas-aligned anatomy. Stimulus-dependent impulse response functions (IRFs) were estimated using AFNI's 3dDeconvolve tool, modeling voxel-specific responses with forepaw stimulus blocks as event regressors. Six estimated motion parameters were included as nuisance regressors. Group-level mean response maps were generated separately for LT-HSV-NOSTIC and LT-HSV-mCherry animals, both before and after 1400 W administration. By performing fMRI in the presence and absence of 1400 W during forelimb stimulation, we identified areas of substantial difference signals in the ipsilateral S1FL, secondary somatosensory cortex (S2), and superior colliculus (SC), representing presumed presynaptic information flow from these regions to POm in this sensory paradigm. Peaks of difference signal observed in LT-HSV-NOSTIC-treated animals were weak or absent in HSV-GFP animals, as indicated both in difference amplitude images (Fig. 6a) and region-of-interest (ROI) averaged values (Fig. 6b). In the NOSTIC animals, the difference amplitudes of 1.2% ± 0.3%, 0.82% ± 0.70%, and 0.48% ± 0.11% were observed in S1, S2, and SC, respectively, and were significant in S1FL and SC with a paired t-test $p \leq 0.027$ ($n = 4$); controls showed no significant differences ($p \geq 0.28$, $n = 3$).

Time courses of fMRI signal change followed the expected response profile in both S1FL and SC, with rapid onset and offset of hemodynamic responses relative to stimulus application and parallel mean time courses observed in the presence vs. absence of 1400 W (Fig. 6c). These data suggest that temporal aspects of NOSTIC-mediated signals are similar to those of endogenous BOLD responses. Although cell populations projecting from SC to POm have not been well studied, we confirmed the ability of an LT-HSV retrograde tracer injected in POm to label cells in SC (Fig. 6d), consistent with the hemogenetic fMRI findings. These results therefore provide further in vivo validation of the hemogentic imaging technique, while also revealing convergent input to the POm region from both cortical and brainstem sources during somatosensory stimulation.

3.2 Image Brain-Wide Inputs to the Striatum Activated During Reward Processing

The striatum serves as a critical neuromodulatory hub within the basal ganglia for decision making, movement control, and reward learning, etc, by integrating sensory, motor, and affective input from remote brain areas. The anatomical connections of the striatum have been studied by applying viral vectors that undergo axonal transport and express fluorescent proteins detectable in postmortem microscopy

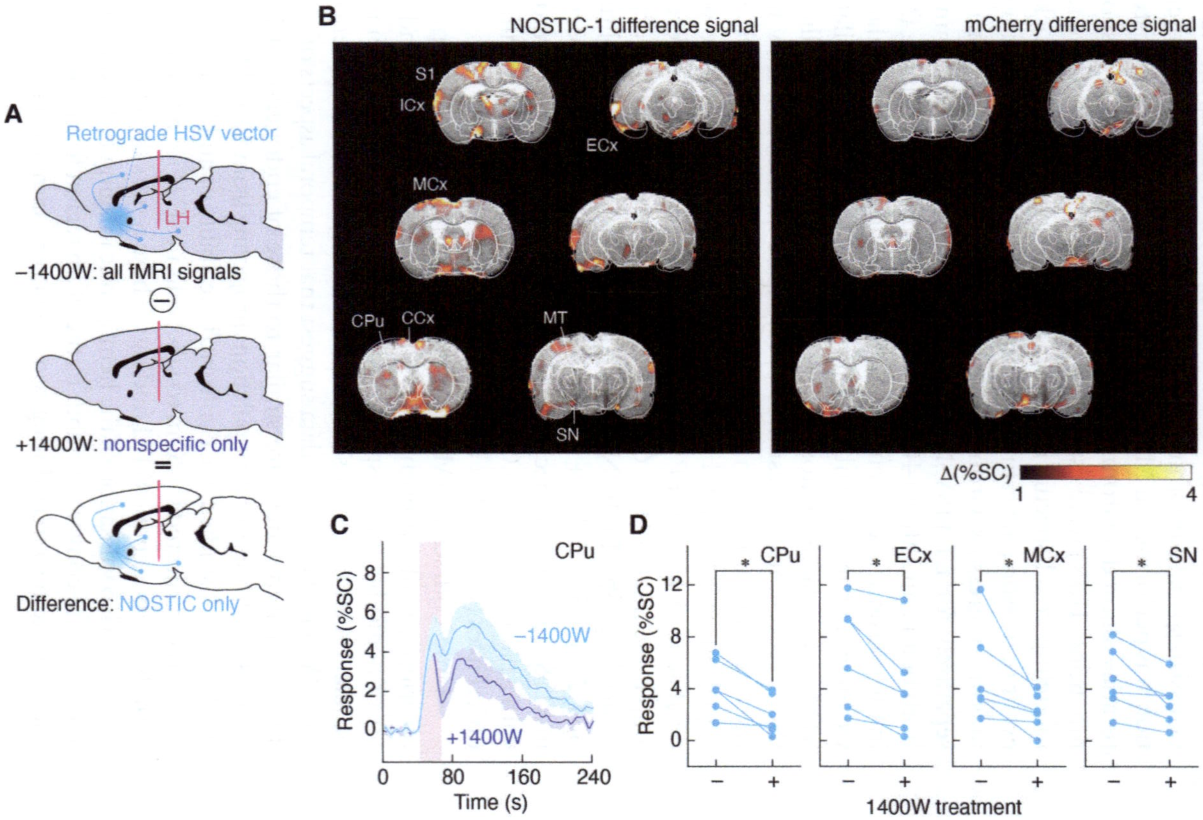

Fig. 7 (**A**) A retrograde LT-HSV vector delivers NOSTIC to neurons presynaptic to the central striatum (cyan). FMRI imaging during medial forebrain bundle stimulation (magenta) without (***top***) and with (***middle***) NOSTIC blocker 1400W[19]. (**B**) Mean amplitude of 1400W suppressed fMRI responses for animals infected with NOSTIC or mCherry viruses. (**C**) Average time courses of fMRI signals in medial striatum (CPu). (mean ± SEM, $\boldsymbol{n}=6$). (**D**) Consistent 1400W-dependent fMRI responses in CPu, entorhinal cortex (ECx), motor cortex (MCx), and substantia nigra (SN) [25]

[37–40]. However, a comprehensive and systematic functional connectivity of striatal afferents is by far challenging. To image functional afferents to the anterior central caudate putamen (aCPu) during reward processing events, the retrograde LT-HSV virus packaging NOSTIC-1 was delivered to two sites with a coordinate of anterior–posterior (AP): −0.5 mm, medial–lateral (ML): 3 mm, dorsal–ventral (DV): 5 and 6 mm, with an injection volume of 3 μL per site to cover most of the aCPu area (Fig. 7A).

Medial forebrain bundle (MFB) stimulation electrode implantation—Implantation of stimulation electrodes. Three weeks after viral injection, rats underwent a second surgery to implant electrodes suitable for deep-brain stimulation of the MFB region of the lateral hypothalamus (LH). Animals were prepared similarly for cranial surgery to expose the desired stimulation site. Bipolar stimulating electrodes were targeted to coordinates (AP: −3.4 mm, ML: 1.7 mm, DV: 8.6 mm). In addition, a plastic headpost was placed to facilitate the positioning of each animal during MRI scanning. After all the items were secured on the skull, dental cement (C&B Metabond, Parkell) was applied to the entire exposed surface area of the skull to hold them rigidly in place. After the surgery, animals were maintained under a heating pad and treated with 0.3 mg/kg slow-release buprenorphine subcutaneously during a convalescence period. Rats were allowed 7 days for recovery before fMRI experiments.

Intracranial self-stimulation tests—To assess the ability of the implanted MFB electrodes to induce rewarding stimulation, rats were evaluated for intracranial self-stimulation behavior in a plexiglass operant chamber (28 × 21 × 21 cm) (Lafayette Instruments). An infrared nose poke sensor was positioned at one end of the operant chamber. Input from the sensor was recorded via a digital input–output interface (National Instruments). Each nose poke triggered a 1-s train of 0.15-mA 1-ms electrical pulses delivered to the rat with a frequency of 60 Hz via its MFB electrode. A minimum of 2 s was required between the onsets of successive nose-poke-triggered rewards. Intracranial self-stimulation was quantified over three 30-s trial periods in each rat before imaging.

In-Scanner MFB stimulation paradigm—Each hemogenetic fMRI session employs a block design consisting of five stimulation cycles. Each cycle consisted of 16-s MFB stimulation followed by a 5-min rest period. During each 16-s stimulation block, eight 1-s trains were delivered, each composed of 1-ms current pulses (0.17 mA) at a frequency of 60 Hz, separated by 1-s inter-train intervals. The stimulation bipolar electrode, implanted in the MFB, was connected to a constant-current stimulus isolator (World Precision Instruments), and the stimulator cable was filtered using a 60-MHz low-pass filter (Mini-Circuits) prior to entry into the MRI room to reduce potential RF artifacts.

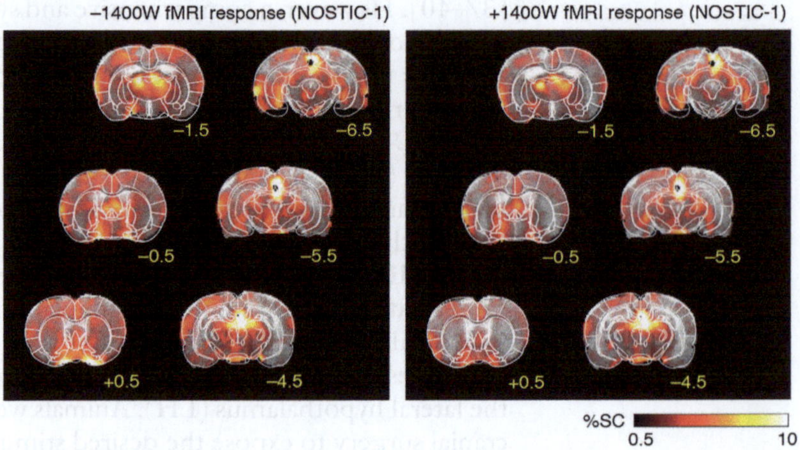

Fig. 8 Average fMRI responses to MFB stimulation in the −1400 W condition (***left***) and +1400 W condition (***right***), among six animals infected with NOSTIC-encoding LT-HSV. Significant responses with ***F*** test $p \leq 0.01$ are shown

MRI imaging and analysis—Animals preparation and maintenance during the MRI scans were the same as described above in Sect. 3.1. MRI data were acquired on a 9.4 T Bruker scanner using a 70-cm-inner-diameter transmit-only linear volume coil (Bruker) for excitation and a 2-cm-diameter receive-only loop surface coil (Doty Scientific) for signal detection, which allowed the MFB stimulation electrode to pass through. To discern NOSTIC inhibitor dependence of stimulus-induced fMRI signals, image series from each animal were acquired first in the absence and then in the presence of systemically delivered 1400 W (bolus injection 25 mg/kg i.v., with imaging starting 30 min after the bolus injection). MRI images were reconstructed and preprocessed as described in Sect. 3.1. Group-level mean response maps were generated separately for HSV-NOSTIC and HSV-mCherry animals, both before and after 1400 W administration (Fig. 7). Post-1400 W responses were normalized to midline amplitudes to account for systemic changes due to the MFB stimulations. Differences of fMRI activation maps were generated by voxel-wise subtracting post-1400 W from pre-1400 W response maps and thresholded at $p \leq 0.01$ (F test). (Fig. 8B). An alternative way to analyze the differential effects of 1400 W and extract the NOSTIC-specific fMRI response is to compute 1400 W-dependent difference maps using a different general linear modeling approach that assumes a fixed stimulus-associated IRF (i.e., hemodynamic BOLD fMRI response) for each voxel in each animal. The CPu IRF computed from the analysis described above was used for this purpose. Input to the regression analysis consisted of group data from all animals imaged in both pre- and post-1400 W experimental periods, normalized by midline responses as above, to account for signal drift. These data were subjected to a standard regression

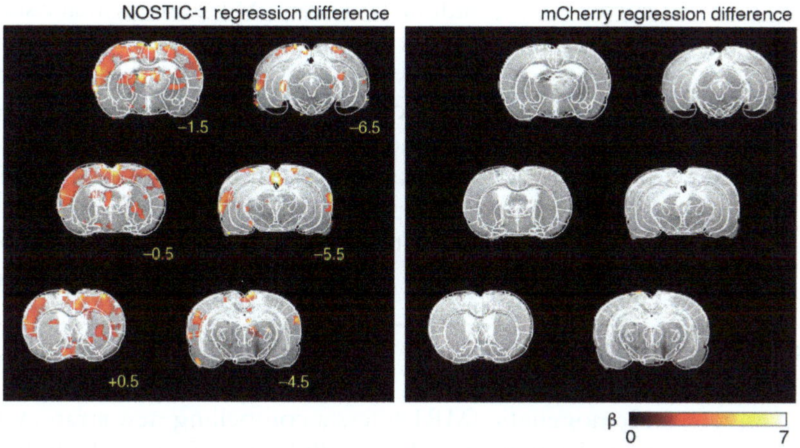

Fig. 9 Group-level general linear model analysis of NOSTIC data. Regression analysis for fMRI signal in animals treated with LT-HSVs encoding NOSTIC-1 (***left***, ***n*** = 6) or mCherry (***right***, ***n*** = 5). In this analysis, regressors for intrinsic and NOSTIC-driven BOLD signals were generated using a spatially invariant hemodynamic response function and applied in conjunction with motion and baseline terms. Maps display regression coefficients for the NOSTIC-specific regressor, thresholded by an ***F*** test $p \leq 10^{-6}$ and overlaid on an anatomical image

analysis that included regressors to account for intrinsic BOLD fMRI signal and NOSTIC-specific BOLD fMRI signal, and six motion correction parameters as nuisance regressors, along with a linear baseline term. The intrinsic BOLD regressor was computed by convolving all stimulus blocks with the CPu IRF, while the NOSTIC regressor was computed by convolving only pre-1400 W stimulus blocks with the CPu IRF, which include the effects of NOSTIC signals. Maps of regression coefficients (β values) for the NOSTIC-specific BOLD signal are presented in Fig. 9.

Across both analyses, animals expressing the NOSTIC construct showed strong 1400 W-dependent NOSTIC-specific fMRI signals in multiple brain regions—including CPu, motor cortex (MCx), entorhinal cortex (ECx), and substantia nigra (SN) (Figs. 7 and 8). These NOSTIC-dependent responses were not only statistically robust but also displayed temporal dynamics closely matching those of endogenous BOLD signals (Fig. 7C), suggesting a common underlying neurovascular mechanism. Importantly, this approach enabled the identification of previously unresolved long-range input pathways to the striatum. In particular, ECx, MCx, and SN emerged as significant presynaptic contributors during MFB reward-related stimulation—findings that were independently validated by c-Fos immunostaining and calcium fiber photometry [25]. These distal inputs complement local activation within CPu and reveal a more integrated view of circuit engagement during reward processing. In contrast, several brain regions that were strongly activated by MFB stimulation (Fig. 8)—including cingulate cortex (CCx), insular cortex (ICx), midline thalamus (MT), and primary somatosensory cortex (S1)—did not exhibit significant NOSTIC signals, emphasizing the method's specificity for presynaptic input activity rather than downstream or correlated responses.

Overall, these findings highlight the power of NOSTIC-based hemogenetic fMRI to selectively and noninvasively visualize input-defined calcium activity patterns in vivo, advancing our ability to dissect functional connectivity with high cellular and circuit specificity.

4 Conclusions

Hemogenetic fMRI offers a compelling new strategy for mapping neural activity with both cellular specificity and whole-brain coverage—an achievement long sought in systems neuroscience. In this chapter, we presented the molecular design, experimental protocols, and applications of NOSTIC-based hemogenetic fMRI as a first-in-class tool to image functionally defined neuronal populations based on their calcium-dependent nitric oxide (NO) production. By leveraging genetically encoded sensors that operate through endogenous neurovascular pathways, this approach enables noninvasive detection of calcium activity from specific cells or circuits across the entire rodent brain.

We demonstrated how NOSTIC-based fMRI could be applied to resolve functional presynaptic inputs to the striatum and thalamus, revealing brain-wide connectivity patterns in response to physiologically relevant stimulation that would otherwise remain hidden with conventional imaging methods. In comparison to anatomical tracers and BOLD fMRI, hemogenetic imaging offered improved specificity and pharmacological resolution, successfully identifying long-range direct afferent pathways—such as cortical and midbrain projections to the striatum—that are not discernible through structural mapping or hemodynamic correlational analysis. These findings, validated using c-Fos immunohistochemistry and calcium photometry, underscore the power of this approach to uncover functionally engaged input sources that go beyond static anatomical connectivity.

One key advantage of hemogenetic fMRI is that it taps into native biological amplification mechanisms to report neural activity, making it particularly effective for detecting signals from sparse or distributed cell populations. The use of NO as a readout provides a high-sensitivity signal that translates intracellular calcium transients into robust, localized vascular responses measurable by MRI. Importantly, the pharmacological isolation of NOSTIC signals (e.g., using 1400W) adds an extra layer of specificity and interpretability.

However, several challenges remain to be addressed. First, the performance of NOSTIC probes may vary depending on the cell type, subcellular localization, variations in viral titer and expression levels. For example, the cytosolic expression of NOSTIC-1 differs from the membrane-tethered localization of endogenous nNOS, potentially altering the kinetics or spatial spread of NO signaling. Systematic comparisons between NOSTIC- and nNOS-expressing cells under matched stimulation conditions—such as with optogenetic activation—will be needed to clarify these differences and quantify probe performance. Second, while subtraction-based analyses (comparing signals before and after NOS inhibition) provide a practical route to isolate NOSTIC-dependent signals, more sophisticated computational frameworks will be essential for extracting specific signals when endogenous and exogenous hemodynamic responses overlap. Future developments in signal deconvolution, regression modeling, and machine learning could enhance the sensitivity and reliability of hemogenetic readouts, especially in complex or pathological brain states where vascular responses may be altered. Third, temporal control of NOSTIC expression remains limited in current protocols. Prolonged or high-level expression of nitric oxide–producing enzymes may introduce cytotoxic effects or perturb baseline vascular tone, particularly in chronic or repeated imaging settings. To address this, we recommend using inducible expression systems—such as the tetracycline-responsive TET-ON or TET-OFF systems [41]—which allow researchers to control the timing and duration of NOSTIC expression with high precision. This strategy enables temporally restricted expression aligned with specific experimental windows and helps mitigate potential long-term side effects associated with continuous NO production.

Despite these limitations, NOSTIC-based hemogenetic fMRI opens new avenues for probing functional connectivity with cellular precision. The technique is well-suited for studying dynamic brain states, behaviorally relevant circuits, and cell-type-specific contributions to neurovascular coupling. With further protein engineering, it may become possible to design next-generation probes with improved signal-to-noise ratio, reduced off-target effects, and multiplexing capability for imaging multiple cell populations simultaneously. Ultimately, hemogenetic fMRI offers a powerful and adaptable framework for mechanistically probing brain function across multiple spatial scales. As advances in probe engineering, delivery technologies, and analytical methodologies continue to progress, we expect hemogenetic fMRI to become an integral part of the neuroimaging repertoire—providing unprecedented insight into how defined cell types and circuits contribute to large-scale brain dynamics with molecular precision.

Conflict of Interest

This study was funded by 2022 NARSAD Young Investigator Grant, Welch Foundation Grant No. I-2148-20230405, Whitehall Foundation Research Grant No. 2024-08-038, and NIH Grant No. R21AG088596. The authors have no conflicts of interest to declare that are relevant to the content of this chapter.

References

1. Ogawa S et al (1990) Brain magnetic resonance imaging with contrast dependent on blood oxygenation. Proc Natl Acad Sci USA 87(24):9868–9872
2. Hillman EM (2014) Coupling mechanism and significance of the BOLD signal: a status report. Annu Rev Neurosci 37:161–181
3. Bandettini PA (2012) Twenty years of functional MRI: the science and the stories. Neuroimage 62(2):575–588
4. Logothetis NK et al (2001) Neurophysiological investigation of the basis of the fMRI signal. Nature 412(6843):150–157
5. Lu H, van Zijl PC (2012) A review of the development of vascular-space-occupancy (VASO) fMRI. Neuroimage 62(2):736–742
6. Detre JA et al (2009) Arterial spin-labeled perfusion MRI in basic and clinical neuroscience. Curr Opin Neurol 22(4):348–355
7. Lin MZ, Schnitzer MJ (2016) Genetically encoded indicators of neuronal activity. Nat Neurosci 19(9):1142–1153
8. Chalfie M et al (1994) Green fluorescent protein as a marker for gene expression. Science 263(5148):802–805
9. Nakai J, Ohkura M, Imoto K (2001) A high signal-to-noise Ca(2+) probe composed of a single green fluorescent protein. Nat Biotechnol 19(2):137–141
10. Zhang Y et al (2023) Fast and sensitive GCaMP calcium indicators for imaging neural populations. Nature 615(7954):884–891
11. Bando Y et al (2019) Genetic voltage indicators. BMC Biol 17(1):71
12. Zhuo Y et al (2024) Improved green and red GRAB sensors for monitoring dopaminergic activity in vivo. Nat Methods 21(4):680–691
13. Jing M et al (2020) An optimized acetylcholine sensor for monitoring in vivo cholinergic activity. Nat Methods 17(11):1139–1146
14. Feng J et al (2024) Monitoring norepinephrine release *in vivo* using next-generation GRAB(NE) sensors. Neuron 112(12):1930–1942.e6
15. Attwell D et al (2010) Glial and neuronal control of brain blood flow. Nature 468(7321):232–243
16. Hamilton NB, Attwell D, Hall CN (2010) Pericyte-mediated regulation of capillary diameter: a component of neurovascular coupling in health and disease. Front Neuroenergetics 2(5):1–14
17. Moon HS et al (2021) Contribution of excitatory and inhibitory neuronal activity to BOLD fMRI. Cereb Cortex 31(9):4053–4067
18. Brown E et al (2026) Striatum-wide dopamine encodes trajectory errors separated from value. Nature. EPub Feb 11, 2026
19. Gagnon L et al (2015) Quantifying the microvascular origin of BOLD-fMRI from first principles with two-photon microscopy and an oxygen-sensitive nanoprobe. J Neurosci 35(8):3663–3675
20. Stefanovic B, Bosetti F, Silva AC (2006) Modulatory role of cyclooxygenase-2 in cerebrovascular coupling. Neuroimage 32(1):23–32
21. Mishra A (2017) Binaural blood flow control by astrocytes: listening to synapses and the vasculature. J Physiol 595(6):1885–1902
22. Stuehr DJ (2004) Enzymes of the L-arginine to nitric oxide pathway. J Nutr 134(10 Suppl):2748S–2751S. discussion 2765S-2767S
23. Chen K, Pittman RN, Popel AS (2008) Nitric oxide in the vasculature: where does it come from and where does it go? A quantitative perspective. Antioxid Redox Signal 10(7):1185–1198
24. Krawchuk MB et al (2020) Optogenetic assessment of VIP, PV, SOM and NOS inhibitory neuron activity and cerebral blood flow regulation in mouse somato-sensory cortex. J Cereb Blood Flow Metab 40(7):1427–1440
25. Ghosh S et al (2022) Functional dissection of neural circuitry using a genetic reporter for fMRI. Nat Neurosci 25(3):390–398
26. Stefanovic B et al (2007) Functional uncoupling of hemodynamic from neuronal response by inhibition of neuronal nitric oxide synthase. J Cereb Blood Flow Metab 27(4):741–754
27. Cinelli MA et al (2020) Inducible nitric oxide synthase: regulation, structure, and inhibition. Med Res Rev 40(1):158–189
28. Zhou L, Zhu DY (2009) Neuronal nitric oxide synthase: structure, subcellular localization, regulation, and clinical implications. Nitric oxide 20(4):223–230

29. Brenman JE et al (1996) Interaction of nitric oxide synthase with the postsynaptic density protein PSD-95 and alpha1-syntrophin mediated by PDZ domains. Cell 84(5):757–767
30. Forstermann U, Sessa WC (2012) Nitric oxide synthases: regulation and function. Eur Heart J 33(7):829–837.837a-837d
31. Lavallee P et al (2005) Feedforward inhibitory control of sensory information in higher-order thalamic nuclei. J Neurosci 25(33):7489–7498
32. Chiaia NL et al (1991) Thalamic processing of vibrissal information in the rat. I. Afferent input to the medial ventral posterior and posterior nuclei. J Comp Neurol 314(2):201–216
33. Yonk AJ et al (2025) Role of posterior medial thalamus in the modulation of striatal circuitry and choice behavior. eLife 13:e98563
34. Li N et al (2014) Study of the spatial correlation between neuronal activity and BOLD fMRI responses evoked by sensory and channelrhodopsin-2 stimulation in the rat somatosensory cortex. J Mol Neurosci 53(4):553–561
35. Zhao F et al (2008) BOLD study of stimulation-induced neural activity and resting-state connectivity in medetomidine-sedated rat. Neuroimage 39(1):248–260
36. Cox RW (1996) AFNI: software for analysis and visualization of functional magnetic resonance neuroimages. Comput Biomed Res 29(3):162–173
37. Mandelbaum G et al (2019) Distinct cortical-thalamic-striatal circuits through the parafascicular nucleus. Neuron 102(3):636–652.e7
38. Hunnicutt BJ et al (2016) A comprehensive excitatory input map of the striatum reveals novel functional organization. eLife 5:e19103
39. Guo Q et al (2015) Whole-brain mapping of inputs to projection neurons and cholinergic interneurons in the dorsal striatum. PLoS One 10(4):e0123381
40. Oh SW et al (2014) A mesoscale connectome of the mouse brain. Nature 508(7495):207–214
41. Das AT, Tenenbaum L, Berkhout B (2016) Tet-on systems for doxycycline-inducible gene expression. Curr Gene Ther 16(3):156–167

Part 2

Multi-scale Optogenetic Assessments of Neural Mechanisms from Cells to Circuits in Intact and Dysfunctional Brains

Check for updates

Chapter 9

Optogenetic Strategies to Dissect Neuronal Contributions to Neurovascular and Neurometabolic Responses

Adiya Rakymzhan, Andrew E. Toader, Mitsuhiro Fukuda and Alberto L. Vazquez

Abstract

Optogenetics has revolutionized the study of neurovascular and neurometabolic interactions by enabling precise, cell-type-specific manipulation of neuronal activity. This technique allows for dissecting the contributions of distinct neuronal populations to cerebral blood flow (CBF) regulation and metabolic load. In this chapter, we outline key methodological considerations for optogenetic investigations into neurovascular and neurometabolic coupling. We discuss the selection of optogenetic actuators, strategies for targeting excitatory and inhibitory neuronal subtypes using transgenic mice and viral vectors, and approaches for delivering optical stimulation in anesthetized and awake preparations. Furthermore, we highlight the importance of controlling for optical and hemodynamic artifacts, designing rigorous experimental controls, and selecting appropriate imaging modalities to assess vascular and metabolic responses. By integrating optogenetics with advanced imaging techniques, researchers can gain detailed mechanistic insight into neurovascular and neurometabolic dynamics underlying brain function in health and disease.

Keywords Neurovascular coupling, Neurometabolism, Optogenetics, Excitatory neurons, Inhibitory neurons, Transgenic mice, Viral vectors, Anesthetized, Awake, Optical artifacts, Control experiments

1 Introduction

To infer and quantify neuronal activity using vascular signals, it is essential to understand the relationship between neuronal activity and cerebral blood flow (CBF) changes. Unlike other tissues in the body, the brain has minimal energy reserves and relies on blood vessels for a continuous supply of oxygen and glucose. Neurons actively regulate local blood supply by directly or indirectly influencing vessel tone, leading to dilation or constriction—a process known as neurovascular coupling. However, the precise relationship between neuronal activity and vascular changes remains elusive, as multiple different types of neurons contribute in varying ways (Table 1). Optogenetics, combined with different genetic cellular targets in transgenic

Kevin Chuen Wing Chan and Seong-Gi Kim (eds.), *Optogenetic Functional Neuroimaging*, Neuromethods, vol. 232, https://doi.org/10.1007/978-1-0716-5178-0_9,

Table 1
Summary of neuronal mediators of CBF

Neuronal subpopulation	Principal neurotransmitter	Neurovascular mediator	Effect on vasculature
Pyr	Glutamate	PGE_2	Vasodilation/ vasoconstriction (high firing rate)
VIP	GABA	VIP	Inconsistent
NOS	GABA	NO	Vasodilation
NPY	GABA	NPY	Vasoconstriction
SST	GABA	NO NPY Substance P	Biphasic (vasodilation then vasoconstriction)
PV	GABA	Substance P	Inconsistent

animals, allows for cell-specific and temporally precise control of discrete subsets of neurons. This advancement has enabled the study of how specific neuronal subpopulations contribute to the regulation of brain blood supply by adjusting local vascular tone.

There is general agreement about the blood flow response to the activation of certain neuronal subtypes. For instance, activation of pyramidal cells or nitric oxide synthase (NOS) cells typically leads to an increase in blood flow, whereas activation of neuropeptide Y (NPY) cells results in decreased blood flow [1–4]. However, studies on other neuronal subtypes, such as vasoactive intestinal peptide (VIP) or parvalbumin (PV) cells, have produced inconsistent results across different research groups [5–8]. These discrepancies have been attributed to several factors, including differences in the measured cortical depth (superficial vs. deep layers), stimulation protocols (short vs. long stimulation periods), brain states (anesthetized vs. awake), and methods of expressing ChR2 (virus injection vs. transgenic mouse). Each of these factors can significantly influence the hemodynamic response, highlighting the complexity of neurovascular coupling and the need for careful experimental design and interpretation in optogenetic studies.

The first studies that assessed the role of different cell types on CBF regulation were conducted in vitro. These intricate experiments relied on neuronal patching to selectively activate and identify individual neurons [9]. Samples from patched neurons were also used to identify their genetic identity. Selective activation of specific neuronal subpopulations became possible with the development and dissemination of optogenetic tools. Although in vitro studies offer precise control over experimental conditions and easier manipulation of specific regions, brain slices lack a more complete physiological context, including normal vascular pressurization, the pulsatile nature of blood supply, natural blood flow, and interactions via longer-range neuronal connections. Thus, depth-dependent studies, which are essential for understanding laminar-specific neurovascular processing, can be difficult in slices. In contrast, in vivo optogenetic studies enable the interrogation of genetically distinct neurons embedded in complex heterogeneous networks and physiological systems, mitigating many if not most of these barriers, and allowing for depth-dependent studies of neurovascular coupling.

Imaging techniques for studying neurovascular coupling, such as laser Doppler flowmetry (LDF), intrinsic optical signal (IOS) imaging, laser speckle contrast imaging (LSCI), two-photon microscopy (2P), functional magnetic resonance imaging (fMRI), and optical coherence tomography (OCT), capture different aspects of vascular responses (e.g., depth capabilities). This variation can also define the choice of opsin, whether it is excited by blue or red light, the light delivery method, depending on the required penetration depth to reach the targeted brain regions. For instance, fMRI offers great imaging depth, but often uses implanted optical fibers to deliver light deep into the tissue. In contrast, IOS imaging, which is more sensitive to superficial tissue signals, typically employs light delivered through the cranial window using fibers, broad beams, or shaped beams. The goal of this type of experiment is to deliver sufficient light to the target and manipulate cells within an area. More recently, patterned light delivery with 2P can be employed to activate pre-identified cells within small 2P illumination volumes for applications where tailored activation of specific cells is desired. These varying methods of light delivery ensure that optogenetic stimulation aligns with the imaging technique's depth sensitivity and the specific research objectives, thereby optimizing the investigation of neurovascular coupling.

This chapter discusses key considerations for optogenetics experiments targeting specific cell types to observe their impact on neurovascular and neuro-metabolic responses in mice. It emphasizes experimental details, including animal preparation and determining light power requirements, to activate neurons expressing optogenetic proteins at various depths. Additionally, it covers the measurement of vascular or hemodynamic responses to optogenetic stimuli with different energies in various mouse models, including naive animals.

2 Materials

2.1 Cell-Type Specific Targeting

Genetic targeting of specific neuronal populations can be achieved through two main approaches. The first uses transgenic animals expressing recombinase enzymes like Cre or Flp under specific promoters. The loxP system is a versatile tool in which genetic sequences flanked by loxP sites (floxed) can be conditionally excised, inverted, or activated by Cre recombinase. This enables precise, cell-specific manipulation of gene expression. Ai animals build on this foundation, containing floxed stop cassettes upstream of reporter or effector genes (e.g., fluorescent proteins, calcium indicators, or optogenetic actuators). The combination of Cre-animals, loxP systems, and Ai animals allows the introduction of genetic constructs into specific targets via breeding. The careful design of Cre, loxP, and Ai mice creates endless possibilities for the types of genetic manipulations that can be generated. The Jackson Laboratory has a convenient guide on breeding strategies for this system. The second approach is the development of viral vectors carrying a recombinase-dependent construct (e.g., floxed or FRT-flanked sequences). These viruses can be injected into the brain, activating the transgene in recombinase-expressing cells. These methods, highly specific and widely used in mice, allow targeting of various neuronal (and other cell) populations through a wide range of mouse Cre lines. This approach utilizes the natural or engineered tissue tropism of viral serotypes in combination with cell-type-specific promoters or enhancers. Serotypes vary in their ability to target specific neuronal or glial populations and can be selected based on the experimental needs. Promoters refine the expression further by ensuring specificity within the target cell population. This approach is especially valuable in nongenetic models or species lacking transgenic lines, such as nonhuman primates. Further details are provided in the following section.

Cre-Driver Lines

Building on the utility of Cre–loxP systems for cell-specific targeting, the combination of viral vector delivery and Cre-driver lines has revolutionized the study of genetically defined cell types, particularly in mouse models. These Cre lines allow selective targeting of specific neuronal subtypes, such as PV or SST inhibitory neurons. Similarly, excitatory neurons can be targeted using CaMKIIα-Cre and Thy1-Cre lines. Beyond neurons, astrocytes and vascular cells, such as pericytes and endothelial cells, can also be selectively targeted using Cre lines like Aldh1l1-Cre for astrocytes or Pdgfrb-Cre for pericytes, facilitating studies on their roles in neurovascular coupling and metabolism. A summary of the available Cre-lines can be found in Table 2. CreER/ERT/ERT2 systems provide an additional layer of temporal control to Cre–loxP recombination. These systems fuse Cre recombinase to a modified estrogen receptor (ER) that remains inactive in the cytoplasm until it binds to Tamoxifen or its metabolites. Upon Tamoxifen administration, the CreER/ERT/ERT2 translocates to the nucleus, allowing temporal activation of floxed genetic constructs. The Jackson Laboratory has a guide for the administration of Tamoxifen to drive recombination in these mice. An interesting side effect of this process is that the dosage of Tamoxifen drives sparse expression (as opposed to strong and dense) in these models.

Viruses and Their Characteristics

While Cre driver lines are a unique advantage of mouse models, viral vector-based approaches significantly expand the unique opportunities for the use of optogenetics or other constructs, enabling opsin

Table 2
Transgenic animals and Cre-recombinase

	Line	Expression	Catalog number
Excitatory	Thy1-cre/creER	Cortex, hippocampus, amygdala	JAX 006143
	CamK2a-cre	Cortex, hippocampus, thalamus, amygdala, cerebellum	JAX 005359
	Scn1a-creER	Cortex, hippocampus, cerebellum, testes	JAX 012641
	Slc17a7-cre	Cortex, olfactory bulb, anterior olfactory nuclei	JAX 037512
	Cux2-creER	Cortex, hippocampus, pancreas	JAX 012551
	Nr5a1-cre	Hypothalamus, adrenal cortex, testes/ovaries, adrenal gland	JAX 022781
	Ntsr1-cre	Cortex, thalamus, hypothalamus, amygdala, gastrointestinal tract	JAX 017266
	Etv1-creER	Cortex, cerebellum, olfactory bulb	JAX 013048
	Rorb-IRES-cre	Cortex, thalamus, retina, suprachiasmatic nucleus (hypothalamus)	JAX 023526
	Rbp4-cre	Cortex, hippocampus, thalamus, hepatocytes, adipose tissue	JAX 037128
Inhibitory	Gad2-cre	Cortex, hippocampus, basal ganglia, amygdala, cerebellum, enteric inhibitory neurons	JAX 010802
	VGAT-IRES-cre	Cortex, hippocampus, basal ganglia, amygdala, cerebellum, retina, enteric inhibitory neurons	JAX 016962
	VIP-cre	Cortex, hippocampus, amygdala, hypothalamus, pancreas, bronchial smooth muscle	JAX 010908
	SST-cre	Cortex, hippocampus, amygdala, basal ganglia, thalamus, hypothalamus, pituitary, pancreas, gastrointestinal tract	JAX 013044
	NOS-cre/creER	Cortex, hippocampus, striatum, hypothalamus, skeletal muscle	JAX 017526
	NPY-cre	Cortex, arcuate and paraventricular nuclei (hypothalamus), amygdala, hippocampus, adrenal medulla, pancreas	JAX 027851
	PV-cre	Cortex, cerebellum, medulla, pons, pallidum, thalamus	JAX 008069
	Calb-cre	Cortex, hippocampus, cerebellum, amygdala, thalamus, pancreas, kidney, muscle	JAX 016835
	CCK-cre	Cortex, hippocampus, amygdala, hypothalamus, ventral tegmental area, pancreas, gastrointestinal tract	JAX 012706
	5HTa3-ires-Flpo	Cortex, hippocampus, amygdala, thalamus, enteric nervous system	JAX 030755
	DOP1Drd1-cre	Prefrontal cortex, striatum, amygdala, hippocampus, thalamus, retina, kidney, adrenal gland	JAX 024860 006410 (D1R)
	DOP2Drd2-cre	Prefrontal cortex, striatum, globus pallidus, substantia nigra (pars reticulata), hippocampus, amygdala, pituitary gland, retina, kidney, pancreas, adrenal gland, gastrointestinal	JAX 036716 030036 (D2R)

expression isolated to injected locations, access to these manipulations in other species, and even broader cell types. The most commonly used viral vectors are lentiviruses (LVs) and adeno-associated viruses (AAVs), both widely applied in rodents and nonhuman primates for introducing optogenetic tools into the adult brain.

When designing an optogenetic study, it's important to understand the differences between recombinant AAVs (rAAVs) and LVs. rAAVs are generally safer because they stay in the nucleus as episomes rather than integrating into the host genome, reducing the risk of disrupting host DNA [10]. They also produce lower immune responses and can spread more widely from the injection site, making them ideal for experiments needing higher levels of expression over larger brain areas [11]. However, high expression levels from rAAVs can lead to risks, such as protein clumping or axonal malformations, which may interfere with the experimental goal [10, 11]. LVs can integrate into the host genome, which can be a safety concern, but they tend to cause fewer aggregates. This makes them a better option for long-term studies, especially in nonhuman primates, where lentiviruses have been successfully used [12].

Selecting the right viral serotype is crucial for targeting specific cell types accurately and efficiently. A serotype refers to the variation in a viral vector's capsid proteins, which determines its ability to infect particular cells or tissues (tissue tropism). AAV9 is commonly used because it can cross the blood–brain barrier and efficiently target neurons and glial cells. This makes it ideal for systemic delivery or intravenous injections in studies requiring widespread expression of optogenetic tools. AAV2 is often preferred for direct injections into the brain due to its strong ability to transduce neurons at the injection site. AAV5 is effective for targeting nonneuronal cells like astrocytes and vascular endothelial cells. When choosing a serotype, the immune response is another important factor. For example, preexisting immunity to AAV2 in some animal models can reduce its effectiveness. In such cases, researchers often use alternative serotypes like AAV8 or AAV1 [13].

Expanding upon the strategies for achieving cell-specific expression in optogenetic studies, the selection of a promoter plays an important role in refining the precision of targeting. Promoters are DNA sequences that drive the transcription of a gene, and their specificity ensures that the optogenetic tools are expressed only in the desired cells [14]. For example, the CamKIIα promoter is widely used to target excitatory pyramidal neurons in the cortex and hippocampus, making it a crucial tool in studies focusing on excitatory neuronal contributions to neurovascular coupling. Similarly, the DLX, Gad67, and VGAT promoters are commonly employed to drive expression in inhibitory neurons, which play critical roles in regulating cortical network activity and cerebral blood flow. The DLX promoter broadly targets GABAergic interneurons with a focus on developmental expression, and Gad67 drives expression in all GABA-synthesizing neurons, whereas VGAT specifically targets synaptically active GABA- and glycine-releasing neurons.

Promoters can also be used to target nonneuronal cells involved in neurovascular coupling. For instance, the GfaABC1D promoter is highly specific for astrocytes, enabling studies of their contributions to vasodilation and neurovascular support. The Tie2 promoter, on the other hand, is effective for driving expression in endothelial cells, facilitating research into vascular dynamics and blood–brain barrier regulation. In some cases, enhancers are added to promoters to increase specificity and expression levels. For example, the CAG promoter is broadly active but can be paired with tissue-specific enhancers to limit expression to neurons or glial cells. Promoter-based approaches do not require maintaining special animal strains and can be applied in nongenetic models. However, viral vector limitations, especially the small packaging size of AAVs (~4.7 kb), restrict the use of large or native promoter sequences. While LVs have a smaller transduction area and lower expression levels than rAAVs, their larger cargo capacity (~8–10 kb) enables delivery of large genes or complex constructs [11, 13, 14].

2.2 Common Optogenetic Modulators

There are different types of optogenetic modulators available, including excitatory and inhibitory opsins (Table 3). The most widely used opsin is Channelrhodopsin-2 (ChR2-H134R), which is activated with blue light (470 nm) [15]. However, red-shifted opsins, such as C1V1 and Chrimson, are activated with orange/red light, allowing deeper light penetration and activation of cells in deeper brain regions [16–18]. The choice of opsins and excitation wavelength is often considered when conducting imaging experiments to avoid or minimize spectral overlap with light sources necessary for imaging. For example, blue-shifted opsins like ChR2 can be used with red-shifted calcium indicators, such as RCaMP1 or RGECO, instead of green-shifted indicators like GCaMP, as blue light will excite both ChR2 and GCaMP, introducing

Table 3
Common optogenetic modulators

Opsin name	Wavelength (nm)	On/off kinetics (ms)	Photocurrent (pA)
ChR2 (H134R)	470	~2/20	~500
ChETA (H134R)	470	~0.5/3	~400
ReaChR	610	~15/20	~400
Chrimson	590	~10/10	~600
vfChrimson	590	~5/5	~800
Chronos	500	~5/5	~700
C1V1	535	~20/30	~300
Jaws	590	~1/10	~300
ArchT	550	~1/10	~400
eNpHR (3.0)	580	~15/60	~500

large artifacts and concerns over continuous stimulation during experiments. Similarly, red-shifted opsins like Chrimson can be used with green-shifted calcium indicators, such as GCaMP.

Additionally, the choice of opsins can depend on their activation/deactivation kinetics and light sensitivity (photocurrent). For example, the ultrafast ChR2 mutant ChETA and the ultrafast, red-shifted vfChrimson mutant of Chrimson are characterized by their rapid kinetics but produce smaller photocurrents [18, 19], requiring higher light power. On the other hand, highly potent opsins like the chimeric ReaChR generate substantial ionic currents but exhibit very slow deactivation [20, 21]. However, one of the latest developments, Chronos, maintains very fast kinetics while also demonstrating relatively high potency [22].

2.3 Animal Preparation

Surgical Procedures and Viral Injections

To perform optogenetic imaging experiments in rodents, especially mice, a surgery is conducted to implant a head holder and cranial window for optical access (Fig. 1). We describe here typical surgical procedures done in mice because of the wider availability of transgenic mouse models. Mice are typically anesthetized with ketamine (75 mg/kg) and xylazine (10 mg/kg). Craniotomy with this cocktail is considered less prone to brain swelling during the procedure. Aseptic procedures are followed during the entire surgery, which usually include three major steps: (1) expose the skull and ensure dryness to install a head-holder or head-plate (e.g., Narishige CF-10); (2) thin or remove skull bone over a location of interest (e.g., parietal bone) and replace the removed bone with glass or glass-assembly (Warner Instruments, Inc.); and (3) inject virus using a nano-injector, if necessary, prior to glass placement. Supplemental oxygen (40% balanced by air) is typically administered via nose cone at 500 mL/min. Body temperature needs to be maintained throughout at 37–38°C using a temperature feedback unit. Surgery consists of removing the skin

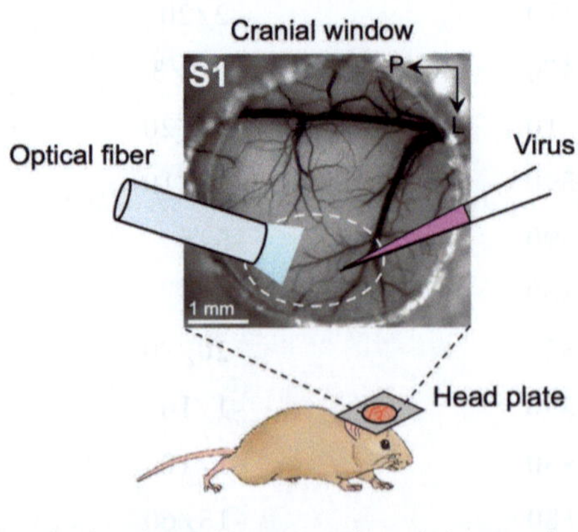

Fig. 1 Typical animal setup for awake head-fixed imaging. Sample image shown

over the skull and exposing the bone until dry. The head plate we use (CP-2, Narishige Inc.) contains an 8 mm round opening that is positioned over the area of interest, and then the plate is cemented onto the skull using VetBond (3 M Inc.) and dental cement. A 4-mm diameter craniotomy is performed over the somatosensory cortex centered around 3 mm lateral and 1.5 mm posterior from the bregma. In contrast to the genetically engineered Thy1-ChR2 or VGAT-ChR2 mouse models, which already express ChR2 throughout the brain, transgenic models expressing Cre-recombinase to target neuronal populations like PV or SST neurons require viral injection to express optogenetic constructs in those cells. In that case, mice are placed under a microscope to visualize virus injection. A glass micropipette (borosilicate glass [OD = 1.50 mm, ID = 0.84 mm, with a filament], tip open diameter = 2 μm, taper length = 0.42 mm) is used to pressure inject (20 psi, 3 s) small amounts of virus at depths of 300 μm and 600 μm into the somatosensory cortex. A small amount of fluorescent dye (e.g., 200 μM Sulphorodhamine-101) can be added to the virus solution (10 μL virus + 1 μL dye) to visualize virus delivery in the targeted locations. Usually, injected volumes of about 0.1 μL per site are sufficient. Many groups, including our own, use a custom cover glass assembly consisting of a 4-mm round cover glass glued onto a 5-mm round cover glass (CS-4R and CS-5R, Warner Instruments Inc.) and cemented on the skull to maintain visual access to the brain.

Acclimation

In our experience, these procedures require about 4 weeks for the mouse to recover and the virally driven expression to stabilize. Mice are given 1–3 weeks to recover from surgery before beginning acclimation to awake head-fixed imaging in the fourth week. A typical acclimation procedure consists of increasing head-fix time from 15 min to at least 1 h either in their cage or in the imaging setup for 5 days.

Experimental Conditions: Anesthetized vs. Awake

Lightly anesthetized experiments are often necessary for experiments where pharmacological manipulations are necessary for the delivery of agents into the brain. However, anesthesia affects neurovascular coupling, modulating the blood flow response (see Fig. 2). In this experiment, we measured CBF using LDF in response to optogenetic stimulation of neurons in Thy1-ChR2 and VGAT-ChR2 mice under both anesthetized and awake conditions. The optogenetic stimulation was delivered using various pulse widths (2 ms, 10 ms, and 30 ms), and the CBF response was compared across states. In awake mice, hemodynamic responses to photostimulation occur earlier and return to baseline faster compared to anesthetized mice. The hemodynamic response in awake mice typically peaks around 2–3 s after stimulation, while anesthetized mice show a lagged time-to-peak of approximately 6–7 s under various anesthetics like ketamine–xylazine and isoflurane. This difference in response timing is likely due to the altered neural activity and vascular reactivity induced by anesthesia. The effects of different anesthetics are summarized in Table 4.

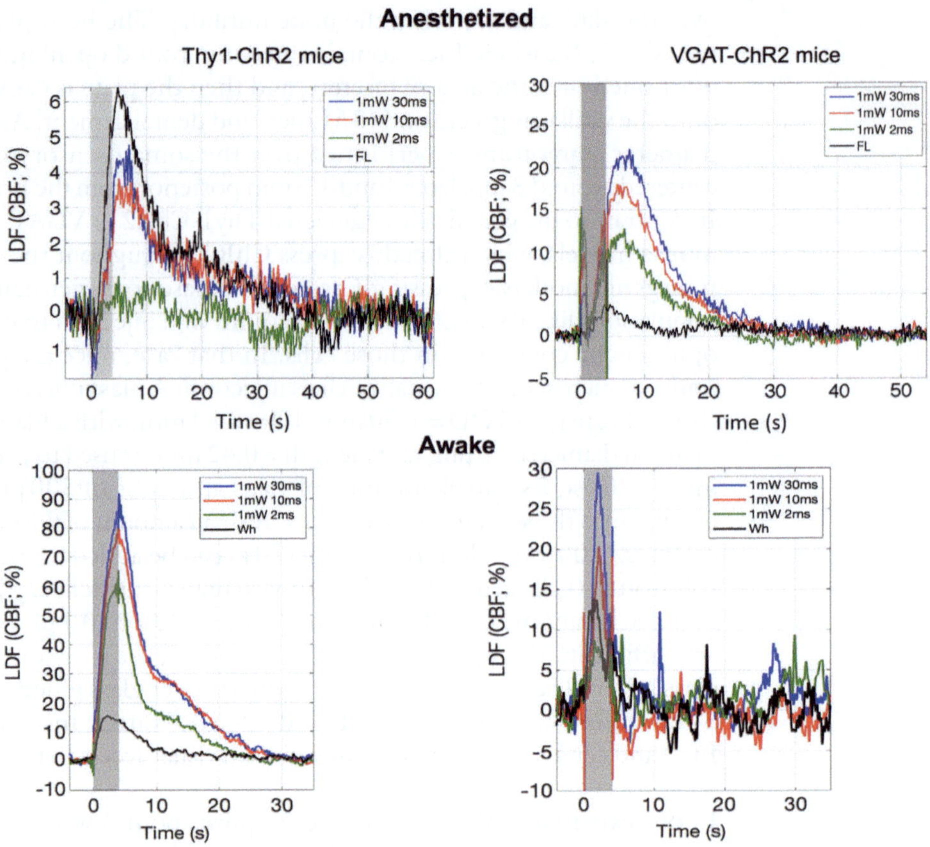

Fig. 2 Hemodynamic responses evoked by optogenetic stimulation in anesthetized vs. awake mice

Table 4
Common anesthetics for NVC experiments

Anesthetic	Mechanism and effects
Isoflurane	activates two-pore potassium channels, GABA agonist; enhances GABAergic inhibition, suppresses excitatory activity; vasodilatory effect, increases CBF.
Ketamine	NMDA antagonist; suppresses excitatory activity, with minimal direct vascular effects.
Medetomidine/dexmedetomidine/ xylazine (often used with ketamine)	α2-adrenergic agonist; reduces sympathetic activity, promotes sedation; direct impact on vascular tone.
Urethane	GABA agonist; enhances GABAergic inhibition and inhibits NMDA receptor activity; reduces vascular tone.
Alpha-chloralose (often used with urethane)	potentiates GABAergic inhibition, with minimal impact on vascular tone.

Interestingly, the magnitude and direction of hemodynamic changes can also differ between awake and anesthetized states. For instance, under 1% isoflurane, photostimulation evoked negative hemodynamic responses similar to those in awake mice, while 1.5% isoflurane resulted in positive responses more typical of functional hyperemia [8]. These findings highlight the importance of considering the arousal state when interpreting hemodynamic data and underscore the complex interplay among neural activity, vascular responses, and anesthesia in the brain.

3 Methods

3.1 Light Delivery

Delivering light to target neurons after successfully introducing optogenetic probes presents a significant challenge, particularly due to the scattering properties of light in biological tissues. To address this, researchers must carefully tailor their approach to align with specific experimental goals, factoring in the light's wavelength, intensity, and scattering behavior, as well as the optical setup employed.

Selecting an appropriate light source is the first step. Options include mercury or xenon bulbs, LEDs, continuous-wave lasers, and ultrafast pulsed lasers for two-photon excitation. Mercury and xenon bulbs provide a broad-spectrum output but require filtering to isolate desired wavelengths. Despite their high power, these bulbs have limited lifespans and need proper disposal. LEDs, by contrast, are more durable, emitting light for tens of thousands of hours, producing less heat, and offering affordability. Lasers generate coherent light, which is advantageous for applications requiring precise light delivery, such as holographic patterning or efficient fiber coupling [10]. Regardless of the source, it is critical to balance the amount of light delivered: too little may fail to activate opsins, while excessive illumination risks damaging cells or triggering nonspecific activation [22]. Additionally, light desensitization of optogenetic tools, observed in ChR2, can occur over time, requiring adjustments during extended experiments [23]. Optical fibers are used to deliver light from the source to the neurons. These fibers, made from glass or plastic, can be used with cranial windows or fiber implants to stimulate neurons in the brain. The numerical aperture (NA) of the fiber is an important factor to consider: a higher NA provides broader illumination with lower intensity, while a lower NA concentrates light more tightly onto a smaller region. Additionally, the choice of fiber material is crucial for specific experimental environments, where, for example, plastic fibers minimize imaging artifacts in MRI experiments. For long-term experiments, fibers can be implanted and connected to an external light source [14].

Light scattering in biological tissue is another important consideration. When light passes through tissue, it is deflected, and the scattered light behaves unpredictably. This means that the light distribution inside the tissue will not be the same as the light emitted from the source. To better understand light propagation, tools, such as electrophysiological recordings, discussed in the following section, can be utilized.

3.2 Electrophysiological Assessment of the Neural Responses to ChR2 Excitation as a Function of Time and Depth

To determine the optimal power and pulse duration of light required to activate neurons expressing optogenetic protein at various depths, it is essential to measure neural activity while delivering photo-stimuli (also called optogenetic stimuli), as light traversal in the brain is limited by both attenuation and scatter, affecting depth and specificity. This can be done relatively easily by electrophysiology. We implemented this approach in the following experiment [24]. An optical fiber and a metal electrode were carefully positioned near the hotspot location while avoiding large surface vessels. The optical fiber, with a core diameter of 125 μm (S-405-HP, ThorLabs, Inc.), delivered the photo-stimulus. A carbon electrode (Carbostar-1 with a tip diameter of 5 μm and resistance of ~1 MΩ; Kation Scientific) was placed adjacent to the fiber, 260 μm deep, at a 60° angle. This setup was allowed to settle for 15 min before starting data acquisition. A power-adjustable, TTL-controlled 473 nm laser diode (CrystaLaser, Inc.) connected to the optical fiber provided the photo-stimulus. Six photo-stimulation parameters were used: 1.0, 0.5, and 0.25 mW pulses (10 ms duration), 1.0 and 0.25 mW pulses (30 ms duration), and 1.0 mW pulses (2 ms duration). Thy1-ChR2 mice, in which ChR2 expression is biased to layer 5 pyramidal neurons and their processes that extend to superficial layers, received single pulses every 3 s for 48 s per parameter set. Measurements were taken in 100 μm depth increments, pausing for 5 min at each new depth. After depth recordings, the electrode was repositioned laterally along the pial surface, avoiding visible surface vessels, and electrophysiological recordings were repeated.

The average current source density was calculated from local field potential (LFP) data to investigate depth-dependent differences in evoked potential activity (Fig. 3A). Photo-stimulation opened channels in the superficial dendrites of layer 5 pyramidal neurons, indicated by a negative deflection in the top 400 μm. The corresponding activity sink was located between 500 and 900 μm, aligning with layer 5 pyramidal neurons. As the photo-stimulus energy decreased, such as reducing the duration from 30 to 2 ms, the magnitude of the evoked field potential also decreased while maintaining the same depth profile (Fig. 3A). The average time-to-first spike was analyzed to examine the progression of evoked spiking activity by depth. At 1 mW and 30 ms, the first spike occurred on average 5.5 ms after

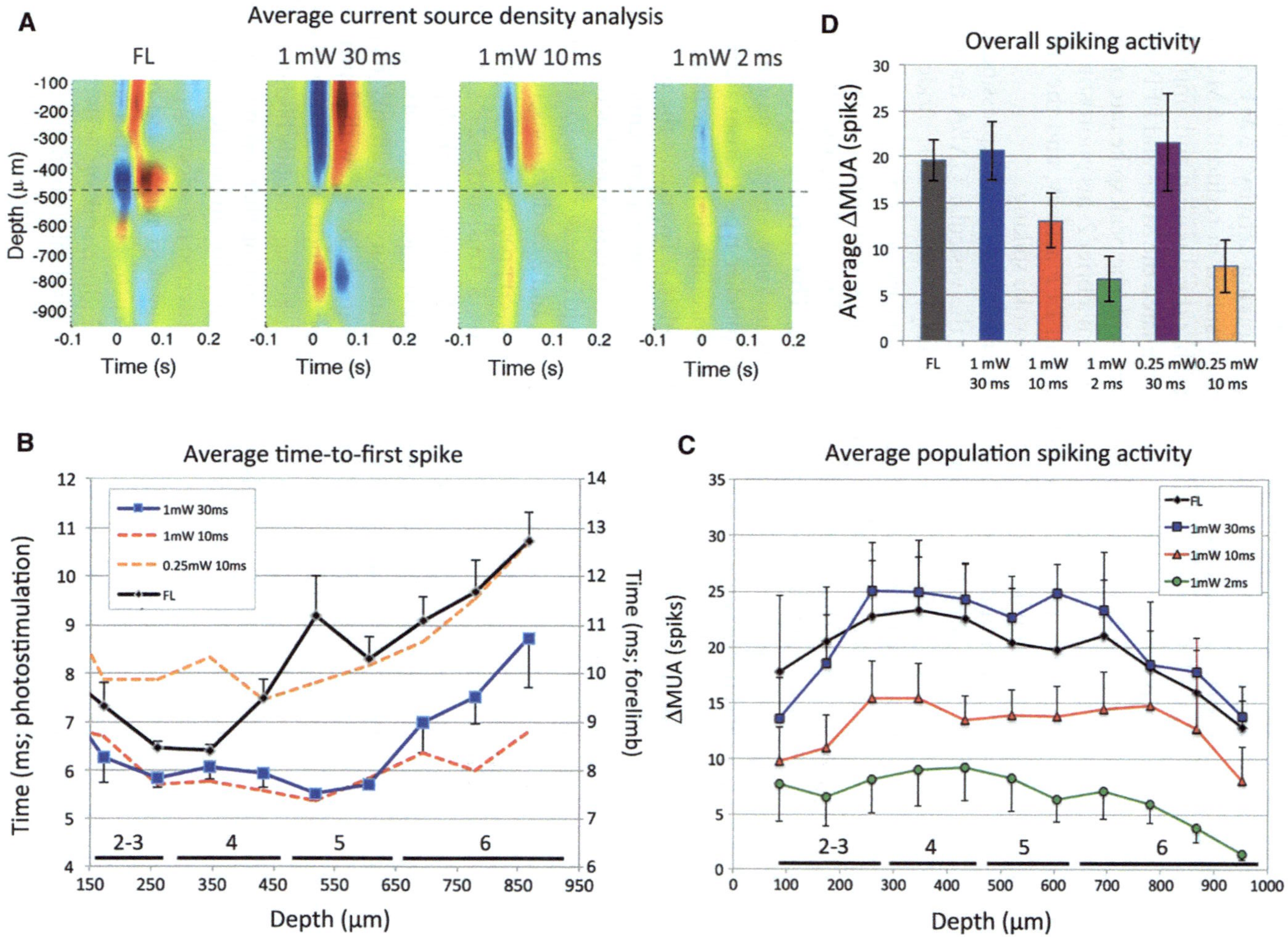

Fig. 3 Electrophysiological assessment of the neural responses to ChR2 excitation in Thy1-ChR2 mice as a function of time and depth

stimulation onset in layer 5, followed by spikes in layers 2-3 and layer 4, then deeper layers (Fig. 3B, blue line). Lower intensity (0.25 mW) increased the time-to-first spike by 2.2 ms. All stimulation parameters evoked significant activity between 173 and 780 μm, with less activity at 86 μm and below 780 μm (Fig. 3C). Photo-stimulation at 30 ms produced the largest changes in spiking activity (ΔMUA), while shorter durations (10 and 2 ms) resulted in reduced ΔMUA (Fig. 3D). These results demonstrate that the duration and amplitude of the photo-stimulus effectively modulated the MUA and LFP activities in layer 5 as well as all other layers, with the peak spike rate occurring in layer 5, as expected. Notably, even a brief 2-ms photostimulation was sufficient to evoke spiking and LFP activity, and the stimulus magnitude influenced the time to evoke spiking.

We conducted the same experiment in mice expressing ChR2 in inhibitory neurons by targeting the vesicular GABA transporter (VGAT) [25]. In contrast to the Thy1-ChR2 mouse model, in this model, ChR2 was expressed in inhibitory neurons across all cortical layers. Because inhibitory neurons intend to inhibit their postsynaptic partners, this experiment can be used to examine the efficiency of photo-stimulation at different depths. This criterion is crucial because different types of inhibitory neurons are known to preferentially reside at different depths. For instance, chandelier-type neurons are typically found in layers 2 and 6, while NPY-expressing neurons are more evenly distributed throughout layers 2 to 6 [26]. To reduce the number of photo-stimulation parameters, all ensuing experiments only used 1 mW photo-stimuli. Photo-stimulation first evoked activity in the most superficial layer (86 μm deep), with average time-to-first spike of 1.4 ms, 1.6 ms, and 1.3 ms for 30, 10, and 2 ms durations, respectively (Fig. 4B). Deeper cortical layers showed progressively latent activity, reaching layer 6 on average at 5.8 ms, 5.4 ms, and 5.2 ms after photo-stimulus onset. The increase in spike latency with depth is likely due to longer times to recruit sufficient photo-detectors, given the reduced light availability. Reducing intensity to 0.25 mW increased time-to-first spike by 2.7 ms. Spike rate decreased with depth, averaging 1.0 ± 0.2 spikes/s at the deepest location for 2-ms stimuli (Fig. 4C). The strongest photo-stimulus evoked significantly higher average activity than shorter pulses ($P<0.05$), showing that spiking activity can be effectively modulated by photo-stimulation amplitude and duration throughout cortical depth. Another criterion for selecting photo-stimulation parameters was ensuring the evoked activity did not surpass the forelimb-evoked activity at any depth. This is important because excessive photo-stimulation could induce spiking activity beyond normal physiological levels. Sensory forelimb stimulation provides a generous upper limit for spiking activity as it activates both excitatory and inhibitory neurons. The chosen photostimulation parameters evoked substantial activity, approaching the levels seen with forelimb stimulation at many depths (Fig. 4C, D).

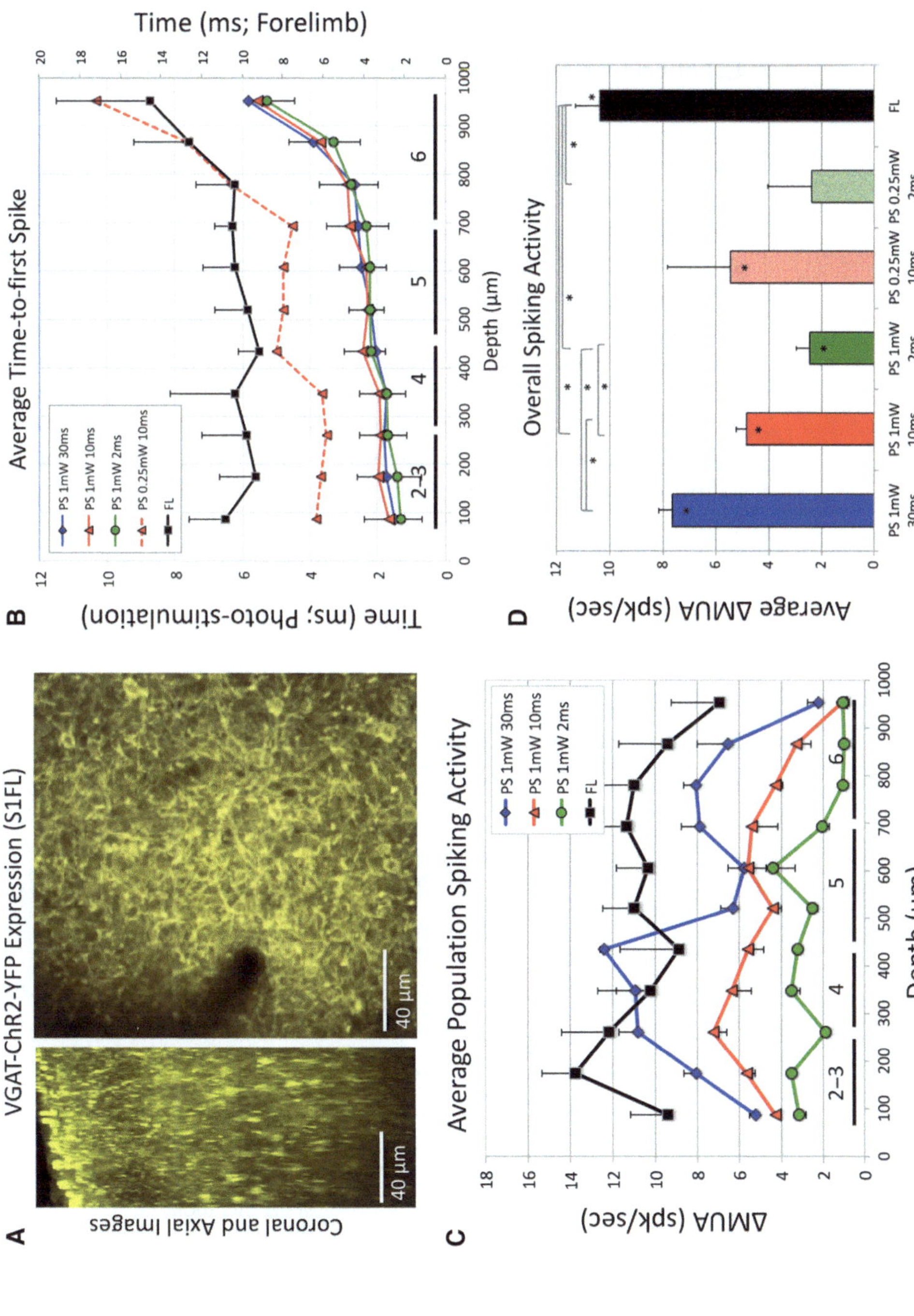

Fig. 4 Electrophysiological assessment of the neural responses to ChR2 excitation in VGAT-ChR2 mice as a function of time and depth

3.3 Imaging Methods

Imaging methods for studying neurovascular and neurometabolic dynamics vary significantly in their scale, spatial, temporal, and depth resolution, each offering unique advantages and limitations (Table 5). Wide-field optical imaging (WFOI) provides large-scale cortical coverage with high temporal resolution but is limited in spatial resolution and depth, capturing signals predominantly from superficial layers. LD offers high temporal resolution for assessing blood flow but lacks spatial resolution, making it more suited for localized measurements. A distinct advantage of LDF is its probe-based configuration with immediate readout of blood flow changes. 2P excels in spatial and depth resolution, enabling detailed cellular and subcellular imaging of vascular and neuronal structures, but with a limited field of view and slower temporal resolution. Functional near-infrared spectroscopy (fNIRS) offers noninvasive depth penetration into cortical regions with moderate temporal resolution but relatively coarse spatial resolution, capturing hemodynamic changes but no direct neuronal activity. Functional magnetic resonance imaging (fMRI) provides whole-brain coverage with reasonable spatial resolution and the ability to probe deep brain structures, though its temporal resolution is limited by the slow hemodynamic response. Positron emission tomography (PET) enables imaging of many physiological processes,

Table 5
Imaging methods

Method	Spatial resolution	Temporal resolution	Depth	Indications
WFOI	High (~50 μm)	High (up to 100 Hz)	Shallow (~200 μm)	Neurovascular, neurometabolic
LDF probe	Moderate (~200 μm)	High (~100 Hz)	Shallow (~500 μm)	Neurovascular
2P	High (~300 nm)	Moderate (~30 Hz)	Moderate (~800 μm)	Neurovascular, structural
OCT	High (~10 μm)	High (up to 1000 Hz)	Moderate (up to 2 mm)	Neurovascular
fNIRS	Low (1 cm)	High (20 Hz)	Moderate (1–2 mm)	Neurovascular, neurometabolic
fMRI	Moderate (~100 μm)	Low (5 Hz)	Any depth	Neurovascular, neurometabolic, structural
PET	Low (~500 μm)	Low (~0.1 Hz)	Any depth	Neurometabolic
pO_2 probe	High (~25 μm)	Moderate (10 Hz)	Moderate (2 mm)	Neurometabolic
Lactate probe	Moderate (~100 μm)	Low (~1 Hz)	Shallow (200 μm)	Neurometabolic

including vascular and metabolic measurements, with whole brain coverage, super sensitivity, and reasonable spatial resolution, but suffers from low temporal resolution and the requirement for radiotracers. Probes, such as pO2 and lactate probes, deliver real-time, localized measurements of oxygen and lactate levels, respectively, offering unparalleled temporal resolution but at the cost of invasive procedures and single-point spatial sampling. Collectively, these methods complement one another, allowing researchers to select the appropriate tool based on the specific scale, resolution, and depth required for their experimental goals. We will discuss the applications of these methods for measuring neurovascular and neurometabolic responses in the next sections.

3.4 Characterization of Hemodynamic Responses to Optogenetic Light Stimuli of Various Power and Pulse Widths in Various Mouse Models

An important property of neurovascular regulation is that the evoked vascular changes are activity-dependent (i.e., directly related to the amount of activity generated in neurons). There are two main neuron types that arise from the classification of their primary neurotransmitter: excitatory and inhibitory. Excitatory neurons comprise about 80% of all cortical neurons and promote activity to their connected network via the release of neurotransmitters like glutamate. In contrast, inhibitory interneurons comprise only 20% of all cortical neurons and suppress the activity of postsynaptic neurons via the release of neurotransmitters like GABA [27, 28]. Several signaling pathways have been identified by which neuronal activity generates local vascular changes. For example, excitatory neurons, the largest neuronal population recruited during sensory-evoked stimulation, increase CBF by releasing vasodilating cyclooxygenase-2 product prostaglandin E2 [29]. This is demonstrated in Fig. 5A, where a 30-ms photo-stimulus in Thy-ChR2 mice produced a CBF response comparable to natural sensory stimuli (forelimb stimulation). Inhibitory neurons, due to their anatomical connections to local cerebral microvasculature and high correlation of their density with vascular length, are also significant in modulating CBF [30]. In VGAT-ChR2 mice, optogenetic stimulation induced a CBF increase similar to that by pyramidal cells, but with a significantly larger amplitude (Fig. 5B). Notably, the response amplitude in both Thy1-ChR2 and VGAT-ChR2 models increased with increasing light pulse duration (2, 10, 30 ms).

One subtype of inhibitory neurons, VIP neurons, has receptors in smooth muscle cells of intracortical microvessels, suggesting a role in regional CBF control [31]. However, the involvement of VIP neurons in neurovascular coupling was later questioned in both in vitro and in vivo studies showing that vasodilation persisted in the presence of the VIP receptor antagonist [32]. Direct optogenetic activation of VIP neurons in awake mice in our experiments did not consistently produce significant changes in CBF, as shown in Fig. 5C [5]. Another inhibitory neuron subtype, SST cells, includes neurons expressing NOS, NPY, and substance-P (SP) receptor [33]. In vivo optogenetic stimulation of cortical SST neurons in awake mice produced a significant increase in CBF, followed by a CBF decrease (Fig. 5E) [2, 5]. This biphasic

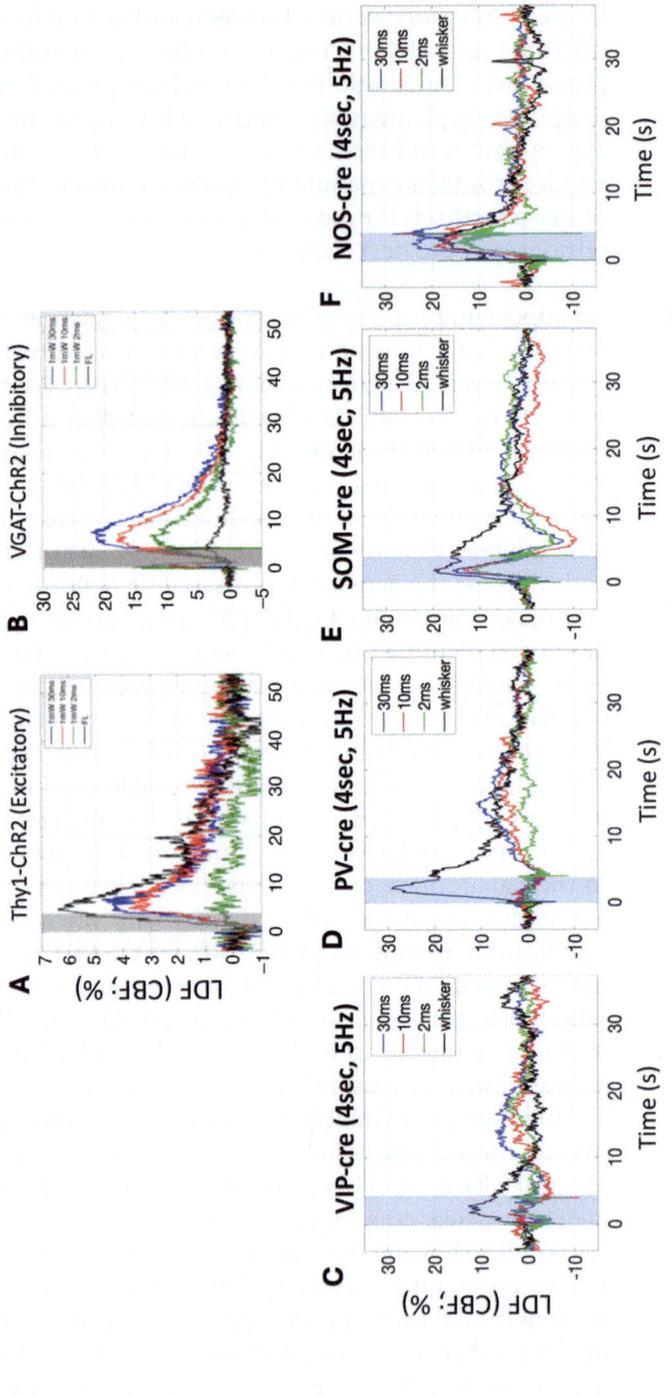

Fig. 5 LDF measurements of CBF responses to optogenetic activation of different neuronal subpopulations

hemodynamic response to photo-stimulus might be explained by NOS and NPY expression in SST neurons, which are known vasodilator and vasoconstrictor, respectively. The constriction phase of the biphasic hemodynamic response induced by activation of GABAergic cells in mice was eliminated in the presence of NPY receptor antagonist, providing evidence that NPY neurons mediate CBF decreases [3]. Optogenetic stimulation of NOS neurons in mice, brief or longer-term, triggered robust fast CBF increases similar to sensory-evoked responses (Fig. 5F) [2, 5]. In contrast, optogenetic stimulation of PV neurons induces delayed, slow CBF increases, aligning with the slow fluctuations observed in resting-state studies, as demonstrated in Fig. 5D [5]. Interestingly, in these Cre mouse models, the response amplitude did not significantly depend on the pulse width of the light stimulus. Thus, some neuronal subtypes have little or no influence over local blood supply while others exert a stronger influence, with optogenetics serving as a crucial tool to elucidate these properties.

In contrast to LDF, which is primarily sensitive to deep cortical layers, 2P offers distinct advantages. In combination with calcium indicators, 2P enables the examination of neuronal contributions at the cellular level, allowing for precise observation of individual neurons. Moreover, 2P enables the visualization and analysis of single blood vessels, offering insights into their specific responses to various stimuli. Perhaps most significantly, this method provides crucial depth information, allowing to investigation of the behavior of individual vessels across different cortical layers. This layer-specific analysis is particularly valuable in understanding the complex interplay between neuronal activity and vascular responses throughout the cortical depth.

We used 2P in a PV-Cre mouse to investigate how PV neuron activity influences arterial dynamics across cortical depth. We examined arterial responses to optogenetic stimulation of PV neurons in both superficial (50–200 μm) and deeper (250–400 μm) cortical layers (Fig. 6A). During the experiments, we simultaneously imaged pial and penetrating arteries while varying the duration and frequency of PV neuron activation. In superficial layers, brief optogenetic stimulation induced fast vasodilation, whereas prolonged stimulation at higher frequencies tended to induce vasoconstriction. In contrast, optogenetic stimulation of PV neurons in deeper layers did not result in fast vasodilation but rather produced delayed, slow vasodilation, particularly with longer stimulation periods. These findings suggest that the vascular response to PV neuron activation exhibits depth-dependent characteristics, aligning with observations of CBF changes measured using LDF and indicating that this effect extends into deeper cortical layers. These experiments were repeated over multiple depth locations in repeated experiments. Devices, such as electro-tunable lenses and acousto-optical deflectors, can be used to redirect the beam to different planes without moving the objective lens. This strategy would be useful when assessing 3D neurovascular changes in experiments without a specific perturbation.

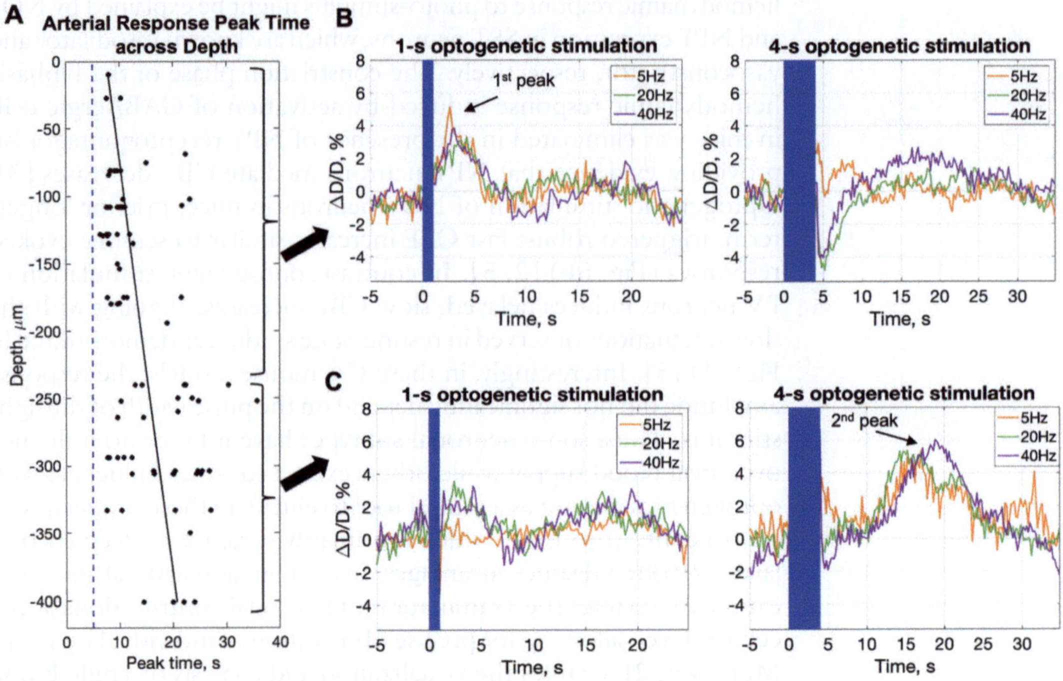

Fig. 6 Optogenetic stimulation of PV neurons induces depth-dependent vascular responses

3.5 Examining Metabolic Differences Between Neuronal Populations

Neurometabolism is thought to be a primary physiological mechanism underlying neurovascular coupling. That is, the blood supply's primary function is to sustain neuronal metabolism through the continuous delivery of oxygen and glucose. Many methods have been proposed to examine oxidative neurometabolism via optical and magnetic properties of oxygenated and deoxygenated hemoglobin. The primary appeal of this strategy stems from the relative noninvasiveness of these measurements. However, this method relies on assumptions that can be difficult to assess or validate, such as the baseline blood oxygen extraction, input oxygen saturation, and immediate consumption in tissue. Measurements sensitive to glucose (or lactate) are less accessible, although several promising methods are maturing and will likely be evaluated in the not-so-distant future. A complicating factor for glucose measurements is that it participates in both glycolysis (in the cytosol) and oxidative metabolism (in mitochondria). In this section, we will describe how optical methods adaptable to optogenetic experiments can be used to reliably calculate changes in the cerebral metabolic rate of oxygen metabolism, or CMRO2.

A model is typically used to study CMRO2 by MRI or intrinsic optical signals because changes in blood oxygenation are highly influenced by changes in CBF. This necessitates measurements of the impact of changes in CBF on oxygen measurements, such that the amount of oxygen removed by blood due to tissue metabolism can be calculated. Several strategies have been devised to generate changes in brain blood flow without altering oxygen metabolism. All of these rely on a transient increase in CBF that is similar to that measured with evoked neuronal activity through the supply of a known vascular agent like low-grade carbon dioxide inhalation (e.g., 5% CO_2 in air). Therefore, a CMRO2 measurement strategy will require that changes in CBF and blood oxygenation are generated so that changes in blood oxygenation in excess of those generated by CBF alone can be used to calculate CMRO2; Fig. 7 shows three different models used to test this principle, and as shown in panel C, CO_2 inhalation generates large increases in CBF but calculated CMRO2 remains unchanged. These data were obtained from awake mice with a nose cone placed by their nose, and the transient changes come from normal neurovascular responses taking place at the LDF measurement location. After examining the reliability of these methods, we used them to calculate CMRO2 changes evoked by optogenetic activation of excitatory and inhibitory neurons in the mouse cortex (Fig. 8).

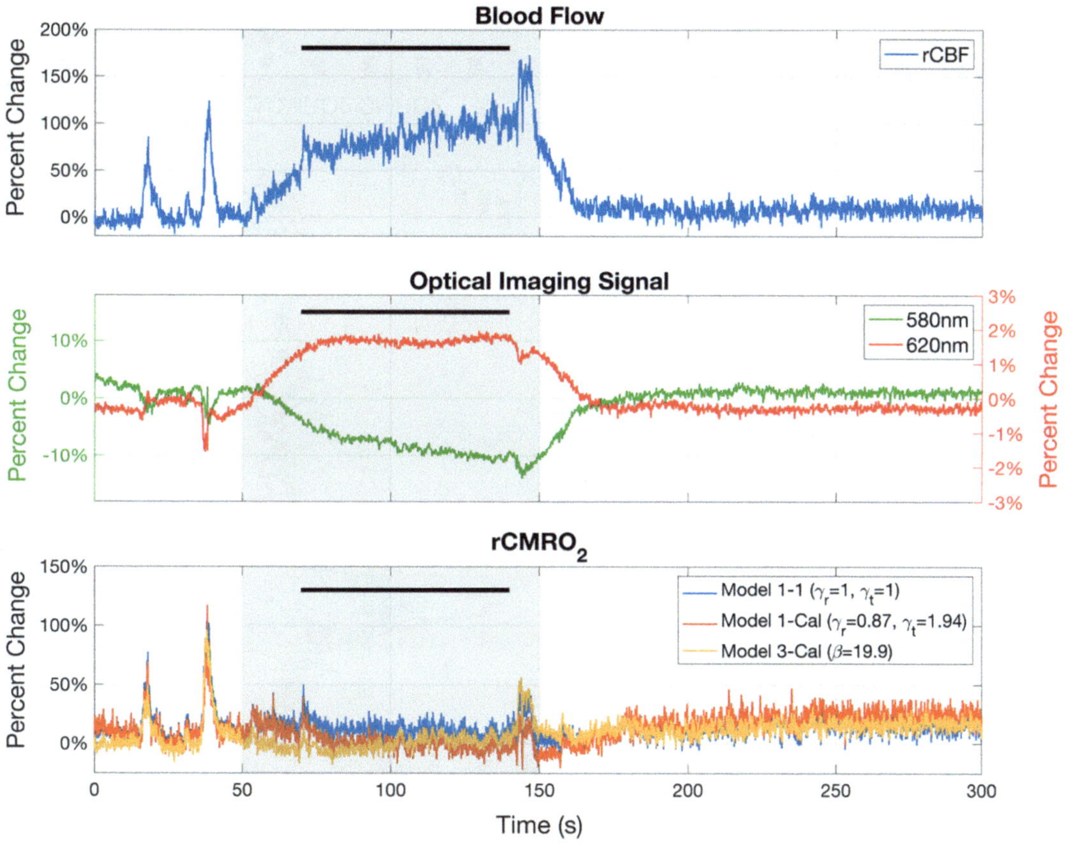

Fig. 7 CBF, WFOI at 580 nm, WFOI at 620 nm, and computed rCMRO2 in one representative animal after 8% CO_2 administration. rCMRO2 was computed using different factors for correcting errors in spatial variations in vascular content (γr, γt)

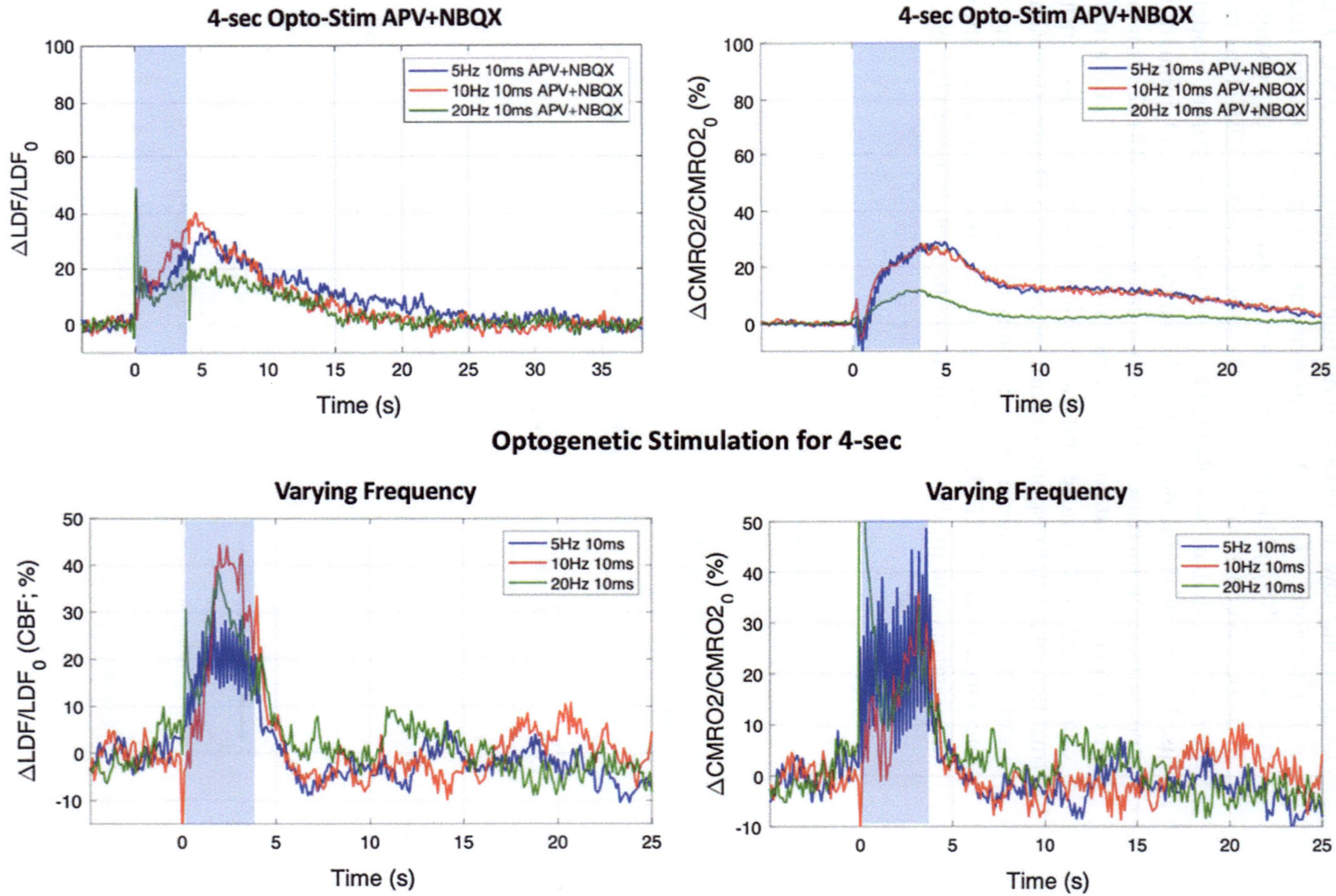

Fig. 8 Changes in cerebral blood flow (left panels) and rCMRO2 (right panels) at different frequencies isolated to excitatory neurons (Thy1-ChR2 mice) were possible using a broad post-synaptic neurotransmission blockade to restrict activity to excitatory neurons expressing ChR2 (top panels). Optogenetic activation of inhibitory neurons (bottom panels) shows similar changes to those from excitatory neurons, although these appear faster. No post-synaptic blockade was used for inhibitory neuron experiments since, in general, their activation does not activate connected neurons

4 Notes

4.1 Light Stimulation Artifacts

To accurately measure hemodynamic responses in optogenetic experiments, it is crucial to remove artifacts induced by light stimulation. One effective strategy involves using a high sampling rate for the hemodynamic signal, which is higher than the light stimulation rate, allowing for the identification of frames without light artifacts (or at least synchronizing the light stimulus to the imaging device start-of-frame trigger). Interspersed frames without artifacts can be retained, while those contaminated by the artifact can be easily discarded by the drastic change in light intensity. Keep in mind that optogenetic constructs are usually conjugated with a fluorescent probe to assess their expression, and that the optogenetic light stimulation light sources can also mildly activate these fluorescent probes. In general, the fluorescence emission is not sufficient to open channels but can be sufficient to be picked up by cameras even when barrier filters are used to block the optogenetic light excitation band [34]. A simple approach to remove light artifacts is to sample the optogenetic experiment outside the brain and use that as a template to remove light artifacts from the brain. This approach relies on either known or consistent and/or repeating sources of variance in the recordings, which can then be removed using techniques, such as principal component analysis (PCA) or linear regression. When using PCA, components with clear artifact-only contributions can be isolated and either zeroed out or removed using linear regression, allowing for recomposition of clean data [35]. Since vascular responses are relatively slow, hemodynamic activity occurring immediately after a short light stimulus (1–2 s) can provide a clean window for observing the hemodynamic changes without the interference of the light artifact.

4.2 Hemodynamic Responses in ChR2 Naive Animals

Hemodynamic responses in ChR2-naive animals (those not expressing optogenetic proteins) are necessary to verify the effects of light stimulation on CBF independent of optogenetic activation. Studies have shown that light stimulation alone can induce changes in CBF and vessel dilation in animals without opsin expression [36]. This phenomenon occurs at light intensities commonly used in optogenetic experiments and can be comparable in magnitude to sensory-evoked responses. Direct and continuous light delivery to brain tissue exceeding 2 mW through cranial windows for periods over 1 s can elicit vascular responses (Fig. 9). The wavelength of light can be an important factor in the hemodynamic response, as longer wavelengths may be more easily absorbed by water, which can lead to localized heating in the tissue. The effect appears to be mediated by a decrease in smooth muscle cell calcium or simply overheating, leading to arteriole dilation. Importantly, these light-induced hemodynamic changes can confound the interpretation of optogenetic

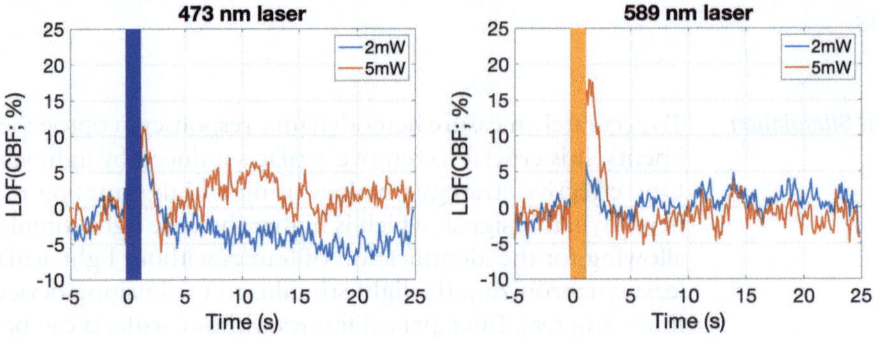

Fig. 9 Changes in CBF produced by delivery of 1-s 473 nm and 589nm laser pulses in awake naive mice

experiments, particularly those investigating neurovascular coupling. This underscores the importance of careful experimental design and appropriate controls in optogenetic studies, especially those relying on blood flow regulation or blood oxygen-level dependent (BOLD) signals, to distinguish between optogenetic activation and direct light effects on vasculature.

Acknowledgments

This work was supported in part by NIH grants R01-NS117515 (AV), R01-NS119404 (AV), and American Heart Association (AHA) predoctoral fellowship 23PRE1019289 (AR).

Conflict of Interest The author has no conflicts of interest to declare that are relevant to the content of this chapter.

References

1. Lacroix A, Toussay X, Anenberg E et al (2015) COX-2-derived prostaglandin E2 produced by pyramidal neurons contributes to neurovascular coupling in the rodent cerebral cortex. J Neurosci 35:11791–11810
2. Lee L, Boorman L, Glendenning E et al (2020) Key aspects of neurovascular control mediated by specific populations of inhibitory cortical interneurons. Cereb Cortex 30:2452–2464
3. Uhlirova H, Kılıç K, Tian P et al (2016) Cell type specificity of neurovascular coupling in cerebral cortex. eLife 5:e14315
4. Perrenoud Q, Rossier J, Férézou I et al (2012) Activation of cortical 5-HT(3) receptor-expressing interneurons induces NO mediated vasodilatations and NPY mediated vasoconstrictions. Front Neural Circuits 6:50
5. Krawchuk MB, Ruff CF, Yang X et al (2020) Optogenetic assessment of VIP, PV, SOM and NOS inhibitory neuron activity and cerebral blood flow regulation in mouse somato-sensory cortex. J Cereb Blood Flow Metab 40:1427–1440
6. Dahlqvist MK, Thomsen KJ, Postnov DD, Lauritzen MJ (2020) Modification of oxygen consumption and blood flow in mouse somatosensory cortex by cell-type-specific neuronal activity. J Cereb Blood Flow Metab 40:2010–2025
7. Vo TT, Im GH, Han K et al (2023) Parvalbumin interneuron activity drives fast inhibition-induced vasoconstriction followed by slow substance P-mediated vasodilation. Proc Natl Acad Sci USA 120:e2220777120
8. Lee J, Stile CL, Bice AR et al (2021) Opposed hemodynamic responses following increased

excitation and parvalbumin-based inhibition. J Cereb Blood Flow Metab 41:841–856
9. Cauli B, Tong X-K, Rancillac A et al (2004) Cortical GABA interneurons in neurovascular coupling: relays for subcortical vasoactive pathways. J Neurosci 24:8940–8949
10. Packer AM, Roska B, Häusser M (2013) Targeting neurons and photons for optogenetics. Nat Neurosci 16:805–815
11. Parr-Brownlie LC, Bosch-Bouju C, Schoderboeck L et al (2015) Lentiviral vectors as tools to understand central nervous system biology in mammalian model organisms. Front Mol Neurosci 8:14
12. Han X (2012) Optogenetics in the nonhuman primate. Prog Brain Res 196:215–233.
13. Haggerty DL, Grecco GG, Reeves KC, Atwood B (2020) Adeno-associated viral vectors in neuroscience research. Mol Ther Methods Clin Dev 17:69–82
14. Yizhar O, Adamantidis A (2018) Cell type-specific targeting strategies for optogenetics. In: Stroh A (Ed) Optogenetics: a roadmap. Springer New York, New York, NY, pp 25–42
15. Boyden ES, Zhang F, Bamberg E et al (2005) Millisecond-timescale, genetically targeted optical control of neural activity. Nat Neurosci 8:1263–1268
16. Yizhar O, Fenno LE, Davidson TJ et al (2011) Optogenetics in neural systems. Neuron 71:9–34
17. Oda K, Vierock J, Oishi S et al (2018) Crystal structure of the red light-activated channelrhodopsin Chrimson. Nat Commun 9:3949
18. Gunaydin LA, Yizhar O, Berndt A et al (2010) Ultrafast optogenetic control. Nat Neurosci 13:387–392
19. Mager T, Lopez de la Morena D, Senn V et al (2018) High frequency neural spiking and auditory signaling by ultrafast red-shifted optogenetics. Nat Commun 9:1750
20. Chaigneau E, Ronzitti E, Gajowa MA et al (2016) Two-photon holographic stimulation of ReaChR. Front Cell Neurosci 10:234
21. Klapoetke NC, Murata Y, Kim SS et al (2014) Independent optical excitation of distinct neural populations. Nat Methods 11:338–346
22. Mardinly AR, Oldenburg IA, Pégard NC et al (2018) Precise multimodal optical control of neural ensemble activity. Nat Neurosci 21:881–893
23. Mattis J, Tye KM, Ferenczi EA et al (2011) Principles for applying optogenetic tools derived from direct comparative analysis of microbial opsins. Nat Methods 9:159–172
24. Vazquez AL, Fukuda M, Crowley JC, Kim S-G (2014) Neural and hemodynamic responses elicited by forelimb- and photo-stimulation in channelrhodopsin-2 mice: insights into the hemodynamic point spread function. Cereb Cortex 24:2908–2919
25. Vazquez AL, Fukuda M, Kim S-G (2018) Inhibitory neuron activity contributions to hemodynamic responses and metabolic load examined using an inhibitory optogenetic mouse model. Cereb Cortex 28:4105–4119
26. Taniguchi H, Lu J, Huang ZJ (2013) The spatial and temporal origin of chandelier cells in mouse neocortex. Science 339:70–74
27. Cauli B, Hamel E (2010) Revisiting the role of neurons in neurovascular coupling. Front Neuroenerg 2:9
28. Rudy B, Fishell G, Lee S, Hjerling-Leffler J (2011) Three groups of interneurons account for nearly 100% of neocortical GABAergic neurons. Dev Neurobiol 71:45–61
29. Lacroix A, Toussay X, Anenberg E et al (2015) COX-2-derived prostaglandin E2 produced by pyramidal neurons contributes to neurovascular coupling in the rodent cerebral cortex. J Neurosci 35:11791–11810
30. Wu Y-T, Bennett HC, Chon U et al (2022) Quantitative relationship between cerebrovascular network and neuronal cell types in mice. Cell Rep 39:110978
31. Fahrenkrug J, Hannibal J, Tams J, Georg B (2000) Immunohistochemical localization of the VIP1 receptor (VPAC1R) in rat cerebral blood vessels: relation to PACAP and VIP containing nerves. J Cereb Blood Flow Metab 20:1205–1214
32. Perrenoud Q, Rossier J, Férézou I et al (2012) Activation of cortical 5-HT(3) receptor-expressing interneurons induces NO mediated vasodilatations and NPY mediated vasoconstrictions. Front Neural Circuits 6:50
33. Yavorska I, Wehr M (2016) Somatostatin-expressing inhibitory interneurons in cortical circuits. Front Neural Circuits 10:76
34. Boyden ES, Zhang F, Bamberg E et al (2005) Millisecond-timescale, genetically targeted optical control of neural activity. Nat Neurosci 8:1263–1268
35. Kozai TDY, Vazquez AL (2015) Photoelectric artefact from optogenetics and imaging on microelectrodes and bioelectronics: new challenges and opportunities. J Mater Chem B Mater Biol Med 3:4965–4978
36. Rungta RL, Osmanski B-F, Boido D et al (2017) Light controls cerebral blood flow in naive animals. Nat Commun 8:14191

Chapter 10

Genetic Tools Meet Functional MRS: Probing Brain Function and Metabolism in Animal Models

Nathalie Just

Abstract

Advances in genetic engineering and metabolic imaging are transforming our ability to probe brain function with unprecedented precision. This chapter explores how the integration of optogenetics, chemogenetics, and genetically encoded biosensors with functional magnetic resonance spectroscopy (fMRS) enables dynamic, cell-type-specific interrogation of brain metabolism in animal models. Building on foundational concepts, such as the astrocyte–neuron lactate shuttle, the malate–aspartate shuttle, and the glutamate–glutamine–GABA cycle, we highlight the limitations of classical frameworks in explaining fast, stimulus-locked metabolic changes captured by modern fMRS techniques. We discuss emerging analysis methods inspired by functional MRI—including glutamate response functions and general linear modeling—that are redefining how fMRS data are interpreted. Finally, we argue that the synergy between genetic control, real-time metabolic monitoring, and computational modeling is ushering in a new era of mechanistic, quantitative neuroenergetics, with profound implications for understanding both normal brain function and metabolic pathology.

Keywords Functional magnetic resonance spectroscopy (fMRS), Optogenetics, Chemogenetics, Neuroenergetics, Brain metabolism models

1 Introduction

Understanding brain function requires an integrative approach that captures both the dynamics of neural circuitry and the underlying neurochemical processes. Over the past two decades, the development of genetically encoded tools, such as optogenetics and chemogenetics, has revolutionized systems neuroscience by allowing precise manipulation of neuronal *activity* in vivo. At the same time, functional magnetic resonance spectroscopy (fMRS) has emerged as a noninvasive method for tracking real-time changes in brain metabolite levels, offering a window into the biochemical underpinnings of neural activity [1, 2]. While functional MRI (fMRI) has traditionally been used to assess circuit-level changes in blood oxygenation (BOLD signal), fMRS provides a complementary approach

Kevin Chuen Wing Chan and Seong-Gi Kim (eds.), *Optogenetic Functional Neuroimaging*, Neuromethods, vol. 232,
https://doi.org/10.1007/978-1-0716-5178-0_10,

by directly measuring concentration changes of metabolites such as glutamate, GABA, lactate, and N-acetylaspartate (NAA). In animal models, where invasive interventions and precise genetic targeting are feasible, the combination of fMRS with genetic tools enables the causal link between specific circuit manipulations and neurochemical outcomes.

This chapter explores the integration of fMRS with optogenetics, chemogenetics, and other biosensors in animal models to investigate brain function. We outline the principles of each technique, discuss experimental strategies for combining them, highlight key applications, and consider both the technical challenges and future directions of this promising approach.

Combining functional magnetic resonance spectroscopy (fMRS) with genetic tools like optogenetics or chemogenetics offers several compelling advantages that allow for a deeper understanding of brain function, neural circuits, and the mechanisms underlying behavior. Optogenetics uses light to precisely control genetically modified neurons that express light-sensitive ion channels, allowing the activation or inhibition of specific neural populations with millisecond precision. Ideally, by combining fMRS with optogenetics, the real-time metabolic changes can be correlated with the immediate effects of specific neuronal activity. Similar to optogenetics, chemogenetics uses engineered receptors (such as DREADDs—designer receptors exclusively activated by designer drugs) to modulate neural activity. The advantage here is that chemogenetics can work on a longer time scale than optogenetics, with drugs activating or deactivating neurons more slowly. This allows for a more gradual and sustained modulation of neural circuits. fMRS can provide a continuous readout of metabolic shifts in response to this prolonged manipulation.

fMRS measures the concentration of neurochemical fluctuations over time, which reflects metabolic activity in the brain [3–6]. By coupling fMRS with opto- or chemogenetic manipulation, the selective tracking of how specific neuronal populations contribute to the brain's metabolic dynamics can be performed. This combination allows for the mapping of metabolic pathways that are activated or inhibited during behavioral tasks, cognitive processes, or sensory inputs. Genetic tools like optogenetics and chemogenetics enable the manipulation of specific neural circuits, providing insight into the function of distinct brain regions and their role in behavior or cognition. fMRS can help measure the downstream metabolic consequences of these manipulations. By combining these methods, it is expected to map out how altering the activity of a specific brain region or network affects the broader brain's metabolic profile and how those changes are associated with behavior or physiological states.

Unlike traditional electrophysiology or invasive techniques that require direct electrical recordings, fMRS is a noninvasive method that provides real-time metabolic data in live subjects (in both animals and humans). This makes it an ideal companion to genetic tools for studying brain function without the need for invasive probes or electrodes. The ability to monitor metabolic changes without disrupting the brain's structure and function allows for more naturalistic studies, where the brain's activity can be modified precisely, while its metabolic response is continuously observed. The combination of fMRS and genetic tools can provide valuable insights into neuroplasticity—how the brain's neural circuits change in response to experience, injury, or disease. This can be particularly useful in understanding recovery mechanisms or therapeutic interventions for neurological disorders. For example, in animal models of neurodegenerative diseases (like Alzheimer's or Parkinson's), one can activate or inhibit specific pathways using opto- or chemogenetics and observe how these manipulations affect the brain's metabolic response, which could reveal new targets for treatment or highlight the brain's compensatory mechanisms in the face of pathology [7, 8]. fMRS alone can show correlations between brain metabolism and behavior [9–11], but it doesn't allow making causal inferences about the relationship between specific neural circuits and the observed metabolic changes [12]. By integrating optogenetics or chemogenetics, researchers can causally manipulate neural circuits and directly observe the resulting metabolic effects. This is crucial for understanding which specific neuronal cell populations or pathways are responsible for certain brain functions and behaviors. The rationale for combining fMRS with genetic tools lies in the ability to generate a precise, controlled, and mechanistic understanding of brain function. This approach allows us to link specific neuronal or glial activity with metabolic changes, investigate neural circuit dynamics, and explore the underlying mechanisms of behavior and disease. It provides a powerful combination for both basic research and the development of novel therapeutic strategies.

2 Materials

2.1 About Genetic Tools

Genetic tools have profoundly advanced our understanding of brain function by enabling the precise exploration of how genes influence the structure, development, and activity of the nervous system. They are critical in manipulating brain function, and in exploring the genetic basis of neural development, behavior, and cognition. Across the past 20 years, significant developments have been made that span a large spectrum of techniques and processes that are currently defining the picture of neuroscience research and promising great advances and enhanced approaches for understanding both the healthy and diseased brain. Among those, we can cite gene editing with CRISPR-Cas9, single-cell RNA sequencing (scRNA-seq), or epigenetic tools allowing to directly target specific genes and their expressions or activation [15, 16]. These key genetic tools have various applications in brain function, such as deactivating genes, correcting gene mutations, modulating gene expression to either study their role in brain development, plasticity, or disease and can serve to characterize brain circuitry, the diversity of neurons and glial cells, how new cell types emerge and mature or to identify molecular changes in brain disorders. Of course, genetic engineering led to the development of transgenic animal models (with modified DNA), which are essential for studying gene function in vivo. By creating mice, rats, and higher-order animal models that express, overexpress, or lack certain genes, researchers can observe how these changes affect brain function and behavior. Transgenic (or knockout) animal models are widely used to study the genetic underpinnings of neurological disorders, such as autism spectrum disorders (ASDs), schizophrenia, and many neurodegenerative diseases (e.g., Alzheimer's, Parkinson's, ALS). With genetic engineering, techniques like fluorescent tagging (for example, green fluorescent protein (GFP) labeling) allow the visualization of gene expression in live cells, which can be used to map brain activity linking it to gene expression, track the expression of critical developmental genes, or investigate disease pathways.

With optogenetics, chemogenetics, or novel biosensors, neuronal and/or glial activities can be controlled using light or pharmacological compounds allowing neural circuit mapping and understanding dysregulated brain activity. These techniques use engineered viruses to deliver genetic material to specific cell populations within the brain. In 2005, Deisseroth's laboratory (along with Ed Boyden and Feng Zhang, who were postdocs at the time) published the seminal paper: "Millisecond-timescale, genetically targeted optical control of neural activity" [17]. This work fundamentally changed the landscape of neuroscience by enabling the causal study of brain circuits in real time with millisecond precision. They demonstrated that Channelrhodopsin-2 (ChR2), a light-sensitive ion channel from algae, could be genetically expressed in mammalian neurons, and that blue light could trigger action potentials. This was the first time light was used to control neuronal firing directly and precisely in intact brain tissue. Deisseroth's group is widely recognized as a pioneer of optogenetics, a revolutionary technique that allows scientists to control specific neurons with light. Deisseroth's group continued to engineer and characterize new opsins to fine-tune optogenetic control (see Table 1). Among these, we can cite Halorhodopsin (NpHR): a yellow light-sensitive chloride pump used for neuronal silencing [18], archaerhodopsin (Arch): another inhibitory opsin offering hyperpolarization via proton pumping; deeper silencing and less phototoxicity [19] and many new opsin variants (ChETA, ChR2 variants) engineered for faster kinetics or different light sensitivities (e.g., red-shifted opsins like C1V1, ReaChR) and step-function opsins (SFOs) that are mutated to remain open longer, allowing long-term activation without continuous light [20].

Table 1
Opsins for optogenetics

Opsin	Function	Light wavelength	Origin	Notable features
ChR2	Excitation (Na^+ influx)	Blue (~470 nm)	Green algae (Chlamydomonas)	First opsin used for neuronal firing (2005)
NpHR	Inhibition (Cl^- influx)	Yellow (~590 nm)	Halobacterium	Silences neurons via chloride pumping
Arch	Inhibition (H^+ pump)	Green-yellow (~530 nm)	Archaea	Strong, sustained silencing with minimal side effects
ChETA	Fast excitation	Blue	Engineered ChR2	High-speed spike fidelity
C1V1/ReaChR	Red-shifted excitation	Orange-red (~590–630 nm)	Engineered variants	Enables deeper brain stimulation with less scattering
SFOs	Prolonged excitation	Blue	ChR2 mutants	Maintains activity with brief light pulse

In 2007, Bryan Roth's group (University of North Carolina, Chapel Hill), on the other hand, sought to create tools that would allow remote, reversible, and cell-type-specific control of neuronal activity—without the need for invasive optics or light delivery. This led to the development of designer receptors exclusively activated by Designer Drugs (DREADDs), modified G protein-coupled receptors (GPCRs) that no longer respond to their natural ligands but can be selectively activated by synthetic, pharmacologically inert compounds. Roth's lab engineered GPCRs to be exclusively activated by clozapine-N-oxide (CNO), a drug that has minimal off-target effects at low doses. DREADDs were introduced in: "Evolving the lock to fit the key to create a family of GPCRs potently activated by an inert ligand." [21]. DREADDs (see Table 2) are expressed in specific cells via viral vectors, often under Cre control. The systemic administration of the designer drug (e.g., CNO or deschloroclozapine (DCZ) for improved brain penetration) activates only the cells with DREADDs—leaving other systems untouched. This system is noninvasive, long-lasting, and highly tunable—ideal for behavioral studies in animals.

Table 2
DREADD types and pathways

Receptor	Coupling	Effect	Applications
hM3Dq	Gq	Neuronal excitation	Activating specific circuits
hM4Di	Gi	Neuronal silencing	Inhibiting or modulating circuits
rM3Ds	Gs	cAMP increase	Modulation of intracellular signaling
KORD	Gi (activated by Salvinorin B)	Silencing	Alternative to CNO-based DREADDs

Table 3
Features of GCaMP and FRET biosensors

Feature	GCaMP	FRET-based biosensors
Signal type	Fluorescence increase	Ratiometric fluorescence
Measures	Ca^{2+} levels	Various signals (Ca^{2+}, cAMP, pH…)
Temporal resolution	Good (ms-s range)	Very good (ms range)
Spatial resolution	High	High
Complexity	Easy to use	More complex setup
Multiplexing	Limited (color overlap)	Better (with proper fluorophores)

Other biosensors, such as GCaMP- and FRET-based biosensors, have revolutionized neuroscience by enabling real-time visualization of brain function in living organisms.

In vivo calcium imaging using GCaMP (Table 3) has allowed researchers to monitor activity from individual neurons or entire networks while animals are awake and behaving. For example, head-fixed mice running on a treadmill while scientists track visual cortex neurons responding to visual stimuli have been extensively used [22, 23]. GCaMP also helps define which neurons are active during specific tasks or sensory inputs (e.g., mapping decision-making circuits in the prefrontal cortex during memory or attention tasks). GCaMP allows for chronic neuroimaging and can be used for investigating learning and plasticity. Finally, it enables correlation of neural activity with specific actions, choices, or emotional states (e.g., *Studying fear responses or reward anticipation in the amygdala and striatum*).

Förster resonance energy transfer (FRET biosensors, [24]) FRET biosensors (Table 3) rely on the energy transfer between two fluorophores to detect molecular interactions or conformational changes inside live cells. Two fluorescent proteins (a donor and an acceptor) are linked via a protein or domain that changes conformation in response to a signal (e.g., calcium, cAMP, kinase activity). When close together (typically <10 nm), energy from the excited donor is transferred to the acceptor, which emits fluorescence at a different wavelength. The FRET ratio changes depending on the activity/state of the molecule of interest. Across the past decade, FRET biosensors have been developed for real-time tracking of calcium transients, cAMP, PKA activity, GTPase signaling, pH, and redox states, and more recently various metabolites, such as glucose, lactate, glutamate, GABA, or acetylcholine. FRET nanosensors can detect subtle and fast biochemical changes and are efficient in exploring the dynamics of signaling pathways. However, calibrations are needed as well as specialized imaging equipment to separate the emissions, and the signal-to-noise ratio may be reduced compared to single fluorophores.

2.2 Using Animal Models with Genetic Tools

Animal models, particularly in neuroscience, are invaluable for studying neural circuitry and brain metabolism. These models, often involving rodents or primates, allow the dissection of complex brain functions and pathologies that would be impossible to investigate in humans directly due to ethical and technical limitations. Animal models allow the manipulation of specific neural circuits or regions using techniques like optogenetics, chemogenetics, or electrophysiological recordings. These methods allow for high precision in activating, inhibiting, or monitoring the activity of individual neurons or neural populations, giving insights into how these circuits contribute to behavior and brain function. In animal models, scientists can selectively target neural pathways to observe their role in specific behaviors or cognitive functions. For example, using optogenetics, researchers can activate or inhibit neural circuits in the prefrontal cortex and observe how changes in this circuit influence behavior, mood, and cognition [13, 14]. In addition, by combining neurochemical analysis with direct manipulation of neural circuits, a better understanding of how neural activity leads to metabolic changes can be reached. Animal models allow for the creation of models that replicate neurological disorders in a controlled environment. These models enable researchers to explore how specific neural circuits or metabolic pathways malfunction in diseases like Parkinson's disease, Alzheimer's disease, epilepsy, and schizophrenia. They provide the ability to study disease progression over time, something that's difficult to do in humans due to ethical constraints. This includes investigating how neural circuitry and brain metabolism change as the disease advances, which is crucial for identifying potential therapeutic targets. With transgenic animals, the effects of specific genes on brain function can be studied. By selectively modifying genes related to neural development, neurotransmission, or metabolism, insights into how these genetic factors shape neural circuits and metabolism can be gained. Animal models allow for the detailed study of behavioral phenotypes (e.g., memory, anxiety, motor skills) and the corresponding metabolic changes in the brain. By altering specific genes or circuits, researchers can correlate these changes with measurable outcomes at the behavioral and metabolic level. Animal models are ideal for investigating neural plasticity. For instance, in rodents, after a stroke or brain injury, how the brain's metabolic response adapts over time and which neural circuits or areas show compensatory activity can now be observed. This helps in understanding recovery mechanisms that could inform rehabilitation strategies in humans. By following animals over time, tracking of the metabolic profile of the brain changes in response to environmental factors or learning can be investigated. This is essential for understanding neuroplasticity and how neural circuits and metabolism adjust based on experience, which has implications for learning, memory, and even recovery from neurological diseases.

3 Methods

3.1 Functional MRS (fMRS): Principles and Applications

What is fMRS, and How Does it Differ from Traditional MRS?

Magnetic resonance spectroscopy (MRS) measures the chemical composition or the concentration of certain metabolites in tissues. With conventional proton MRS (^{1}H-MRS), the most common observed metabolites in the rodent brain are N-acetyl aspartate (NAA), a marker of neuronal health, glutamate (Glu) and γ-aminobutyric acid (GABA), the main excitatory and inhibitory neurotransmitters within the brain as well as creatine (Cr) and phosphocreatine (PCr) or glucose (Glc) and lactate (lac) to measure energy metabolism. In rodents, up to 22 metabolite concentrations have been reliably measured at high field strength. An exhaustive review of rodent MRS can be found in [25]. The power of MRS techniques has been abundantly demonstrated over the past two decades: MRS allows researchers to track brain chemistry over time in the *same animal*, enabling within-subject designs and reducing variability. The metabolites detected (e.g., Glu, GABA, NAA, Cr) are the same as in humans, which makes rodent MRS highly translational, especially for modeling human diseases like depression, schizophrenia, Alzheimer's disease, epilepsy, and many others. MRS is ideal for detecting subtle biochemical changes before structural changes are evident [26], allowing neurochemical fingerprinting that gives a unique profile of metabolic status in specific regions of the rodent brain, such as the cortex, the striatum, the thalamus, and the hippocampus [27]. In short, rodent MRS is a powerhouse for neuroscience—enabling precise, repeatable, and meaningful insights into brain metabolism from real-time tracking of neurochemical effects of pharmacological agents, from studies on the metabolism of transgenic rodent models, or from exploration of how brain chemistry evolves across the lifespan.

Functional MRS (fMRS) techniques appeared almost at the same time as blood oxygen level dependent functional MRI (BOLD fMRI) at the beginning of the 90s, with groundbreaking studies in the human visual cortex from [28], revealing for the first time significant task-related changes in PCr and Lactate. It took, however, another 15 years to further demonstrate the validity and reliability of this technique using higher spectral resolution at 7T [29, 30]. The term "fMRS" is nowadays accepted but remains debated among the MR community, with many researchers preferring to talk about combined fMRI–MRS studies or simply dynamic MRS. Nevertheless, what makes fMRS unique compared to traditional MRS is the use of dedicated paradigms of stimulation, allowing a better understanding of neurometabolic changes occurring in the healthy or diseased brain [31, 32]. While traditional MRS provides a snapshot of the brain's chemical composition, fMRS adds the temporal component by tracking dynamic shifts in neurochemical levels. This technique has emerged as an essential tool in neuroscience, particularly for understanding the biochemical underpinnings of brain function, behavior, and disease in animal models.

3.2 Technical Challenges and Considerations in Animals

fMRS acquisitions are gradually becoming complementary to fMRI acquisitions, providing crucial insights into various functions of the normal and altered human brain. The number of fMRS studies in the human brain has significantly increased over the past decade, demonstrating the power of this technique for investigating the complexities of brain activity [1, 3, 9, 33–36].

fMRS applications in rodents have been more challenging essentially due to the remaining use of anesthesia which induces baseline changes, and the lack of signal-to-noise ratio (SNR) in small single voxel studies [2]. High-field MR scanners (7T or greater) are commonly used to obtain sufficient SNR and spatial resolution, which are crucial for detecting subtle changes in metabolite levels. The metabolites detected (e.g., Glu, GABA, NAA, Cr) are the same as in humans, making it of high translational relevance. However, fMRS is typically performed on the rodent brain, where the relatively small size of the brain presents both challenges and opportunities for high-resolution imaging (see Table 4). Nevertheless, rodent fMRS brings unique opportunities to interpret the underlying events linking neuronal activity to neuroenergetics. Compared to human fMRS, preclinical fMRS benefits from high flexibility with possibilities for repeating experiments under various paradigms, therefore, challenging the reproducibility of experiments. In addition, appropriate controls or sham animals can be used, enabling precise experimental

Table 4
Challenges in animal fMRS

Challenge	Notes
Small brain size	Requires high spatial resolution and SNR
Motion artifacts	Especially in awake animals
Anesthesia effects	Anesthesia alters baseline metabolism
Low concentration changes	Metabolite changes are often small (~1%–5%)
Data analysis	Complex fitting (e.g., LCModel) + time series modeling

control. Finally, the advent of genetic tools has literally revolutionized neuroscience, enabling the selective and specific targeting of cell populations. Optogenetics, chemogenetics, and other biosensors can be used in combination with fMRS techniques to deliver controlled stimuli or interventions that enable causal studies and emphasize the necessity of performing fMRS in rodent models. These possibilities bring higher orders of interpretation on neurovascular or neurometabolic coupling mechanisms that currently cannot be reached in the human brain but can be efficiently transferred to humans.

3.3 fMRS Paradigms in Rodents

In rodents, task-based fMRS was employed to study neurotransmitter dynamics during controlled activities. These activities may include sensory stimuli, such as electrical whisker stimulation [5], electrical forepaw stimulation [37], visual cortex [38, 39], or drug exposure [40]. The main goal is to observe how some regions of the brain's metabolic profile change in response to functional demands.

- **Sensory stimulation:** for instance, electrical whisker pad stimulation was used to evoke glutamate and lactate release in the primary somatosensory barrel field cortex (S1BF), allowing the tracking of neurochemical responses during sensory processing in the rat brain [5]. This first experiment replicated the seminal results found in the human visual cortex at 7T [29, 30], demonstrating the feasibility of conducting ^{1}H-fMRS in the rodent brain at 9.4T. Furthermore, this first experiment confirmed findings in the human visual cortex: the positive BOLD responses in S1BF accompanied by increases in Glu and Lac concentrations, suggested increased oxidative and glycolytic metabolism in agreement with findings in the human brain [4, 12]. These findings were further confirmed in mice [38] upon visual cortex stimulation with an enhanced temporal resolution. These experiments allowed for the first time the definition of a glutamate response function (GRf) paralleling the use of the hemodynamic response function (HRf) in BOLD fMRI studies. The GRf is emerging as a promising concept enabling similar processing steps as those used for fMRI processing [31, 38].

- **Pharmacological interventions**: administration of drugs (e.g., ketamine, amphetamines, GABA agonists, or antagonists) or neuromodulatory agents (e.g., serotonergic compounds) during fMRS scanning should allow the tracking of real-time changes in neurotransmitter levels, facilitating the investigation of drug effects on brain chemistry. This intervention was experimented with for the first time by Seuwen et al. [40] in epilepsy mouse models upon bicuculline injection. Authors also demonstrated the feasibility of functional spectroscopic imaging in rodents and notably the importance of getting metabolic responses independent of hemodynamic contamination. Notably, metabolite concentrations may be better indicators of neural activity than BOLD or cerebral blood volume (CBV) in cases of neurovascular coupling failure.
- **Optogenetics and chemogenetics**: ^{1}H-fMRS was successfully combined with optogenetic techniques, called ofMRS, where specific neurons are activated or inhibited using light [37]. This experiment demonstrated the feasibility of combining optogenetics and ^{1}H-fMRS for the functional investigation of the forelimb somatosensory cortex (S1FL), revealing similar Glu concentration increases to electrical stimulation of S1FL but potentially differing neurovascular coupling mechanisms. On the other end, chemogenetics and ^{1}H-fMRS were also combined, demonstrating increases in NAA concentrations in total contrast to previously acquired ^{1}H-fMRS data showing no changes in NAA concentrations [41–43]. While these controversies need to be discussed and understood (see below), the combination of genetic tools with fMRS is expected to provide a precise causal link between neurotransmitter changes and specific neuronal circuits.

All these fMRS applications in animal models were performed under anesthesia, which remains one of the biggest challenges of the technique since anesthetic effects are expected to significantly influence the quantification of neurochemical profiles [2, 44]. In addition, most fMRS studies in rodents were performed with prolonged paradigms of stimulation leading to habituation issues [32]. More recently, event-related paradigms were tested in conjunction with ^{1}H-fMRS [31], leading to a larger effect size on metabolite concentrations, increased temporal resolution, and decreased habituation effects. An earlier trial of an event-related paradigm was performed during ^{1}H-fMRS and ^{13}C-fMRS of the visual cortex of tree-shrews [39], demonstrating no significant changes in Glu and Lac concentrations but significant changes in glucose and PCr to Cr concentrations while demonstrating strong coupling mechanisms between neurotransmission and TCA-cycle in both neurons and astrocytes under low isoflurane concentrations (0.7%).

3.4 Optogenetics, Chemogenetics, and Other Biosensors: Precision Control of Neural Activity Supporting fMRS

To date, the number of fMRS studies in rodents remains marginal. Again, this can be ascribed to the use of anesthesia, poor SNR, and temporal resolution, although the complexity of the design of protocols involving genetic tools might also be difficult. Nevertheless, the past decade has seen an increasing number of studies using optogenetics and chemogenetics to probe the underlying neurochemical properties of neural activities, paving the way to more efficient and rewarding future fMRS studies, which remain the only means of measuring and quantifying metabolic concentrations in vivo and noninvasively. The combination of optogenetics/chemogenetics and neuroenergetics is one of the most exciting intersections in systems neuroscience and metabolism. It allows researchers to *causally manipulate* specific neural circuits or glial cells and observe the real-time energetic consequences of that activity, especially using tools like fMRS in rodents. Unfortunately, to date, studies using fMRS are rare. For example, optogenetics gives precise control over the activity of genetically defined cell types using light. Neuroenergetics studies how the brain generates, uses, and recycles energy—primarily in the form of ATP, glucose, lactate, and oxygen metabolism [45–48].

By combining them, one can ask: "What is the metabolic cost of activating this neuron type or brain region?" "How do astrocytes respond metabolically to neuronal demand?" "How are energy pathways like the ANLS or MAS dynamically regulated by activity?" Bringing optogenetics, chemogenetics, and neuroenergetics together opens up powerful ways to control, measure, and understand how neural circuits consume and manage energy. Because metabolism is not just a byproduct of activity, it's a key player in *how* neural activity is sustained, regulated, and adapted, several key questions need to be answered:

- How much energy do excitatory vs. inhibitory neurons require?
- Do astrocytes provide lactate *only* in response to activity—or proactively?
- What happens to mitochondrial metabolism when certain circuits are upregulated?

Let us explore with examples of groundbreaking studies investigating the metabolism of neural activity with precision, how genetic tools can be integrated to investigate brain energy metabolism at cell-type resolution, and how they illuminate pathways like ANLS, MAS, and neurovascular coupling.

Nikos Logothetis's research demonstrated that local field potentials (LFPs), which represent the summed synaptic activity, correlate more strongly with the BOLD signal measured by fMRI than do the spiking activities of individual neurons [49]. This finding has been pivotal in interpreting fMRI data in terms of underlying neural processes. Exploring how neurons encode information, he has shown that the timing of neuronal spikes relative to the phase of ongoing network oscillations carries significant information about sensory stimuli. This phase-of-firing coding suggests a temporal dimension to how the brain processes complex inputs. Phase-of-firing coding is not energetically neutral: it may be more metabolically demanding than simple rate coding, because precise spike timing requires high-fidelity synaptic integration, which demands tight ion homeostasis (Na^+, K^+, Ca^{2+}). In addition, fast oscillations (like gamma) require synchronized interneuron networks, which are energy-intensive (GABAergic fast-spiking interneurons are metabolically expensive), and temporal coding efficiency might reduce the number of spikes, but increase the precision and background synaptic activity needed to maintain timing accuracy. Using optogenetics, which gives you precise spatial and temporal control of specific cell populations, Newman et al. [50] introduced the "optoclamp," a closed-loop optogenetic system that maintains neuronal firing rates despite disruptions in glutamatergic or GABAergic signaling. They demonstrated the ability to decouple firing activity from neurotransmitter fluctuations, highlighting the precision of optogenetic control over neural circuits.

Neuroenergetics and neurovascular coupling (NVC) are tightly interconnected processes that ensure the brain's metabolic demands are met through dynamic regulation of blood flow. Neural activity leads to energy consumption, particularly glucose and oxygen, and the brain compensates with localized increases in cerebral blood flow—a phenomenon central to functional imaging methods like BOLD fMRI. Neuroenergetics refers to how neurons and glial cells manage and consume energy during activity, involving key pathways, such as glycolysis, the ANLS, and mitochondrial oxidative phosphorylation. NVC, on the other hand, is the process by which changes in neural activity lead to adjustments in cerebral blood flow, mediated by signaling molecules like nitric oxide, prostaglandins, and lactate. Optogenetics and chemogenetics allow for precise, cell-type-specific control over neural and glial activity. These tools have revealed that NVC is not only solely dependent on neuronal spiking but also involves astrocytic signaling and metabolic intermediates. For instance, astrocytic activation via optogenetics can independently elicit increases in blood flow, even in the absence of direct neuronal activation, suggesting a significant metabolic and vascular role for glial cells [51]. Conversely, optogenetic stimulation of specific neuronal populations (e.g., excitatory vs. inhibitory neurons) has shown distinct patterns of vascular responses, emphasizing the differential contributions of neural subtypes to metabolic signaling and blood flow regulation [52]. In addition, several more recent studies have significantly advanced our understanding of the cellular mechanisms underlying neurovascular coupling, highlighting the critical roles of specific GABAergic interneurons in regulating cerebral blood flow in response to neural activity. Studies [53–55] utilized optogenetics to selectively activate interneuron subtypes, including parvalbumin

(PV), somatostatin (SOM), vasoactive intestinal peptide (VIP), and nNOS-expressing neurons. These studies revealed that different interneuron classes have distinct effects on vascular responses, with nNOS-expressing neurons notably contributing to vasodilation. They also explored how inhibitory neuron activity influences hemodynamic responses [54]. The findings indicated that activation of inhibitory neurons can lead to significant changes in blood flow, emphasizing their role in NVC and the importance of considering inhibitory circuits in functional imaging interpretations. Other works identified a subset of cortical GABAergic neurons expressing both neuronal nitric oxide synthase (nNOS) and tachykinin receptor 1 (Tacr1) [56]. These neurons, located primarily in cortical layers 5/6, project locally and across cortical areas. Optogenetic activation of Tacr1 neurons induced vasodilation, mirroring responses to sensory stimulation, suggesting their pivotal role in mediating NVC through nitric oxide signaling. Wakabayashi et al. [57] also conducted experiments to activate GABAergic neurons in the ventral tegmental area (VTA), revealing that such activation disrupts responses to reward-predictive cues. The findings suggest that modulating GABAergic activity in the VTA can influence glutamatergic signaling pathways involved in reward processing used chemogenetics. Together, these findings underscore the importance of metabolic signaling pathways in linking neural activity to hemodynamic changes. Optogenetics and chemogenetics provide essential tools to dissect the respective contributions of neurons, astrocytes, and vascular cells to both energy metabolism and vascular dynamics, enabling a deeper understanding of the physiological basis of brain imaging signals and potential mechanisms underlying neurodegenerative and metabolic brain disorders.

Other groundbreaking studies used laconic, a genetically encoded FRET-based biosensor designed to monitor intracellular lactate dynamics in real time [58]. Laconic utilizes the bacterial transcription factor LldR, flanked by fluorescent proteins mTFP and Venus, to detect lactate concentrations ranging from 1 μM to 10 mM. This sensor has been instrumental in various studies exploring cellular metabolism and lactate signaling. Using two-photon microscopy in live mice, Mächler et al. [59] observed a lactate gradient from astrocytes to neurons, supporting the ANLS hypothesis. This study demonstrated that astrocytes have higher basal lactate levels and can supply lactate to neurons as an energy substrate. Laconic's combination with ofMRI proved difficult due to important hemodynamic confounds [60]. Accurate calibration of laconic in vivo is crucial for reliable measurements. A study highlighted potential pitfalls in calibration procedures [61], such as the influence of rapid lactate-pyruvate interconversion and ongoing metabolism, which can affect the interpretation of lactate gradients between astrocytes and neurons. The link between laconic studies and fMRS lies in their shared goal: understanding real-time lactate dynamics in the brain during neural activity. While they operate at different scales and use different methods, combining insights from both can provide a multi-scale, cross-validated picture of brain energy metabolism without BOLD contamination. Laconic could confirm whether fMRS-detected lactate changes arise from neurons, astrocytes, or both. Moreover, the ANLS model could be refined by testing it at both cellular and mesoscopic levels.

4 Notes

4.1 Functional Metabolic Models in the Brain: Concepts, Challenges, and New Directions

Metabolic models, such as the astrocyte-neuron lactate shuttle (ANLS) [62], the malate-aspartate shuttle (MAS) [3, 63], and the glutamate-glutamine-GABA cycle, form the backbone of our current understanding of how brain energy metabolism is linked to neural activity. These frameworks describe how neurons and astrocytes cooperatively maintain energy homeostasis and neurotransmitter cycling, and they serve as theoretical scaffolds for interpreting data from fMRS.

- The ANLS model suggests that astrocytes, upon glutamate uptake, engage in glycolysis and produce lactate, which is shuttled to neurons as a preferred oxidative substrate. This model implies a functional coupling between synaptic activity and astrocytic energy metabolism.
- The MAS plays a key role in intracellular redox balance [3, 63], allowing reducing equivalents (NADH) from glycolysis to feed into the mitochondrial respiratory chain, especially critical in neurons with high oxidative demands.
- The glutamate–glutamine–GABA cycle [39, 64, 65] captures the fast, energetically expensive recycling of neurotransmitters between astrocytes and neurons. This cycle highlights the centrality of astrocytes not only in neurotransmitter clearance but also in supporting neuronal metabolism.

Despite their importance, these models are often not sufficient to explain dynamic metabolic changes detected by modern fMRS techniques. Time-resolved fMRS can now track transient fluctuations in glutamate, glutamine, lactate, and aspartate in response to specific stimuli. However, there is a lack of quantitative, cell-type resolved, and temporally accurate kinetic models that incorporate features, such as compartmental transport, enzymatic regulation, and neuron-glia interactions [66].

Toward Dynamic, Testable Metabolic Models: New Analytical and Genetic Tools

Recent innovations are beginning to address this gap by applying systems-level and computational modeling approaches to fMRS data. These include: the development of the glutamate response function (GRF), akin to the hemodynamic response function (HRF) used in BOLD-fMRI, which enables the parametric modeling of stimulus-evoked metabolite changes in time. The adaptation of fMRI analytical pipelines (e.g., SPM, FSL, AFNI) to fMRS datasets allows researchers to apply general linear models, temporal filtering, and statistical mapping to metabolite dynamics. This offers new opportunities for hypothesis-driven testing of metabolic models, statistical comparison between conditions, and integration with multimodal data. Combined with machine learning and Bayesian inference, these tools are paving the way for data-driven discovery of metabolic network behavior, especially when combined with multi-metabolite time series.

4.2 The Critical Role of Genetic Tools in Building and Validating Models

The integration of genetic tools with fMRS is a game-changer for both constructing and validating new metabolic models: optogenetics allows for precise, cell-type-specific stimulation of defined neural populations (e.g., excitatory neurons, GABAergic interneurons, or astrocytes), enabling the isolation of metabolic responses attributable to specific circuit elements. Chemogenetics (e.g., DREADDs) enables prolonged or subtle modulation of neural activity, ideal for studying steady-state metabolic shifts in the absence of strong hemodynamic artifacts. FRET-based biosensors, including genetically encoded indicators for $NADH/NAD^+$, ATP, lactate, glutamate, and pH, offer real-time intracellular readouts that can be directly correlated with MR spectroscopy findings. These tools provide ground-truth validation for MR-derived metabolite dynamics and help refine model parameters with subcellular resolution. Together, these technologies allow researchers to experimentally perturb specific pathways, record the resulting fMRS profiles, and iteratively refine metabolic models based on empirical data. For example, selective activation of astrocytes via chemogenetics, combined with fMRS readouts of lactate and glutamate, can directly test predictions of the ANLS model under physiologically relevant conditions. With the synergy of advanced fMRS methods, computational modeling, and precise genetic manipulation, neuroscience is on the verge of a paradigm shift: from correlative metabolic observations to mechanistic, testable, and dynamic models of brain energy metabolism. This integration will be key to unraveling the cellular basis of neurometabolic diseases, identifying novel therapeutic targets, and understanding the energetic logic of the functioning brain.

4.3 Combining Genetic Tools with fMRS: A Synergistic Approach

A central conceptual framework of this chapter is the integration of circuit-specific neuromodulation with neurochemical readouts, offering a dynamic window into the functional architecture of brain metabolism. Recent advances in tools, such as optogenetics and chemogenetics allow for precise, cell-type, and projection-specific activation or inhibition of neuronal populations. In rodents, when these manipulations are paired with fMRS, unprecedented access to the causal links between circuit activity and neurochemical fluxes can be gained. This approach moves beyond static or correlational measurements, enabling mechanistic interrogation of neuroenergetic processes within behaviorally relevant or physiologically defined time windows. Such a framework offers a powerful strategy for elucidating the metabolic signatures of distinct neuronal ensembles, bridging the gap between cellular activity and whole-brain imaging signals.

Experimental Design Considerations

Our study "Probing activation-induced neurochemical changes using optogenetics combined with functional magnetic resonance spectroscopy: a feasibility study in the rat primary cortex" [37] explored for the first time the integration of optogenetic stimulation with fMRS to monitor neurochemical dynamics in vivo. One of the main issues when combining optogenetics and fMRS in rodents is the complexity of the setup for conducting MR-compatible studies.

Choice of Opsins

For this feasibility study, experiments were conducted using the red-shifted chimeric excitatory opsin C1V1 TT [14] to target specifically and selectively all excitatory neurons in the primary somatosensory forelimb cortex (S1FL) of Fisher rats. Adeno-associated virus particles (AAV2) containing a CamKII-C1V1TT-mcherry construct (rAAV2/CamKII-C1V1_Ts-mCherry (from addgene) were injected in S1FL (anterior–posterior: +0.2 mm; left–right: 2.4 mm; dorso–ventral: 1.2 mm; and an angle of 35°) 4 weeks or more before fMRI–fMRS experiments took place. Using an optic fiber (OFS) implanted on the day of MRI experiments, S1FL was excited through the delivery of green light at 552 nm using a powerful laser.

A Complex Setup for ^{1}H-fMRS Conducted at 9.4 T (Fig. 1)

The setup for experiments was installed in a dedicated 9.4 T animal Bruker scanner and included mechanical ventilation through direct intubation of each rat and a dedicated stereotactic device to firmly secure the head with a bite bar and ear bars. To conduct experiments, a 90 mm volume coil (Rapid Biomedical GMBH, Germany) for transmission and a 20 mm in diameter surface coil for reception (Bruker, Germany) were used. Monitoring of both respiration and temperature was performed using a pneumatic sensor placed underneath the rat's body and a rectal probe, respectively. Electrical stimulation via the insertion of electrodes between the digits of the forepaw of the rat was also conducted. Stimulation paradigms for electrical and optogenetic stimulations were delivered through a LABVIEW interface and directly triggered by the scanner. All these details can be found with more information in the paper [37] and are described in Fig. 2. The animal was initially anesthetized with 1.5–2% isoflurane in air and oxygen (70/30) for surgeries and installation purposes. Sedation using medetomidine and isoflurane (0.25–1%) was induced when the animal was in the scanner.

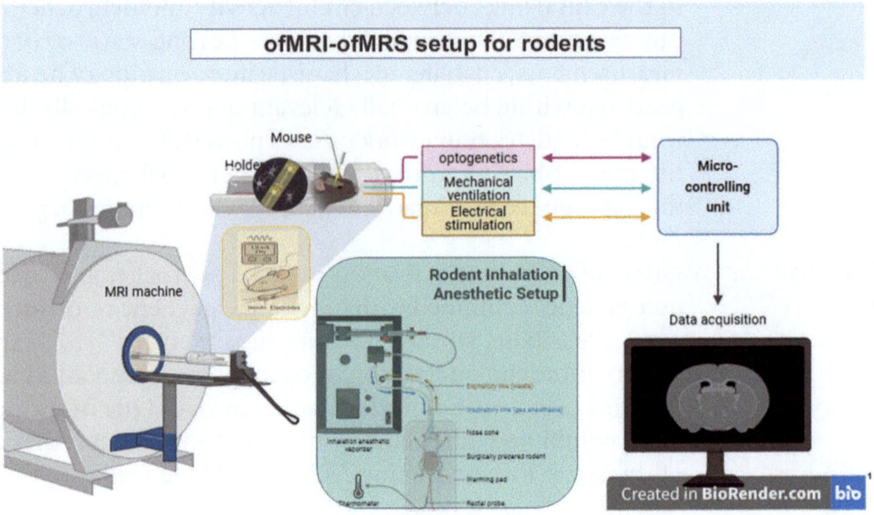

Fig. 1 A complex setup for ofMRI/ofMRS experiments in rodent animal models. The setup includes optogenetic stimulation, electrical stimulation of the forepaw, intubation of the rodent, and physiological monitoring

For these experiments, the correct and fast installation of the animal in its cradle is crucial for the outcome of the experiments. Dedicated and repeated gestures, as well as taping of cables, were necessary to preserve the OF implantation during the installation of the animal.

ofMRI and ofMRS (Figs. 2 and 3)

Anatomical images were acquired using a RARE T2-weighted sequence. BOLD-fMRI acquisition was performed using GRE-EPI (TR/TE = 1000/18 ms; 80 × 80 matrix; 16 slices). Electrical and optogenetic (green laser, 552 nm, <22 mW/mm^2) stimulations were delivered in a 10s-OFF/10s-ON/10s-OFF block paradigm. Data were preprocessed with SPM12 and analyzed using the general linear model to generate statistical T-maps ($p<0.01$).

^{1}H-fMRS was performed using a STEAM sequence (TR = 4 s, TE = 2.6 ms, mixing time = 10 ms). A small voxel of interest (VOI) (1.1 × 2.5 × 3 mm^3) was positioned in the activated region of the cortex. VAPOR water suppression and outer volume saturation were applied to optimize signal quality. FASTMAP shimming ensured water linewidths below 18 Hz; rats not meeting this criterion were excluded.

The stimulation protocol included: optogenetic and sensory stimulations were performed in a 2.5-min OFF/5-min ON/2.5-min OFF cycle. Each rat underwent four repetitions per stimulation type (optogenetic and sensory), alternating the order between animals. A 10-min pause was included between runs. Control experiments were performed in five rats without in non-opsin-expressing rats and dead rats to evaluate artifacts and

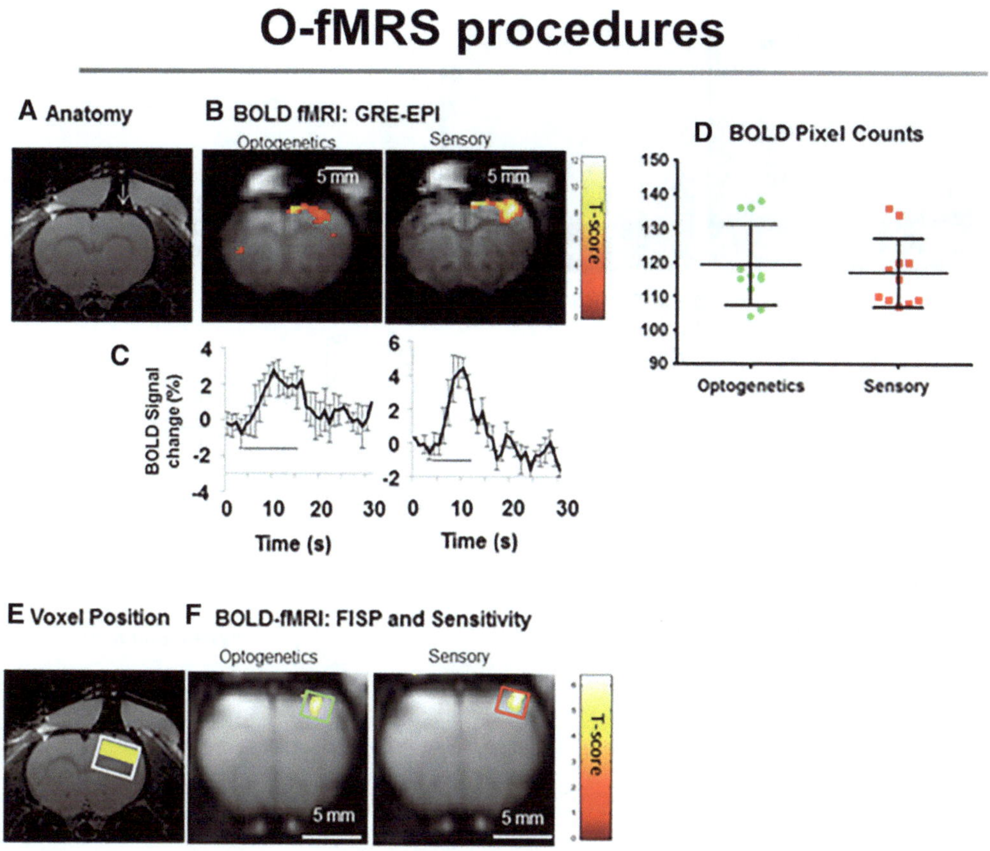

Fig. 2 ofMRS procedures: description of a typical optogenetic functional magnetic resonance imaging (o-fMRI) session. (**A**) Rats expressing the opsin C1V1 underwent a second craniotomy under isoflurane anesthesia in order to implant an optic fiber (OF) above primary somatosensory forelimb cortex (S1FL) on the day of 1H-fMRS [white arrow on rapid imaging with refocused echoes (RARE) image]. (**B**) After overall shimming with MAPSHIM, the presence of BOLD responses in S1FL upon optogenetic (9 Hz, 10 ms, 10 mW) and sensory stimulations (9 Hz, 10 ms, 1.2 mA) (10 s ON-20 s OFF.; 10 min) was verified using a conventional gradient echo echo planar imaging (GRE-EPI) sequence. BOLD T-value maps were thus obtained successively with both types of stimulation. (**C**) Averaged BOLD responses were obtained for each type of stimulation and (**D**) the number of activated pixels in S1FL was not different (mean voxel counts_standard deviation: 119_12 for optogenetic vs. 117_10 for sensory stimulation; $t=0.459$; $d=10$; $p=0.7$, Student paired t test) (**E**) The $3\times3\times3\,mm^3$ for further FASTMAP shimming (open white box) as well as the acquisition voxel of interest (VOI) (yellow box) are depicted on RARE T2-weighted images. (**F**) To avoid missing the core of the BOLD activation when positioning the OF for 1H-fMRS acquisitions and to limit partial volume effects and lipid contamination, high spatially resolved fast imaging with steady state free precession images (matrix $=256\times128$, $78\times156\,\mu m^2$) were acquired successively during optogenetic stimulation and electrical forepaw stimulation. These images were useful to position the FASTMAP VOI for shimming (27 μL, white box) and the acquisition VOI (8 μL, yellow box in f) over the anatomical RARE images and to obtain maximal sensitivity to BOLD-induced T2* effects. Reproduced with permission from [37]

photothermal effects. Spectral quality was assessed by signal-to-noise ratio (SNR), calculated using the NAA peak. Unsuppressed water spectra were used for eddy current correction and absolute metabolite quantification.

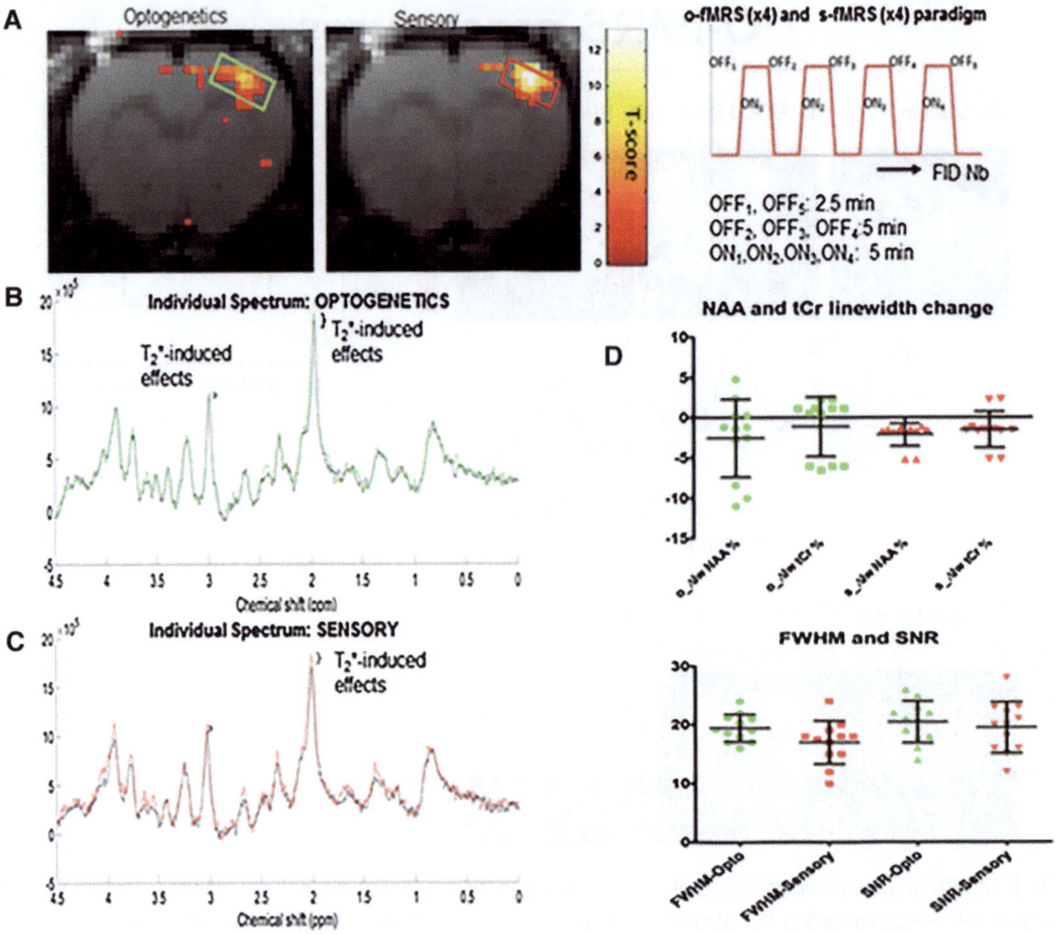

Fig. 3 Description of a typical ¹H functional magnetic resonance spectroscopy (¹H-fMRS) session. (**A**) Example of T-value BOLD maps overlaid onto rapid imaging with refocused echoes anatomical images with the 8 μL voxel of interest (VOI) for further optogenetic fMRS (ofMRS, green VOI) and sensory fMRS (s-fMRS, red VOI). A 5-min ON–5-min OFF paradigm of stimulation repeated for 45 min (acquisitions of 650 free induction decay, FIDs) was used for o-fMRS and s-fMRS using interleaved acquisitions. To increase the signal-to-noise ratio (SNR) of functional spectra, four acquisitions were performed per type of stimulation, separated by a 10-min waiting time. Examples of single run overlaid stimulated and rest spectra (600FIDs) for (**B**) optogenetics (green; [SNR = 18_1]) and (**C**) sensory paradigms (red; [SNR = 18_1]). BOLD-T2* induced effects can be observed as NAA and tCr peak heights resulting from linewidth narrowing. (**D**) BOLD-T2*induced effects induced NAA and tCr linewidth narrowing that were reported as relative percentage changes for both optogenetic and sensory stimulations ($n = 12$) (oΔlwNAA and oΔlwtCr: changes in NAA and tCr linewidths during optogenetic stimulation; ΔlwNAA and sΔlwtCr: changes in NAA and tCr linewidths during sensory stimulation). Reproduced with permission from [37]

5 Results

5.1 Feasibility and Quality

Successful acquisition of ¹H-MRS spectra without significant susceptibility artifacts was achieved, provided the optical fiber was inserted carefully and there were no bleeding or scars. Laser stimulation at calibrated parameters (22 mW/mm^2, 9 Hz) did not produce thermal artifacts or nonspecific metabolic changes (Fig. 4).

5.2 Neurochemical Changes

Glutamate (Glu) levels increased during both optogenetic (+8.8%) and sensory (+9.9%) stimulation. Glutamine (Gln) levels decreased significantly (−17% to −18%). Aspartate (Asp) decreased significantly during sensory (−23%) but not optogenetic stimulation. No significant lactate changes were observed, suggesting no anaerobic glycolysis under either condition. Changes in PCr/Cr ratio indicated energy metabolism shifts during stimulation (Fig. 5).

Water linewidths remained under 15 Hz, and spectra had high signal-to-noise ratios (SNR ~20). VOI placement was optimized using BOLD fMRI to ensure accurate targeting of active regions.

6 Discussion

6.1 Feasibility of Combining Optogenetics and ^{1}H-fMRS

The study successfully demonstrated the feasibility of combining optogenetic stimulation with proton functional magnetic resonance spectroscopy (^{1}H-fMRS) in vivo in rats. This is significant because optogenetics allows for cell-type-specific stimulation, while fMRS enables noninvasive monitoring of neurochemical dynamics, offering a powerful synergy for investigating brain function. Key technical achievements included: a precise placement of a superficial optic fiber (layer 1 of S1FL), which minimized susceptibility artifacts, and a reliable shimming (water linewidth <18 Hz) in the majority of animals, ensuring spectral quality. In addition, no photothermal artifacts were detected at the tested laser power (<22 mW/mm^2), and control animals confirmed the specificity of neurochemical responses to opsin-mediated activation.

6.2 Spectral Quality and Experimental Rigor

High-quality spectra were consistently obtained with good SNR (~20–38) and stable linewidths (FWHM ~17–19 Hz). The use of repetitive block paradigms for stimulation and the careful alternation of conditions per animal ensured robustness and minimized order effects. The implementation of BOLD-fMRI to guide VOI placement in S1FL enhanced spatial accuracy for detecting localized metabolic changes, optimizing sensitivity to neurotransmitter shifts.

6.3 Metabolic Changes Reflect Functional Activation

Both optogenetic and sensory stimulation produced similar and significant changes in key metabolites: increased glutamate (Glu): (+8.8% optogenetic; +9.9% sensory), indicating increased excitatory neurotransmission but decreased glutamine (Gln): (−17% optogenetic; −18% sensory), suggesting enhanced Glu–Gln cycling, likely driven by glutamatergic neuron activation. Aspartate (Asp) decreased significantly with sensory but not optogenetic stimulation, possibly reflecting broader cell recruitment (e.g., astrocytic or inhibitory neurons) during electrical stimulation. These results are

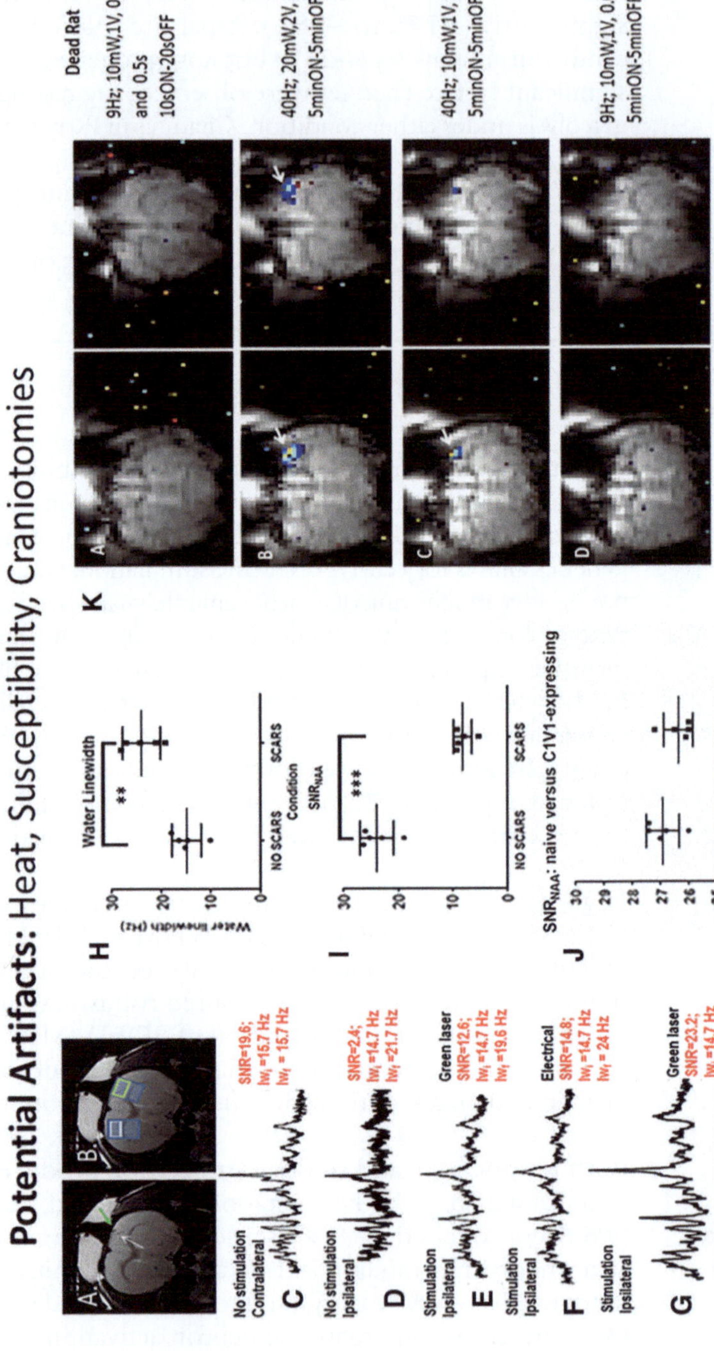
Potential Artifacts: Heat, Susceptibility, Craniotomies
A
B
No stimulation
Contralateral
C
SNR=19.6;
lw$_i$ =15.7 Hz
lw$_f$ = 15.7 Hz
No stimulation
Ipsilateral
D
SNR=2.4;
lw$_i$ =14.7 Hz
lw$_f$ =21.7 Hz
Stimulation
Ipsilateral
E
Green laser
SNR=12.6;
lw$_i$ =14.7 Hz
lw$_f$ =19.6 Hz
Stimulation
Ipsilateral
F
Electrical
SNR=14.8;
lw$_i$ =14.7 Hz
lw$_f$ = 24 Hz
Stimulation
Ipsilateral
G
Green laser
SNR=23.2;
lw$_i$ =14.7 Hz
lw$_f$ = 16 Hz
Chemical shift (ppm)
H
Water Linewidth
Water linewidth (Hz)
**
NO SCARS
SCARS
Condition
I
SNR$_{NAA}$
***
NO SCARS
SCARS
J
SNR$_{NAA}$: naive versus C1V1-expressing
SNRnoOF-noopsin
SNROF-C1V1
K
Dead Rat
9Hz; 10mW,1V, 0.01
and 0.5s
10sON-20sOFF
40Hz; 20mW,2V, 0.01s
5minON-5minOFF
40Hz; 10mW,1V, 0.01s
5minON-5minOFF
9Hz; 10mW,1V, 0.01s
5minON-5minOFF

Fig. 4 Potential artifacts in ofMRS studies: susceptibility effects and SNR: (**A**) The scar left by C1V1 injection (white arrow) and the presence of the optical fiber (green arrow) superficially implanted (100 μm) in primary somatosensory forelimb cortex (S1FL) can be detected in anatomical T2-weighted rapid imaging with refocused echoes images. (**B**) Rectangles represent the respective voxels of interest for FASTMAP shimming (blue, $3 \times 3 \times 3\,mm^3$) and 1 H functional magnetic resonance spectroscopy (1H-fMRS) acquisitions (white and green, $1.1 \times 2.5 \times 3\,mm^3$). (**C**) 1H-MR spectrum (512 free induction decays) acquired in the contralateral S1FL voxel of interest (VOI) (white rectangle, no craniotomy) of the rat brain, showing high spectral quality and high SNR. The water linewidth remained stable during the MRS acquisition, as shown by the initial water linewidth (lwi) of an unsuppressed water spectrum acquired prior to 1H-fMRS and the final water linewidth (lwf) of an unsuppressed water spectrum acquired after 1H-fMRS. (**D**) By contrast, in the ipsilateral S1FL VOI of the same rat (green voxel [optic fiber (OF) + C1V1 opsin], the presence of scars strongly degraded the water linewidth and SNR during MRS acquisitions prior to stimulation, as well as (**E**) during optogenetic stimulation and (**F**) during electrical forepaw stimulation (**G**). This SNR degradation can be specifically attributed to the scars, as MRS acquisitions in a different rat presenting no traces of scars in S1FL demonstrated high quality under optogenetic stimulation. All the rats were under similar physiological conditions. (**H**) The water linewidth was compared in five animals ($n = 5$) with and without scars, showing a significant increase in the presence of scars ($t = 4.17$; $df = 8$; $p = 0.0031$; Student *t* test). (I) Comparison of SNR with and without scars showing a significant decrease of SNR in case of scars or micro-bleedings (mean SNR (no scar) = 23.9_3 vs. mean SNR (scar) = 8_2; $t = 10.16$; $df = 8$; $p < 0.0001$). (**J**) Comparison of SNR in naive and C1V1TT-expressing rats without scars showing no significant difference (SNRnoOF-noOpsin = 26.9_0.6 vs. SNROFC1V1 = 66.4_0.5, $t = 1.623$; $df = 8$; $p = 0.1432$, unpaired *t* test, $n = 5$). (**K**) Examples of heating artifact mapping in a dead rat for different parameters of optogenetic stimulation. Reproduced with permission from [37]

consistent with previous studies and support the notion that Glu and Gln shifts serve as biomarkers of local neuronal activation and energy metabolism.

6.4 Energetic Demands and PCr/Cr Ratio

In control animals (non-opsin expressing), decreases in phosphocreatine (PCr) and increases in creatine (Cr) were observed during both electrical and light stimulation. This likely reflects a shift toward ATP regeneration from PCr to meet energy demands. Interestingly, a reduction in the PCr/Cr ratio was also seen in non-expressing animals during light stimulation, suggesting that photodilation or nonspecific heating might alter energy metabolism even in the absence of opsins, although no significant BOLD or neurochemical artifacts were detected.

6.5 Temporal Dynamics and Limitations

Time-resolved metabolite measurements revealed transient Glu increases that may not fully return to baseline between blocks. This suggests potential cumulative or adaptive changes with repeated stimulation, a factor that should be carefully controlled in future protocols. However, the study also includes limitations: a modest sample size and voxel volume (8 μL), which constrain detection sensitivity for low-concentration metabolites like GABA or lactate. While superficial fiber placement minimized artifacts, it restricts stimulation depth; future studies using miniaturized or multiphoton tools may address this. The lack of histological validation for all animals may limit the interpretation of opsin expression variability.

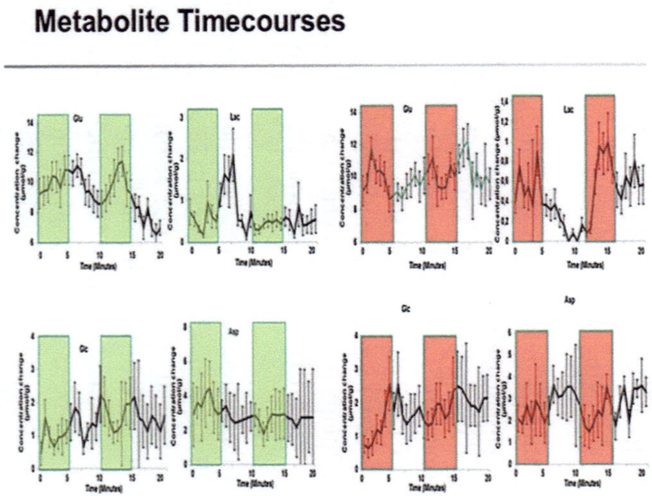

CMR Glc-Optogenetics	CMR Glc-Sensory
0.68 ± 0.06 µmol/g/min	0.47 ± 0.17 µmol/g/min

Condition	Electrical	OPTICAL	Electrical	OPTOGENETICS
	Rest versus Stimulation	Rest versus Stimulation	C1V1 Rest versus Stimulation	C1V1 Rest versus Stimulation
Cr (µmol/g)	4.2± 6 %versus 5.0±4%	4± 5 % versus 5.3±6%	4.1± 4 % versus 2.4± 13%	2.6± 19% versus 3.9± 6%
PCr (µmol/g)	5.6±5% versus 4.0± 5%	5.8± 6versus 2.0±11%	3.2± 5% versus 4.7± 6%	5.5%± 8 % versus 4.3± 5%

$$PCr^{2-} + MgADP^{-} + H^{+} \underset{k_{ATP-PCr}}{\overset{k_{PCr-ATP}}{\rightleftarrows}} Cr + MgATP^{2-}$$

(k: creatine kinase rate constant)

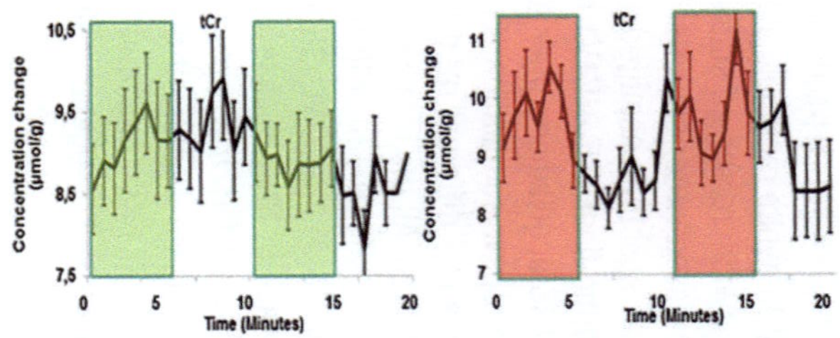

Fig. 5 Example of metabolite time courses for glutamate, lactate, aspartate, and glucose for optogenetics and sensory stimulation of S1FL of the rat. The consumption rate of glucose (CMR) was compared between the two types of stimulation. The time courses for TCr were also added, as well as the quantification for Cr and PCr for different conditions. Explanations can be found in [37]

6.6 Conclusion

This study provided a proof-of-concept for the integration of optogenetics with ^{1}H-fMRS in rodents, offering a powerful methodology to probe cell-type-specific neuroenergetics in vivo. The robust neurochemical readouts of Glu and Gln in response to selective activation pave the way for:

- Exploring functional roles of specific neural populations.
- Testing neuroenergetic hypotheses of brain function and disease.
- Informing the development of targeted neuromodulation therapies.

Future research should focus on expanding the approach to deeper brain regions, enhancing spectral sensitivity, and applying the method in disease models to uncover pathophysiological mechanisms with cell-level specificity. This study demonstrated for the first time the viability of combining optogenetics with ^{1}H-fMRS in vivo in rats. It provided a cell-type-specific window into brain metabolism during functional activation. The results pave the way for using this approach to study neuroenergetics and validate brain energy metabolism theories.

The integration of genetic control over neural activity (via optogenetics or chemogenetics) with metabolic readouts (via ^{1}H-fMRS) represents a transformative advancement in systems neuroscience. This combination is important for:

- Precisely stimulating specific neuronal populations (e.g., excitatory or inhibitory neurons), bypassing the complexity of conventional stimulation techniques that activate diverse cell types.
- Directly measuring neurochemical consequences (e.g., changes in glutamate, glutamine, and aspartate) in real time, within a localized brain region, provides insights into the energetic cost of activity at the cell-type level.
- Disentangling neurovascular from neurochemical responses, a long-standing challenge in interpreting BOLD-fMRI, by linking molecular metabolism with controlled neural input.
- Testing and validating neuroenergetic theories, such as the glutamate–glutamine cycle and the role of astrocytes in fueling neuronal activity, with unprecedented specificity and temporal resolution (see Fig. 6).

This approach bridges the gap between functional, cellular, and biochemical levels of brain function, providing a mechanistic understanding of how genetically defined circuits drive metabolic processes.

The future of functional MR spectroscopy (fMRS) in rodents is set to be revolutionized by emerging technologies that drastically enhance sensitivity, allowing finer temporal and spatial resolution of neurochemical dynamics. Two key advancements—^{2}H-fMRS and hyperpolarized ^{13}C-fMRS—offer promising avenues for overcoming the traditional SNR limitations of ^{1}H-fMRS.

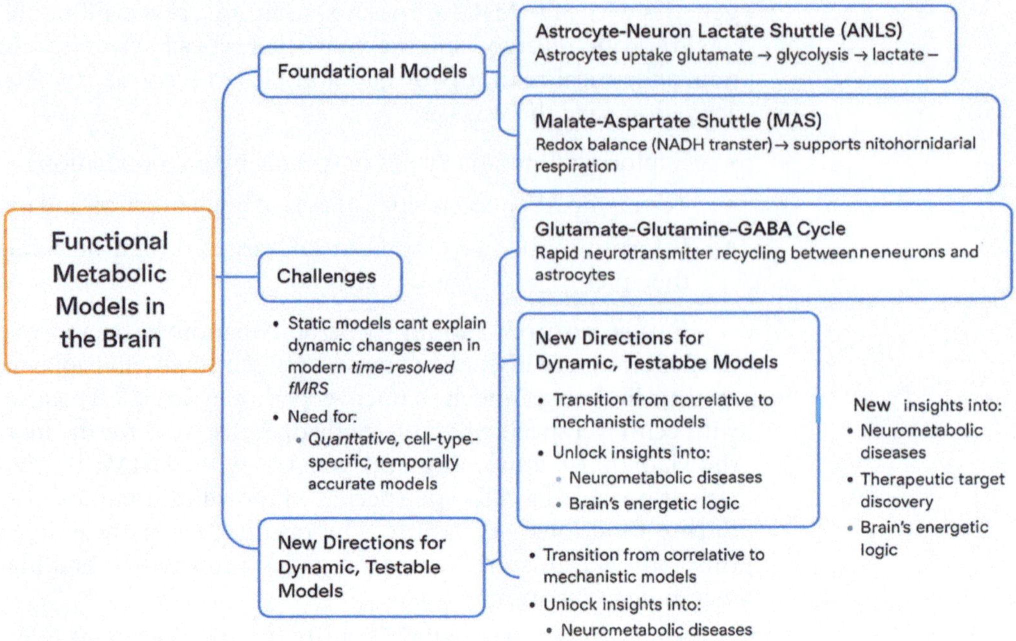

Fig. 6 New directions, new concepts for fMRS modeling

^{2}H-labeled substrates (e.g., [6,6'-^{2}H$_2$]-glucose) can be administered to trace glycolytic and oxidative metabolism in real time. In the future, ^{2}H-fMRS could offer a direct, nonradioactive way to quantify metabolic fluxes, such as the tricarboxylic acid (TCA) cycle and glutamate–glutamine cycling. When combined with optogenetics, ^{2}H-fMRS may enable cell-type-specific interrogation of energy usage and metabolic flexibility under controlled neural activation. Due to low background signal and broader chemical shift dispersion, ^{2}H-fMRS improves spectral clarity and quantification in small animal models. On the other end, hyperpolarization via dynamic nuclear polarization (DNP) should enhance the MR signal of ^{13}C-labeled substrates by 10,000-fold, allowing detection of low-abundance metabolites within seconds. This technique can track real-time enzymatic conversions (e.g., pyruvate to lactate, alanine, or bicarbonate), giving insight into neuronal vs. astrocytic metabolic contributions. When integrated with genetic stimulation tools, hyperpolarized fMRS could become a cutting-edge way to identify how specific cell types alter fuel preference, redox state, and energy metabolism on a second-by-second basis. Coupling these ultra-sensitive techniques with genetic precision (e.g., optogenetics, DREADDs) promises the dissection of metabolic roles for excitatory, inhibitory, or glial cells and the tracking of metabolic fingerprints of genetically defined circuits during behavior, sleep, learning, or disease states. These technologies will pave the way for whole-brain metabolic mapping, enhanced cell-type resolution, and real-time measurements of neurometabolic adaptation. They will also support high-throughput screening of therapeutic interventions targeting metabolic dysfunction in models of neurodegeneration, epilepsy, or psychiatric disorders.

In essence, combining ^{2}H and hyperpolarized ^{13}C-fMRS with genetic control mechanisms will supercharge the capabilities of rodent neuroscience, transforming fMRS from a complementary tool into a primary modality for cellular-resolution brain energetics. Ultimately, the fusion of genetic specificity with noninvasive metabolic and functional imaging will redefine our understanding of the living brain, making animal models an even more powerful platform for translational neuroscience and neurotherapeutic development [20].

Acknowledgments

The author acknowledges support from the Lundbeck Foundation (Experiment grant, grant no. R370-2021-402) to N.J. The author also acknowledges support from the Laboratory for Functional Imaging and Spectroscopy (LIFMET) from the EPFL, Lausanne, Switzerland

Conflict of Interest The author has no conflict of interest to disclose.

References

1. Mangia S, Simpson IA, Vannucci SJ, Carruthers A (2009) The in vivo neuron-to-astrocyte lactate shuttle in human brain: evidence from modeling of measured lactate levels during visual stimulation. J Neurochem 109 (Suppl 1):55–62
2. Just N (2021) Proton functional magnetic resonance spectroscopy (^{1}H-fMRS) in rodents: an early review of current status, difficulties, and future perspectives. NMR Biomed 34(5):e4254
3. Mangia S, Tkác I, Gruetter R, Van de Moortele PF, Maraviglia B, Ugurbil K (2007) Sustained neuronal activation raises oxidative metabolism to a new steady-state level: evidence from ^{1}H NMR spectroscopy in the human visual cortex. J Cereb Blood Flow Metab 27(5):1055–1063
4. Schaller B, Xin L, O'Brien K, Magill AW, Gruetter R (2014) Are glutamate and lactate increases ubiquitous to physiological activation? A ^{1}H functional MR spectroscopy study during motor activation in human brain at 7 Tesla. Neuroimage 93:138–145
5. Just N, Xin L, Frenkel H, Gruetter R (2013) Characterization of sustained BOLD activation in the rat barrel cortex and neurochemical consequences. Neuroimage 74:343–351. https://doi.org/10.1016/j.neuroimage.2013.02.039
6. Mullins PG (2018) Towards a theory of functional magnetic resonance spectroscopy (fMRS): a meta-analysis and discussion of using MRS to measure changes in neurotransmitters in real time. Scand J Psychol 59(2):91–103
7. Conde-Berriozabal S, Sitjà-Roqueta L, García-García E, García-Gilabert L, Sancho-Balsells A, Fernández-García S, Rodriguez-Urgellés E, Giralt A, Castañé A, Rodríguez MJ, Alberch J, Masana M (2025) Differential impact of optogenetic stimulation of direct and indirect pathways from dorsolateral and dorsomedial striatum on motor symptoms in Huntington's disease mice. Exp Neurol 383:114991. https://doi.org/10.1016/j.expneurol.2024.114991
8. Ji G, Presto P, Kiritoshi T, Chen Y, Navratilova E, Porreca F, Neugebauer V (2024) Chemogenetic manipulation of amygdala kappa opioid receptor neurons modulates amygdala neuronal activity and neuropathic pain behaviors. Cells 13(8):705. https://doi.org/10.3390/cells13080705
9. Ip IB, Emir UE, Lunghi C, Parker AJ, Bridge H (2021) GABAergic inhibition in the human visual cortex relates to eye dominance. Sci Rep 11(1):17022. https://doi.org/10.1038/s41598-021-95685-1
10. Ip IB, Berrington A, Hess AT, Parker AJ, Emir UE, Bridge H (2017) Combined fMRI-MRS

acquires simultaneous glutamate and BOLD-fMRI signals in the human brain. Neuroimage 155:113–119. https://doi.org/10.1016/j.neuroimage.2017.04.030
11. Eisenstein T, Furman-Haran E, Tal A (2023) Increased cortical inhibition following brief motor memory reactivation supports reconsolidation and overnight offline learning gains. Proc Natl Acad Sci USA 120(52):e2303985120. https://doi.org/10.1073/pnas.2303985120
12. Bednařík P, Tkáč I, Giove F, Eberly LE, Deelchand DK, Barreto FR, Mangia S (2018) Neurochemical responses to chromatic and achromatic stimuli in the human visual cortex. J Cereb Blood Flow Metab 38(2):347–359. https://doi.org/10.1177/0271678X17695291
13. Kim CK, Adhikari A, Deisseroth K (2017) Integration of optogenetics with complementary methodologies in systems neuroscience. Nat Rev Neurosci 18(4):222–235. https://doi.org/10.1038/nrn.2017.15
14. Yizhar O, Fenno LE, Prigge M, Schneider F, Davidson TJ, O'Shea DJ, Sohal VS, Goshen I, Finkelstein J, Paz JT, Stehfest K, Fudim R, Ramakrishnan C, Huguenard JR, Hegemann P, Deisseroth K (2011) Neocortical excitation/inhibition balance in information processing and social dysfunction. Nature 477(7363):171–178. https://doi.org/10.1038/nature10360
15. Li C, Fleck JS, Martins-Costa C, Burkard TR, Themann J, Stuempflen M, Peer AM, Vertesy Á, Littleboy JB, Esk C, Elling U, Kasprian G, Corsini NS, Treutlein B, Knoblich JA (2023) Single-cell brain organoid screening identifies developmental defects in autism. Nature 621(7978):373–380. https://doi.org/10.1038/s41586-023-06473-y
16. Wang ZH, Chen L, Li W, Chen L, Wang YP (2022) Mitochondria transfer and transplantation in human health and diseases. Mitochondrion 65:80–87. https://doi.org/10.1016/j.mito.2022.05.002
17. Boyden ES, Zhang F, Bamberg E, Nagel G, Deisseroth K (2005) Millisecond-timescale, genetically targeted optical control of neural activity. Nat Neurosci 8(9):1263–1268. https://doi.org/10.1038/nn1527
18. Zhang F, Wang L, Boyden ES, Deisseroth K (2007) Channelrhodopsin-2 and the optogenetic control of neural activity. Nature 446(7136):633–639. https://doi.org/10.1038/nature05844
19. Chow BY, Prigge M, Pullen MW et al (2010) High-performance optogenetic excitable neurons and their use in systems neuroscience. Neuron 67(5):698–708. https://doi.org/10.1016/j.neuron.2010.08.025
20. Nagel G, Ollig D, Fuhrmann M et al (2003) Channelrhodopsin-2, a directly light-activated cation channel. Proc Natl Acad Sci USA 100(24):13940–13945. https://doi.org/10.1073/pnas.1936192100
21. Armbruster BN, Li X, Pausch MH, Herlitze S, Roth BL (2007) Evolving the lock to fit the key to create a family of G protein–coupled receptors potently activated by an inert ligand. Proc Natl Acad Sci USA 104(12):5163–5168. https://doi.org/10.1073/pnas.0700293104
22. Fritsche M, Montijn JS, Dreosti E (2024) Ensemble coding and context dependence in the mouse visual cortex. eLife 12:RP89986. https://doi.org/10.7554/eLife.89986
23. Musall S, Kaufman MT, Juavinett AL, Gluf S, Churchland AK (2021) Single-trial neural dynamics are dominated by richly varied movements. Nat Commun 12:4029. https://doi.org/10.1038/s41467-021-24311-5
24. Imani M, Mohajeri N, Rastegar M, Zarghami N (2021) Recent advances in FRET-based biosensors for biomedical applications. Anal Biochem 630:114323. https://doi.org/10.1016/j.ab.2021.114323
25. Lanz B, Abaei A, Braissant O, Choi IY, Cudalbu C, Henry PG, Gruetter R, Kara F, Kantarci K, Lee P, Lutz NW, Marjańska M, Mlynárik V, Rasche V, Xin L, Valette J, Experts' Working Group on Magnetic Resonance Spectroscopy in the Rodent Brain (2024) Magnetic resonance spectroscopy in the rodent brain: experts' consensus recommendations. NMR Biomed e4325. https://doi.org/10.1002/nbm.4325
26. Lei H, Berthet C, Hirt L, Gruetter R (2009) Evolution of the neurochemical profile after transient focal cerebral ischemia in the mouse brain. J Cereb Blood Flow Metab 29(4):811–819. https://doi.org/10.1038/jcbfm.2009.8
27. Duarte JM, Do KQ, Gruetter R (2014) Longitudinal neurochemical modifications in the aging mouse brain measured in vivo by ^1H magnetic resonance spectroscopy. Neurobiol Aging 35(7):1660–1668. https://doi.org/10.1016/j.neurobiolaging.2014.01.135
28. Sappey-Marinier D, Calabrese G, Fein G, Hugg JW, Biggins C, Weiner MW (1992) Effect of photic stimulation on human visual cortex lactate and phosphates using ^1H and ^{31}P magnetic resonance spectroscopy. J Cereb Blood Flow Metab 12(4):584–592. https://doi.org/10.1038/jcbfm.199
29. Mangia S, Tkác I, Gruetter R, Van de Moortele PF, Maraviglia B, Uğurbil K (2007a) Sustained neuronal activation raises oxidative metabolism to a new steady-state level: evidence from ^1H NMR spectroscopy in the human visual cortex. J Cereb Blood Flow Metab 27(5):1055–1063. https://doi.org/10.1038/sj.jcbfm.9600401
30. Mangia S, Tkác I, Logothetis NK, Gruetter R, Van de Moortele PF, Uğurbil K (2007b) Dynamics of lactate concentration and blood

oxygen level-dependent effect in the human visual cortex during repeated identical stimuli. J Neurosci Res 85(15):3340–3346. https://doi.org/10.1002/jnr.21371
31. Koolschijn RS, Clarke WT, Ip IB, Emir UE, Barron HC (2023) Event-related functional magnetic resonance spectroscopy. Neuroimage 276:120194. https://doi.org/10.1016/j.neuroimage.2023.120194
32. Sonnay S, Duarte JMN, Just N (2017) Lactate and glutamate dynamics during prolonged stimulation of the rat barrel cortex suggest adaptation of cerebral glucose and oxygen metabolism. Neuroscience 346:337–348. https://doi.org/10.1016/j.neuroscience.2017.01.034
33. Rogan M, Friend AT, Rossetti GM, Edden R, Mikkelsen M, Oliver SJ, Macdonald JH, Mullins PG (2022) Hypoxia alters posterior cingulate cortex metabolism during a memory task: a (1)H fMRS study. Neuroimage 260:119397. https://doi.org/10.1016/j.neuroimage.2022.119397
34. Jelen LA, Lythgoe DJ, Jackson JB, Howard MA, Stone JM, Egerton A (2021) Imaging brain Glx dynamics in response to pressure pain stimulation: a (1)H-fMRS study. Front Psychiatry 12:681419. https://doi.org/10.3389/fpsyt.2021.681419
35. Bell TK, Craven AR, Hugdahl K, Noeske R, Harris AD (2023) Functional changes in GABA and glutamate during motor learning. eNeuro 10:ENEURO.0356-20.2023. https://doi.org/10.1523/ENEURO.0356-20.2023
36. de Joode NT, van den Heuvel OA, Koster M, Clarke WT, van Balkom AJLM, Schrantee A, Vriend C (2024) Glutamate dynamics and BOLD response during OCD symptom provocation in the lateral occipital cortex: a 7 Tesla fMRI-fMRS study. J Affect Disord 367:416–425. https://doi.org/10.1016/j.jad.2024.08.216
37. Just N, Faber C (2019) Probing activation-induced neurochemical changes using optogenetics combined with functional magnetic resonance spectroscopy: a feasibility study in the rat primary cortex. J Neurochem 150:402–419. https://doi.org/10.1111/jnc.14762
38. Ligneul C, Fernandes FF, Shemesh N (2021) High temporal resolution functional magnetic resonance spectroscopy in the mouse upon visual stimulation. Neuroimage 234:117973. https://doi.org/10.1016/j.neuroimage.2021.117973
39. Sonnay S, Poirot J, Just N, Clerc AC, Gruetter R, Rainer G, Duarte JMN (2018) Astrocytic and neuronal oxidative metabolism are coupled to the rate of glutamate–glutamine cycle in the tree shrew visual cortex. GLIA 66(3):477–491. https://doi.org/10.1002/glia.23257
40. Seuwen A, Schroeter A, Grandjean J, Rudin M (2015) Metabolic changes assessed by MRS accurately reflect brain function during drug-induced epilepsy in mice in contrast to fMRI-based hemodynamic readouts. Neuroimage 120:55–63. https://doi.org/10.1016/j.neuroimage.2015.07.004
41. Baslow MH, Cain CK, Sears R, Wilson DA, Bachman A, Gerum S, Guilfoyle DN (2016) Stimulation-induced transient changes in neuronal activity, blood flow and N-acetylaspartate content in rat prefrontal cortex: a chemogenetic fMRS-BOLD study. NMR Biomed 29(12):1678–1687. https://doi.org/10.1002/nbm.3629
42. Just N (2025). Modulating interhemispheric connectivity through chemogenetic stimulation: a fMRI-fMRS study. ResearchSquare preprint. https://doi.org/10.21203/rs.3.rs-xxx (Placeholder DOI)
43. Mangia S, Tkac I (2008) Dynamic relationship between neurostimulation and N-acetylaspartate metabolism in the human visual cortex: evidence that NAA functions as a molecular water pump during visual stimulation. J Mol Neurosci 35(2):245–246. https://doi.org/10.1007/s12031-008-9037-y
44. Just N, Hoehn M (2020). To intubate or not? Balancing anesthesia in rodent fMRI: strategies to mitigate confounding effects. Cereb Cortex. https://doi.org/10.1093/cercor/bhz209 (Placeholder DOI)
45. Yellen G (2018) Lactate in the brain: from metabolic end-product to signaling molecule. J Cell Biol 217(7):2235–2246. https://doi.org/10.1083/jcb.201803152
46. Magistretti PJ, Allaman I (2018) Lactate in the brain: from metabolic end-product to signaling molecule. Nat Rev Neurosci 19(4):235–249. https://doi.org/10.1038/nrn.2018.19
47. Magistretti PJ, Allaman I (2015) A cellular perspective on brain energy metabolism and functional imaging. Neuron 86(4):883–901. https://doi.org/10.1016/j.neuron.2015.03.035
48. DiNuzzo M, Giove F, Maraviglia B, Mangia S (2017) Computational Flux Balance Analysis Predicts that Stimulation of Energy Metabolism in Astrocytes and their Metabolic Interactions with Neurons Depend on Uptake of K+ Rather than Glutamate. Neurochem Res 202-216. https://doi.org/10.1007/s11064-016-2048-0. Epub 2016 Sep 14. PMID: 27628293; PMCID: PMC5283516
49. Logothetis NK (2002) The neural basis of the blood-oxygen-level-dependent functional magnetic resonance imaging signal. Philos Trans R Soc Lond B Biol Sci 357(1424):1003–1037. https://doi.org/10.1098/rstb.2002.1114

50. Newman JP, Fong MF, Millard DC, Whitmire CJ, Stanley GB, Potter SM (2015) Optogenetic feedback control of neural activity. eLife 4:e07192. https://doi.org/10.7554/eLife.07192
51. Takata K, Nakayama H, Saitoh Y, Miyamoto K, Nagano I, Yamasaki R, Iwamoto M, Takeuchi T (2018) Glial cell differentiation and glutamate metabolism in the developing mouse brain. GLIA 66(9):2013–2023. https://doi.org/10.1002/glia.23454
52. Uhlirova H, Kılıç K, Tian P, Thunemann M, Desjardins M, Saisan PA, Sakadžić S, Ness TV, Mateo C, Cheng Q, Weldy KL, Razoux F, Vandenberghe M, Cremonesi JA, Ferri CG, Nizar K, Sridhar VB, Steed TC, Abashin M, Fainman Y, Masliah E, Djurovic S, Andreassen OA, Silva GA, Boas DA, Kleinfeld D, Buxton RB, Einevoll GT, Dale AM, Devor A (2016) Cell type specificity of neurovascular coupling in cerebral cortex. eLife 5:e14315. https://doi.org/10.7554/eLife.14315
53. Krawchuk MB, Ruff CF, Yang X, Ross SE, Vazquez AL (2020) Optogenetic assessment of VIP, PV, SOM and NOS inhibitory neuron activity and cerebral blood flow regulation in mouse somato-sensory cortex. J Cereb Blood Flow Metab 40(7):1427–1440. https://doi.org/10.1177/0271678X19870105
54. Vazquez AL, Fukuda M, Kim SG (2018) Inhibitory neuron activity contributions to hemodynamic responses and metabolic load examined using an inhibitory optogenetic mouse model. Cereb Cortex 28(11):4105–4119. https://doi.org/10.1093/cercor/bhy225
55. Moon HS, Jiang H, Vo TT, Jung WB, Vazquez AL, Kim SG (2021) Contribution of excitatory and inhibitory neuronal activity to BOLD fMRI. Cereb Cortex 31(9):4053–4067. https://doi.org/10.1093/cercor/bhab068
56. Ruff CF, Juarez Anaya F, Dienel SJ, Rakymzhan A, Altamirano-Espinoza A, Couey JJ, Fukuda M, Watson AM, Su A, Fish KN, Rubio ME, Hooks BM, Ross SE, Vazquez AL (2024) Long-range inhibitory neurons mediate cortical neurovascular coupling. Cell Rep 43(4):113970. https://doi.org/10.1016/j.celrep.2024.113970
57. Wakabayashi KT, Feja M, Baindur AN, Bruno MJ, Bhimani RV, Park J, Hausknecht K, Shen RY, Haj-Dahmane S, Bass CE (2019) Chemogenetic activation of ventral tegmental area GABA neurons, but not mesoaccumbal GABA terminals, disrupts responding to reward-predictive cues. Neuropsychopharmacology 44(2):372–380. https://doi.org/10.1038/s41386-018-0097-6
58. San Martin A et al (2013) A genetically encoded FRET lactate sensor and its use to detect the Warburg effect in single cancer cells. PLoS One 8(2):e57712. https://doi.org/10.1371/journal.pone.0057712
59. Mächler P, Wyss MT, Elsayed M, Stobart J, Gutierrez R, von Faber-Castell A, Kaelin V, Zuend M, San Martín A, Romero-Gómez I, Baeza-Lehnert F, Lengacher S, Schneider BL, Aebischer P, Magistretti PJ, Barros LF, Weber B (2016) In vivo evidence for a lactate gradient from astrocytes to neurons. Cell Metab 23(1):94–102. https://doi.org/10.1016/j.cmet.2015.10.010
60. Lambers H, Wachsmuth L, Thomas D, Boumezbeur F, Hoesker V, Pradier B, Faber C (2022) Fiber-based lactate recordings with fluorescence resonance energy transfer sensors by applying an magnetic resonance-informed correction of hemodynamic artifacts. Neurophotonics 9(3):032212. https://doi.org/10.1117/1.NPh.9.3.032212
61. Dienel GA, Rothman DL (2024) In vivo calibration of genetically encoded metabolite biosensors must account for metabolite metabolism during calibration and cellular volume. J Neurochem 168(5):506–532. https://doi.org/10.1111/jnc.15775
62. Pellerin L, Magistretti PJ (1994) Glutamate uptake into astrocytes stimulates aerobic glycolysis: a mechanism coupling neuronal activity to glucose utilization. Proc Natl Acad Sci USA 91(22):10625–10629. https://doi.org/10.1073/pnas.91.22.10625
63. Andersen JV, Markussen KH, Jakobsen E, Schousboe A, Waagepetersen HS, Rosenberg PA, Aldana BI (2021) Glutamate metabolism and recycling at the excitatory synapse in health and neurodegeneration. Neuropharmacology 196:108719. https://doi.org/10.1016/j.neuropharm.2021.108719
64. Hertz L, Rothman DL (2016) Glucose, lactate, beta-hydroxybutyrate, acetate, GABA, and succinate as substrates for synthesis of glutamate and GABA in the glutamine-glutamate/GABA cycle. Adv Neurobiol 13:9–42. https://doi.org/10.1007/978-3-319-45096-4_2
65. Patel AB, de Graaf RA, Mason GF, Rothman DL, Shulman RG, Behar KL (2005) The contribution of GABA to glutamate/glutamine cycling and energy metabolism in the rat cortex in vivo. Proc Natl Acad Sci USA 102(15):5588–5593. https://doi.org/10.1073/pnas.0501703102
66. Lea-Carnall CA, El-Deredy W, Stagg CJ, Williams SR, Trujillo-Barreto NJ (2023) A mean-field model of glutamate and GABA synaptic dynamics for functional MRS. Neuroimage 266:119813. https://doi.org/10.1016/j.neuroimage.2022.119813

Check for updates

Chapter 11

Optogenetic Functional MRI Interrogation of Brain-Wide Neural Oscillations and Behavioral Relevance

Xunda Wang

Abstract

One grand challenge of the twenty-first century is to gain a holistic understanding of neural circuits and their spatiotemporally coordinated patterns of neural activities that underlie brain functions. Neural oscillatory activities in the brain, particularly in the thalamus and cortex, are widely considered to mediate critical brain functions, such as memory consolidation during spindle activities (i.e., 0.5–3 s brief oscillatory events at 7–15 Hz). However, it remains unclear where such oscillatory activities propagate across the brain and, more importantly, their sites of multi-scale actions to mediate sensory or cognitive functions. This barrier precludes our existing frameworks from describing the exact roles of these oscillatory activities in functional processes. Here, we provide a multi-scale multimodal neuroimaging protocol that integrates small animal optogenetic fMRI with electrophysiological, behavioral, genetic, and molecular assays. This approach permits comprehensively interrogating where and how spontaneous neural oscillations causally act across brain-wide networks to support brain functions at the systems level.

Keywords Multi-scale, Multimodal, fMRI, Optogenetics, Neural oscillation, Brain-wide

1 Introduction

The mammalian brain is a highly sophisticated biological information processing system, comprised of billions of interconnected neuronal populations organized into numerous local circuits and extensive brain-wide networks. It dynamically orchestrates neural activity across diverse temporal and spatial dimensions, whether at rest or during active behaviors, leading to the emergence of specific spatiotemporal patterns of brain activity [1, 2]. Particularly noteworthy are the spontaneous oscillatory activities generated by rhythmic interactions among localized neural populations. These neural oscillations propagate across interlinked circuits and can trigger dynamic patterns across the entire brain during critical phases of information transmission, integration, storage, consolidation, and execution. Since the invention of noninvasive electroencephalography (EEG) in 1929, the prominence of these neural oscillations has drawn significant

Kevin Chuen Wing Chan and Seong-Gi Kim (eds.), *Optogenetic Functional Neuroimaging*, Neuromethods, vol. 232, https://doi.org/10.1007/978-1-0716-5178-0_11,

research attention [2–4]. Over the last century, extensive studies have aimed to elucidate the role of these neural oscillations in supporting essential brain functions, including perception, cognition, emotion, and motor control. This has led to the identification of a hierarchical nesting organization of neural oscillations—such as slow oscillations, spindles, and sharp-wave ripples—now recognized as a fundamental feature for long-range functional integration within large-scale brain networks [5–10].

Existing studies of neural oscillations have primarily relied on passive observation techniques, such as scalp and intracranial cortical electrophysiology, or local depth recordings, including the use of electrophysiological activity-triggered functional magnetic resonance imaging (fMRI) analysis [11]. However, existing methodologies struggle to elucidate the precise relationship between neural oscillations observed in different regions and single or multiple specific initiations or sites. Consequently, these approaches are inherently limited in their ability to clarify how neural oscillations initiated at a specific local circuit propagate and target various downstream structures brain-wide. Moreover, we still lack a systems-level understanding of the specific multi-scale actions of brain-wide coordinated neural oscillations concerning brain functionality. It is imperative to develop an integrative framework that traces the evolution of neural oscillations from their local circuitry initiation to their long-range propagation and network coordination, while also examining their causal impact on behavior and underlying cellular neuroplasticity. Currently, progress in this area is impeded by the absence of cross-scale multimodal causal probing technologies that can spatiotemporally and precisely initiate neural oscillations, effectively visualize their downstream targets, and directly map the functional consequences throughout the brain.

Hence, we developed a methodology protocol that integrates small animal whole-brain large-scale fMRI, fMRI-guided multisite meso/microscale neuronal ensemble recording, multimodal neuroimaging-calibrated optogenetic emulation of spontaneous neural impulses and oscillations, behavioral assays, and fMRI-guided multisite profiling of microscale activity-dependent neuroplasticity [11–21] (Fig. 1). This integrative multi-scale multimodal causal probing approach addresses the limitations of previous methodologies, allowing us to systematically and causally interrogate the multi-scale actions of neural oscillations across brain-wide networks. Here, a guide for implementing this multi-scale multimodal causal neural probing protocol is provided. Specifically, the procedures in this protocol include stereotactic surgeries for viral injection and optical fiber implantation, histological confirmation of virus expression, animal preparation for multimodal neuroimaging, MRI acquisition procedure, MRI scanner-synchronized optogenetic and sensory stimulation, fMRI-guided multisite electrophysiology recording, behavioral relevance testing, animal model induction, and fMRI-guided multisite profiling of microscale activity-dependent neuroplasticity.

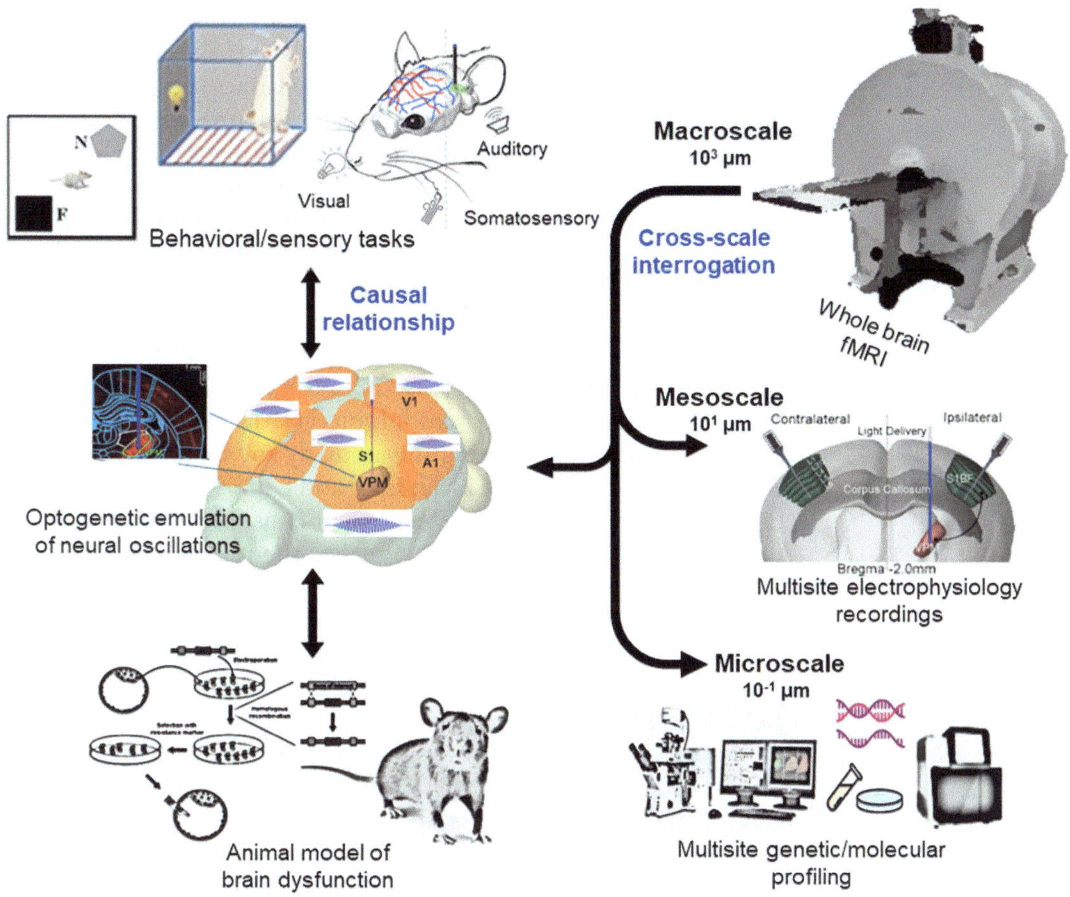

Fig. 1 Illustration of integrative multimodal neuroimaging for multiscale causal interrogation of brain-wide neural oscillations and behavioral relevance

1.1 Basic Theory of fMRI and Blood Oxygenation Level Dependent Signal

In recent decades, fMRI has been established as a robust noninvasive technique for probing large-scale functional brain networks and their activities in both human and animal models. It can monitor neural activities across the entire brain, whether in response to stimulations/tasks or during rest, with millimeter-level spatial resolution and ~1–2 s range temporal resolution. Since its introduction in 1990 [22], task-based fMRI has been employed extensively to study various sensory, motor, language, and cognitive functions [11, 12, 17, 23–29]. In parallel, immense efforts have been made in exploring the intricate, intrinsic functional networks of the brain at rest through resting-state fMRI (rsfMRI) [12, 13, 30–32]. By analyzing regional correlations of low-frequency temporal fluctuations in fMRI signals, this method facilitates the mapping of extensive whole-brain networks [31, 33]. The experimental simplicity of rsfMRI has led to numerous studies aimed at examining dynamics of brain connectivity across developmental stages [34], aging [35], various pathologies [36], and learning processes [37].

To date, the majority of fMRI studies rely on measuring the blood oxygenation level dependent (BOLD) signal across time, which reflects changes in vascular hemodynamics that correlate with neural activities [38]. Although the exact neurovascular coupling mechanism giving rise to BOLD signals is not yet fully elucidated, the prevailing understanding indicates that BOLD dynamics are linked to changes in blood oxygenation level, cerebral blood flow (CBF), and cerebral blood volume (CBV) resulting from fluctuation of neural metabolism in specific brain regions. Specifically, deoxygenated hemoglobin (dHb) in red blood cells acts as an endogenous paramagnetic contrast agent that can attenuate MR signal and alter the T2* time-constant. Increased neural activities consume oxygen released by the blood oxyhemoglobin (Hb), resulting in a higher concentration of dHb both intravascularly and extravascularly. This is accompanied by an increase in CBF and CBV, with enhanced CBF typically overcompensating the oxygen consumption increase and thereby amplifying the BOLD signal as the relative concentration of Hb rises against dHb. As such, the BOLD contrast utilizes the differential magnetic susceptibility of oxygenated versus deoxygenated blood [39]. This allows for the detection and characterization of shifts in the local distribution of Hb and dHb through large-scale BOLD fMRI.

With the rapid advent of high-field strength magnets to 7–17 Tesla, small animal fMRI studies predominantly employ echo-planar imaging (EPI) sequences to achieve rapid whole brain imaging at a submillimeter (0.2–0.5 mm) resolution [40–44]. In particular, after a small flip angle (50–60°) RF excitation pulse, gradient-echo EPI (GE-EPI) rapidly reverses the frequency-encoding gradient and applies a train of blipped phase-encoding gradient to sample the entire K-space at 1–2 s temporal resolution. GE-EPI uses a relatively short TE optimized for different field strengths (9–25 ms) to maximize sensitivity to T2* changes driven by the field inhomogeneities/magnetic susceptibility of dHb concentration alterations. Therefore, it provides a favorable balance between spatial and temporal resolution, both essential attributes for effective fMRI investigations.

Pioneering work by Logothetis and colleagues has demonstrated that the BOLD signal exhibits a high correlation with local field potentials (LFPs) [45, 46], which predominantly reflect a weighted summation of postsynaptic signal inputs to a neural ensemble. Therefore, an appreciable sum of evoked synaptic potentials is necessary to elicit a robust BOLD response in a particular brain region. Furthermore, variations in correlated spontaneous LFPs across different brain regions critically contribute to the dynamics of regional correlations of intrinsic BOLD signal fluctuations observed in fMRI connectivity analyses. Although discrepancies in the temporal and spatial resolution between BOLD and LFP signals can result in differences in brain signal organization, the brain-wide coordination of low-frequency neural activities (i.e., <20 Hz) shows a strong correspondence with BOLD activities, typically under 1 Hz [47–51]. Therefore, integrating electrophysiological recordings with fMRI can further enhance our capacity to investigate neural oscillations across the entire brain [5, 7, 45, 52–54].

1.2 Integration of fMRI and Optogenetics

Optogenetics combines optical and genetic methods to manipulate specific neuronal populations within the brain through precisely designed spatiotemporal excitatory or inhibitory modulation events [55–57]. Upon expressing light-sensitive opsins in genetically targeted neurons, researchers can achieve reversible neuromodulation at millisecond timescales and micrometer spatial resolution by delivering light via optical fibers. With its capability for delivering precise causal control while minimizing invasiveness, optogenetics has opened up numerous new landscapes for neuroscience, such as enabling the exploration of causal relationships between regional activations within and across neural circuits and influencing behaviors in both healthy subjects and those with neuropsychiatric disorders [56, 58, 59].

The combination of optogenetics with fMRI, thus allows for the mapping of large-scale network activities controlled causally through the stimulation or inhibition of specific local neuronal populations [16, 17, 60–74]. This integrative approach facilitates the simultaneous (1) perturbation of the long-range networks by spatiotemporally precise and cell-type-specific optogenetic control of local neuronal populations, and (2) monitoring of brain-wide neural activities evoked by local optogenetic perturbation. Consequently, optogenetic fMRI has emerged as a powerful tool for mapping the brain-wide downstream effects of cell-type or projection-specific manipulations, and for characterizing spatiotemporal response dynamics and interactions within the long-range networks.

Furthermore, several recent studies have highlighted the substantial potential to initiate local neural activities that mimic spontaneous neural oscillations while concurrently monitoring their brain-wide spatiotemporal propagation across multiple relevant brain regions [11–13, 16–18, 20]. While previous electrophysiological investigations suggested the involvement of numerous brain regions, their limited spatial coverage and resolution hindered direct interrogation. The integration of whole-brain fMRI mapping with optogenetics offers a powerful method for examining these dynamics. However, several challenges must be addressed to successfully induce and monitor neural activities that emulate brain-wide spontaneous neural oscillations across multiple aspects. Key challenges include: (1) precise expression of light-sensitive opsins and accurate light delivery within the targeted neuron populations; (2) optimization of temporal-frequency stimulation protocols suitable for emulating spontaneous neural oscillations; (3) development of effective multimodal neuroimaging protocols. The subsequent chapters will demonstrate detailed methodologies to address these challenges and advance a flexible optogenetic fMRI approach for studying brain-wide neural oscillations, using thalamic spindle oscillation as an example, and their multi-scale actions across brain-wide networks.

2 Materials

2.1 Animal Subjects and Virus Packaging

All animal experiments were approved by the University of Hong Kong's Committee on the Use of Live Animals in Teaching and Research (CULATR). Groups of age-matched male Sprague Dawley rats (6–7 weeks, 150–220 g during virus injection; 10–12 weeks, 250–350 g during fiber implantation) were used in this study. Animals were housed (individually, if after implantation surgeries) under a 12 h light/dark cycle with access to food and water ad libitum. We did not consider sex in our experiments, as no consensus has been made in the field on the sex differences of neural oscillations and their functions.

Recombinant adeno-associated virus (AAV) vectors were serotyped with AAV5 coat proteins and produced by the vector core at the University of North Carolina at Chapel Hill, Chapel Hill, NC. For excitatory thalamic nuclei, such as the ventral posteromedial nucleus (VPM) and the medial dorsal nucleus (MD), AAV5-CaMKIIα-ChR2-H134R-mCherry was used. The thalamic reticular nucleus (TRN) consists solely of inhibitory neurons, which are GABAergic in nature. We used AAV5-hSyn-ChR2-H134R-mCherry that expresses ChR2 in GABAergic-expressing neurons via the human synapsin (hSyn) promoter. These viral vectors (e.g., AAV5-CaMKIIα-ChR2-H134R-mCherry and AAV5-hSyn-ChR2-H134R-mCherry) and their usages are well characterized and commercially available from the Vector Core at the University of North Carolina (www.med.unc.edu/genetherapy/vectorcore).

3 Methods

3.1 Stereotactic Surgery for Viral Injection and Fiber Implantation

All stereotactic surgeries for viral injection were performed with rats at 6–7 weeks of age. Rats were anesthetized with an intraperitoneal bolus injection of a ketamine (90 mg/kg) and xylazine (40 mg/kg) mixture after 3%–5% isoflurane induction. The eyes were treated with ointment to avoid dehydration. Successful induction of anesthesia was confirmed by the absence of palpebral reflex (eye blink), the absence of foot withdrawal following pressure to the footpad, and no startle from tail pinch. Buprenorphine (0.05 mg/kg) was administered subcutaneously, and Emla ointment will be applied to the affected ear skin areas before fixing the rat skull in the stereotaxic frame with a tooth bar and intra-aural positioning ear bars. A small-animal heating system was used to maintain animal rectal temperature at ~37.0°C. The surgical site was shaved, and the remaining hair was removed via depilatory cream to expose the scalp with an area of 3 cm in diameter. A surgical scrub on the skin was conducted using betadine and rinsing with 70% ethanol, repeated twice, for a total of three times before making a midline 2 cm scalp incision and exposing the skull by retracting skin with hemostats. The stereotaxic frame was centered at the Bregma point, and the site of craniotomy was marked according to those reference points and the brain atlas. Following a small craniotomy (0.5–1 mm diameter) with a hand-held drill, viral injections were performed at two depths in VPM (−3.6 mm posterior to Bregma, +3.0 mm medial-lateral right hemisphere, −6.0 and −6.3 mm from the surface of dura), MD (−2.8 mm posterior to Bregma, +0.9 mm medial-lateral right hemisphere, −5.2 and −5.5 mm from the surface of dura), or in TRN (−2.7 mm posterior to Bregma, +3.8 mm medial-lateral right hemisphere, −5.1 and −5.4 mm from the surface of dura). A total of 3 μL of viral constructs (1.5 μL at each depth) were delivered through a 5-μL syringe and 33-gauge beveled needle at 150 nL/min with an injection pump. The injection needle was held in place for 5 min at each depth and was slowly retracted from the brain to ensure sufficient diffusion and to avoid any backflow. Scalp incision was sutured and disinfected, and animals were kept on a heating pad until recovery from anesthesia. Buprenorphine (0.05 mg/kg, subcutaneous) was administered postinjection twice daily for 72 h to minimize discomfort. Enrofloxacin was also administered orally for 72 h to minimize infection and inflammation postsurgery. Animals recovered for 4–6 weeks before conducting optical fiber implantation, MRI, electrophysiology, and behavioral experiments.

For fiber implantation, other procedures were the same as mentioned above for injection, while surgery was performed under 1.2%–2.0% isoflurane to implant an opaque custom-made plastic optical fiber cannula ($d = 450$ μm) at VPM/MD/TRN with subsequent high-resolution anatomical MRI verification under 1.0%–1.2% isoflurane maintenance [12–14, 16, 61]. Animals were only stabilized at 2% isoflurane during the first 15 min of surgery for craniotomy to minimize the duration of exposure to high-dose anesthesia. The animals were subsequently maintained with 1.2%–1.5% isoflurane for the remaining 15–30 min surgical procedure (i.e., dura removal and optical fiber implantation). The animals were then maintained under 1.0%–1.2% isoflurane during MRI setup and anatomical scans. Note that the fiber cannula was made opaque using heat-shrinkable sleeves to prevent light leakage during optogenetic stimulation [12–14, 16, 61]. The plastic fiber was first bent to create an arc with a 4 mm radius and secured permanently in place with UV glue. The fiber tip surface was beveled (45°) to facilitate fiber insertion and minimize injury to brain tissue. Two to three MR-compatible plastic screws were inserted into the skull and secured by UV glue. Dental cement was applied to fix the fiber cannula on the skull without air pockets. The optical fiber tip was typically situated in the center of the VPM/MD/TRN nucleus through stereotactic implantation and verified by high-resolution anatomical MRI. The spatial spread from the fiber tip for 473 nm blue light is rather small (200 μm and 350 μm at 50% and 10% of initial light intensity, respectively) and has little or no spreading in the backward and lateral directions (Fig. 2) [75], confining the light delivery within the targeted VPM/MD/TRN nucleus. Buprenorphine (0.05 mg/kg) was also given twice a day for three consecutive days as postoperation analgesics. After surgeries, animals observed not to be eating or showing a decrease in body weight will be offered a food nutrient supplement until weight gain is observed. Animals will be euthanized when showing as much as 20% weight loss.

3.2 Animal Preparation for fMRI Experiments

Endotracheal intubation was conducted with prior application of one drop of 2% lidocaine to the cords to minimize discomfort before MRI experiments. The animals were mechanically ventilated at a rate of 60 breaths per minute (180–350 mL/min) with ~1.0% isoflurane in room-temperature air using a ventilator (TOPO, Kent Scientific, Torrington, CT, USA). During all fMRI experiments, animals were placed on a plastic holder, and their heads were fixed with a tooth bar and intra-aural positioning ear bars. Custom-made earplugs were used to minimize acoustic noise. Continuous physiological monitoring was performed using sensors of an MRI-compatible system (SA Instruments, Stony Brook, NY, USA). Rectal temperature was maintained at ~37.0°C using a water circulation system. Vital signs were within normal physiological ranges (rectal temperature: 36.5–37.5°C, heart rate: 350–420 beats/min with <10 beats/min within-scan variation, breathing: ~60 breaths/min and not synchronized to optogenetic stimulation, oxygen saturation: >95%) throughout the length of the experiments [12, 13, 16]. End-tidal CO_2 (1.5%–3.5%) levels were also monitored. Water-based eye drops were applied to avoid dehydration. The eyes of the animals were then blindfolded in optogenetic fMRI experiments to avoid undesired visual stimulation [14, 64, 69]. Animal eyes were kept open with tapes in the visual fMRI (vfMRI) experiments. Our previous studies have demonstrated that such a preparation protocol with eyes either blindfolded [14] or uncovered [16] showed no evoked neural activities and fMRI responses in the visual pathway during light delivery to VPM for optogenetic stimulation in opsin-free control animals.

3.3 MRI Scanner-Synchronized Optogenetic and Sensory Stimulation

An Arduino programming board was used to synchronize the scanner trigger and the lasers/stimulators for optogenetic and sensory stimulation. Computers and stimulation systems were kept outside

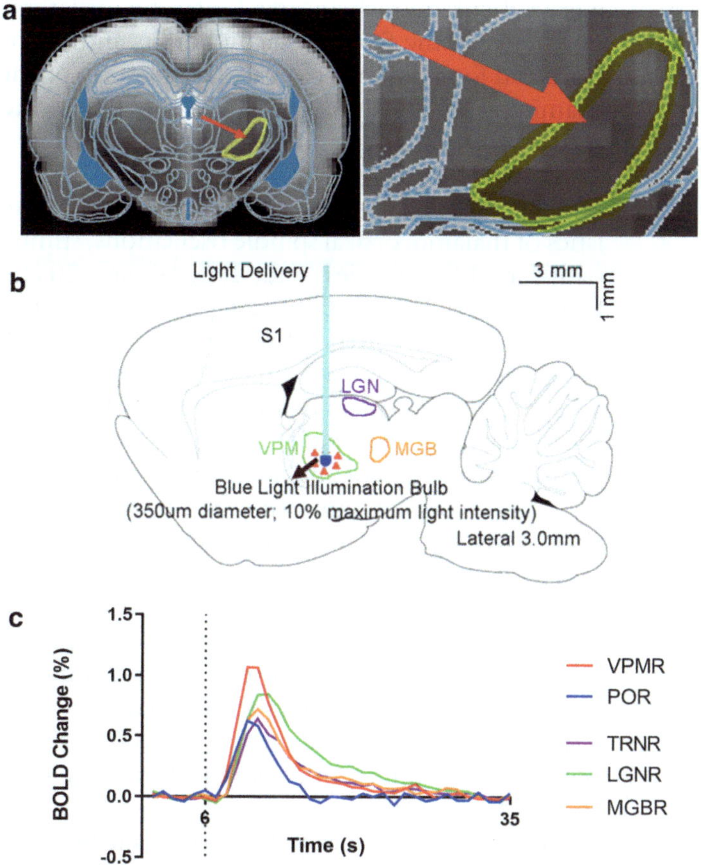

Fig. 2 Confirmation of spatially precise optogenetic stimulation of the targeted thalamic nucleus. (**a**) Averaged 2D T2-weighted anatomical MRI images of the VPM fiber implanted animal brains overlaid with the Paxinos atlas. The whole field-of-view image showed the perpendicular optical fiber implantation tract (left), and the enlarged image showed that the fiber tip was located at the middle of VPM (right). Fiber tip location is indicated by the red arrow, where the fiber tract (perpendicular continuous darkened pixels) was discontinued by a lighter pixel. VPM is demarcated by the yellow line. Image resolution is 0.25 mm^2 isotropic. (**b**) The light delivery procedure used in our experiments only directly activates the targeted thalamic nucleus. (**c**) Averaged BOLD signal profiles extracted from different thalamic nuclei (right brain hemisphere) showing the delayed responses to 8 Hz 24-pulse stimulation in non-somatosensory thalamic nuclei (TRN, LGN, and MGB)

the magnet. Optical patch cables (5–10 m) and electrode wires (4 m) delivered light and electrical pulses into the bore of the scanner. For optogenetic stimulation, blue light was delivered using a 473 nm diode-pumped solid-state laser measured before scanning as 8 mW at the fiber-tip (450 μm, NA = 0.5) corresponding to a light intensity of 40 mW/mm^2. For visual stimulation, blue light was delivered binocularly using a 473 nm DPSS laser via a separate optical patch cable. The emitted light was measured before the start of scanning as 0.5 mW at the fiber tip (400 μm, NA = 0.39). Laser stimulation protocols varied depending on the fMRI experiment parameters. For somatosensory forepaw stimulation, two needle electrodes were subcutaneously inserted into the ipsilateral/right forepaw: one between the first and second digits and the other between the third and fourth digits, typical placements in previous fMRI studies examining somatosensory forepaw functions. The electrodes were secured using surgical tape, and the stimulus effectiveness was first qualitatively confirmed by digit twitching.

To examine the brain-wide propagation and targeting characteristics of thalamocortical spindle oscillations, stimulation pulse trains with varied frequencies (e.g., 4, 8, 14, and 20 Hz for 24-pulse) or lengths (e.g., 2-, 4-, 8-, 16-, 24-, and 96-pulse for 8 Hz) were used (light intensity = 8 mW: 40 mW/mm^2; 10 ms or 63 ms pulse width). We covered slow and fast spindle frequencies (8 and 14 Hz, respectively), and frequencies below or above (4 and 20 Hz, respectively) the typical range of spindle frequency. Lengths for 8 Hz stimulations were chosen within or above the typical range of spindle length (i.e., 0.5, 1, 2, 3, or >3 s, equal to 4, 8, 16, 24, or >25 cycles). Two sessions were acquired for each stimulation pulse train paradigm. The sequence of varied stimulation paradigms was randomized across animals. For each session, the same stimulation pulse train was repeated every 30 s. The interval between two consecutive stimulation pulse trains was set as 30 s to ensure that the density of the evoked spindle-like activities was within the natural density range of spontaneous spindle oscillations (2–12 events/min) [76–78]. Note that identical optogenetic stimulation paradigms were employed for VPM, MD, and TRN experiments.

To examine whether and how optogenetically evoked spindle-like activities act on brain-wide functional responses to memory consolidation-associated visual stimulation (i.e., memory representation), vfMRI experiments were conducted 8 days before (PRE) and 1 day after (POST) animals underwent the same habituation, acquisition, and consolidation phases as in the fear conditioning experiments. For optogenetic stimulation, an 8 Hz 24-pulse paradigm delivered once every 30 s was used, as it is the most effective paradigm in evoking brain-wide spindle-like activities. For visual stimulation, 10 s 5 Hz light flash was employed (light intensity = 0.5 mW: 50% duty cycle). Blue light flashing at 5 Hz was delivered via an optical fiber placed 3.5 cm in front of the eyes for binocular visual stimulation in vfMRI. The vfMRI paradigm used consisted of six blocks of 10 s visual stimulation and 30 s rest. To examine effects of optogenetically evoked spindle-like activities on general somatosensory or cross-modal somatosensory-visual processing, binocular visual stimulation or ipsilateral electrical stimulation were presented at 4 Hz in a block design paradigm (30 s pulse-off followed by 20 blocks of 5 s pulses-on and 25 s pulses-off, 25 ms pulse width for visual stimulation, a constant voltage of 8 V with 3 ms square wave pulses for forepaw stimulation). Sixteen sessions (eight for PRE and eight for POST) were acquired in total for each animal.

To examine the effects of optogenetically evoked spindle-like activities on brain-wide long-range rsfMRI connectivity, a total of eight sessions were acquired for each animal, four before (PRE) and four after (POST) optogenetic stimulation. For each session of optogenetic stimulation (OG-On), the 8 Hz 24-pulse stimulation paradigm was employed.

3.4 MRI Acquisition Procedures

All MRI experiments were performed on a 7 T MRI scanner (ParaVision v5/v6, PharmaScan 70/16, Bruker Biospin GmbH, Ettlingen, Germany) using a transmit-only birdcage coil in combination with an actively decoupled receive-only surface coil. A single-channel receive-only surface coil was used for the optogenetic fMRI, sensory fMRI, and rsfMRI experiments. The animal was placed in the magnet, with its brain located at the center of the magnetic field. Scout and anatomical RARE T2-weighted (T2W) images were first acquired for accurate positioning and reference with a field of view (FOV) = $32 \times 32\ mm^2$, matrix = 256×256, RARE factor = 8, echo time (TE) = 36 ms, and repetition time (TR) = 4200 ms. Sixteen contiguous 1.0-mm slices were positioned in the transverse orientation according to the rat brain atlas (38–40) to cover the majority of the brain. A field map covering the whole brain was acquired, and local shimming of the rat brain was conducted. All fMRI, sensory fMRI, and rsfMRI data were obtained at the same geometry as the anatomical reference T2W images, using a single-shot gradient-echo echo-planar imaging (GE-EPI) sequence with field of view (FOV) = $32 \times 32\ mm^2$, matrix = 64×64 (resolution = $0.5 \times mm^2$), flip angle (FA) = 56° (optogenetic and sensory fMRI) or 50° (rsfMRI), echo time (TE) = 20 ms, repetition time (TR) = 1000 ms (optogenetic and sensory fMRI) or 750 ms (rsfMRI), and interleaved slice acquisition. A total of 600 s volumes were collected during each fMRI or rsfMRI session, while 270 s volumes were acquired for each sensory fMRI session. These procedures ensured a sufficient temporal signal-to-noise ratio at >50 levels across cortical regions.

3.5 Multisite Electrophysiology Recordings

Electrophysiology recordings were performed under the same anesthesia protocols and similar physiological conditions as in the fMRI experiments to confirm the successful induction of thalamocortical spindle oscillations and their brain-wide propagation. Local field potential (LFP) and multiunit activity (MUA) recordings at the thalamic nuclei (VPM/MD/TRN) were conducted to confirm the evoked spindle-like neural activities. An optrode was constructed by a fiber cannula and a single-channel electrode and was implanted, followed by identical procedures as described for optogenetic fMRI/rsfMRI experiments. Only 8 and 14 Hz 24-pulse, and 8 Hz 4-, 8-, and 16-pulse stimulations evoked rhythmic burst-like MUA activities that corroborated previously recorded spontaneous and optogenetically evoked spindle oscillations [10, 79–81] within typical frequencies and lengths (7–15 Hz, 0.5–3 s, and 4–25 cycles). Both spontaneous and induced spindle oscillations were detected as 7–15 Hz, 0.5–3 s, and 4–25 cycles, brief oscillatory events that exhibit distinct spindle-shaped signal envelopes (i.e., wide in the middle and tapering at both ends) in LFP waveforms [76, 81, 82]. Co-occurrences of spontaneous spindle oscillations (7–15 Hz) and slow oscillations (<2 Hz) were also detected. Other recording locations were then strategically chosen based on our fMRI findings of brain-wide cross-modal targets, particularly at subcortical and deep cortical regions, of thalamically evoked activities at spindle frequencies and lengths (e.g., under 8 Hz 24-cycle stimulation). Craniotomies were made, and dura matter was removed with reference to ipsilateral primary somatosensory cortex barrel field (S1BF, −1.6 mm anterior to Bregma, +5.4 mm ML), amygdala (Amg, −3.0 mm anterior to Bregma, + 5.0 mm ML), medial prefrontal cortex (mPFC, +3.0 mm anterior to Bregma, +0.6 mm ML), and retrosplenial cortex and superior colliculus (RS and SC, −6.4 mm anterior to Bregma, +1.0 mm ML). Amg, RS, and SC were chosen to confirm the detection (i.e., using BOLD fMRI mapping) of deep and/or small regions as targets of thalamocortical spindle oscillations. Among them, SC represented the subcortical sensorimotor regions, while Amg and deep subregions in RS (granular area) and mPFC (prelimbic area) represented the remaining subcortical and deep cortical limbic regions, respectively. mPFC, the prominent anterior projection target of MD [83], was chosen to echo RS (posterior cortical limbic region) for the purpose of verifying the robust cross-modal recruitment of cortical limbic regions over anterior and posterior brain.

To examine robust recurrent activities over multiple cortical layers (II/III to V/VI) evolving with the spindle-shaped profile at ipsilateral S1BF, a linear microelectrode silicon array (16 recording channels equally spaced at 100 μm; 1.5 MΩ impedance; NeuroNexus Technologies, Ann Arbor, MI) was inserted perpendicular to the cortical surface approximately 2.5 mm below the dura matter with micromanipulators (Narishige, Amityville, NY, USA). For other regions, single-channel electrodes with 2.0 MΩ and 10-μm tip diameter were implanted (Amg: −8 mm DV; mPFC: −3 mm DV; RS: −2.0 mm DV; SC: −3 mm DV). S1BF recordings were acquired using a multichannel neurophysiology recording system (sampling frequency: 24 kHz; notch-filter: 50 Hz, 100 Hz, and 150 Hz; Synapse v86, Tucker Davis Technologies/TDT, Alachua, FL). Synchronized laser stimulation was controlled by the same system, and light pulses were recorded simultaneously with the neural data. Electrophysiological data from VPM/MD/TRN, Amg, mPFC, RS, and SC were acquired with a 32-channel OpenEphys data acquisition system (digitized at 30 kHz and bandpass filtered between 0.1 kHz and 8 kHz). The experimental paradigms were the same as those used in optogenetic fMRI experiments, and stimulations were controlled by the TDT system.

3.6 Visual-Somatosensory Associative Fear Conditioning Experiments

To examine the functional effects of evoked brain-wide spindle oscillations on memory consolidation, the memory performances of age-matched optogenetic (OG), Sham, and Naive normal animal groups were tested using fear conditioning experiments with combined sequential vfMRI experiments. Before OG or Sham animals underwent fear conditioning experiments, chronic fiber implantation was performed using similar procedures that were utilized for the MRI experiments. Animals recovered for 1 week and subsequently underwent 15 min of animal handling and fiber acclimatization for 1 week in the habituation phase. Naive animals also underwent the same habituation phase.

Memory acquisition and extinction were performed in the day phase using a startle and fear conditioning system (Panlab Harvard Apparatus, MA, USA). During the acquisition phase, each animal was first placed in a black chamber with a grid floor. After 2 min of exploration and adaptation, the animal received four pairs of 10 s, 5 Hz light flashes (CS, conditioned stimulus) co-terminated with a 1 s, 0.6 mA foot shock (US, unconditioned stimulus) at randomized intervals of 75 to 135 s for the acquisition of visual-somatosensory associative fear memory. The animal was moved back to the home cage after 2 min of rest. Based on the previously reported critical window for spindle-associated memory consolidation as 2–6 h after acquisition [78, 84, 85], we stimulated the OG and Sham animals at ~3 h after memory acquisition. An 8 Hz 24-pulse stimulation paradigm was used as the most effective paradigm in evoking brain-wide spindle oscillations. Both the light flash and optogenetic stimulation were controlled by the same system used in electrophysiology experiments. Twenty-four hours following memory acquisition, all groups underwent the extinction phase. Animals were placed in a white chamber with flat metal floors, a different context from the one received during the acquisition phase. After 2 min of exploration, they received 45 periodic repetitions of 10 s, 5 Hz light flashes (at intervals of 10 s) in the absence of foot shock. After each acquisition or extinction phase, the chambers were cleaned with 70% ethanol and allowed to fully dry to minimize olfactory cues.

For animals that underwent vfMRI experiments, baseline vfMRI sessions (PRE) were acquired 1 day before the habituation phase, while the other vfMRI sessions (POST) were acquired during the same phase as memory extinction in the pure behavioral experiments. The vfMRI animals underwent the same habituation, acquisition, and consolidation phases as the animals in the pure behavioral experiments. Freezing levels during each CS presentation were reported for memory acquisition. Animals met the same outlier exclusion criteria as in pure behavioral experiments, or displayed weak baseline vfMRI activations, were excluded from further analysis. Comparisons of freezing levels during acquisition were done by two-way ANOVA with post hoc Bonferroni-corrected *t* tests to ensure successful memory acquisitions in vfMRI animals and to validate that any between-conditions differences (i.e., POST vs. PRE) in the vfMRI results were learning/memory-dependent.

3.7 Accelerated Aging Animal Model, Genetic, Histological, and Synaptic Morphological Profiling

Spindle oscillations change with age and have been proposed as a marker of aging-related memory deficits [86–88]. To further investigate the causal actions of spindle oscillations in critical functions, such as memory, it is essential to examine the effects of aging in disrupting brain-wide spindle propagation and memory, and to explore the potential therapeutic effects of boosting brain-wide spindle propagation on aging-associated memory deficits. VPM optogenetically transfected animals were treated with daily injection of D-galactose (50 mg/kg, subcutaneous) for 8 weeks to induce an accelerated aging animal model [89] before chronic fiber implantation, behavioral, and fMRI experiments. Aging-associated memory deficits were examined by visual-somatosensory associative fear conditioning experiments. Guided by fMRI results, brain tissues in multiple key regions were extracted immediately after experiments and were analyzed using the next-generation-sequencing (NGS) method to identify memory-related gene expressions significantly regulated with OG compared to Sham in control animals. Further analyses, including quantitative polymerase chain reaction (qPCR), focused on identifying the alteration of these gene expressions in aging animals and their normalization after boosting brain-wide spindle propagation. Histological and synaptic morphological profiling were conducted to confirm the outcome of such causal relationships.

To confirm the specific expression of ChR2-mCherry in the VPM/MD excitatory neurons and TRN inhibitory neurons, histology was performed according to the previously published procedure [12, 16]. Upon completion of experiments, a selected number of optogenetically transfected animals were anesthetized with pentobarbital and then transcardially perfused with ice-cold 4% paraformaldehyde (PFA) in phosphate-buffered saline (PBS). The brains were equilibrated in 20% sucrose in PBS at 4°C overnight. Axial sections (40 μm) were prepared on a freezing microtome (model 860, AO Scientific Instruments). Consecutive sections were mounted and examined with a laser confocal microscope (Carl Zeiss LSM780). For immunohistochemistry, free-floating sections were processed with 5% normal goat serum and 0.3% Triton X-100 in PBS with primary antibodies against rabbit polyclonal to CaMKIIα (1:400; Abcam) at 4°C for 24 h. After washing with PBS, sections were then incubated for 2 h at room temperature with secondary antibodies Alexa Fluor 647 conjugate goat anti-rabbit IgG and Alexa Fluor 488 conjugate goat anti-guinea pig IgG (both 1:500; Molecular Probe). Slices were then washed and mounted using FluoroShield mounting medium with DAPI (Abcam). Double or triple immunofluorescence was assessed with a laser confocal microscope (Carl Zeiss LSM780). To confirm the effects of spindle propagation on gene expression, selected genes were also stained. The number of cells expressing these genes across key brain regions, informed by fMRI mapping, was then analyzed. Last, for synaptic morphological profiling, quantitative analysis of dendritic spine density was conducted. Secondary and tertiary apical dendritic segments parallel to the section surface were imaged using a Nikon C2 confocal microscope (Japan) with a 60× oil objective. Z-stack imaging was performed with 0.1-μm step intervals along the *z* axis. The images were imported, calibrated, and manually traced using Imaris 8.1 software.

4 Notes

4.1 Ensuring Favorable Brain States During Experiments

Spindle oscillations occur during sleep and at other states, such as under drowsy, awake quiescence, and light anesthesia states [52, 79, 90, 91]. Since optogenetic fMRI experiments were conducted under anesthesia, it is important to maintain a brain state in which spontaneous spindle oscillations occur naturally [11]. Converging evidence indicates that the influence of anesthesia on characteristics of spindle oscillations is dosage- and stage-dependent [92–95]. For example, spindle oscillations are suppressed under deep anesthesia stages (e.g., high-dose isoflurane at >1.5%) but are slightly decreased under light anesthesia (e.g., low-dose isoflurane at 0.5%–1.0%) [52, 92, 94, 96, 97]. Further, the basic characteristics of spindle oscillations, such as frequency, length, and brain-wide responses, are largely preserved under light anesthesia states, such as with low dosage isoflurane (0.5%–1.0%) [98, 99], dexmedetomidine plus low dosage isoflurane [96], remifentanil [52], urethane or ketamine plus xylazine [81, 100, 101], and barbiturate [101–103]. We controlled this by carefully optimizing our anesthesia induction and maintenance procedures, i.e., shortened exposure to high-dose isoflurane during induction and applied low-dose anesthesia exposure for maintenance during experiments (detailed in Sects. 3.1 and 3.2).

Further, this was confirmed by examining S1BF LFP activities [11]. Spontaneous thalamocortical spindle oscillations were detected at S1BF recordings with an established amplitude and length thresholding based automatic spindle detection algorithm [76, 104] followed by visual inspection. The detected spontaneous thalamocortical spindle oscillations in our recordings exhibit well-documented spindle characteristics, such as bursts of oscillatory activities in the spindle frequency range (7–15 Hz); spindle-shaped signal envelopes (i.e., wide in the middle and tapers at both ends), which are especially prominent in the 7–15 Hz bandpass filtered LFP signals; and 0.5–3 s in length (a majority of them are ~1 s) [76, 82, 104]. Note that, as expected, we detected the co-occurrences of spontaneous spindle oscillations (7–15 Hz) and slow oscillations (< 2 Hz). The densities of spindle oscillations and slow oscillations under 1.0% isoflurane in the baseline period (i.e., prior to any optogenetic stimulation) were 5.34 ± 0.89 events/min (via automatic detection) and 12–36 events/min (via visual inspection), respectively, comparable to 2–12 spindle events/min and 10–50 slow oscillation events/min measured in rodent/monkey brains during natural slow-wave or NREM sleep [52, 78, 79, 91]. We also found that the density of spontaneous spindle oscillations was similar to baseline (5.88 ± 0.83 vs. 5.34 ± 0.89 events/min) when examining periods after 8 Hz 24-pulse stimulation (post, 27 s period after each 3 s optogenetic stimulation pulse train). Further, stimulations did not alter the length and amplitude of spontaneous spindle oscillations (723.0 ± 28.2 ms vs. 672.5 ± 26.9 ms and 98.20 ± 12.28 μV vs. 96.01 ± 9.20 μV for post- vs. pre-stimulation conditions). The density, length, and amplitude of spontaneous spindle oscillations were also confirmed to be unaffected across different light anesthesia levels of 0.0%–1.5% isoflurane. These validation procedures are critical to ensure that spontaneous spindle oscillations were not diminished and altered in our optogenetic stimulation experiment under light anesthesia conditions.

During visual specifically, when animals had rest for ~3 h in the consolidation phase and were observed to be asleep (with eyes closed and curled-up body posture [105, 106]) or immobile/quiet resting (i.e., quiescent state) for a ~30 min period, periodic 8 Hz 24-pulse stimulations were delivered once every 30 s for 40 min. The behavioral state of the animal during stimulation was maintained comparable to the drowsy, asleep, and awake quiescent states in previous literature, whereby spontaneous spindle oscillations were detected [52, 79, 90, 91]. Similar stimulation protocols were shown to successfully avoid the disruption of sleep [78, 79, 91].

4.2 Careful Development of Suitable Temporal-Frequency Stimulation Protocols

To ensure our induced activities emulate spontaneous spindles oscillations from multiple aspects, we carefully controlled the stimulation parameters as in the five optogenetic spindle studies by others so far [78–81, 91], where spindle oscillations were evoked from the thalamus. Our stimulation parameters were also designed with reference to the temporal characteristics of spontaneous spindle oscillations, such as frequency (7–15 Hz), length (0.5–3 s and 4–25 cycles), and occurrence density (2–12 events/min). Previous optogenetic spindle studies [78–81, 91] demonstrated the similarities of evoked spindle-like activities to spontaneous spindle oscillations, mainly through the examination of representative LFP/EEG waveforms [78–81, 91] and their spectra/spectrogram [78–81], and comparison of behavioral outcomes [78, 79, 91], so to generalize their evoked neural activities as spindle oscillations. We adopted a similar approach to demonstrate that the key features of our optogenetically evoked spindle-like activities were similar, which include their electrophysiological characteristics at local thalamic and remote cortical/subcortical regions, and their functional/behavioral effects on memory consolidation. Further, unlike the previous optogenetic spindle studies, we performed additional analyses on the optogenetically evoked spindle-like activities and showed their similarities to spontaneous spindle oscillations in terms of co-occurring slow oscillations and recurrent activities across multiple cortical layers [76, 82, 104, 107, 108]. We also confirmed that the brain-wide propagation characteristics of evoked spindle-like activities were not affected by light anesthesia and reduced optogenetic stimulation light intensities.

The intensity of our optogenetic light pulse (40 mW/mm^2) is much lower than that utilized in previous studies (76–650 mW/mm^2) [78–81, 91]. Further, each of our optogenetic stimulation events was delivered via a pulse train consisting of 10 ms pulses, not via a long pulse continuous throughout the whole stimulation period (e.g., in total, 24 10 ms pulses, not a continuous 3 s pulse, for the 8 Hz 24-pulse stimulation paradigm). Optogenetic stimulation with similar intensity and duty cycle caused no tissue damage based on our previous optogenetic fMRI studies [12, 16, 109]. In addition, although we did not directly measure the local temperature rise in this study, it was estimated to be <0.1 °C based on the previous literature [110]. Such careful control of stimulation parameters is essential for avoiding tissue heating or other undesired confounds.

In this study, we did not explicitly encode spindle shape in our stimulation paradigm. Occurrence of spindle shape depends on activity recurrence, facilitation, and synchronization in the initial spindle generation phase, and subsequent activity desynchronization in the later termination phase of the spindle events. Therefore, spindle shape does not solely depend on the stimulation design but is due to the interplay between the stimulation/input and the inherent response properties of the thalamocortical circuitry.

5 Conclusion

This integrative multi-scale multimodal neuroimaging protocol centered around optogenetic fMRI provides enormous potential. For example, this powerful approach has been demonstrated to enable systematically elucidating the multi-scale spatiotemporal dynamics and functional mechanisms underlying the emergence of intrinsic spontaneous neural oscillations in local meso/microscale brain circuits, their long-range brain-wide propagation properties, and hierarchical cross-scale interactions with brain-wide networks in specific functions and behaviors [11, 15, 20, 21]. Ideally, future studies should include experiments performed in animals without anesthesia to completely overcome the confounding effects. It would also be ideal to monitor the sleep–wake state of animals during behavioral experiments. However, even though several awake or sleep mouse fMRI studies have been reported in recent years [71, 111–114], unanesthetized rat fMRI experiments are technically challenging because stable fixation of the rat during MRI scanning remains technically difficult and is unreliable in our experience, as the rat head cannot be restricted readily, severely degrading fMRI quality and detection sensitivity. Further, brutal head immobilization could induce stress. Future studies could be conducted with simultaneous fMRI-electrophysiology recordings and optogenetic perturbation. Ongoing efforts have adopted this approach to further ensure the observed brain-wide network activities on fMRI mapping and multisite electrophysiology recordings are not merely parallel phenomena [113, 115–117]. However, care should be taken not to damage brain tissues, which will be utilized for further genetic and molecular profiling.

To further examine the brain-wide spatiotemporal coordination of slow oscillation, spindles, and ripples, future studies could extend our protocol to target the events where these oscillations are coupled and uncoupled. In our previous study [11], we did not detect hippocampal ripple activity in our LFP recordings during optogenetic stimulation. The purpose of the electrophysiology recording experiments was primarily to confirm the brain-wide spindle propagation characteristics observed in fMRI experiments. Recording locations were chosen based on our fMRI findings of brain-wide cross-modal targets of the optogenetically evoked spindle-like activities. We did not record hippocampal LFPs in our experiments because of the following considerations: (1) the relatively weaker fMRI responses in the hippocampus compared to those of the selected recording regions (BOLD signal amplitude of $<0.5\%$ vs. $\geq 1.0\%$), which could be due to weaker propagation of evoked spindle-like activities to hippocampus; and (2) the challenge with placing electrode probes at too many locations for simultaneous recording. Meanwhile, the recording electrodes in the previous study were not located in the regions receiving direct anatomical projections from the hippocampus. Therefore, our recordings were not optimized for detecting hippocampal ripples, and thus were not able to sensitively detect ripple activity during our evoked spindle-like activities.

We also have not yet explicitly encoded the parameters of slow oscillation in the stimulation paradigm [11], as it has been demonstrated that evoking spindle-like activities also triggers the co-occurrence of slow oscillations [78]. Furthermore, since the brain activities of the animals were under a state similar to slow-wave or NREM sleep states in our experiments, stimulation of the sensory thalamocortical circuits could generate slow oscillations with high probability [118]. This is likely because the generation of these activities relies on similar thalamocortical circuits and brain states [10, 119]. Slow oscillations are typically initiated from the cortex [9, 10, 120]. In our study [11], however, they are the subsequent secondary activities that are induced after the spindle-like activities that are evoked at the thalamus propagate to the cortex. Although the co-occurrence of slow oscillations and spindle activities is an important feature, it is not an indispensable aspect of spontaneous spindle activities, as they do not always co-occur (e.g., usually this phenomenon only represents 15%–40% of detected spindle events) and have different origins [9, 10, 78, 119, 121]. Ideally, future studies focusing on couplings of spindle-slow oscillation-ripples should simultaneously initiate or perturb them from their different origins.

In conclusion, the multi-scale multimodal causal neural probing protocol overcomes the limitations of single-modality studies, enabling systematic causal interrogation of complex brain functional architecture governing brain-wide actions of neural oscillations. In the coming future, this approach will facilitate the identification of multi-scale brain functional architecture biomarkers for brain disorders and preclinical development of novel precision intervention strategies leveraging these mechanisms.

Conflict of Interest

The author has no conflicts of interest to declare that are relevant to the content of this chapter.

References

1. Raichle ME, Mintun MA (2006) Brain work and brain imaging. Annu Rev Neurosci 29:449–476.
2. Raichle, ME (2015) The restless brain: how intrinsic activity organizes brain function. Philos Trans R Soc Lond B Biol Sci 370(1668):20140172.
3. Berger H (1929) Über das elektrenkephalogramm des menschen. Eur Arch Psychiatry Clin Neurosci 87(1):527–570.
4. Buzsaki G (2006) Rhythms of the Brain. Oxford University Press
5. Logothetis NK (2015) Neural-event-triggered fMRI of large-scale neural networks. Curr Opin Neurobiol 31:214–222.
6. Staresina BP et al (2015) Hierarchical nesting of slow oscillations, spindles and ripples in the human hippocampus during sleep. Nat Neurosci 18(11):1679–1686
7. Florin E, Watanabe M, Logothetis NK (2015) The role of sub-second neural events in spontaneous brain activity. Curr Opin Neurobiol 32:24–30.
8. Buzsaki G, Logothetis N, Singer W (2013) Scaling brain size, keeping timing: evolutionary preservation of brain rhythms. Neuron 80(3):751–764.
9. Mak-McCully RA et al (2017) Coordination of cortical and thalamic activity during non-REM sleep in humans. Nat Commun 8:15499.
10. Steriade M, McCormick DA, Sejnowski TJ (1993) Thalamocortical oscillations in the sleeping and aroused brain. Science 262(5134):679–685.
11. Wang X et al (2023) Functional MRI reveals brain-wide actions of thalamically-initiated oscillatory activities on associative memory consolidation. Nat Commun 14(1):2195.
12. Chan RW et al (2017) Low-frequency hippocampal–cortical activity drives brain-wide resting-state functional MRI connectivity. Proc Natl Acad Sci USA 114(33):E6972–E6981.
13. Wang X et al (2019) Thalamic low frequency activity facilitates resting-state cortical interhemispheric MRI functional connectivity. Neuroimage 201:115985.
14. Leong ATL et al (2021) Neural activity temporal pattern dictates long-range propagation targets. Neuroimage 235:118032.
15. Wang X et al (2022) Optogenetically-evoked brain-wide spindles alleviate associative memory dysfunction in aging animal. In: ISMRM 2022 Annual Meeting. Virtual Conference. p4668.

16. Leong AT et al (2016) Long-range projections coordinate distributed brain-wide neural activity with a specific spatiotemporal profile. Proc Natl Acad Sci USA 113(51):E8306–E8315.
17. Leong ATL et al (2023) Hippocampus modulates vocalizations responses at early auditory centers. Neuroimage 270:119943.
18. Ma T et al (2024) Olfactory cortical outputs recruit and shape distinct brain-wide spatiotemporal networks. eLife RP101189.
19. Liu J et al (2022) Astrocyte dysfunction drives abnormal resting-state functional connectivity in depression. Sci Adv 8(46):eabo2098.
20. Liu J et al (2025) Brain-wide spatiotemporally distinct traveling waves drive anxiety-like behaviors in mice. Adv Sci 12(40):e11867.
21. Wang X et al (2023) Brain-wide fMRI connectivity and regional genetic modulations underlying optogenetically-evoked spindles in rescuing memory decline in aging. In: ISMRM 2023 Annual Meeting, Toronto. p1306.
22. Ogawa S et al (1990) Brain magnetic resonance imaging with contrast dependent on blood oxygenation. Proc Natl Acad Sci USA 87(24):9868–9872.
23. Belliveau J et al (1991) Functional mapping of the human visual cortex by magnetic resonance imaging. Science 254(5032):716–719.
24. Desmond JE et al (1995) Functional MRI measurement of language lateralization in Wada-tested patients. Brain 118(6):1411–1419.
25. D'esposito M et al (1995) The neural basis of the central executive system of working memory. Nature 378(6554):279.
26. Cheung MM et al (2012) BOLD fMRI investigation of the rat auditory pathway and tonotopic organization. Neuroimage 60(2):1205–1211.
27. Lau Cet al (2011) BOLD responses in the superior colliculus and lateral geniculate nucleus of the rat viewing an apparent motion stimulus. Neuroimage 58(3):878–884.
28. Chan KC et al (2010) Functional MRI of postnatal visual development in normal and hypoxic–ischemic-injured superior colliculi. Neuroimage 49(3):2013–2020.
29. Gao PP et al (2015) BOLD fMRI study of ultrahigh frequency encoding in the inferior colliculus. Neuroimage 114:427–437.
30. Biswal B et al (1995) Functional connectivity in the motor cortex of resting human brain using echo-planar MRI. Magn Reson Med 34(4):537–541.
31. Fox MD, Raichle ME (2007) Spontaneous fluctuations in brain activity observed with functional magnetic resonance imaging. Nat Rev Neurosci 8(9):700–711.
32. Zhou IY et al (2014) Brain resting-state functional MRI connectivity: morphological foundation and plasticity. Neuroimage 84:1–10.
33. Dosenbach NUF, Raichle ME, Gordon EM (2025) The brain's action-mode network. Nat Rev Neurosci 26(3):158-168.
34. Chan RW et al (2015) Structural and functional brain remodeling during pregnancy with diffusion tensor MRI and resting-state functional MRI. PLoS One 10(12):e0144328.
35. Damoiseaux JS et al (2007) Reduced resting-state brain activity in the "default network" in normal aging. Cereb Cortex 18(8):1856–1864.
36. Drysdale AT et al (2017) Resting-state connectivity biomarkers define neurophysiological subtypes of depression. Nature Med 23(1):28–38.
37. Taubert M et al (2011) Long-term effects of motor training on resting-state networks and underlying brain structure. Neuroimage 57(4):1492–1498.
38. Kim SG, Ogawa S (2012) Biophysical and physiological origins of blood oxygenation level-dependent fMRI signals. J Cereb Blood Flow Metab 32(7):1188–1206.
39. Kim SG, P Bandettini (2023) Principles of BOLD functional MRI. In: SH Faro et al (eds) Functional Neuroradiology: Principles and Clinical Applications. Springer, US, p 461-472.
40. Schmitt F, Stehling MK, Turner R, (2012) Echo-planar imaging: theory, technique and application. Springer Science & Business Media
41. Grandjean J et al (2020) Common functional networks in the mouse brain revealed by multi-centre resting-state fMRI analysis. Neuroimage 205:116278.
42. Grandjean J et al (2023) A consensus protocol for functional connectivity analysis in the rat brain. Nature Neurosci 26(4):673–681.
43. Vrooman RM et al (2025) fMRI data acquisition and analysis for task-free, anesthetized rats. Nat Protoc 20(6):1393–1412.
44. Klohs J, Chen WC, Araki R (2025) Advanced preclinical functional magnetic resonance imaging of the brain. npj Imaging 3(1):27.
45. Logothetis NK et al (2001) Neurophysiological investigation of the basis of the fMRI signal. Nature 412(6843):150–157.
46. Logothetis NK, Wandell BA (2004) Interpreting the BOLD signal. Annu Rev Physiol 66:735–769.
47. Schölvinck ML et al (2010) Neural basis of global resting-state fMRI activity. Proc Natl Acad Sci USA 107(22):10238–10243.
48. Wang L et al (2012) Electrophysiological low-frequency coherence and cross-frequency

coupling contribute to BOLD connectivity. Neuron 76(5):1010–1020.
49. Matsui T, Murakami T, Ohki K (2016) Transient neuronal coactivations embedded in globally propagating waves underlie resting-state functional connectivity. Proc Natl Acad Sci USA 113(23):6556–6561.
50. Ma Y et al (2016) Resting-state hemodynamics are spatiotemporally coupled to synchronized and symmetric neural activity in excitatory neurons. Proc Natl Acad Sci USA 113(52):E8463–E8471.
51. Pan WJ et al (2013) Infraslow LFP correlates to resting-state fMRI BOLD signals. Neuroimage 74:288–297.
52. Logothetis NK et al (2012) Hippocampal-cortical interaction during periods of subcortical silence. Nature 491(7425):547–553.
53. Ramirez-Villegas JF et al (2021) Coupling of hippocampal theta and ripples with pontogeniculooccipital waves. Nature 589(7840):96-102.
54. Schabus M et al (2007) Hemodynamic cerebral correlates of sleep spindles during human non-rapid eye movement sleep. Proc Natl Acad Sci USA 104(32):13164–13169.
55. Boyden ES et al (2005) Millisecond-timescale, genetically targeted optical control of neural activity. Nat Neurosci 8(9):1263–1268.
56. Zhang F et al (2007) Circuit-breakers: optical technologies for probing neural signals and systems. Nat Rev Neurosci 8(8):577.
57. Deisseroth K (2011) Optogenetics. Nat Methods 8(1):26–29.
58. Häusser M (2014) Optogenetics: the age of light. Nature Methods 11(10):1012.
59. Tye KM, Deisseroth K (2012) Optogenetic investigation of neural circuits underlying brain disease in animal models. Nat Rev Neurosci 13(4):251–266.
60. Lee JH et al (2010) Global and local fMRI signals driven by neurons defined optogenetically by type and wiring. Nature 465(7299):788–792.
61. Leong ATL et al (2019) Optogenetic fMRI interrogation of brain-wide central vestibular pathways. Proc Natl Acad Sci USA 116(20):10122–10129.
62. Weitz AJ et al (2019) Thalamic input to orbitofrontal cortex drives brain-wide, frequency-dependent inhibition mediated by GABA and zona incerta. Neuron 104(6):1153–1167.e4p
63. Lee HJ et al (2016) Activation of direct and indirect pathway medium spiny neurons drives distinct brain-wide responses. Neuron 91(2):412–424.
64. Bernal-Casas D et al (2017) Studying brain circuit function with dynamic causal modeling for optogenetic fMRI. Neuron 93(3):522–532.e5p
65. Liu J et al (2015) Frequency-selective control of cortical and subcortical networks by central thalamus. eLife 4:e09215.
66. Weitz AJ et al (2015) Optogenetic fMRI reveals distinct, frequency-dependent networks recruited by dorsal and intermediate hippocampus stimulations. Neuroimage 107:229–241.
67. Liang Z et al (2015) Mapping the functional network of medial prefrontal cortex by combining optogenetics and fMRI in awake rats. Neuroimage 117:114–123.
68. Grandjean J et al (2019) A brain-wide functional map of the serotonergic responses to acute stress and fluoxetine. Nat Commun 10(1):350.
69. Jung WB et al (2022) Dissection of brain-wide resting-state and functional somatosensory circuits by fMRI with optogenetic silencing. Proc Natl Acad Sci USA 119:4.
70. Menon V et al (2023) Optogenetic stimulation of anterior insular cortex neurons in male rats reveals causal mechanisms underlying suppression of the default mode network by the salience network. Nat Commun 14(1):866.
71. Zou Y et al (2024) Cell-type-specific optogenetic fMRI on basal forebrain reveals functional network basis of behavioral preference. Neuron 112(8):1342-1357.
72. Grimm C et al (2024) Tonic and burst-like locus coeruleus stimulation distinctly shift network activity across the cortical hierarchy. Nat Neurosci 27(11):2167–2177.
73. Yang C et al (2022) Medial prefrontal cortex and anteromedial thalamus interaction regulates goal-directed behavior and dopaminergic neuron activity. Nat Commun 13(1):1386.
74. Ferenczi EA et al (2016) Prefrontal cortical regulation of brainwide circuit dynamics and reward-related behavior. Science 351(6268):aac9698.
75. Yizhar O et al (2011) Optogenetics in neural systems. Neuron 71(1):9–34.
76. Warby SC et al (2014) Sleep-spindle detection: crowdsourcing and evaluating performance of experts, non-experts and automated methods. Nat Methods 11(4):385–392.
77. Andrillon T et al (2011) Sleep spindles in humans: insights from intracranial EEG and unit recordings. J Neurosci 31(49):17821–17834.
78. Latchoumane C-F et al (2017) Thalamic spindles promote memory formation during sleep through triple phase-locking of cortical, thalamic, and hippocampal rhythms. Neuron 95(2):424–435.

79. Kim A et al (2012) Optogenetically induced sleep spindle rhythms alter sleep architectures in mice. Proc Natl Acad Sci USA 109(50):20673–20678.
80. Halassa MM et al (2011) Selective optical drive of thalamic reticular nucleus generates thalamic bursts and cortical spindles. Nat Neurosci 14(9):1118–1120.
81. Bartho P et al (2014) Ongoing network state controls the length of sleep spindles via inhibitory activity. Neuron 82(6):1367–1379.
82. Fernandez LMJ, Luthi A (2020) Sleep spindles: mechanisms and functions. Physiol Rev 100(2):805–868.
83. Vertes RP, Linley SB, Hoover WB (2015) Limbic circuitry of the midline thalamus. Neurosci Biobehav Rev 54:89–107.
84. Boyce R et al (2016) Causal evidence for the role of REM sleep theta rhythm in contextual memory consolidation. Science 352(6287):812–816.
85. Tudor JC et al (2016) Sleep deprivation impairs memory by attenuating mTORC1-dependent protein synthesis. Sci Signal 9(425):ra41–ra41.
86. Purcell SM et al (2017) Characterizing sleep spindles in 11,630 individuals from the National Sleep Research Resource. Nat Commun 8(1):15930.
87. Mander BA, Winer JR, Walker MP (2017) Sleep and human aging. Neuron 94(1):19–36.
88. Helfrich RF et al (2018) Old brains come uncoupled in sleep: slow wave-spindle synchrony, brain atrophy, and forgetting. Neuron 97(1):221–230.e4p
89. Ali T et al (2015) Melatonin attenuates D-galactose-induced memory impairment, neuroinflammation and neurodegeneration via RAGE/NF-KB/JNK signaling pathway in aging mouse model. J Pineal Res 58(1):71–85.
90. Sarasso S et al (2014) Hippocampal sleep spindles preceding neocortical sleep onset in humans. Neuroimage 86:425–432.
91. Bandarabadi M et al (2020) A role for spindles in the onset of rapid eye movement sleep. Nat Commun 11(1):5247.
92. Brown EN, Lydic R, Schiff ND (2010) General anesthesia, sleep, and coma. N Engl J Med 363(27):2638–2650.
93. Brown EN, Purdon PL, Dort CJVan (2011) General anesthesia and altered states of arousal: a systems neuroscience analysis. Annu Rev Neurosci 34:601–628.
94. Hagihira S (2015) Changes in the electroencephalogram during anaesthesia and their physiological basis. Br J Anaesth 115(Suppl 1):i27–i31.
95. Purdon PL et al (2015) Clinical electroencephalography for anesthesiologists: part I: background and basic signatures. Anesthesiology 123(4):937–960.
96. Benveniste H et al (2017) Anesthesia with dexmedetomidine and low-dose isoflurane increases solute transport via the glymphatic pathway in rat brain when compared with high-dose isoflurane. Anesthesiology 127(6):976–988.
97. Li D et al (2013) Cross-frequency coupling during isoflurane anaesthesia as revealed by electroencephalographic harmonic wavelet bicoherence. Br J Anaesth 110(3):409–419.
98. Stitt I et al (2015) Intrinsic coupling modes reveal the functional architecture of cortico-tectal networks. Sci Adv 1(7):e1500229.
99. Tolner EA et al (2012) Subplate neurons promote spindle bursts and thalamocortical patterning in the neonatal rat somatosensory cortex. J Neurosci 32(2):692–702.
100. Magill PJ et al (2004) Brain state–dependency of coherent oscillatory activity in the cerebral cortex and basal ganglia of the rat. J Neurophysiol 92(4):2122–2136.
101. Contreras D, Steriade M (1996) Spindle oscillation in cats: the role of corticothalamic feedback in a thalamically generated rhythm. J Physiol 490(1):159–179.
102. Contreras D et al (1997) Spatiotemporal patterns of spindle oscillations in cortex and thalamus. J Neurosci 17(3):1179–1196.
103. Contreras D et al (1996) Control of spatiotemporal coherence of a thalamic oscillation by corticothalamic feedback. Science 274(5288):771–774.
104. Fernandez LM et al (2018) Thalamic reticular control of local sleep in mouse sensory cortex. eLife 7:e39111.
105. van Betteray JN, Vossen JM, Coenen AM (1991) Behavioural characteristics of sleep in rats under different light/dark conditions. Physiol Behav 50(1):79–82.
106. Sawangjit A et al (2018) The hippocampus is crucial for forming non-hippocampal long-term memory during sleep. Nature 564(7734):109–113.
107. Sirota A et al (2003) Communication between neocortex and hippocampus during sleep in rodents. Proc Natl Acad Sci USA 100(4):2065–2069.
108. Hagler DJ Jr et al (2018) Heterogeneous origins of human sleep spindles in different cortical layers. J Neurosci 38(12): 3013–3025.
109. Leong ATLet al (2019) Optogenetic fMRI interrogation of brain-wide central vestibular pathways. Proc Natl Acad Sci USA 116(20):10122–10129.
110. Owen SF, Liu MH, Kreitzer AC (2019) Thermal constraints on in vivo

optogenetic manipulations. Nat Neurosci 22(7):1061-1065.
111. Chen X et al (2020) Sensory evoked fMRI paradigms in awake mice. Neuroimage 204:116242.
112. Han Z et al (2019) Awake and behaving mouse fMRI during Go/No-Go task. Neuroimage 188:733–742.
113. Yu Y et al (2023) Sleep fMRI with simultaneous electrophysiology at 9.4 T in male mice. Nat Commun 14(1):1651.
114. Hike D et al (2025) High-resolution awake mouse fMRI at 14 tesla. eLife 13:RP95528.
115. Tu W, Zhang N (2022) Neural underpinning of a respiration-associated resting-state fMRI network. eLife 11:e81555.
116. Tu W, Cramer SR, Zhang N (2024) Disparity in temporal and spatial relationships between resting-state electrophysiological and fMRI signals. eLife 13:RP95680.
117. Xie L et al (2025) Brain-wide resting-state fMRI network dynamics elicited by activation of single thalamic input. Nat Commun 16:11247.
118. Stroh A et al (2013) Making waves: initiation and propagation of corticothalamic Ca2+ waves in vivo. Neuron 77(6): 1136–1150.
119. Niethard N et al (2018) Cortical circuit activity underlying sleep slow oscillations and spindles. Proc Natl Acad Sci USA 115(39):E9220–E9229.
120. Timofeev I et al (2000) Origin of slow cortical oscillations in deafferented cortical slabs. Cereb Cortex 10(12):1185–1199.
121. Kim J, Gulati T, Ganguly K (2019) Competing roles of slow oscillations and delta waves in memory consolidation versus forgetting. Cell 179(2):514–526.e13p

Chapter 12

Understanding Resting-State Functional Connectivity Using Opto-fMRI

Xu Han, Lauren N. Beloate and Nanyin Zhang

Abstract

Resting-state functional connectivity (RSFC) is a fundamental aspect of brain organization that is traditionally assessed using resting-state functional magnetic resonance imaging (fMRI). Combining this method with optogenetics (opto-fMRI) offers a powerful approach to causally investigate RSFC by allowing precise modulation and monitoring of local neural activity along with brain-wide fMRI signals. Here, we describe key experimental procedures for implementing opto-fMRI, including awake rodent imaging, opsin expression strategies, and calcium signal recording, as well as preprocessing of calcium and fMRI signals. We then discuss the applications of opto-fMRI in uncovering cell type-specific contributions to brain network activity and inferring causality in RSFC.

Keywords Optogenetics, fMRI, Opto-fMRI, Resting-state functional connectivity, Non-anesthetized fMRI, Calcium imaging, Causal connectivity

1 Introduction

fMRI is a non-invasive technique widely used in neuroscience to measure and map brain activity. It detects blood-oxygen-level-dependent (BOLD) signals, which are based on the concept of neurovascular coupling (NVC). NVC describes the process in which neural activity increases intracellular Ca^{2+} in neurons and astrocytes, leading to the release of vasodilators like nitric oxide, which dilate local blood vessels [1, 2]. This leads to local cerebral blood flow (CBF), cerebral blood volume (CBV), and cerebral metabolic rate of oxygen ($CMRO_2$) increases, which, in turn, raises the ratio of oxygenated hemoglobin (Hb) to deoxygenated hemoglobin (dHb), lowering local magnetic susceptibility and thus increasing the fMRI signal intensity.

fMRI is a versatile tool that offers insights into brain function through a variety of approaches. One prominent application is task-based fMRI, where specific cognitive, sensory, or motor tasks are

Kevin Chuen Wing Chan and Seong-Gi Kim (eds.), *Optogenetic Functional Neuroimaging*, Neuromethods, vol. 232, https://doi.org/10.1007/978-1-0716-5178-0_12,

used to map brain regions activated in response to external stimuli. Complementing this, resting-state fMRI (rsfMRI) explores intrinsic brain activity in the absence of tasks, often focusing on functional connectivity (FC)—a widely used approach to examine the temporal correlations of the BOLD signal between different brain regions. By revealing how different parts of the brain communicate and synchronize their activity over time, FC analyses provide key insights into the brain's intrinsic networks. Beyond these applications, fMRI is also a powerful tool for investigating brain dynamics, network topology, and disease mechanisms. It supports advanced analysis approaches such as models of functional integration, dynamic functional connectivity analysis, and causal modeling methods like dynamic causal modeling (DCM). These techniques provide deeper insights into how brain regions interact, how connectivity changes over time, and how causal relationships between regions can be inferred, continually expanding our understanding of the brain's complexity.

Despite its widespread use, fMRI also has several important limitations. The BOLD signal is only a proxy of neural activity, and NVC can be influenced by factors such as age, arousal state, and drugs, and may vary across different brain areas and cell populations. Therefore, fMRI might not be able to accurately reflect neural activity. Additionally, fMRI cannot differentiate between specific neuronal subtypes, which is crucial for dissecting the complexities of brain circuits. Furthermore, the slow hemodynamic response and relatively low temporal resolution make it difficult to accurately determine the timing and causal relationship of neural activity across brain regions. Therefore, to obtain a more comprehensive understanding of brain activity, fMRI is often combined with other techniques such as electrophysiology and optogenetics. These methods provide higher temporal resolution, more direct measures of neural activity, and/or the ability to target specific cell types, complementing fMRI's ability to map large-scale brain networks. Optogenetics, in particular, enables precise modulation of neuronal subpopulations by expressing diverse opsin variants in specific cell types in vivo, while also allows the monitoring of neural activity through the recording of associated Ca^{2+} signals.

When used as neural modulators, optogenetic tools enable researchers to causally probe brain networks by stimulating targeted neurons and observing the resulting BOLD responses both locally and in distant projection areas [3, 4]. The first use of opto-fMRI was reported by Lee et al. [4] in the rodent sensorimotor system. They observed BOLD fMRI responses in the primary motor cortex (M1) and downstream thalamus following optogenetic activation of excitatory cells in anesthetized rats. Similarly, Desai et al. [5] showed that BOLD fMRI responses to optogenetic stimulation of excitatory neurons in the primary somatosensory cortex (S1) were comparable to those elicited by external somatosensory stimuli in awake mice. There are multiple advantages of pairing optogenetics, as opposed to other neuromodulatory techniques, with fMRI. For instance, optogenetics can be used to activate or inhibit neural activity with high temporal precision, allowing for rapid transition from one neural state to another. Opto-fMRI also allows for simultaneous modulation of multiple brain regions through multi-channel fibers, enabling the exploration of complex interactions and connectivity within neural circuits [6]. Additionally, unlike electrophysiological and lesion-based interventions, optogenetics enables investigations of brain networks with cell-type specificity, surpassing the capabilities of electrical stimulation by avoiding antidromic activation and minimizing confounding effects from nearby cell types.

When used as neural indicators, optogenetic tools like GCaMP, a genetically encoded calcium indicator (GECI), enable real-time imaging of neuronal activity at the cellular level. In contrast to other neural indicators, such as chemical Ca^{2+} dyes and electrophysiology, GCaMP allows for longitudinal, cell-type-specific monitoring of calcium dynamics and overcomes the challenges of electrophysiology-fMRI integration, where fMRI noise can interfere with neural spiking detection. Combining GCaMP with fMRI provides crucial insights into how different brain cell types contribute to NVC underlying the BOLD signal [7, 8]. Moreover, with calcium signals providing the ground truth for neural activity, GCaMP reveals how specific brain cell populations correlate with and influence overall FC, deepening our understanding of brain network dynamics [9, 10].

Taken together, opto-fMRI offers a unique opportunity to both modulate and monitor neural activity across multiple scales, from local circuits to broader network-level interactions. It allows researchers to address critical questions that fMRI alone cannot answer, such as the causal effects of specific neural connections, the dynamics of brain state transitions, and the mechanisms underlying the BOLD signal. In this chapter, we will cover the basics of optogenetics and opto-fMRI techniques, with a focus on their application to studying RSFC.

2 Material

2.1 Animals

Long-Evans rats (weighing 250–350 g, Charles River Laboratories, Wilmington, MA) were used for opto-fMRI studies. Prior to surgery, all rats were acclimated to the animal facility for at least one week to reduce physiological and psychological stress.

2.2 Surgery

1. Ketamine/xylazine mixture: 40 mg/kg ketamine and 12 mg/kg xylazine in sterile saline, used to maintain anesthesia.
2. Baytril solution: Enrofloxacin, 2.5 mg/kg, used as a broad-spectrum antibiotic.

3. Dexamethasone solution: 0.5 mg/kg, used to reduce inflammation.
4. Buprenorphine solution: 1.0 mg/kg, used for long-lasting analgesia.
5. Bupivacaine solution: 4 mg/kg, used for local analgesia.
6. Stereotaxic frame.
7. Surgical microscope.
8. Pulse oximeter.
9. PhysioSuite monitoring system with an integrated heating pad module.
10. Robotic microinjection system.
11. Micromotor drill, used to create small craniotomies on the skull.
12. Brass screws, used to secure implants by providing a stable base for dental cement; fMRI-compatible.
13. Hamilton syringe connected to a pulled glass micropipette.
14. Fiber optic cannula (core = 400 μm, NA = 0.5).
15. Metabond dental cement.

2.3 Optical Setup for GCaMP Experiment

1. LED light sources: 405 nm and 465 nm
2. Mini-cube (Doric Lenses Inc., Quebec, Canada) that combines two LED light sources (405 nm and 465 nm) via dichroic mirrors and routes them through a single optical path.
3. Low-autofluorescence optical patch cord with 400 μm core diameter, numerical aperture (NA) of 0.48, and a length of 7 m (Doric Lenses Inc., Quebec, Canada).
4. Microscope objective (40×, 0.65 NA) that focuses the emission light onto the photoreceiver.
5. Photomultiplier (MiniSL-30035, Sensl, Ireland) that converts the collected fluorescence signal to an electric signal.
6. Amplifier and bandpass filter (0.3–1 kHz).
7. NI-DAQ board and LabVIEW software, used to record GCaMP fluorescence signals and synchronize TTL signals from the LED and the fMRI scanner.

3 Methods

3.1 Expression of Opsins and GCaMP Via Viral or Transgenic Methods

Optogenetics offers precise control over neuronal activity by utilizing viral vector technology or transgenics to express light-sensitive proteins that modulate membrane potentials in the cells in which they are expressed. Channelrhodopsin-2 (ChR2) is a commonly used tool for inducing membrane depolarization. It is a light-gated non-specific cation channel permeable to Na^+, K^+, and Ca^{2+}, and is maximally activated by blue light [11] (Fig. 1A). Several ChR2 variants have been engineered to fulfill different functionalities. For example, ChETA (ChR2/E123T) and ChIEF (ChR2/L132C/T159S) are designed for high-frequency stimulation (>200 Hz) [12, 13], and red-activatable channelrhodopsin (ReaChR) responds optimally to red light (590–630 nm), which is less absorbed by blood and penetrates deeper into tissues, making it ideal for transcranial activation [14]. The stabilized step-function opsin (SSFO) is a double mutant of ChR2 (D156A/C128S) that possesses two distinct features: high stability and subthreshold depolarization [15]. When briefly illuminated with blue light, the SSFO can remain activated for up to 30 min and can be rapidly switched off with yellow light [15]. In addition to excitatory opsins, optogenetic tools for neural inhibition include Halorhodopsin (NpHR), a chloride pump activated by yellow light, and Archaerhodopsin (Arch), an outward proton pump triggered by green light, which induces hyperpolarization (Fig. 1A).

GCaMP is a GECI that has become a widely used tool in neuroscience research for in vivo monitoring of neuronal activity. It consists of a circularly permutated green fluorescent protein (GFP) fused to calmodulin and M13 [16]. Calcium ions play an essential role in processes such as neurotransmitter release, synaptic plasticity, and other neuronal functions [17]. When calcium ions bind to calmodulin, it undergoes a conformational change, which alters the fluorescence intensity of GFP (Fig. 1B). This change in fluorescence allows for the detection of fluctuations in calcium concentration within neurons, reflecting underlying neural activity.

To achieve targeted opsin expression in specific brain regions, several methods are available. One commonly used approach in rats is the viral delivery system, utilizing vectors such as adeno-associated virus (AAV) and lentivirus. AAVs are particularly favored due to their ability to provide stable, long-term expression [18] and being only mildly immunogenic [19]. They can accommodate approximately 4.8 kilobases (kb) of DNA, including both the transgene and the necessary regulatory elements [20]. Notably, AAV serotypes 1, 2, 5, 7, 8, and 9 all show a strong preference for neurons in many studies, making them highly suitable for neural targeting [20]. The most straightforward approach is through the use of promoter-driven AAVs, such as injecting AAVs with the CaMKIIα promoter to selectively target excitatory neurons in cortical regions. Alternatively, a second AAV driven by a promoter that encodes Cre recombinase can be used in combination with an AAV carrying a double-floxed inverted open reading frame (AAV-DIO). This approach allows for transduction of the viral genome to highly specific cell populations through the promoter driving the Cre recombinase, with additional specificity provided by the promoter in the AAV-DIO construct.

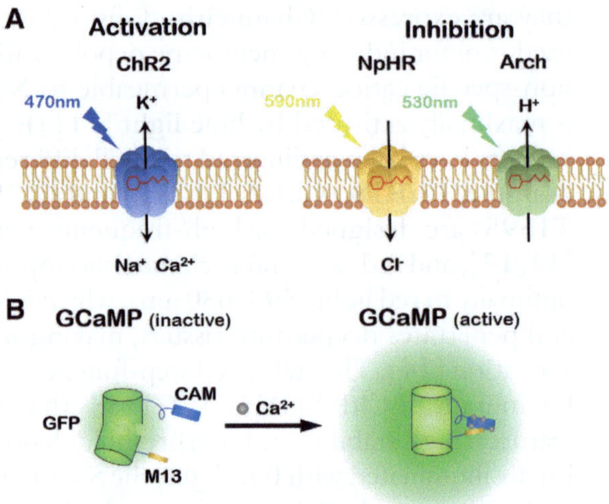

Fig. 1 Commonly used optogenetic tools (**A**) *Left*: ChR2 is a light-gated cation channel that, when activated, causes membrane depolarization. Right: NpHR and Arch are light-gated chloride (Cl^-) and proton (H^+) channels, respectively, whose activation results in membrane hyperpolarization. (**B**) The binding of calcium ions to calmodulin triggers conformational changes in the GCaMP protein, leading to an increase in fluorescence emission. ChR2 channelrhodopsin-2, NpHR halorhodopsin, Arch archaerhodopsin

When Cre-expressing mouse or rat lines are available, cell-type-specific opsin expression can be achieved through either genetic crosses using the Cre-LoxP system or by viral injection of AAV-DIO [21, 22]. This combination of methods and the broad neuronal tropism of AAV serotypes enables researchers to precisely target and manipulate specific neural circuits, ensuring accurate and controlled experimental outcomes. In addition, some transgenic mouse lines that express opsins in specific cell types are also available. For example, Zhao et al. [23] established transgenic lines that express ChR2 under the control of promoters specific to GABAergic, cholinergic, and serotonergic neurons, as well as parvalbumin-expressing neurons.

To integrate optogenetics with fMRI, several technical aspects need to be considered, as described below.

3.2 Non-Anesthetized Rodent fMRI

Although there has been significant progress in the human fMRI field, fMRI studies in animals, especially rodents, open up more research avenues to be explored, as a variety of neural recording and more invasive neuromodulation techniques can be more easily applied. These translational models, in turn, allow for further progress in understanding the human brain. However, one major obstacle that limits the use of rsfMRI in animals is the difficulty of immobilizing them during scanning. Although the use of anesthesia during imaging is useful for many applications, especially for motion reduction, there are some limitations. For instance, anesthesia alters neural activity and dramatically changes the organization of networks [24–26]. Along with a change of neural activity, chemicals used for anesthesia repress activity in astrocytes and, therefore, alter NVC [27, 28]. Commonly used anesthetic drugs in animals also affect CBF and CBV directly, thereby altering the BOLD signal. For example, isoflurane and propofol are known as vasodilators [29, 30], while medetomidine can cause vasoconstriction [31]. As a result, the hemodynamic response function (HRF), which describes the shape of the BOLD signal over time following neural activation, is usually slower in response and lower in amplitude in the anesthetized state compared to the awake state [32]. Moreover, anesthesia can significantly alter FC patterns, with some connections present only in the awake state and disappearing under anesthesia, such as the anticorrelated relationship between the infralimbic cortex (IL) and amygdala [24], and the intrinsic connectivity within the motor cortices [33]. Compared to the awake state, anesthesia results in a global reduction in FC, and this effect was shown to be dose-dependent [34]. Therefore, when possible, imaging in non-anesthetized animals is crucial to avoid misinterpretation of the BOLD signal and to ensure that the rsfMRI result can be accurately compared with behavioral results.

The major obstacle for non-anesthetized imaging is the motion of the animals due to the loud noise (~120 dB) generated by the scanner, inadequate restraining, and discomfort caused by the restrainer. To solve this problem, previous studies have shown that acclimation of the animal to a similar scanning environment can significantly reduce motion and stabilize physiological conditions, such as heart and respiration rates [32, 35, 36]. Additionally, various designs of restrainers have been created to immobilize the animals [37–39]. In our acclimation protocol [40–42], the animal is first anesthetized using isoflurane to allow for preparation. Once anesthetized, the animal is placed into an in-lab custom-designed, 3D-printed restrainer (Fig. 2A) [43]. The rat's upper incisors are secured into an apparatus with a fixed bite bar, designed in different versions to accommodate various rat sizes. This apparatus also integrates a nose cone and a pipe that delivers isoflurane for continued anesthesia. Two pairs of adjustable side bars are located at the cheeks to secure the head position and minimize movement. An additional piece is then pressed over the nose to further prevent movement. To reduce discomfort from mechanical restraint, 2% lidocaine paste is applied to the contact area with the bars. The four limbs are positioned in specially designed sockets, and the tail is placed in a customized tube. Additionally, the lower body is gently restrained at the waist, and a bar is placed over the shoulders to prevent movement. The restrainer is also designed to secure our 3-channel coil and fit the core of our 7 T fMRI scanner. The entire preparation process typically takes less than 20 min, after which the isoflurane is discontinued. Once the animal regains consciousness, the restrainer containing the animal is placed

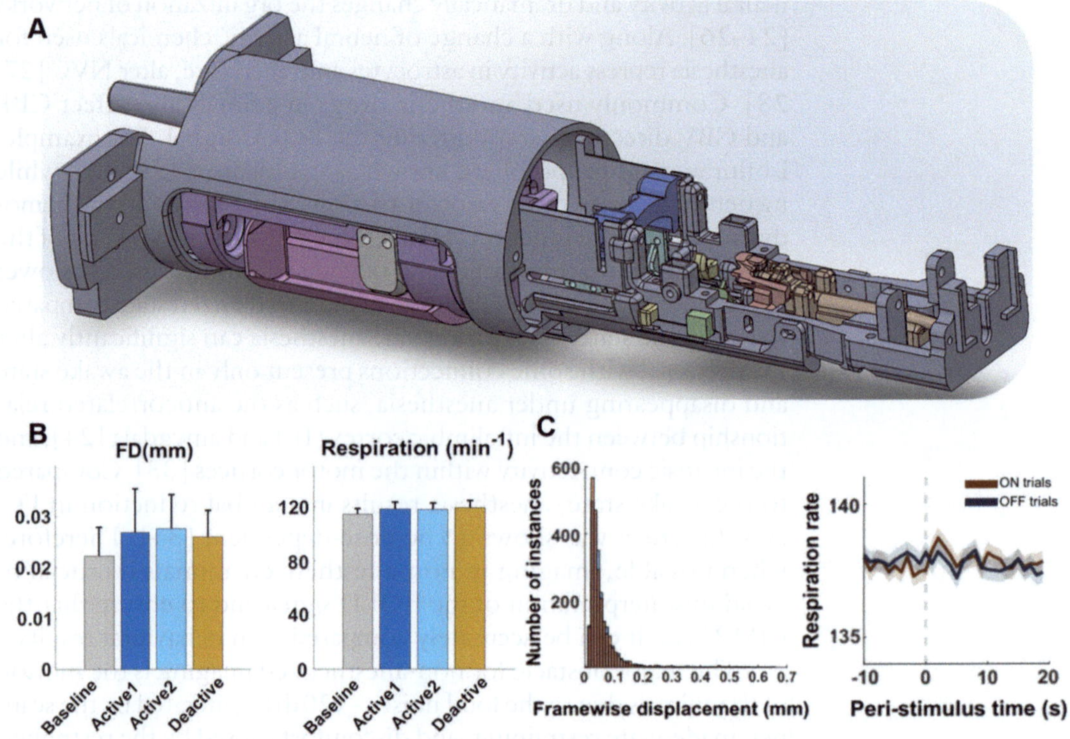

Fig. 2 Custom restrainer design and acclimation results. (**A**) Computer-aided design (CAD)-rendered depiction of the custom restraining device used for awake rat imaging and acclimation. Adapted from Cramer et al. [43]. (**B**) Head motion level and respiration rate under different activation statuses of SSFO expressed in the dorsal dentate gyrus. (Adapted from Han et al. [41].) (**C**) *Left*: frame-wise displacement (FD) distribution during 1–15 s post light stimulation to the eye for ON and OFF trials. *Right*: respiration rate during ON and OFF light stimulation periods. Adapted from Zhang et al. 2022

in an unlit, sound-attenuating chamber to continue the acclimation. Throughout each acclimation session, audio recordings of fMRI pulse sequences are played at a comparable dB level to familiarize the animal with the sounds associated with imaging. The acclimation period spans 7 days, with incremental increases in exposure time: 15 min on Day 1, gradually increasing to 60 min by Day 4, and maintaining 60-min sessions for the remaining days (i.e., 15, 30, 45, 60, 60, 60, and 60 min from Day 1 to Day 7). This acclimation procedure, combined with the use of our specially designed restrainer, ensures minimal head motion and stable respiration, both with and without light stimulation (Fig. 2B, C).

3.3 Intracranial Virus Injection and Fiber Optic Cannula Implantation

Aseptic stereotaxic surgeries are conducted to enable intracranial virus injection and fiber optic cannula implantation. The procedure begins with the induction of anesthesia using 5% isoflurane in a sealed chamber. Once anesthetized, the rat receives intramuscular (IM) injections of ketamine (40 mg/kg) and xylazine (12 mg/kg) to maintain deep anesthesia. The head is then shaved with an electric clipper, and the rat is given subcutaneous injections of Baytril (2.5 mg/kg) as an antibiotic, dexamethasone (0.5 mg/kg) as an anti-inflammatory, and buprenorphine (1.0 mg/kg) for long-term analgesia. The rat's scalp is infiltrated with bupivacaine (4 mg/kg) for local analgesia. The rat is then transferred to a stereotaxic platform (Fig. 3) and secured using ear bars and a nose clamp to stabilize the head. The surgical site is cleaned alternately with iodine and isopropyl alcohol, and eye lubricant is applied to protect the corneas during the procedure. Deep anesthesia is maintained throughout the surgery, with isoflurane available as needed to address any signs of light anesthesia, such as a positive toe-pinch reflex or a sudden increase in heart rate. A midline incision is made along the scalp to expose the skull, and the area is cleaned with saline and a cotton swab. Excess fluid is removed using a suction tube until bregma and lambda are clearly visible. For accurate anatomical targeting, the head position is adjusted to ensure that the anteroposterior (AP) and mediolateral (ML) axes of the skull are aligned with the corresponding plane of the

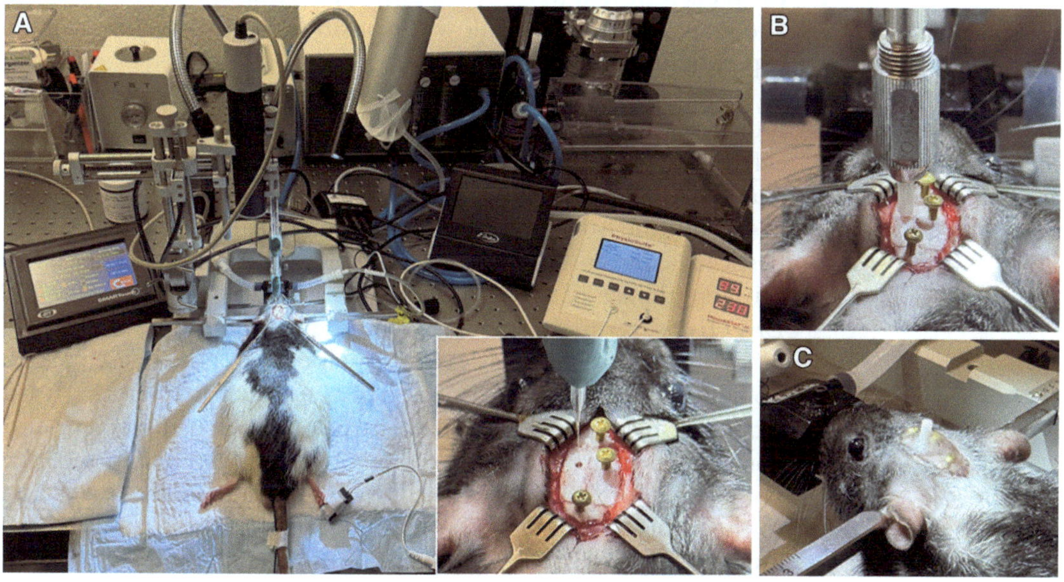

Fig. 3 Overview of surgical procedures for intracranial virus injection and optic cannula implantation. (**A**) The rat is mounted onto a stereotaxic device, with ear bars and a tooth bar securely stabilizing the head. Body temperature is monitored via a rectal probe, while heart rate and blood oxygen levels are continuously tracked. A craniotomy is performed at the skull location corresponding to the targeted brain site, determined by stereotaxic coordinates. A syringe fitted with a thin glass pipette tip is guided into the target brain region using the stereotaxic arm. The injection rate is precisely controlled by a microsyringe pump (Hamilton Company, Reno, NV). (**B**) An optical cannula is inserted into the targeted brain region using stereotaxic guidance. (**C**) Dental cement (Parkell Inc.) is applied over the skull to secure the optical cannula and anchor the implant

stereotaxic instrument. Dorsoventral (DV) positions of bregma and lambda are also adjusted to make sure they lie on the same horizontal plane. Using a surgical microscope, craniotomies are carefully drilled at the target locations. Drilling is performed slowly until the dura mater becomes visible, which is then gently sliced open with a 25 G needle. Three additional holes are drilled at different locations on the skull—front, middle, and back—to anchor brass screws. These screws provide a firm base for the dental cement, which in turn helps to stabilize the implantation by securing it more effectively to the skull. For the virus injection, a pulled glass micropipette connected to a microsyringe (Hamilton Company, Reno, NV) (Fig. 3A) is loaded with the viral vector solution, and injection into the target region is maintained at a flow rate of 50–100 nL/min. To ensure proper diffusion and prevent backflow, the micropipette is left in place for additional 5 min post-injection. After the injection, a fiber optic cannula is carefully advanced to the injection site for optogenetic stimulation (Fig. 3B). Metabond dental cement (Parkell Inc.) is then applied over the skull to secure the cannula and extend over the surrounding skin, helping to protect the wound area and prevent the rat from scratching it (Fig. 3C). Following the surgery, the rat is returned to its home cage for at least four weeks to ensure adequate recovery and allow for opsin expression before starting experimental procedures.

3.4 Dual-Wavelength Fiber Photometry for GCaMP Signal Acquisition

A dual-wavelength fiber photometry system is implemented to record calcium-dependent and calcium-independent fluorescence signals from GCaMP. Excitation is provided by two modulated light sources at 465 nm and 405 nm, corresponding to calcium-sensitive and isosbestic (calcium-insensitive) wavelengths, respectively (Fig. 4). These wavelengths are alternated at 400 Hz with a 40% duty cycle for each LED to prevent spectral overlap and ensure temporal separation of signals [10]. The dual-excitation approach effectively corrects non-specific signal fluctuations, such as signal drift and motion artifacts. The two excitation channels are combined using a mini-cube containing dichroic mirrors. A low-autofluorescence mono-fiber optic patchcord (400 μm core, NA = 0.48) is used to deliver light to the implanted fiber optic cannula targeting the brain region of interest. The emitted fluorescent light from the implanted brain region passes through the mini-cube via another dichroic mirror, then, with a microscope objective (40×, NA = 0.65), focuses onto a photodetector. The analog signal from the photoreceiver is then amplified and bandpass filtered (0.3–1 kHz) before digitization using an NI-DAQ board (sampling rate: 10 kHz). TTL signals from the two LEDs and the MRI scanner are simultaneously recorded to synchronize GCaMP photometry with fMRI acquisition (Fig. 4).

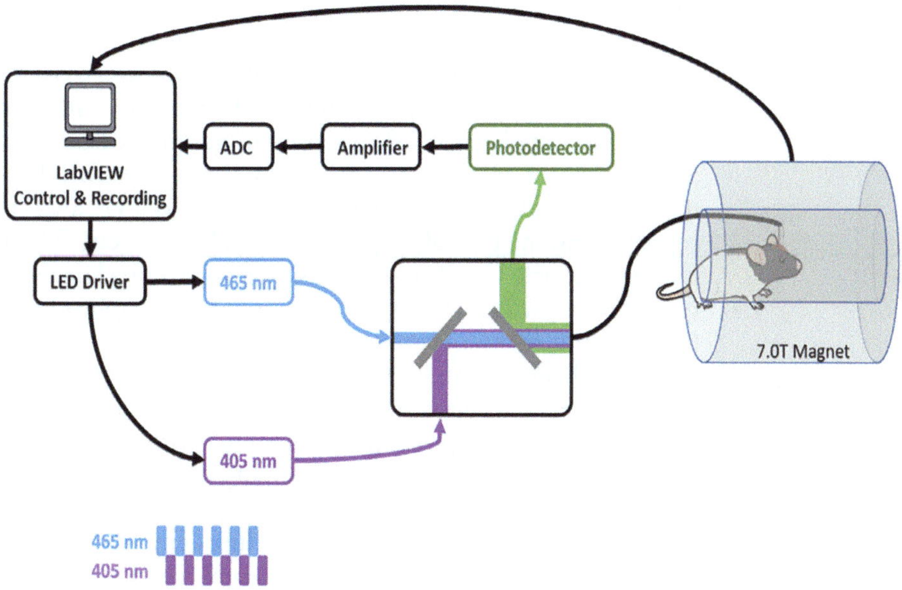

Fig. 4 Schematic of the dual-wavelength optical system used for simultaneous GCaMP calcium imaging and fMRI recordings. GCaMP and calcium-independent fluorescence signals were alternately excited by 465 nm and 405 nm LEDs, respectively. The recorded signals were amplified and sampled at 10 kHz using an NI-DAQ board. A custom-written LabVIEW tool (National Instruments Corp.) controls fMRI data acquisition, stimulation, and detection of fluorescence light and optional recordings of physiological monitoring data. Adapted from Ma et al. [10]

3.5 GCaMP Signal Processing

Proper signal preprocessing is essential to extract meaningful neural activity and minimize artifacts. Since calcium dynamics of interest typically occur in low-frequency ranges [44], a low-pass filter with a 20 Hz cutoff is applied. This step removes high-frequency noise, such as electronic or motion-induced artifacts, while preserving the biologically relevant calcium signals. Baseline drift, often caused by factors like slow motion or photobleaching, is corrected using the adaptive iteratively reweighted Penalized Least Squares (airPLS) algorithm. This method adaptively estimates the baseline by penalizing deviations from smoothness while iteratively reweighting the signal to handle peaks and noise effectively [45]. The corrected calcium-dependent signal at 465 nm is then normalized by subtracting the calcium-independent signal at 405 nm through linear regression, yielding a normalized percentage change relative to the baseline ($\Delta F/F$).

Identifying calcium transients requires distinguishing true neuronal signal spikes from those induced by artifacts. This process considers both the amplitude and temporal profile of the GCaMP signal to differentiate genuine calcium events from non-neuronal fluctuations, such as motion artifacts, which exhibit distinct patterns.

To ensure accurate identification, local peak segments with amplitudes exceeding a certain threshold (e.g., one standard deviation of the signal) are extracted. These segments are pooled across the dataset and subjected to k-means clustering. The cluster with the highest occurrence rate is assumed to represent genuine calcium transients, as these events are expected to be the most frequent. The average temporal profile of this cluster is then used as a GCaMP template, representing the typical shape of a true calcium event. This template is applied to the GCaMP signal using a cross-correlation procedure to identify signals that closely match it in both amplitude and shape. Overlapping segments are excluded to ensure clarity, with only the segment showing the highest peak amplitude retained in cases of overlap [10]. This approach provides a robust framework for identifying calcium peaks, ensuring consistency with established GCaMP dynamics and expected neuronal activity patterns.

3.6 fMRI Data Preprocessing

For rsfMRI preprocessing, the procedures employed in our lab are largely in line with those commonly employed in other rodent rsfMRI labs [46, 47]. To further enhance performance, our lab has also systematically evaluated various preprocessing methods. The BOLD signal is frequently affected by non-neuronal noise sources, including both physiological factors—such as head motion, respiration, and cardiac cycles—and non-physiological factors like MR system instability. These nuisance signals can introduce artifacts into the rsfMRI time series, potentially skewing the quantification of RSFC. In Ma et al. [10], simultaneously recorded calcium signals, as direct measures of neuronal activity, were utilized to validate the effectiveness of fMRI preprocessing pipelines. White matter (WM), cerebrospinal fluid (CSF), and global signal (GS) regression have been shown to improve spatial specificity in BOLD-based seed correlation maps (Fig. 5A). Liu et al. [42] provide a comprehensive discussion comparing the performance of various nuisance regression methods by examining FC matrices and RSFC maps for the insula and anterior cingulate area (ACA), demonstrating that an optimized preprocessing pipeline can effectively mitigates artifacts while preserving meaningful FC patterns. As illustrated in Fig. 5B, the preprocessing pipeline begins with the calculation of FD, followed by the exclusion of volumes with excessive motion. The remaining fMRI images were then manually co-registered to an anatomical template using rigid-body transformation, followed by motion correction with SPM12. Next, independent component analysis (ICA) is applied to identify non-neuronal artifacts. Time courses corresponding to noise-independent components, motion parameters, and nuisance signals from WM and CSF are regressed out. Finally, the data are spatially smoothed and temporally bandpass filtered to enhance signal quality and optimize the detection of the BOLD signal.

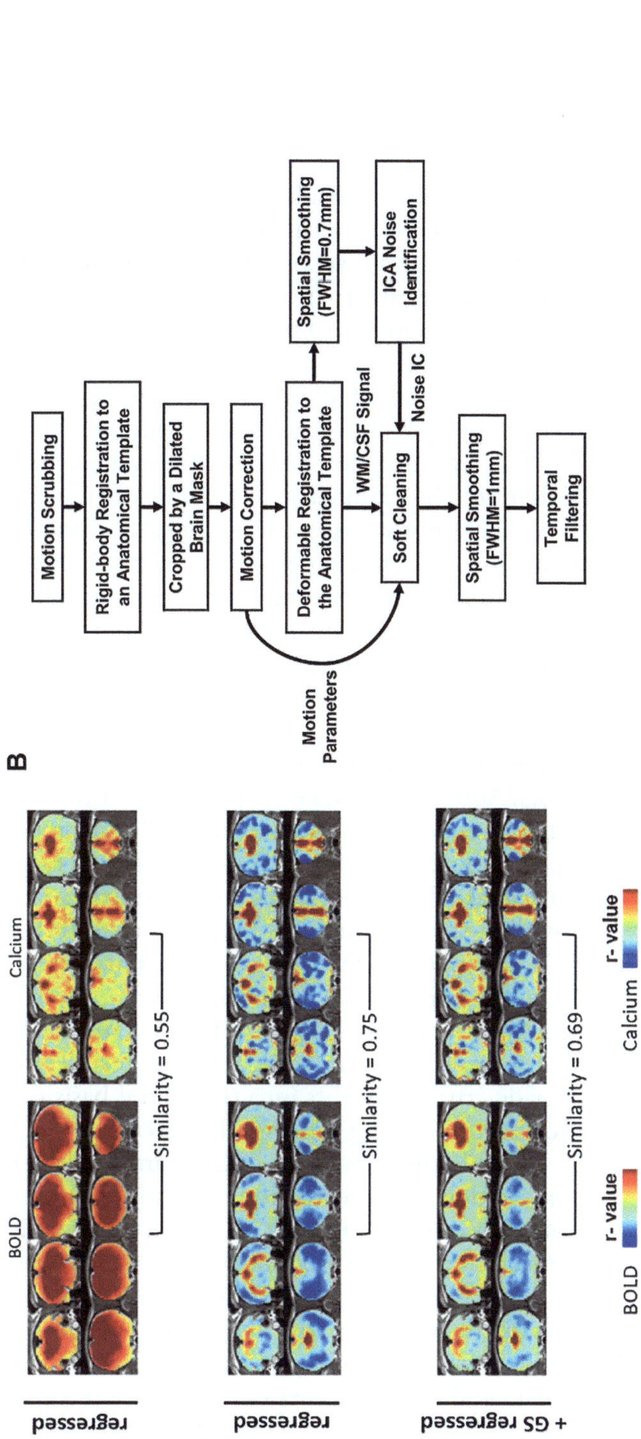

Fig. 5 fMRI data preprocessing methods. **(A)** Comparison of RSFC maps derived using BOLD or calcium signal in the hippocampus between different fMRI data preprocessing methods. Adapted from Ma et al. [10]. **(B)** Optimized preprocessing pipeline of the fMRI signal. Adapted from Liu et al. [42]

3.7 Simultaneous Calcium Imaging and fMRI

To directly assess the neural basis of BOLD signal fluctuations and FC, GECIs, particularly GCaMP, were used to record neuronal activity in vivo. GCaMP-based calcium signals were acquired using fiber photometry, synchronized with fMRI acquisition, allowing simultaneous measurement of cellular-level activity and hemodynamic responses. This approach addresses limitations of traditional BOLD fMRI, which measures neural activity indirectly through changes in blood supply and is susceptible to physiological noise (e.g., cardiac and respiratory fluctuations). Although electrophysiology provides direct measurements of neural activity, it is sensitive to MRI-related interference and lacks cell-type specificity. Calcium imaging via GECIs overcomes these challenges by enabling stable, cell-type-specific recordings with high temporal resolution.

The significance of GCaMP-based calcium imaging was first highlighted by Liang et al. [7], where calcium signals were recorded using fiber photometry with an optical fiber implanted in the superior colliculus (SC) and synchronized with fMRI acquisition. During visual stimulation, a strong coupling was observed between the calcium transients and BOLD responses (Fig. 6A). Leveraging the cell-type specificity of optogenetics, Schlegel et al. [48] revealed distinct coupling relationships between BOLD signals and calcium transients in neurons and astrocytes under paw stimulation, highlighting the importance of differentiating cell-specific contributions to the BOLD signal. Additionally, Ma et al. [10] demonstrated significant coupling between neural calcium spikes in the dorsal hippocampus (dHP) and BOLD signals during resting-state fMRI (Fig. 6B). They further correlated calcium-predicted BOLD signals with whole-brain fMRI, revealing strong couplings between local dHP calcium signals and BOLD activity in the medial prefrontal cortex (mPFC), posterior cingulate cortex (PCC), and hippocampus. These spatiotemporal coactivation patterns align with the default mode network (DMN), a large-scale brain network active at rest. Together, these studies provide critical insights into the neural basis of BOLD signals and their corresponding functional networks.

Despite their coupling, calcium and BOLD signals do not always convey identical information, reflecting differences in their neural basis. For example, Lake et al. [49] found that FC strength derived from calcium and BOLD signals can vary between intrahemispheric cortical regions and even reverse polarity in interhemispheric regions. This discrepancy may arise from inhibitory interhemispheric activity, where increased inhibitory signaling results in an elevated BOLD signal but decreased excitatory calcium activity. Furthermore, Zhang et al. [50] reported a phenomenon where visual stimulation trials consistently elicited positive calcium transients, yet the corresponding BOLD responses in the visual regions varied in effect size and occasionally appeared negative across different trials. They hypothesized that this could be explained by ongoing whole-brain BOLD activity, which represents continuous intrinsic neural and physiological

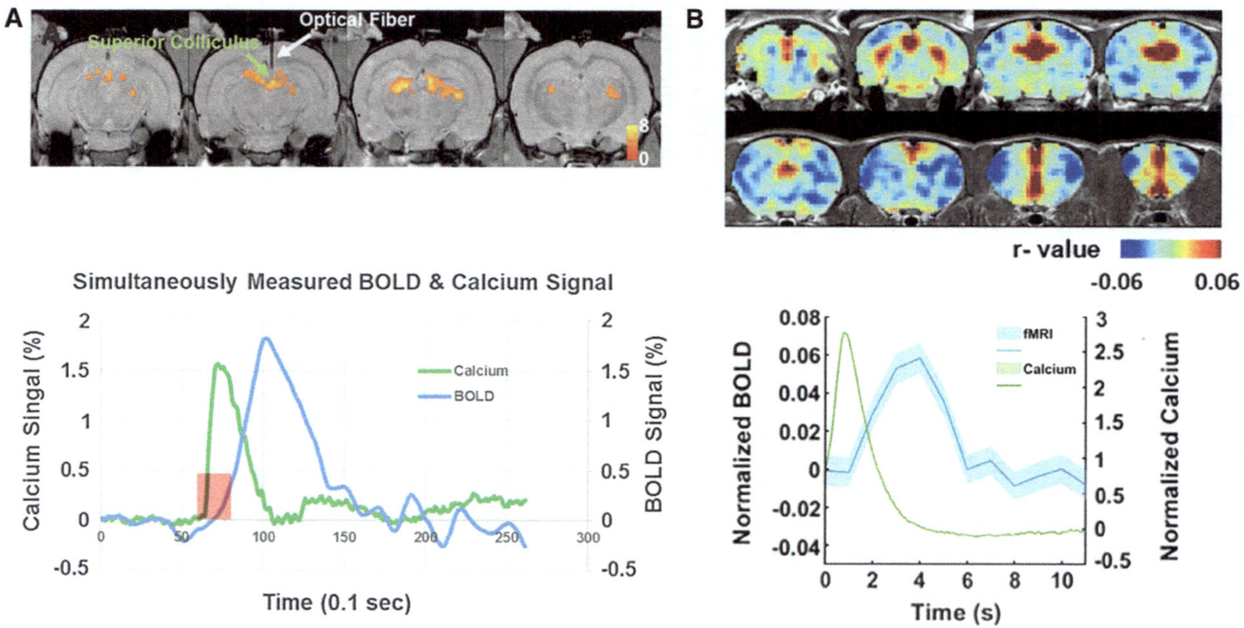

Fig. 6 Correlation between calcium signal and BOLD signal. (**A**) *Upper*: activation map during visual stimulation. *Lower*: averaged calcium and BOLD signals in SC. Adapted from Liang et al. [7]. (**B**) *Upper*: co-activation map generated from the calcium signal predicted BOLD signal. *Lower*: averaged calcium and BOLD signals of spontaneous calcium peaks. Adapted from Ma et al. [10]

processes independent of external stimuli. By using a high-dimensional model that accounted for both stimulation-evoked responses and ongoing activity across responding and non-responding brain regions, they improved coherence between GCaMP and BOLD signals and reduced cross-trial BOLD variability by approximately 30%. Similarly, Wang et al. [9] explored the contribution of astrocytic calcium signals to evoked BOLD responses in the cortex. They discovered that evoked astrocytic calcium signals were linked to positive BOLD responses, whereas intrinsic astrocytic activity correlated with negative BOLD responses. These two distinct gliovascular effects occurred simultaneously in a periodic manner, providing a potential explanation for the variability in BOLD responses to identical stimuli. These studies underscore the necessity of understanding the neural basis underlying complex BOLD fMRI results.

Beyond its role in elucidating BOLD signals, calcium imaging also captures other neurological phenomena, such as calcium slow wave events (<1 Hz). These large-scale, low-frequency oscillations propagate across the brain during deep sleep or under anesthesia. Using combined calcium imaging and fMRI, Schwalm et al. [51] demonstrated that slow calcium wave events are uniquely associated with cortex-wide BOLD activation under light anesthesia. Notably, this relationship was specific to slow wave events and did not extend to other frequency bands. In addition, the coupling between higher calcium frequency (2–3 Hz) and the low frequency (<0.5 Hz) also serves as an indicator of different dynamics of brain states [47]. For instance, under an uncoupled state, a stronger correlation was observed between pons BOLD signals and pupil dynamics, suggesting that arousal pathways originating in the pons are more tightly linked to global cortical activity during such states. In all, calcium imaging provides more information when studying brain function by providing high temporal resolution and being able to label specific cell types.

3.8 Methods to Derive Causal Connectivity Using Opto-fMRI

Opto-fMRI is a powerful technique for investigating how activity in specific brain regions induces global RSFC changes. By selectively modulating neural activity, this method enables researchers to observe how signals propagate through brain circuits and influence downstream regions. For instance, Liang et al. [52] demonstrated that optogenetic stimulation of the IL subregion of the mPFC evoked positive BOLD responses across widespread brain regions (Fig. 7A). These included nearby areas such as the insula, orbital cortex (OC), and ACC, as well as more distal regions like the subiculum, bed nucleus of the stria terminalis (BST), habenula (Hb), mediodorsal thalamus (MD), and hypothalamus. Notably, all regions with increased BOLD activity received direct anatomical projections from the IL [53], underscoring the accuracy of opto-fMRI in identifying functional brain circuits. This neural modulation, however, can extend beyond directly connected targets. Zhang et al. [50] showed that

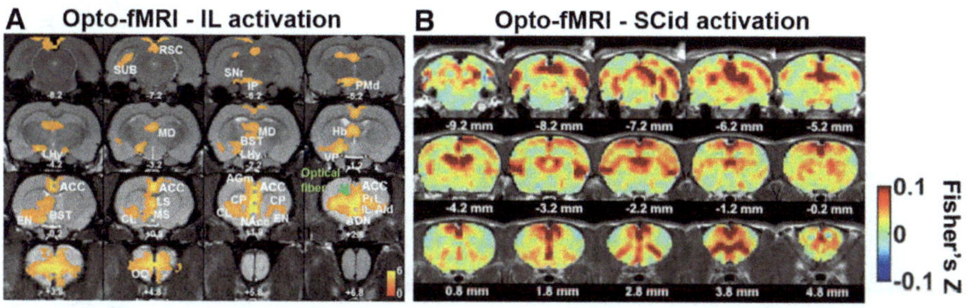

Fig. 7 Averaged activation maps in response to optogenetic stimulation. (**A**) The averaged activation map of IL activation in the awake rats. Adapted from Liang et al. [52]. (**B**) The averaged activation map to SCid activation in the awake rats. Adapted from Zhang et al. [50]

optogenetic stimulation of CaMKII neurons in the intermediate and deep layers of the superior colliculus (SCid) induced BOLD activation not only in direct downstream targets, such as the inferior colliculus, several thalamic regions, and the globus pallidus (GP), but also in areas without direct connections, including the ACC, OC, and motor and auditory cortices (Fig. 7B). These broader effects are likely mediated by indirect pathways, with the thalamus serving as a relay and possibly involving the visual cortex. Beyond soma activation, opto-fMRI can also target axon terminals, triggering both antidromic propagation and postsynaptic activation. For example, Lee et al. [54] expressed AAV-CaMKII-ChR2 in the ACC and stimulated axon terminals in the periaqueductal gray (PAG). This led to positive BOLD responses in both the PAG and the upstream ACC, highlighting the potential of opto-fMRI to uncover complex connectivity patterns, including those involving distant or upstream regions. The cell-type specificity of opto-fMRI provides unique insights into distinct neural pathways. Lee et al. [55] used opto-fMRI to investigate the direct (D1-MSN) and indirect (D2-MSN) pathways of the basal ganglia, which are associated with movement initiation and suppression, respectively.

Beyond these experimental approaches, several analytical methods are commonly applied to infer causality in opto-fMRI studies. DCM is a hypothesis-driven approach that tests predefined models of connectivity. It incorporates intrinsic connectivity and the modulatory effects of optogenetic stimulation, comparing multiple models to identify the best fit [56, 57]. In contrast, Multivariate Dynamical Systems (MDS) offers an exploratory framework for uncovering dynamic interactions within brain networks without prior assumptions [58]. Granger causality has been proposed as a method to infer directional interactions in fMRI data based on temporal precedence [59]. Its application, however, faces challenges due to the low temporal resolution of BOLD signals and variability in neurovascular coupling delays. Other methods, such as psychophysiological interaction (PPI), structural equation modeling (SEM), and Bayesian network analysis,

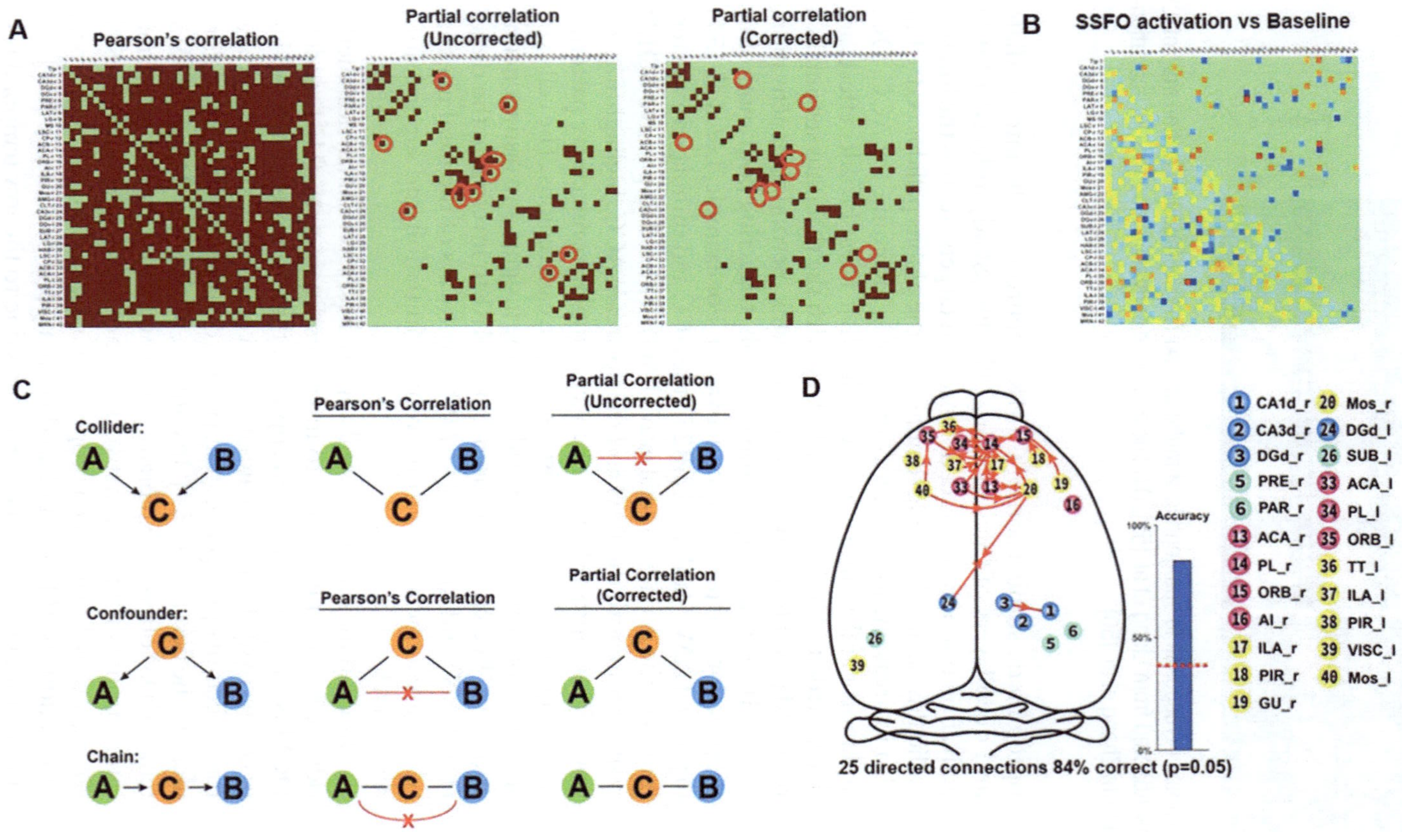

Fig. 8 Causal inferences based on Pearson's correlation and partial correlation. (**A**) Baseline connectivity Matrices without SSFO activation. Showing Pearson's correlation, "uncorrected" partial correlation, and "corrected" partial correlation, where pseudo partial correlations are eliminated. (**B**) ROI-wise partial correlation comparisons between the SSFO activated condition and baseline. (**C**) Nodes and edges illustrate colliders, confounders, and chains. (**D**) Inferred causal relationships. The bar graph shows the accuracy of the inferred causal relationships compared with anatomical tracing data. Red dashed lines in the accuracy bars are the correction rate from a random guess. Adapted from Han et al. [41]

are also employed in task-dependent studies. These approaches may not fully capture the intrinsic dynamics of resting-state network, where activity is not driven by specific external stimuli and causal relationships are harder to infer.

In Han et al. [41], SSFO-based opto-fMRI was combined with a partial correlation approach to infer causal relationships specifically within resting state brain-wide networks. SSFO activation was induced by brief blue light stimulation, leading to prolonged enhancement of neuronal excitability without directly driving spikes. Following stimulation of the right dorsal dentate gyrus (rdDG), altered spontaneous BOLD activity was observed locally, along with brain-wide changes in RSFC. To infer directional connectivity, partial correlation analyses were applied to identify direct connections by controlling for the influence of other variables. Comparing partial correlation matrices between baseline and SSFO-activated conditions revealed causal relationships within the resting-state network (Fig. 8). This approach demonstrates how integrating opto-fMRI with advanced analytical methods can uncover the directional flow of information in resting-state brain networks.

4 Notes

In ChR2-based opto-fMRI, the power of the excitation light is a critical factor. Continuous light illumination can result in tissue heating, which may increase neuron firing rates [60]. Prolonged heating can also cause tissue damage [61]. Additionally, heating-induced artifacts in fMRI, such as altered relaxation times [62], can significantly affect the BOLD signal. To mitigate these risks, it is important to maintain the light power at a level sufficient to achieve experimental goals while minimizing adverse effects. Using lower light intensities and duty cycles, where illumination is delivered intermittently rather than continuously, can help limit cumulative heating and reduce phototoxicity. Additionally, it is best practice to conduct control experiments without ChR2 to ensure that no heating-related artifacts are present [50].

5 Conclusion

In conclusion, the integration of optogenetics and fMRI, or opto-fMRI, provides a powerful means to dissect the neural mechanisms underlying BOLD signals, enabling precise, cell-type-specific manipulations and allowing for direct comparisons between neural activity and hemodynamic responses. The studies discussed here highlight the potential of opto-fMRI to unravel complex causal relationships within brain networks during both task-related and resting-state activities.

Additionally, calcium imaging serves as a crucial ground truth for neural activity, addressing fundamental questions that traditional methods cannot, such as the neural basis of BOLD signals across different cell types and conditions, and the role of ongoing activity in task-based fMRI. Furthermore, it aids in refining the preprocessing pipeline for fMRI data. As this research advances, it will ultimately lead to more accurate interpretations of brain function and deepen our understanding of neurological and psychiatric disorders.

Conflict of Interest

The author has no conflicts of interest to declare that are relevant to the content of this chapter.

References

1. Attwell D, Buchan AM, Charpak S, Lauritzen M, MacVicar BA, Newman EA (2010) Glial and neuronal control of brain blood flow. Nature 468(7321):232–243
2. Brew N, Walker D, Wong FY (2014) Cerebral vascular regulation and brain injury in preterm infants. Am J Physiol Regul Integr Comp Physiol 306(11):R773–R786
3. Bernal-Casas D, Lee HJ, Weitz AJ, Lee JH (2017) Studying brain circuit function with dynamic causal modeling for optogenetic fMRI. Neuron 93(3):522–532. e5
4. Lee JH, Durand R, Gradinaru V, Zhang F, Goshen I, Kim DS et al (2010) Global and local fMRI signals driven by neurons defined optogenetically by type and wiring. Nature 465(7299):788–792
5. Desai M, Kahn I, Knoblich U, Bernstein J, Atallah H, Yang A et al (2011) Mapping brain networks in awake mice using combined optical neural control and fMRI. J Neurophysiol 105(3):1393–1405
6. Sych Y, Chernysheva M, Sumanovski LT, Helmchen F (2019) High-density multi-fiber photometry for studying large-scale brain circuit dynamics. Nat Methods 16(6):553–560
7. Liang Z, Ma Y, Watson GDR, Zhang N (2017) Simultaneous GCaMP6-based fiber photometry and fMRI in rats. J Neurosci Methods 289:31–38
8. Tesler F, Linne ML, Destexhe A (2023) Modeling the relationship between neuronal activity and the BOLD signal: contributions from astrocyte calcium dynamics. Sci Rep 13(1):6451
9. Wang M, He Y, Sejnowski TJ, Yu X (2018) Brain-state dependent astrocytic Ca^{2+} signals are coupled to both positive and negative BOLD-fMRI signals. Proc Natl Acad Sci USA [Internet]. [cited 2024 May 10];115(7). Available from: https://pnas.org/doi/full/10.1073/pnas.1711692115
10. Ma Z, Zhang Q, Tu W, Zhang N (2022) Gaining insight into the neural basis of resting-state fMRI signal. Neuroimage 250:118960
11. Nagel G, Szellas T, Huhn W, Kateriya S, Adeishvili N, Berthold P et al (2003) Channelrhodopsin-2, a directly light-gated cation-selective membrane channel. Proc Natl Acad Sci USA 100(24):13940–13945
12. Gunaydin LA, Yizhar O, Berndt A, Sohal VS, Deisseroth K, Hegemann P (2010) Ultrafast optogenetic control. Nat Neurosci 13(3):387–392
13. Lin JY, Lin MZ, Steinbach P, Tsien RY (2009) Characterization of engineered channelrhodopsin variants with improved properties and kinetics. Biophys J 96(5):1803–1814
14. Lin JY, Knutsen PM, Muller A, Kleinfeld D, Tsien RY (2013) ReaChR: a red-shifted variant of channelrhodopsin enables deep transcranial optogenetic excitation. Nat Neurosci 16(10):1499–1508
15. Yizhar O, Fenno LE, Prigge M, Schneider F, Davidson TJ, O'Shea DJ et al (2011) Neocortical excitation/inhibition balance in information processing and social dysfunction. Nature 477(7363):171–178
16. Chen TW, Wardill TJ, Sun Y, Pulver SR, Renninger SL, Baohan A et al (2013) Ultrasensitive fluorescent proteins for imaging neuronal activity. Nature 499(7458):295–300
17. Augustine GJ, Santamaria F, Tanaka K (2003) Local calcium signaling in neurons. Neuron 40(2):331–346

18. Tenenbaum L, Chtarto A, Lehtonen E, Velu T, Brotchi J, Levivier M (2004) Recombinant AAV-mediated gene delivery to the central nervous system. J Gene Med 6(S1):S212–S222
19. Arjomandnejad M, Dasgupta I, Flotte TR, Keeler AM (2023) Immunogenicity of recombinant adeno-associated virus (AAV) vectors for gene transfer. BioDrugs 37(3):311–329
20. Haggerty DL, Grecco GG, Reeves KC, Atwood B (2020) Adeno-associated viral vectors in neuroscience research. Mol Ther Methods Clin Dev 17:69–82
21. Lee JY, You T, Woo CW, Kim SG (2022) Optogenetic fMRI for brain-wide circuit analysis of sensory processing. IJMS 23(20):12268
22. Hui Y, Zheng X, Zhang H, Li F, Yu G, Li J et al (2022) Strategies for targeting neural circuits: how to manipulate neurons using virus vehicles. Front Neural Circuits 16:882366
23. Zhao S, Ting JT, Atallah HE, Qiu L, Tan J, Gloss B et al (2011) Cell type–specific channelrhodopsin-2 transgenic mice for optogenetic dissection of neural circuitry function. Nat Methods 8(9):745–752
24. Liang Z, King J, Zhang N (2012) Intrinsic organization of the anesthetized brain. J Neurosci 32(30):10183–10191
25. Liang Z, Liu X, Zhang N (2015) Dynamic resting state functional connectivity in awake and anesthetized rodents. Neuroimage 104:89–99
26. Smith JB, Liang Z, Watson GDR, Alloway KD, Zhang N (2017) Interhemispheric resting-state functional connectivity of the claustrum in the awake and anesthetized states. Brain Struct Funct 222(5):2041–2058
27. Thrane AS, Rangroo Thrane V, Zeppenfeld D, Lou N, Xu Q, Nagelhus EA et al (2012) General anesthesia selectively disrupts astrocyte calcium signaling in the awake mouse cortex. Proc Natl Acad Sci USA 109(46):18974–18979
28. Institoris A, Vandal M, Peringod G, Catalano C, Tran CH, Yu X et al (2022) Astrocytes amplify neurovascular coupling to sustained activation of neocortex in awake mice. Nat Commun 13(1):7872
29. Koenig HM, Wang Q, Albrecht RF (1994) Role of nitric oxide and endothelium in rat pial vessel dilation response to isoflurane. Anesth Analg 79(5):886-891
30. Gragasin FS, Davidge ST, Tsui BCH (2009) The potential use of intralipid to minimize propofol's cardiovascular effects. Can J Anesth 56(2):170–171
31. Fukuda M, Vazquez AL, Zong X, Kim S (2013) Effects of the α_2-adrenergic receptor agonist dexmedetomidine on neural, vascular and BOLD fMRI responses in the somatosensory cortex. Eur J Neurosci 37(1):80–95
32. Gao YR, Ma Y, Zhang Q, Winder AT, Liang Z, Antinori L et al (2017) Time to wake up: studying neurovascular coupling and brain-wide circuit function in the un-anesthetized animal. Neuroimage 153:382–398
33. Peltier SJ, Kerssens C, Hamann SB, Sebel PS, Byas-Smith M, Hu X (2005) Functional connectivity changes with concentration of sevoflurane anesthesia. Neuroreport 16(3):285-288
34. Lu H, Zuo Y, Gu H, Waltz JA, Zhan W, Scholl CA et al (2007) Synchronized delta oscillations correlate with the resting-state functional MRI signal. Proc Natl Acad Sci USA 104(46):18265–18269
35. King JA, Garelick TS, Brevard ME, Chen W, Messenger TL, Duong TQ et al (2005) Procedure for minimizing stress for fMRI studies in conscious rats. J Neurosci Methods 148(2):154–160
36. Dopfel D, Zhang N (2018) Mapping stress networks using functional magnetic resonance imaging in awake animals. Neurobiol Stress 9:251–263
37. Khubchandani M, Mallick HN, Jagannathan NR, Mohan Kumar V (2003) Stereotaxic assembly and procedures for simultaneous electrophysiological and MRI study of conscious rat. Magn Reson Med 49(5):962–967
38. Ludwig R, Bodgdanov G, King J, Allard A, Ferris CF (2004) A dual RF resonator system for high-field functional magnetic resonance imaging of small animals. J Neurosci Methods 132(2):125–135
39. Tiefenbach J, Shannon L, Lobosky M, Johnson S, Chan HH, Byram N et al (2024) A novel restrainer device for acquistion of brain images in awake rats. Neuroimage 289:120556
40. Tu W, Ma Z, Ma Y, Dopfel D, Zhang N (2021) Suppressing anterior cingulate cortex modulates default mode network and behavior in awake rats. Cereb Cortex 31(1):312–323
41. Han X, Cramer SR, Zhang N (2022) Deriving causal relationships in resting-state functional connectivity using SSFO-based optogenetic fMRI. J Neural Eng 19(6):066002
42. Liu Y, Perez PD, Ma Z, Ma Z, Dopfel D, Cramer S et al (2020) An open database of resting-state fMRI in awake rats. Neuroimage 220:117094
43. Cramer SR, Han X, Chan DCY, Neuberger T, Zhang N (2025) Neuroimaging model of visceral manipulation in an awake rat. J Neurosci 45(9):e1317242024
44. Zhang Y, Rózsa M, Liang Y, Bushey D, Wei Z, Zheng J et al (2023) Fast and sensitive GCaMP calcium indicators for imaging neural populations. Nature 615(7954):884–891

45. Zhang ZM, Chen S, Liang YZ (2010) Baseline correction using adaptive iteratively reweighted penalized least squares. Analyst 135(5):1138
46. Yu Y, Qiu Y, Li G, Zhang K, Bo B, Pei M et al (2023) Sleep fMRI with simultaneous electrophysiology at 9.4 T in male mice. Nat Commun 14(1):1651
47. Pais-Roldán P, Takahashi K, Sobczak F, Chen Y, Zhao X, Zeng H et al (2020) Indexing brain state-dependent pupil dynamics with simultaneous fMRI and optical fiber calcium recording. Proc Natl Acad Sci USA 117(12):6875–6882
48. Schlegel F, Sych Y, Schroeter A, Stobart J, Weber B, Helmchen F et al (2018) Fiber-optic implant for simultaneous fluorescence-based calcium recordings and BOLD fMRI in mice. Nat Protoc 13(5):840–855
49. Lake EMR, Ge X, Shen X, Herman P, Hyder F, Cardin JA et al (2020) Simultaneous cortex-wide fluorescence Ca^{2+} imaging and whole-brain fMRI. Nat Methods 17(12):1262–1271
50. Zhang Q, Cramer SR, Ma Z, Turner KL, Gheres KW, Liu Y et al (2022) Brain-wide ongoing activity is responsible for significant cross-trial BOLD variability. Cereb Cortex 32(23):5311–5329
51. Schwalm M, Schmid F, Wachsmuth L, Backhaus H, Kronfeld A, Aedo Jury F et al (2017) Cortex-wide BOLD fMRI activity reflects locally-recorded slow oscillation-associated calcium waves. eLife 6:e27602
52. Liang Z, Watson GDR, Alloway KD, Lee G, Neuberger T, Zhang N (2015) Mapping the functional network of medial prefrontal cortex by combining optogenetics and fMRI in awake rats. Neuroimage 117:114–123
53. Hoover WB, Vertes RP (2007) Anatomical analysis of afferent projections to the medial prefrontal cortex in the rat. Brain Struct Funct 212(2):149–179
54. Lee JY, You T, Lee CH, Im GH, Seo H, Woo CW et al (2022) Role of anterior cingulate cortex inputs to periaqueductal gray for pain avoidance. Curr Biol 32(13):2834–2847. e5
55. Lee HJ, Weitz AJ, Bernal-Casas D, Duffy BA, Choy M, Kravitz AV et al (2016) Activation of direct and indirect pathway medium spiny neurons drives distinct brain-wide responses. Neuron 91(2):412–424
56. Friston KJ (2011) Functional and effective connectivity: a review. Brain Connect 1(1):13–36
57. Stephan KE, Friston KJ (2010) Analyzing effective connectivity with functional magnetic resonance imaging. WIREs Cogn Sci 1(3):446–459
58. Ryali S, Shih YYI, Chen T, Kochalka J, Albaugh D, Fang Z et al (2016) Combining optogenetic stimulation and fMRI to validate a multivariate dynamical systems model for estimating causal brain interactions. Neuroimage 132:398–405
59. Roebroeck A, Formisano E, Goebel R (2005) Mapping directed influence over the brain using Granger causality and fMRI. Neuroimage 25(1):230–242
60. Stujenske JM, Spellman T, Gordon JA (2015) Modeling the spatiotemporal dynamics of light and heat propagation for in vivo optogenetics. Cell Rep 12(3):525–534
61. Podgorski K, Ranganathan G (2016) Brain heating induced by near-infrared lasers during multiphoton microscopy. J Neurophysiol 116(3):1012–1023
62. Christie IN, Wells JA, Southern P, Marina N, Kasparov S, Gourine AV et al (2013) fMRI response to blue light delivery in the naïve brain: implications for combined optogenetic fMRI studies. Neuroimage 66:634–641

Chapter 13

Dissection of Brain-Wide Functional Neurosensory Circuits by fMRI with Optogenetic Silencing

Won Beom Jung and Seong-Gi Kim

Abstract

Understanding the functional organization of brain networks requires clearly dissecting the contributions of both long-range and local circuits to information processing. This chapter describes a methodological framework that combines functional MRI (fMRI) with optogenetic focal silencing to systematically determine the relative contributions of different circuits to fMRI responses at both whole-brain and laminar levels. Using cerebral blood volume (CBV)–weighted fMRI together with optogenetic activation of local GABAergic neurons within the mouse somatosensory network, sensory-driven input circuits were effectively separated from downstream output circuits. Selective suppression of excitatory outputs allowed us to characterize the specific roles of spinothalamic (ST), thalamocortical (TC), corticothalamic (CT), corticocortical (CC), and intracortical (IC) circuits in shaping somatosensory processing. Furthermore, laminar-specific fMRI analyses differentiated TC and CC input layers, revealing circuit-dependent processing mechanisms that could not be resolved through conventional brain-wide analyses. These findings provide a strategy for investigating brain-wide network organization with circuit-level specificity, potentially offering insights into functional reorganization associated with neuroplasticity, disease, and learning. By complementing traditional electrophysiological and microscopic methods, this integrative approach significantly advances our understanding of functional connectivity and information flow within the brain.

Keywords fMRI, Optogenetic silencing, Neuronal inhibition, Sensory processing, Brain-wide networks, Laminar-specific analysis

1 Introduction

The investigation of brain networks has traditionally relied upon anatomical tracing, electrophysiological recordings, and optical imaging techniques, each possessing unique strengths but also inherent limitations. Anatomical tracing studies offer detailed structural insights, clearly delineating neuronal connectivity patterns across different brain regions [1, 2]. However, structural information alone does not directly indicate whether connections are functionally active or dynamically modulated during sensory processing or cognitive tasks. Electrophysiological approaches, such as single-unit and multiunit

Kevin Chuen Wing Chan and Seong-Gi Kim (eds.), *Optogenetic Functional Neuroimaging*, Neuromethods, vol. 232, https://doi.org/10.1007/978-1-0716-5178-0_13,

recordings, allow direct measurement of neuronal firing with excellent temporal resolution but are typically restricted in spatial coverage, limiting observations primarily to localized neural circuits [3, 4]. Similarly, optical imaging techniques, including calcium imaging, enable cellular-level monitoring across neuronal populations but remain constrained by limited penetration depth and field of view [5, 6], hindering comprehensive analyses of large-scale brain-wide interactions.

Functional magnetic resonance imaging (fMRI) has significantly advanced neuroscience by enabling noninvasive, brain-wide mapping of neuronal activity through the indirect measurement of hemodynamic responses [7]. This imaging approach effectively bridges structural neuroanatomy with dynamic functional interactions, allowing researchers to investigate brain-wide activation patterns in response to external stimuli (evoked fMRI) [8, 9], as well as intrinsic connectivity patterns during resting states (resting-state fMRI) [10, 11]. These advances have enhanced our understanding of how distinct brain regions dynamically interact. Despite these achievements, interpreting functional networks measured by fMRI remains challenging due to the hierarchical and reciprocal architecture of neuronal circuits, including ascending spinothalamic (ST) and thalamocortical (TC), descending corticothalamic (CT), as well as corticocortical (CC) and intracortical (IC) pathways [12]. The convergence of these interconnected circuits contributes to the observed fMRI signals, complicating the precise attribution of responses to specific neural pathways. Thus, determining the relative contributions and specific roles of these neuronal circuits is crucial, yet challenging, for the accurate interpretation of brain function.

To address these limitations, integrating fMRI with optogenetic modulation has emerged as an effective methodological approach. Optogenetics allows for precise spatial and temporal control of neuronal activity by using light-sensitive opsins to selectively activate or silence genetically defined neuronal populations [13]. Optogenetic activation of excitatory neurons facilitates direct observation of excitatory outputs and their downstream pathways [14, 15]. Conversely, optogenetic silencing (achieved by excitatory opsins in inhibitory neurons [16] or inhibitory opsins in excitatory neurons [17]) effectively suppresses excitatory output from the target site to downstream signaling pathways [18–20], and the downregulated neuronal activity can determine the strength of interregional communication under baseline conditions [21]. Therefore, optogenetic fMRI can offer a unique opportunity to investigate functional processes across multiple scales, from local circuits to network level [22–26].

Specifically, combining evoked fMRI with optogenetic silencing is advantageous for selectively suppressing downstream activity (output-driven circuits) without compromising the upstream and/or collateral inputs from other brain regions (input-driven circuits). This approach allows researchers to precisely determine the contributions of downstream circuits to functional networks by directly comparing fMRI responses obtained with and without targeted neuronal inhibition [27–29]. In this chapter, we will describe methodological strategies for dissecting functional circuits by combining fMRI with optogenetic cortical silencing, using the mouse somatosensory network as a model system. Our approach will include analyses at both brain-wide and laminar-specific levels, providing detailed insights into the functional organization of neuronal circuits.

2 Materials

2.1 Strategies for Optogenetic fMRI

To effectively perform fMRI with optogenetic silencing, selecting an appropriate mouse model is critical. The cortex predominantly consists of excitatory pyramidal neurons (80%–90%) and a smaller population of inhibitory interneurons (10%–20%) [30]. Optogenetic silencing can be achieved through two strategies [18–20]: directly inhibition of excitatory neurons by expressing inhibitory opsins such as Arch (e.g., AAV-CaMKII-Arch) in excitatory pyramidal neurons; or indirect inhibition through activation of inhibitory interneurons using excitatory opsins such as channelrhodopsin-2 (ChR2).

Experimental evidence supports that indirect suppression via excitatory opsins in inhibitory interneurons, particularly using VGAT-ChR2 mice (which express ChR2 selectively in GABAergic neurons), is highly effective [31]. ChR2-mediated photoinhibition in transgenic mice robustly reduces neuronal firing (>80% reduction in spike rates) at remarkably low laser power (0.5–1.5 mW), substantially below the thresholds associated with tissue heating and nonspecific neural activity artifacts. In contrast, alternative opsins such as Arch or Jaws require approximately ten-fold higher laser power to achieve comparable neuronal silencing. Higher laser powers or continuous illumination can induce heating-related artifacts, resulting in pseudo-fMRI responses [32]. Such artifacts may manifest as positive signals due to changes in arteriolar smooth muscle tone [33] or negative signals arising from frequency shifts unrelated to neuronal activity [34, 35]. Therefore, careful verification is essential to differentiate true neuronal responses from heating-induced artifacts. To mitigate heating-related concerns, optogenetic stimulation parameters should be carefully selected, favoring pulsed stimulation protocols at moderate laser powers. Under these conditions, effective neuronal activation or suppression can be achieved with minimal heating risks. Given these considerations, VGAT-ChR2 mice offer significant methodological advantages for optogenetic silencing in fMRI studies by efficiently suppressing excitatory neuronal activity at low laser power [16, 31], minimizing the potential for heating-related artifacts.

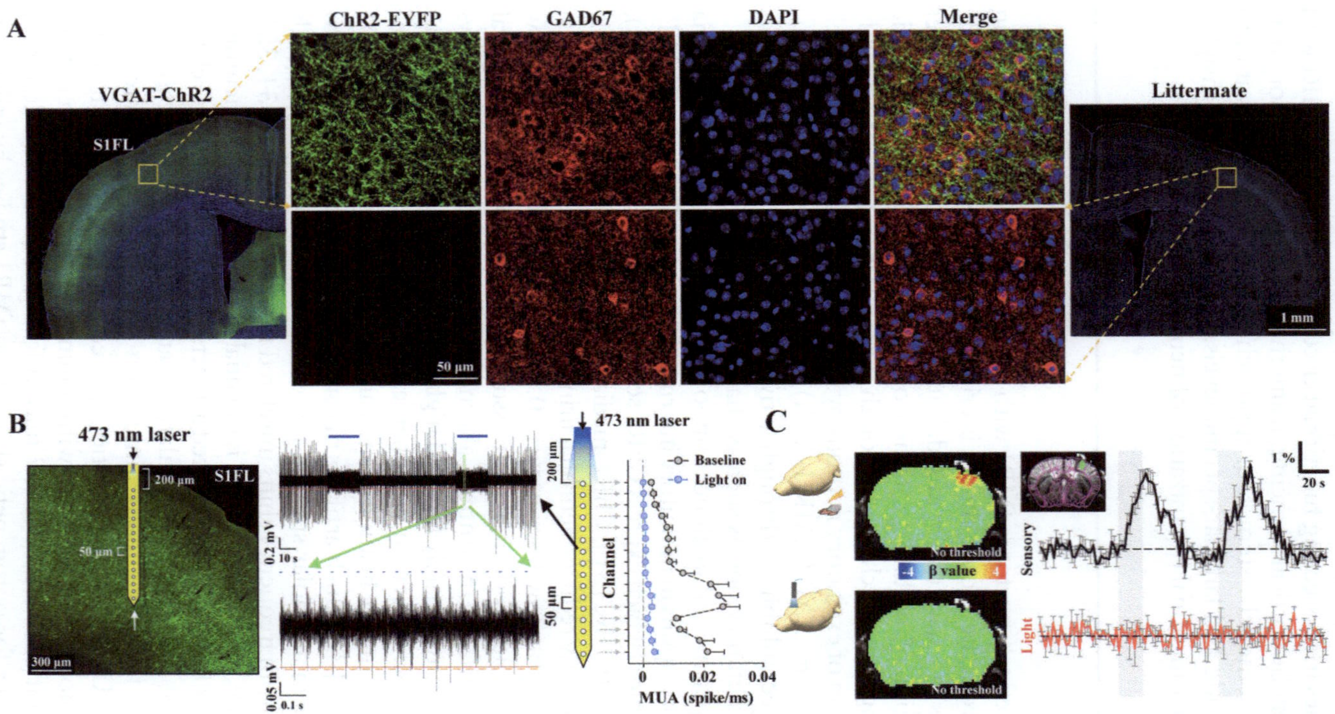

Fig. 1 Characterization of the VGAT-ChR2 mouse model for optogenetic-fMRI circuit analysis. (**A**) Transgenic-specific expression of ChR2-EYFP in the primary somatosensory cortex (S1FL) of a VGAT-ChR2-positive mouse compared to a VGAT-ChR2-negative littermate mouse. ChR2-EYFP (green) colocalizes with the interneuron marker GAD67 (red), indicating selective expression in cortical GABAergic neurons. Cell nuclei were visualized with DAPI (blue). (**B**) Optogenetic silencing of spontaneous activity in excitatory neurons within S1FL. Using a 16-channel optoelectrode (50 μm spacing, 1-mm depth insertion), photostimulation was applied to activate inhibitory neurons at the cortical surface. Multiunit activities (MUAs) were significantly reduced during two consecutive 20-s photostimulation periods (blue bars). Expanded views show brief increases in MUA during photostimulation pulses (blue dots), reflecting enhanced inhibitory neuron activity, with nearly complete suppression of spontaneous activity between pulses (yellowish-orange lines). Photostimulation effectively suppressed spontaneous neuronal activity across all cortical layers (200–1000 μm depth). (**C**) Evaluation of potential artifactual fMRI responses to light stimulation in naïve mice. Control experiments using interleaved forepaw (FP) stimulation and cortical photostimulation were conducted. While clear sensory-evoked fMRI responses occurred during FP stimulation in the S1FL (black time trace), photostimulation alone produced no significant fMRI signal changes (red time trace). Thus, fMRI artifacts due to tissue heating from photostimulation were negligible. Gray vertical bars indicate the 20-s stimulus period. Reprinted from [27]

For mapping brain-wide somatosensory networks by fMRI with optogenetic silencing, VGAT-ChR2 mice expressing light-sensitive opsin proteins in GABAergic interneuron populations were used in this study (see Fig. 1A for expression of ChR2 in inhibitory neurons). The suppression of neural activity was confirmed by recording multiunit activities (MUAs) with a 16-channel optoelectrode with 50 μm spacing between channels (A1×16-5 mm-50-177-OA16LP, NeuroNexus, Ann Arbor, MI, USA) under the blue light stimulation (473 nm, 3 mW, 10-ms duration, and 20 Hz). During 20 s of photostimulation to the cortical surface of the S1FL, an increase in stimulation-induced inhibitory neuronal activity mostly led to suppression of spontaneous activity across all cortical depths (Fig. 1B). With 3-mW optogenetic stimulation of VGAT-ChR2, the spatial spread of inhibition was expected to be >1 mm from the fiber center [31], effectively covering cortical areas.

To confirm whether the light irradiance used can induce heating-related fMRI artifacts, fMRI studies in naïve mice included interleaved sensory-evoked and optogenetic stimulations under the same photostimulation conditions via an optical fiber cannula (105 μ m inner core diameter, NA = 0.22, time-averaged light power of 69.3 mW/mm^2; Thorlabs, Newton, NJ, USA). These stimulus conditions did not induce heating-related fMRI artifacts in naïve mice (Fig. 1C), confirming that fMRI responses to optogenetic stimulation can genuinely reflect modulation of underlying neural activity.

2.2 Design of Stimulus Conditions

To dissect functional circuits using fMRI, we designed three stimulus conditions: (1) optogenetic cortical silencing, (2) simultaneous sensory and optogenetic stimulation, and (3) sensory-evoked stimulation (Fig. 2).

(1) Silencing cortical excitatory neurons by optogenetically activating ChR2-expressing inhibitory neurons suppresses local excitatory recurrent circuits and outputs to downstream pathways (Opto in Fig. 2), leading to a decreased fMRI signal in both local and downstream networked areas (deactivation).

(2) Simultaneous sensory stimulation during cortical silencing can preserve sensory-evoked inputs to the thalamus and cortex via upstream spinothalamic (ST) and thalamocortical (TC) pathways while suppressing downstream activity (Comb in Fig. 2). Since inhibitory neural activation itself can induce a hemodynamic response at the stimulation site and in downstream regions, this effect should be removed. It is assumed that the common photostimulation-driven hemodynamic response is eliminated by calculating the difference between optogenetic fMRI responses observed with and without sensory stimulation (Comb − Opto = Diff in Fig. 2). Consequently, the remaining signal reflects the contribution of upstream ST and TC pathways in the absence of local recurrent circuits and downstream projections from the target region.

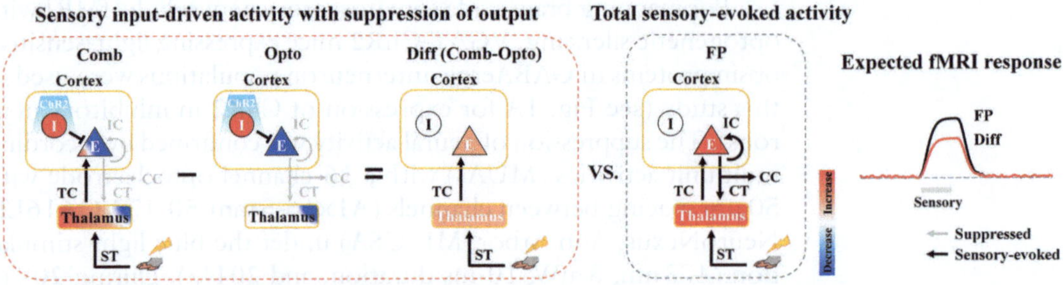

Fig. 2 Strategy for dissecting sensory circuits using fMRI with optogenetic silencing. Schematic illustration of the experimental design for dissecting sensory-evoked, long-range, and local circuit activity using fMRI with optogenetic cortical inhibition. In the simplified circuit, excitatory cortical neurons (E) interact with local inhibitory interneurons (I), receive thalamocortical (TC) inputs from the thalamus, and drive output activity through intracortical (IC), corticocortical (CC), and corticothalamic (CT) pathways. Optogenetic activation of ChR2-expressing inhibitory neurons suppresses excitatory neuron outputs, effectively silencing downstream CT, CC, and IC projections (Opto). When forepaw (FP) stimulation is simultaneously applied (Comb), it activates upstream spinothalamic (ST) and TC inputs to the thalamus and cortex. By subtracting the fMRI response to optogenetic inactivation alone (Opto) from the combined condition (Comb), the difference (Diff = Comb − Opto) separates upstream input-driven activity, excluding direct contributions from inhibitory neuron activation. Comparing this difference (Diff) with the total sensory-evoked response (FP) determines the relative contributions of downstream CT, CC, and IC circuits. Blue-to-red color scale represents relative signal change from baseline (decrease to increase)Reprinted from [27]

(3) By comparing the difference signal (Diff) with sensory-evoked fMRI responses (FP in Fig. 2), the contribution of downstream suppression on fMRI responses can be determined (Diff vs. FP in Fig. 2), allowing us to infer functional causality for long-range and local processing in somatosensory networks. While similar circuit-analysis approaches using sensory stimulation with and without local silencing have been used in electrophysiological studies [3, 4], fMRI combined with optogenetic inhibition provides a powerful tool to analyze brain-wide long-range circuits at the population level.

2.3 Stimulations for Functional Contrast

The stimulation parameters were controlled using a pulse generator (Master 9; World Precision Instruments, Sarasota, FL, USA). For optogenetic stimulation, blue light was delivered to the target cortex via a fiber-optic cable coupled to a 473-nm diode-pumped solid-state laser (MBL-III-473, Changchun New Industries Optoelectronics Tech. Co., Ltd, Changchun, China). The output power was calibrated to 3 mW at the tip of the optic fiber, measured by a power meter (PM100D, Thorlabs, USA). Light pulses were applied at 20 Hz with a pulse width of 10 ms through a 105 μm diameter fiber. To prevent undesired activation of the visual pathway due to light leakage inside the magnet bore, the connection between the fiber-optic cable and the implanted cannula was covered with heat-shrinkable sleeves, and the animal's eyes were shielded with a biocompatible silicone elastomer (Kwik-Sil, World Precision Instruments, Sarasota, FL, USA).

For somatosensory stimulation, the left forepaw (FP) were electrically stimulated using a pair of 30 G needle electrodes inserted subcutaneously under the plantar skin. Electrical pulses were delivered using a stimulus isolator (ISO-Flex, AMPI, Jerusalem, Israel) with a current intensity of 0.5 mA, a pulse width of 0.5 ms, and a frequency of 4 Hz triggered by the pulse generator. Each functional trial consisted of a 60-s prestimulus, 20-s stimulus, 60-s interstimulus interval, 20-s stimulus, and 60-s poststimulus period.

3 Methods

3.1 Animal Preparation

For fMRI with cortical inactivation in the somatosensory network, a stereotactic surgical procedure was performed to implant fiber-optic cannulas in mice at 6–7 weeks of age for optogenetic fMRI. The mice were anesthetized by an intraperitoneal (IP) bolus injection of a mixture of ketamine (Yuhan, Korea) and xylazine (Rompun®, Bayer, Korea) (100 mg/kg and 10 mg/kg, respectively) and fixed in a stereotaxic frame (SR-AM, Narishige, Tokyo, Japan) for surgery. Meloxicam (1 mg/kg) was administered subcutaneously to provide pain relief and reduce inflammation.

After making an incision on the scalp, the skull was cleaned and dried. A burr hole (~100 μm diameter) was carefully made by thinning the skull over the target area. The optical fiber cannula was then slowly inserted into the right S1FL (AP: −0.2 mm relative to bregma, ML: +2.2 mm, DV: +0.5 mm), M1 (AP: +0.05 mm, ML: +1.1 mm, DV: +0.25 mm) and S2 (AP: −1.1 mm, ML: +4.2 mm, DV: +0.9 mm) areas (see Fig. 3B-i for targeting sites) at a rate of 1 μm/s using a micromanipulator (SMX-model, Sensapex, Oulu, Finland). Once in place, a biocompatible silicone elastomer was applied to enclose the fiber implantation site, followed by a thin layer of dental cement (SB, Sun-Medical Co., Shiga, Japan) to secure the fiber onto the skull [36]. After a recovery period of at least 2 weeks, mice underwent optogenetic fMRI experiments.

3.2 MRI Data Acquisition

Functional MRI datasets were acquired using a 15.2 T MRI scanner equipped with an actively shielded gradient system (6-cm diameter) capable of operating at a maximum strength of 100 G/cm and a rise time of 110 μs (Bruker BioSpec, Billerica, MA, USA). A customized surface coil with an inner diameter of 15 mm was used for both radiofrequency (RF) transmission and reception, positioned centrally over the somatosensory cortex, with the brain placed near the isocenter of the magnet. Field inhomogeneity was minimized by performing global shimming, followed by local shimming using the FASTMAP protocol within an ellipsoidal volume covering the cerebrum (ParaVision 6, Bruker BioSpin). The ultrahigh field strength of

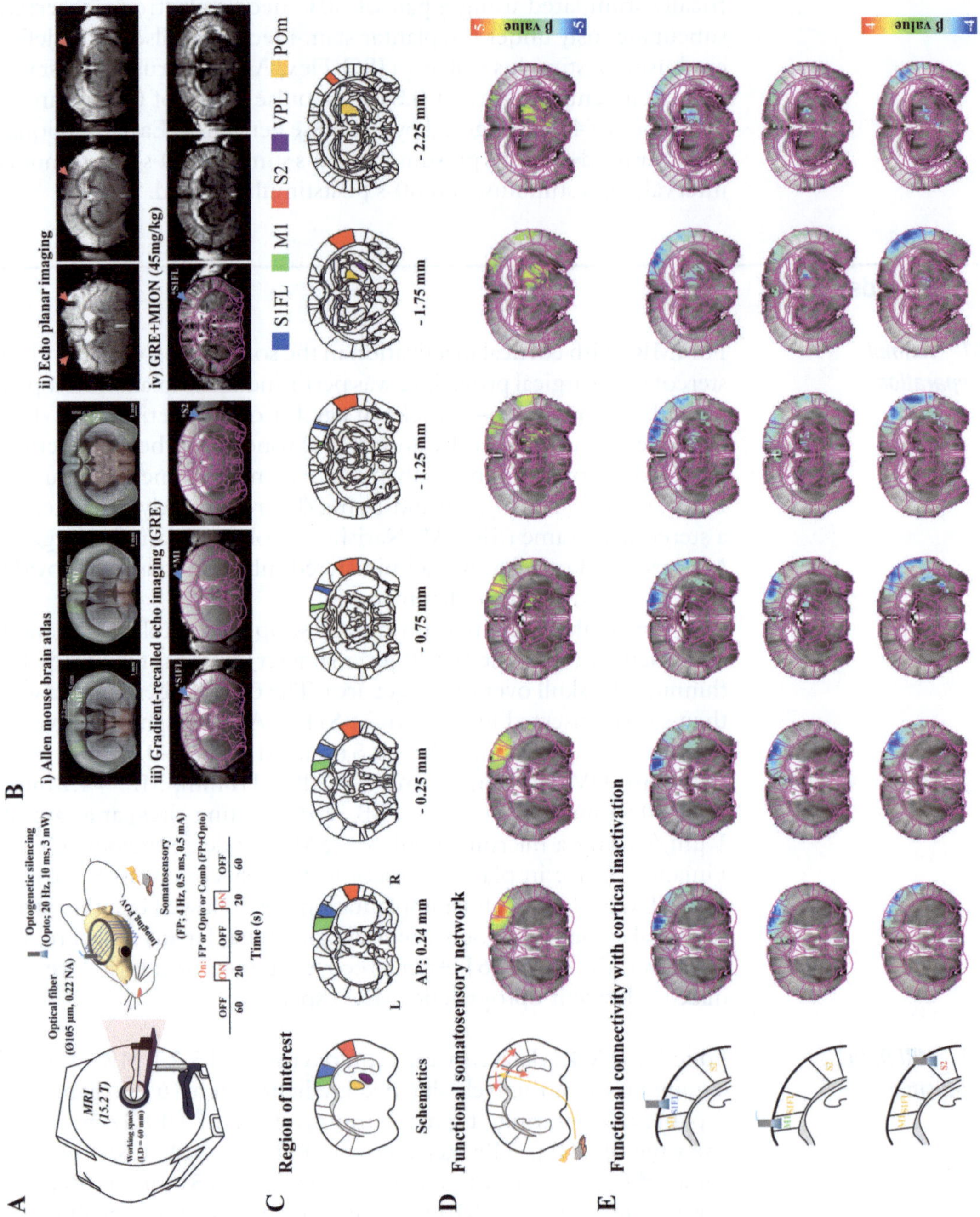
A
Optical fiber (Ø105 μm, 0.22 NA)
Optogenetic silencing (Opto; 20 Hz, 10 ms, 3 mW)
MRI (15.2 T)
Working space (I.D = 60 mm)
Imaging FOV
Somatosensory (FP: 4 Hz, 0.5 ms, 0.5 mA)
On: FP or Opto or Comb (FP+Opto)
OFF ON OFF ON OFF
60 20 60 20 60
Time (s)
B
i) Allen mouse brain atlas
ii) Echo planar imaging
iii) Gradient-recalled echo imaging (GRE)
iv) GRE+MION (45mg/kg)
C
Region of interest
S1FL
M1
S2
VPL
POm
Schematics
L
R
AP: 0.24 mm
- 0.25 mm
- 0.75 mm
- 1.25 mm
- 1.75 mm
- 2.25 mm
D
Functional somatosensory network
β value
E
Functional connectivity with cortical inactivation

15.2 T significantly improves sensitivity, potentially allowing precise detection of fMRI responses even from small or sparsely activated neuronal populations. This capability is particularly beneficial for studying whole-brain networks in mice [37], providing the high spatial resolution required for the small mouse brain while overcoming inherent limitations related to the low sensitivity of the fMRI signal.

CBV-Weighted fMRI

Gradient-echo (GE) echo-planar imaging (EPI), commonly used in BOLD fMRI, is highly susceptible to artifacts, such as distortion and signal drops caused by implanted optical fibers, particularly at long TEs and high field strengths. These artifacts (Fig. 3B-ii) can be minimized by adopting gradient-echo imaging with a short TE (Fig. 3B-iii). Our imaging strategy was designed to maximize functional sensitivity and specificity while mitigating susceptibility artifacts. Therefore, we adopted a CBV-weighted fMRI approach using monocrystalline iron oxide nanoparticles (MIONs; Feraheme, AMAG Pharmaceuticals, Waltham, MA, USA) as a contrast agent [38]. TE was set to 3 ms to match the measured tissue (3.44 ± 0.08 ms), with a 45 mg/kg dose of MION administered (Fig. 3B-iv) to enhance functional CBV sensitivity and specificity while minimizing confounding BOLD signals.

CBV-weighted fMRI data were acquired using the fast low-angle shot (FLASH) sequence with the following parameters: TR/TE = 50/3 ms, flip angle = 15°, FOV = 15×7.5 mm^2, matrix size = 96×48, partial Fourier = 1.2 along the phase-encoding direction, spatial resolution = $156 \times 156 \times 500$ μm^3, six contiguous coronal slices covering somatosensory areas, and a temporal resolution of 2 s.

Fig. 3 Experimental fMRI setup and brain-wide activity mapping using CBV-weighted fMRI with optogenetic silencing. (**A**) Schematic of multislice CBV-weighted fMRI at 15.2T during forepaw (FP) stimulation, cortical inactivation alone (Opto), and simultaneous cortical inactivation with FP stimulation (Comb). (**B**) Imaging quality control for optogenetic fMRI with fiber placement. (*i*) The fiber tips were positioned in S1FL, M1, and S2, overlaid on the corresponding Allen Mouse Brain Atlas. (*ii*) Image artifacts from susceptibility-induced distortion and signal dropout near the implant fiber targeting the S1FL in gradient-echo echoplanar imaging (EPI) (red arrows) were minimized (*iii*) by the adoption of gradient-echo imaging with a short TE of 3 ms (blue arrows). (*iv*) A superparamagnetic monocrystalline iron oxide nanoparticle (MION; 45 mg/kg) was injected to enhance functional sensitivity and specificity. (**C**) Regions of interest (ROIs) within the somatosensory network were defined based on the Allen Mouse Brain Atlas. Slice coordinates relative to bregma are labeled. R, right; L, left. (**D**) Brain-wide fMRI activation maps during FP stimulation demonstrate robust responses in contralateral somatosensory regions, including cortical areas (S1FL, M1, and S2) and thalamic nuclei (VPL and POm), as observed in the FP paradigm in Fig. 2. (**E**) Functional connectivity maps during cortical inactivation (Opto paradigm in Fig. 2) show ipsilateral deactivation across somatosensory cortical areas and thalamic nuclei, in response to optogenetic silencing of the S1FL, M1, and S2. Reprinted from [27]

To investigate functional circuits with optogenetic silencing, fMRI data were acquired under three designed stimulus conditions (Sect. 2.2) in separate groups targeting cortical areas: S1FL, M1, and S2 (Sect. 3.1). For each stimulus condition, 15 fMRI trials were performed per animal, with data acquired in an interleaved manner and an intertrial interval longer than 3 min.

During the experiments, mice were anesthetized using a mixture of ketamine and xylazine (100/10 mg/kg for initial IP and 25/1.25 mg/kg intermittent IP injections every 45–50 min) [39] while spontaneously breathing through a nose cone delivering an oxygen and air (1:4 ratio) at 1 L/min (SAR-1000, CWE, Ardmore, OK, USA or TOPO, Kent Scientific Corporation, Torrington, CT, USA), maintaining an oxygen saturation level >90%. Physiological parameters, including electrocardiograms and motion-sensitive respiratory signals, were continuously monitored (Model 1030, Small Animal Instruments Inc., Stony Brook, USA, for fMRI experiments). The body temperature was maintained at 37 ± 0.5°C using a warm-water heating system with a rectal thermometer.

3.3 Mapping Brain-Wide Networks by Sensory-Evoked and Focal Silencing

For fMRI mapping and voxel-wise analysis, multiple processing steps were performed using functions from the Analysis of Functional Neuroimages package (AFNI) [40], FMRIB Software Library (FSL) [41], FreeSurfer [42], Advanced Normalization Tools (ANTs) [43], and MATLAB codes (MathWorks, Natick, USA). The processing included the construction of a study-specific template, the generation of fMRI maps, and statistical analyses.

(1) Due to limited slice coverage in our imaging protocol, a study-specific template was constructed to spatially normalize individual animal fMRI data into a common brain space. First, intensity nonuniformity was corrected using bias field estimation (FreeSurfer nu_correct), followed by semiautomatic brain extraction from the temporal average of all fMRI images for each animal (FSL BET, AFNI 3dAutomask). Second, individual fMRI images were aligned with each other to identify the most representative image requiring minimal transformation within the population (FSL tbss_2_reg). This image was then used as a reference for initial spatial normalization. Third, all fMRI images were registered to this reference image using both linear and nonlinear transformations (ANTs antsRegistrationSyN), and the averaged dataset was subsequently used to construct the mouse brain template. For anatomical localization, this study-specific template was spatially registered to the Allen Mouse Brain Atlas [1] (Fig. 3C).

(2) The fMRI response maps for individual animals were generated through preprocessing and general linear model (GLM) analyses. Preprocessing steps included slice timing correction (AFNI 3dTshift), realignment to correct minor head motions

(AFNI 3dvolreg), linear signal detrending (AFNI 3dDetrend), and time course normalization based on the baseline period. Averaged fMRI trials for each stimulus condition were spatially normalized to the study-specific template (ANTs antsRegistrationSyN). Spatial smoothing was applied using a Gaussian kernel with a full width at half-maximum (FWHM) of one pixel (AFNI 3dmerge) to reduce misalignment and enhance functional detectability. GLM analysis (AFNI 3dDeconvolve) was performed using design matrices incorporating stimulus blocks convolved with a data-driven hemodynamic response function (HRF), derived by fitting the animal-averaged time course in S1FL during sensory stimulation to a two-gamma variate function (MATLAB lsqcurvefit).

(3) Following GLM analysis, group-averaged fMRI maps were generated for sensory-evoked and optogenetic silencing conditions (S1FL, M1, and S2) using a one-sample t-test with nonparametric inference. Statistical significance was determined using threshold-free cluster enhancement (TFCE)-corrected $p < 0.05$ (FSL randomize). Standardized regression coefficients (β values) were mapped to quantify fMRI response magnitudes.

An increase in CBV leads to an increase in the amount of iron oxide within the voxel, thereby decreasing fMRI signals. To ensure consistency in interpretation, the original CBV-weighted fMRI signal was inverted to align with CBV responses. Somatosensory stimulation induced robust CBV increases in the contralateral somatosensory network (Fig. 3D). Brain-wide activation sites were observed in cortical regions, including the S1FL, M1, and S2, as well as thalamic nuclei, including first-order ventral posterolateral nucleus (VPL) and higher-order posterior medial thalamic nucleus (POm). To dissect the functional circuits underlying fMRI responses, optogenetic silencing was applied to S1FL, M1, and S2 in conjunction with somatosensory stimulation. Cortical silencing via optogenetic activation of inhibitory neurons resulted in CBV reductions in networked cortical and subcortical sites (Fig. 3E). Inactivation of S1FL induced negative CBV changes in the ipsilateral S1, M1, S2, and thalamic nuclei (VPL, POm), as well as in the striatum. Similar patterns of CBV suppression were observed following M1 and S2 inactivation (Fig. 3E). These fMRI networks exhibit topological consistency with anatomical neural tracing studies [1, 2].

3.4 Dissection of functional network at a Brain-Wide Level

To investigate the functional circuits underlying sensory-evoked responses, we examined the relative roles of spinothalamic (ST), thalamocortical (TC), corticocortical (CC), corticothalamic (CT), and intracortical (IC) pathways (Fig. 4A). To separate somatosensory input-driven responses without downstream feedforward and local contributions from the inactivation site, fMRI data from cortical inactivation with and without FP stimulation were subtracted

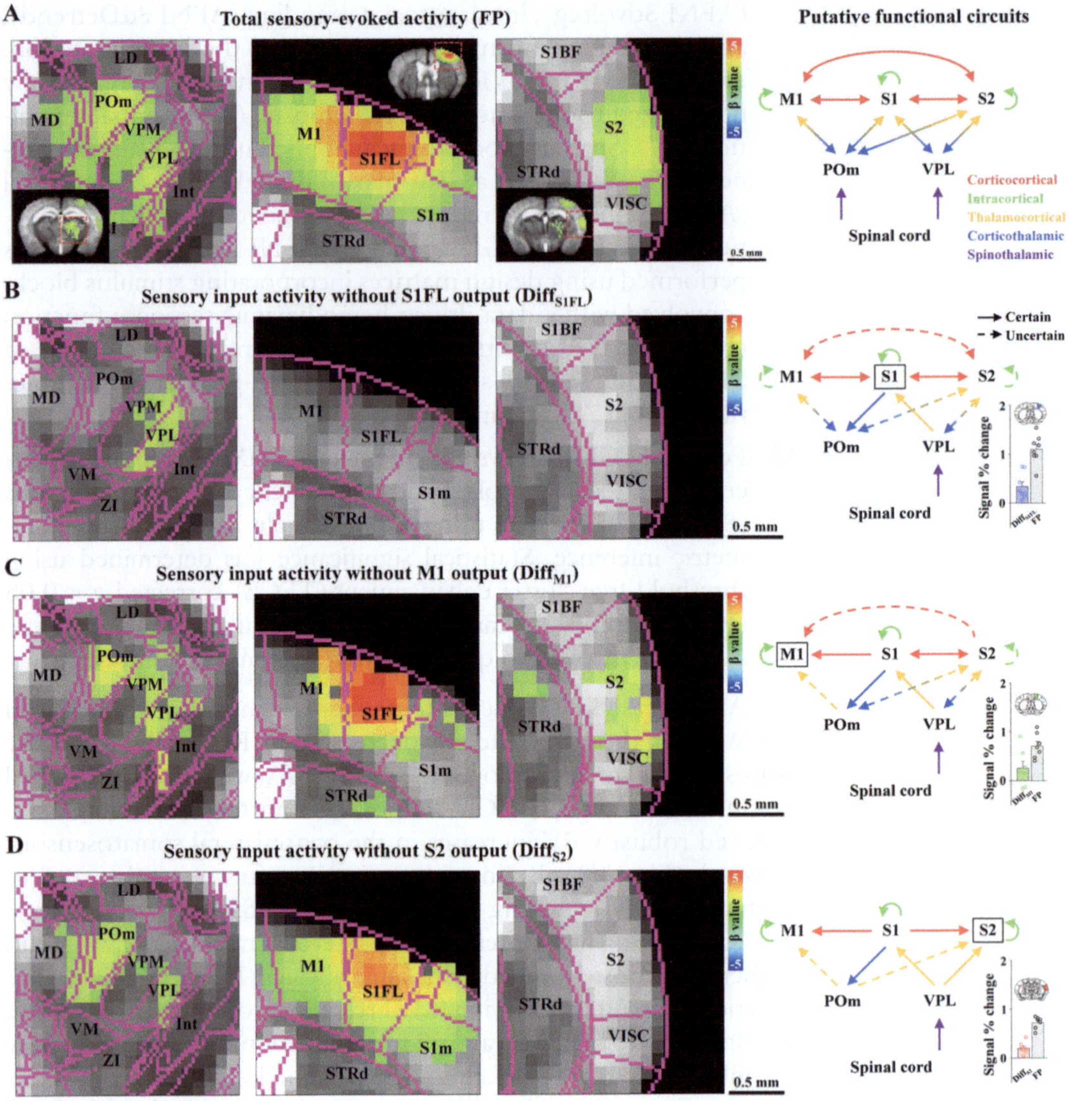

Fig. 4 Dissection of somatosensory-driven fMRI responses with cortical silencing at a brain-wide level. (**A**) fMRI mapping of total sensory-evoked responses. FP stimulation induced widespread CBV increases in thalamic (VPL, POm) and cortical (S1FL, M1, S2) areas. These responses reflect a reciprocally connected functional network including ascending ST and thalamocortical (TC) inputs, along with descending CT, CC, and local IC circuits. (**B**) Input-driven activity without S1FL output. With S1FL silenced, VPL activity remained intact, consistent with direct spinal input, while POm responses were abolished, indicating dominant dependence on S1FL. M1 and S2 responses were mostly reduced, confirming major feedforward CC input from S1FL. Residual activity in S1FL (~31%) suggests that ~69% of the total response (inset graph) results from IC amplification of long-range inputs. (**C**) Input-driven activity without M1 output. Silencing M1 did not affect thalamic and S1FL responses, indicating minimal feedback M1 projections. S2 responses also remained unchanged, suggesting a negligible role for the M1 → S2 pathway. In M1, ~36% of the response was input-driven, with the remaining ~64% attributed to local IC amplification (inset graph). (**D**) Input-driven activity without S2 output. S2 silencing had no effect on thalamic or upstream cortical activity (S1FL, M1), indicating limited feedback projections from S2. Within S2, ~28% of the response was driven by long-range inputs (mainly from S1FL, some from VPL), and

(Comb−Opto). The resulting difference maps ($Diff_{S1FL}$, $Diff_{M1}$, and $Diff_{S2}$ in Fig. 4B–D) were obtained with a significance threshold of uncorrected $p < 0.01$ using a paired t-test (FSL randomize).

Sensory Input-Driven Activity Without S1FL Output

To determine the contribution of S1FL to thalamic and cortical responses, total sensory-evoked activity was compared with input-driven activity under S1FL output suppression (Fig. 4B).

In the thalamic nuclei, the CBV response in the VPL during FP stimulation closely matched this activity ($Diff_{S1FL}$), suggesting VPL possibly receives inputs from the spinal cord (and/or S2) rather than the S1FL. Conversely, POm responses were completely abolished, indicating that S1FL input is the dominant driver of POm responses. S1FL also played a crucial role in modulating cortical responses in M1 and S2. Somatosensory-induced fMRI responses in M1 included CC inputs from S1FL and S2, TC inputs from the POm, and IC activity. When S1FL was silenced, M1 responses were mostly suppressed, highlighting the dominant contribution of the S1FL-to-M1 corticocortical projection. Similarly, functional inputs to S2 could originate from monosynaptic TC circuits from the VPL, direct CC projections from S1FL, disynaptic pathways through M1 (S1FL → M1 → S2) and POm (S1FL → POm → S2), and IC activity. The contribution of the TC projection, calculated by the difference in signal changes without and with S1FL silencing, constituted only 25% of the total S2 response, whereas the remaining 75% of inputs originated from S1FL. This indicates that S2 activity is mainly driven by feedforward inputs from S1FL.

Additionally, somatosensory-driven S1FL activity contained both long-range inputs (TC, CC) and local recurrent activity (IC). Thus, the contribution of IC activity can be determined from the difference between long-range inputs and total somatosensory activity induced by forepaw stimulation. When IC activity in S1FL was suppressed ($Diff_{S1FL}$), a positive CBV response persisted, though its magnitude was reduced to approximately 31% of the total stimulus-driven response (inset graph in Fig. 4B). This finding suggests that input-driven responses to S1FL (TC and/or CC inputs) are significantly amplified through local recurrent activity within the region.

Together, these findings underscore the crucial role of S1FL in integrating and propagating somatosensory information across thalamic and cortical regions. While suppressing S1FL output provided key insights into its functional contributions, fully dissecting

Fig. 4 (continued): ~72% originated from local IC circuits. During S1FL silencing, only ~25% of S2 activity remained, suggesting the dominant role of S1FL in driving S2 responses. LD lateral dorsal nucleus, MD mediodorsal nucleus, VM ventral medial nucleus, Int internal capsule, ZI zona incerta. S1m, primary somatosensory area of the mouth, and STRd, striatum of the dorsal region. S1BF, primary somatosensory area of the barrel field; VISC, visceral area, and STRd, striatum of the dorsal region (Scale bar, 0.5 mm). Reprinted from [27]

the processing of the somatosensory network requires additional investigation into sensory input-driven activity under the suppression of M1 and S2. By comparing fMRI responses across these different conditions, we can further determine the specific roles of each cortical region in shaping sensory information flow and network-wide interactions.

Sensory Input-Driven Activity Without M1 Output

To investigate the contribution of M1 to functional somatosensory networks, we examined changes in sensory-evoked activity when M1 output was selectively suppressed (Fig. 4C). M1 suppression did not affect activity in thalamic nuclei (VPL and POm), indicating that these regions receive their primary inputs from other sources. This supports the hypothesis that the VPL primarily receives spinal cord inputs, functioning as a key relay for spinal sensory information, while the POm predominantly processes signals from S1FL. Although S2 may contribute to inputs to these thalamic nuclei, current findings do not clearly determine its significance.

Based on fMRI data acquired after M1 silencing, we found that CC projections from M1 to S1FL were negligible during FP stimulation. This finding suggests that while M1 primarily receives signals from S1FL, its feedback contribution to S1FL is minimal under anesthesia [44, 45]. Notably, input-driven M1 responses accounted for approximately 36% of the total somatosensory-evoked fMRI response in M1 (Inset graph in Fig. 4C), indicating significant intrinsic amplification within M1, similar to S1FL. Additionally, suppression of M1 did not change responses in S2, suggesting the M1 → S2 pathway plays a negligible role, and S2 activation is predominantly driven by direct inputs from S1FL or other sources, rather than by M1-mediated projections.

Sensory Input-Driven Activity Without S2 Output

To further clarify the functional role of S2 within somatosensory networks, we examined the relative contributions of long-range projections and local circuit activity during S2 silencing. Despite anatomical connections to VPL, POm, S1FL, and M1 [46], silencing S2 did not significantly affect fMRI responses in these regions, indicating a limited modulatory role of S2 on thalamic and cortical processing, compared to S1FL.

Comparative analyses of sensory-evoked S2 responses with and without S2 silencing ($Diff_{S2}$) indicated that long-range synaptic inputs accounted for approximately 28% of the total signal change, while the remaining 72% of the response originated from local IC circuits (Inset graph in Fig. 4D). Notably, residual responses (approximately 25%) observed in S2 during S1FL suppression ($Diff_{S1FL}$) may reflect direct VPL inputs, while the remaining 75% originate from S1FL. These findings suggest that S2 activity during FP stimulation primarily results from feedforward inputs from S1FL, supplemented by minor contributions from VPL. These inputs are further amplified by recurrent IC activity, underscoring the importance of local processing in shaping S2 responses to somatosensory stimulation.

Overall, these brain-wide fMRI results help determine the relative contributions of feedforward projections, feedback connections, and local circuits to sensory-evoked processing. However, to further refine our interpretation, we conducted additional analyses at the laminar-specific level.

3.5 Resolving Circuit Interactions by Laminar-Specific fMRI

Conventional BOLD contrast in fMRI is highly sensitive to draining veins, resulting in poor specificity for cortical layers, while the CBV-weighted fMRI approach using a superparamagnetic contrast agent provides an effective alternative, enhancing both functional sensitivity and spatial specificity for sites of neuronal activity [47]. Therefore, to achieve a better understanding of circuit-level contributions, we investigated layer-dependent input processing within cortical areas, including the S1FL, M1, and S2.

For laminar-specific analyses, cortical areas were flattened using radial projections perpendicular to cortical edges (MATLAB improfile; [48]), and depth profiles were resampled using bicubic interpolation to achieve a nominal resolution of 78 μm [23]. Profiles were averaged across a 0.78 mm width in S1FL and M1 (see Fig. 5A) and a 1.56 mm width in S2 (see Fig. 5E). Group-averaged percent change maps were then generated from individual animal fMRI data (Fig. 5B, E). Laminar boundaries for each cortex were defined according to cortical thickness distribution derived from the Allen mouse brain atlas [1].

First, we examined whether TC and CC input layers could be identified based on cortical depth-dependent CBV responses. In S1FL, somatosensory-evoked CBV responses peaked at layer 4 (L4), indicating predominant TC input (Fig. 5B, C). Conversely, M1 responses peaked at layer 2/3 (L2/3), reflecting CC input rather than TC input from POm (Fig. 5B, D). These findings are consistent with the expected laminar organization of TC and CC pathways [49]. Cortical suppression in the S1FL, M1, and S2 induced negative CBV changes, particularly evident in CC input layers of networked regions (Fig. 5B–D). Notably, M1 suppression led to peak responses in L2/3 and L5A of S1FL with a small dip at the ~150-μm-thick L4 (Fig. 5C, green), indicating a high spatial specificity of CBV responses with a hemodynamic point spread function (PSF) of <150 μm FWHM. These laminar-specific fMRI findings showed that cortical CBV responses peak at synaptic input layers, supporting the utility of CBV-weighted fMRI in resolving layer-specific activation patterns.

Next, we investigated synaptic inputs from S1FL to S2, which occur either through direct corticocortical (CC) projections (S1FL → S2) or indirectly via cortico-thalamo-cortical (CTC) pathways via POm (S1FL → POm → S2) [50]. The question was whether laminar CBV analysis could clearly separate direct CC inputs from indirect CTC inputs. Since direct CC projections from S1FL predominantly

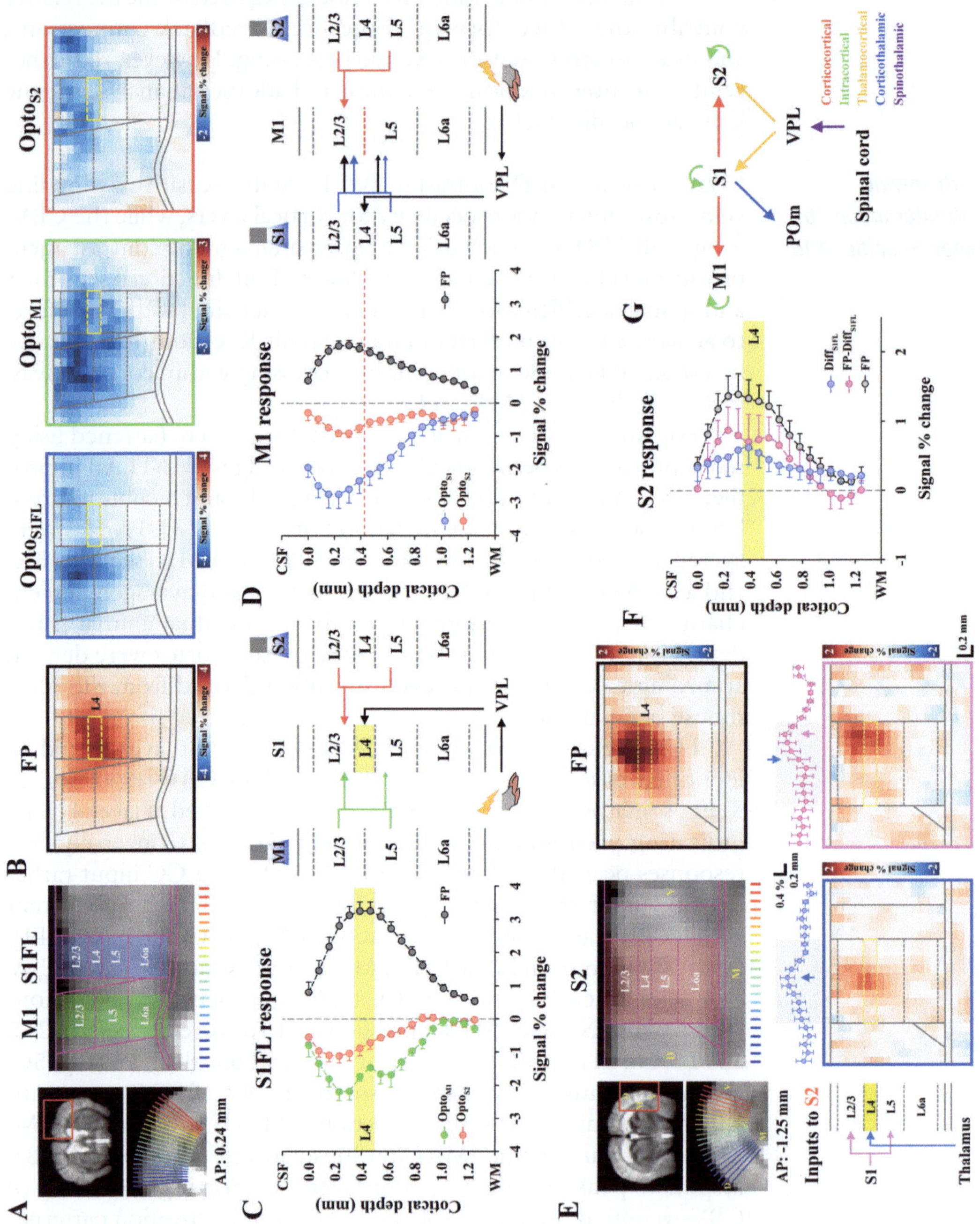
A
M1
S1FL
L2/3
L4
L5
L6a
AP: 0.24 mm
B
FP
OptoS1FL
OptoM1
OptoS2
Signal % change
C
S1FL response
Cotical depth (mm)
CSF
WM
Signal % change
OptoM1
OptoS2
FP
D
M1 response
OptoS1
S1
S2
VPL
E
S2
AP: -1.25 mm
Inputs to S2
Thalamus
F
DiffS1FL
FP-DiffS1FL
G
S2 response
POm
Spinal cord
Corticocortical
Intracortical
Thalamocortical
Corticothalamic
Spinothalamic

target L2/3 and L5, while CTC inputs primarily project to L4 (Fig. 5E, schematic of "Inputs-to-S2") [49–51], laminar-specific CBV responses provided an effective way to differentiate these pathways. Given the spatial specificity of CBV responses (PSF <150 μm, Fig. 5C), the peak locations of CBV signals indicate the cortical sources of inputs. During somatosensory stimulation, S2 showed broad CBV responses across L2/3 and L4, suggesting different input sources (FP in Fig. 5E, black profile in Fig. 5F). To separate the TC input originating from VPL, we obtained the difference between responses with S1FL inactivation with and without FP stimulation ($Diff_{S1FL}$ in Fig. 5E, blue profile in Fig. 5F). Subsequently, we determined the direct CC inputs from S1FL to S2 (FP-$Diff_{S1FL}$) by subtracting the VPL-driven input ($Diff_{S1FL}$) from the total sensory-evoked response (FP). The results indicated two spatially separable input sources within S2: TC inputs from VPL peaked at L4 in dorsal S2, whereas CC inputs from S1FL showed double peaks at L2/3 and L5 in ventral S2, clearly identifying direct CC projections (FP-$Diff_{S1FL}$, Fig. 5E, pink profile in Fig. 5F). Utilizing laminar-specific CBV fMRI with cortical silencing, we successfully differentiated TC from direct CC inputs to S2, providing novel insights into the laminar organization of sensory processing (Fig. 5G).

Fig. 5 (continued) Laminar fMRI mapping of input-specific circuits in the somatosensory cortex. (**A**) Cortical flattening for laminar fMRI analysis of S1FL and M1. To examine depth-dependent fMRI responses, the S1FL and M1 regions were upsampled and linearized using radially projecting lines perpendicular to the cortical surface. Laminar boundaries were defined based on thickness distributions. (**B**) Cortical depth-dependent fMRI activity. Somatosensory-evoked CBV responses were localized to L4 in S1FL and to L2/3 in M1. Optogenetic silencing resulted in notable suppression at L2/3 in both regions, indicating that these layers are sensitive to input-driven CC modulation. (**C–D**) CBV profiles and corresponding circuit diagrams. (**C**) In S1FL, sensory-evoked responses peaked at L4 (black line), consistent with TC input. M1 silencing (green) induced responses at L2/3 and L5, while S2 silencing (red) led to the largest changes at L2/3. Since M1 does not project to L4 of S1FL, the sharp peak suggests the PSF of laminar response is less than 150 μm (FWHM). (**D**) In M1, the largest CBV responses consistently occurred at L2/3 across all stimulus conditions. (**E–F**) Separation of TC and S1FL-derived CC/CTC inputs in S2. To dissect the sources of input to S2, layer-specific responses were analyzed after cortical flattening. Total sensory-evoked responses (black) were decomposed into TC (blue) and CC (pink) components. TC inputs ($Diff_{S1FL}$) projected to L4, while direct and indirect S1FL inputs (FP − $Diff_{S1FL}$) showed double peaks at L2/3 and L5, supporting laminar specificity of feedforward cortical inputs. (**G**) Putative functional circuits underlying CBV-fMRI responses. Expanding upon prior brain-wide findings, layer-specific fMRI mapping revealed how functional circuits are organized within the somatosensory network as measured by fMRI. Reprinted from [27]

4 Conclusions

Our fMRI approach, combining brain-wide and laminar-specific analyses with focal optogenetic silencing, systematically dissected the in vivo contributions of sensory-driven, long-range, and local recurrent circuits to fMRI responses. The high spatial specificity of CBV-weighted fMRI enabled the mapping of layer-specific processing, thereby bridging macroscale network activity with microscale circuit dynamics. This methodology significantly enhances our understanding of sensory information processing across cortical and thalamic networks. Brain-wide fMRI with circuit-level dissection offers a powerful framework for investigating functional reorganization in response to neuropathological conditions and learning processes. In addition, it complements traditional electrophysiological and microscopic approaches by providing insights into large-scale functional connectivity and information flow.

Acknowledgments

This research was supported by the Institute for Basic Science (IBS-R015-D1) and the Korea Brain Research Institute basic research program (25-BR-02-02, 25-BR-03-02, and 25-BR-04-01) in Korea.

Conflict of Interest The author has no conflicts of interest to declare that are relevant to the content of this chapter.

References

1. Oh SW, Harris JA, Ng L, Winslow B, Cain N, Mihalas S, Wang Q, Lau C, Kuan L, Henry AM, Mortrud MT, Ouellette B, Nguyen TN, Sorensen SA, Slaughterbeck CR, Wakeman W, Li Y, Feng D, Ho A, Nicholas E, Hirokawa KE, Bohn P, Joines KM, Peng H, Hawrylycz MJ, Phillips JW, Hohmann JG, Wohnoutka P, Gerfen CR, Koch C, Bernard A, Dang C, Jones AR, Zeng H (2014) A mesoscale connectome of the mouse brain. Nature 508(7495):207–214. https://doi.org/10.1038/nature13186
2. Harris JA, Mihalas S, Hirokawa KE, Whitesell JD, Choi H, Bernard A, Bohn P, Caldejon S, Casal L, Cho A, Feiner A, Feng D, Gaudreault N, Gerfen CR, Graddis N, Groblewski PA, Henry AM, Ho A, Howard R, Knox JE, Kuan L, Kuang X, Lecoq J, Lesnar P, Li Y, Luviano J, McConoughey S, Mortrud MT, Naeemi M, Ng L, Oh SW, Ouellette B, Shen E, Sorensen SA, Wakeman W, Wang Q, Wang Y, Williford A, Phillips JW, Jones AR, Koch C, Zeng H (2019) Hierarchical organization of cortical and thalamic connectivity. Nature 575(7781):195–202. https://doi.org/10.1038/s41586-019-1716-z
3. Diamond ME, Armstrong-James M, Budway MJ, Ebner FF (1992) Somatic sensory responses in the rostral sector of the posterior group (POm) and in the ventral posterior medial nucleus (VPM) of the rat thalamus: dependence on the barrel field cortex. J Comp Neurol 319(1):66–84. https://doi.org/10.1002/cne.903190108
4. Lee S, Kruglikov I, Huang ZJ, Fishell G, Rudy B (2013) A disinhibitory circuit mediates motor integration in the somatosensory cortex. Nat Neurosci 16(11):1662–1670. https://doi.org/10.1038/nn.3544
5. Vanni MP, Chan AW, Balbi M, Silasi G, Murphy TH (2017) Mesoscale mapping of mouse cortex reveals frequency-dependent cycling between distinct macroscale functional modules. J Neurosci 37(31):7513–7533. https://doi.org/10.1523/jneurosci.3560-16.2017
6. Hillman EM (2007) Optical brain imaging in vivo: techniques and applications from animal to man. J Biomed Opt 12(5):051402. https://doi.org/10.1117/1.2789693
7. Kim SG, Ogawa S (2012) Biophysical and physiological origins of blood oxygenation

level-dependent fMRI signals. J Cereb Blood Flow Metab 32(7):1188–1206. https://doi.org/10.1038/jcbfm.2012.23
8. Ogawa S, Tank DW, Menon R, Ellermann JM, Kim SG, Merkle H, Ugurbil K (1992) Intrinsic signal changes accompanying sensory stimulation: functional brain mapping with magnetic resonance imaging. Proc Natl Acad Sci USA 89(13):5951–5955. https://doi.org/10.1073/pnas.89.13.5951
9. Bandettini PA, Wong EC, Hinks RS, Tikofsky RS, Hyde JS (1992) Time course EPI of human brain function during task activation. Magn Reson Med 25(2):390–397. https://doi.org/10.1002/mrm.1910250220
10. Biswal B, Yetkin FZ, Haughton VM, Hyde JS (1995) Functional connectivity in the motor cortex of resting human brain using echo-planar MRI. Magn Reson Med 34(4):537–541. https://doi.org/10.1002/mrm.1910340409
11. Fox MD, Raichle ME (2007) Spontaneous fluctuations in brain activity observed with functional magnetic resonance imaging. Nat Rev Neurosci 8(9):700–711. https://doi.org/10.1038/nrn2201
12. Logothetis NK (2008) What we can do and what we cannot do with fMRI. Nature 453(7197):869–878. https://doi.org/10.1038/nature06976
13. Deisseroth K (2015) Optogenetics: 10 years of microbial opsins in neuroscience. Nat Neurosci 18(9):1213–1225. https://doi.org/10.1038/nn.4091
14. Sato TK, Häusser M, Carandini M (2014) Distal connectivity causes summation and division across mouse visual cortex. Nat Neurosci 17(1):30–32. https://doi.org/10.1038/nn.3585
15. Kim CK, Adhikari A, Deisseroth K (2017) Integration of optogenetics with complementary methodologies in systems neuroscience. Nat Rev Neurosci 18(4):222–235. https://doi.org/10.1038/nrn.2017.15
16. Zhao S, Ting JT, Atallah HE, Qiu L, Tan J, Gloss B, Augustine GJ, Deisseroth K, Luo M, Graybiel AM, Feng G (2011) Cell type–specific channelrhodopsin-2 transgenic mice for optogenetic dissection of neural circuitry function. Nat Methods 8(9):745–752. https://doi.org/10.1038/nmeth.1668
17. Chow BY, Han X, Dobry AS, Qian X, Chuong AS, Li M, Henninger MA, Belfort GM, Lin Y, Monahan PE, Boyden ES (2010) High-performance genetically targetable optical neural silencing by light-driven proton pumps. Nature 463(7277):98–102. https://doi.org/10.1038/nature08652
18. Reinhold K, Lien AD, Scanziani M (2015) Distinct recurrent versus afferent dynamics in cortical visual processing. Nat Neurosci 18(12):1789–1797. https://doi.org/10.1038/nn.4153
19. Wiegert JS, Mahn M, Prigge M, Printz Y, Yizhar O (2017) Silencing neurons: tools, applications, and experimental constraints. Neuron 95(3):504–529. https://doi.org/10.1016/j.neuron.2017.06.050
20. Babl SS, Rummell BP, Sigurdsson T (2019) The spatial extent of optogenetic silencing in transgenic mice expressing channelrhodopsin in inhibitory interneurons. Cell Rep 29(5):1381–1395. e1384. https://doi.org/10.1016/j.celrep.2019.09.049
21. Huo Y, Chen H, Guo ZV (2020) Mapping functional connectivity from the dorsal cortex to the thalamus. Neuron 107(6):1080–1094. e1085. https://doi.org/10.1016/j.neuron.2020.06.038
22. Lee JH, Durand R, Gradinaru V, Zhang F, Goshen I, Kim DS, Fenno LE, Ramakrishnan C, Deisseroth K (2010) Global and local fMRI signals driven by neurons defined optogenetically by type and wiring. Nature 465(7299):788–792. https://doi.org/10.1038/nature09108
23. Jung WB, Im GH, Jiang H, Kim SG (2021) Early fMRI responses to somatosensory and optogenetic stimulation reflect neural information flow. Proc Natl Acad Sci USA 118(11):e2023265118. https://doi.org/10.1073/pnas.2023265118
24. Moon HS, Jiang H, Vo TT, Jung WB, Vazquez AL, Kim SG (2021) Contribution of excitatory and inhibitory neuronal activity to BOLD fMRI. Cereb Cortex 31(9):4053–4067. https://doi.org/10.1093/cercor/bhab068
25. Vo TT, Im GH, Han K, Suh M, Drew PJ, Kim SG (2023) Parvalbumin interneuron activity drives fast inhibition-induced vasoconstriction followed by slow substance P-mediated vasodilation. Proc Natl Acad Sci USA 120(18):e2220777120. https://doi.org/10.1073/pnas.2220777120
26. Kim S, Moon HS, Vo TT, Kim CH, Im GH, Lee S, Choi M, Kim SG (2023) Whole-brain mapping of effective connectivity by fMRI with cortex-wide patterned optogenetics. Neuron 111(11):1732–1747. e1736. https://doi.org/10.1016/j.neuron.2023.03.002
27. Jung WB, Jiang H, Lee S, Kim SG (2022) Dissection of brain-wide resting-state and functional somatosensory circuits by fMRI with optogenetic silencing. Proc Natl Acad Sci USA 119(4):e2113313119. https://doi.org/10.1073/pnas.2113313119
28. Lee JY, You T, Lee CH, Im GH, Seo H, Woo CW, Kim SG (2022) Role of anterior cingulate cortex inputs to periaqueductal gray for pain avoidance. Curr Biol 32(13):2834–2847. e2835. https://doi.org/10.1016/j.cub.2022.04.090
29. Dinh TNA, Moon HS, Kim SG (2024) Separation of bimodal fMRI responses in mouse somatosensory areas into V1 and non-V1 contributions. Sci Rep 14(1):6302. https://doi.org/10.1038/s41598-024-56305-w
30. Meyer HS, Schwarz D, Wimmer VC, Schmitt AC, Kerr JN, Sakmann B, Helmstaedter M (2011) Inhibitory interneurons in a

cortical column form hot zones of inhibition in layers 2 and 5A. Proc Natl Acad Sci USA 108(40):16807–16812. https://doi.org/10.1073/pnas.1113648108
31. Li N, Chen S, Guo ZV, Chen H, Huo Y, Inagaki HK, Chen G, Davis C, Hansel D, Guo C, Svoboda K (2019) Spatiotemporal constraints on optogenetic inactivation in cortical circuits. eLife 8:e48622. https://doi.org/10.7554/eLife.48622
32. Schmid F, Wachsmuth L, Albers F, Schwalm M, Stroh A, Faber C (2017) True and apparent optogenetic BOLD fMRI signals. Magn Reson Med 77(1):126–136. https://doi.org/10.1002/mrm.26095
33. Rungta RL, Osmanski BF, Boido D, Tanter M, Charpak S (2017) Light controls cerebral blood flow in naive animals. Nat Commun 8:14191. https://doi.org/10.1038/ncomms14191
34. Christie IN, Wells JA, Southern P, Marina N, Kasparov S, Gourine AV, Lythgoe MF (2013) fMRI response to blue light delivery in the naïve brain: implications for combined optogenetic fMRI studies. Neuroimage 66:634–641. https://doi.org/10.1016/j.neuroimage.2012.10.074
35. Chen X, Sobczak F, Chen Y, Jiang Y, Qian C, Lu Z, Ayata C, Logothetis NK, Yu X (2019) Mapping optogenetically-driven single-vessel fMRI with concurrent neuronal calcium recordings in the rat hippocampus. Nat Commun 10(1):5239. https://doi.org/10.1038/s41467-019-12850-x
36. Shim HJ, Im GH, Jung WB, Moon HS, Dinh TNA, Lee JY, Kim SG (2022) Protocol for mouse optogenetic fMRI at ultrahigh magnetic fields. STAR Protoc 3(4):101846. https://doi.org/10.1016/j.xpro.2022.101846
37. Jung WB, Shim HJ, Kim SG (2019) Mouse BOLD fMRI at ultrahigh field detects somatosensory networks including thalamic nuclei. Neuroimage 195:203–214. https://doi.org/10.1016/j.neuroimage.2019.03.063
38. Kim SG, Harel N, Jin T, Kim T, Lee P, Zhao F (2013) Cerebral blood volume MRI with intravascular superparamagnetic iron oxide nanoparticles. NMR Biomed 26(8):949–962. https://doi.org/10.1002/nbm.2885
39. Shim HJ, Jung WB, Schlegel F, Lee J, Kim S, Lee J, Kim SG (2018) Mouse fMRI under ketamine and xylazine anesthesia: robust contralateral somatosensory cortex activation in response to forepaw stimulation. Neuroimage 177:30–44. https://doi.org/10.1016/j.neuroimage.2018.04.062
40. Cox RW (1996) AFNI: software for analysis and visualization of functional magnetic resonance neuroimages. Comput Biomed Res 29(3):162–173. https://doi.org/10.1006/cbmr.1996.0014
41. Smith SM, Jenkinson M, Woolrich MW, Beckmann CF, Behrens TE, Johansen-Berg H, Bannister PR, De Luca M, Drobnjak I, Flitney DE, Niazy RK, Saunders J, Vickers J, Zhang Y, De Stefano N, Brady JM, Matthews PM (2004) Advances in functional and structural MR image analysis and implementation as FSL. Neuroimage 23(Suppl 1):S208–S219. https://doi.org/10.1016/j.neuroimage.2004.07.051
42. Fischl B (2012) FreeSurfer. Neuroimage 62(2):774–781. https://doi.org/10.1016/j.neuroimage.2012.01.021
43. Avants BB, Tustison NJ, Song G, Cook PA, Klein A, Gee JC (2011) A reproducible evaluation of ANTs similarity metric performance in brain image registration. Neuroimage 54(3):2033–2044. https://doi.org/10.1016/j.neuroimage.2010.09.025
44. Suzuki M, Larkum ME (2020) General anesthesia decouples cortical pyramidal neurons. Cell 180(4):666–676. e613. https://doi.org/10.1016/j.cell.2020.01.024
45. Keller AJ, Roth MM, Scanziani M (2020) Feedback generates a second receptive field in neurons of the visual cortex. Nature 582(7813):545–549. https://doi.org/10.1038/s41586-020-2319-4
46. Liao CC, Yen CT (2008) Functional connectivity of the secondary somatosensory cortex of the rat. Anat Rec (Hoboken) 291(8):960–973. https://doi.org/10.1002/ar.20696
47. Jin T, Kim SG (2008) Cortical layer-dependent dynamic blood oxygenation, cerebral blood flow and cerebral blood volume responses during visual stimulation. Neuroimage 43(1):1–9. https://doi.org/10.1016/j.neuroimage.2008.06.029
48. Shih YY, Chen YY, Lai HY, Kao YC, Shyu BC, Duong TQ (2013) Ultra high-resolution fMRI and electrophysiology of the rat primary somatosensory cortex. Neuroimage 73:113–120. https://doi.org/10.1016/j.neuroimage.2013.01.062
49. Ni J, Chen JL (2017) Long-range cortical dynamics: a perspective from the mouse sensorimotor whisker system. Eur J Neurosci 46(8):2315–2324. https://doi.org/10.1111/ejn.13698
50. Theyel BB, Llano DA, Sherman SM (2010) The corticothalamocortical circuit drives higher-order cortex in the mouse. Nat Neurosci 13(1):84–88. https://doi.org/10.1038/nn.2449
51. Lee CC, Sherman SM (2008) Synaptic properties of thalamic and intracortical inputs to layer 4 of the first- and higher-order cortical areas in the auditory and somatosensory systems. J Neurophysiol 100(1):317–326. https://doi.org/10.1152/jn.90391.2008

Check for updates

Chapter 14

Understanding Brain-Wide Network by Combining Optogenetics with Genetic and Connectivity Atlases

Kai-Hsiang Chuang

Abstract

Extensive mapping of axonal connections, cell types, and gene expression has provided rich data about the brain's complex organization. When combined with whole-brain functional connectivity imaging, these maps have shed light on the association between brain structure and function. However, it remains unclear how these structural elements support information flow and processing in neural circuits, and how they influence behavior. Combining these multimodal imaging with cell-specific optogenetic stimulation allows in vivo interrogation of a specific circuit to understand the functional and causal roles of these cellular and connectomic substrates on brain network dynamics. This article describes methods for conducting such experiments and comparing functional, connectomic, cell-type, and genomic data in mice. These combined tools could provide new insights into the brain's structure–function relationship, circuit mechanisms of brain disorders, and the development of potential interventions.

Keywords Brain connectome, Neuromodulation, Functional brain imaging, Functional connectivity, Resting-state network, Omics

1 Introduction

Understanding the structural organization and functional dynamics of the brain network is fundamental for decoding information processing, uncovering neural mechanisms of behavior [1–3], predicting disease processes [4–7], and guiding therapeutic interventions [8, 9]. Noninvasive functional brain imaging, such as functional magnetic resonance imaging (fMRI), has provided exquisite information on the network organization and dynamics in development, aging, and disorders in humans [10, 11]. The most commonly used fMRI technique relies on the blood oxygenation level-dependent (BOLD) signal, which reflects regional changes in the cerebral blood volume, blood flow, and metabolic rate of oxygen that follows neural activation [12]. By modeling this neurovascular response with a hemodynamic response function (HRF), neural activity induced by a specific task performance or spontaneous activity without an explicit task (typically

Kevin Chuen Wing Chan and Seong-Gi Kim (eds.), *Optogenetic Functional Neuroimaging*, Neuromethods, vol. 232,
https://doi.org/10.1007/978-1-0716-5178-0_14,

referred to as resting state) can be estimated and located. The connectivity between brain regions can then be estimated from the statistical relationship between regional signal dynamics during either task or resting-state conditions [13, 14]. The task-evoked activity is typically used to infer "effective connectivity" that describes how the activity of a brain region influences the response in other regions, thus reflecting the information flow. The resting-state activity, on the other hand, is typically used to determine "functional connectivity" (FC) that describes the synchrony between brain regions. Both methods have been widely used to identify system-level neural pathways and network interactions that underpin cognitive performance in healthy conditions or impairment in disorders.

As the fMRI signal is a complex function of hemodynamic changes with individually and regionally heterogeneous HRF, the exact neural activity that underlies the network dynamics remains unclear. Furthermore, how macroscopic neural dynamics are constrained by and relate to the underlying axonal, cellular, and genetic organization and affected by their disruptions in disorders is difficult to resolve. Animal imaging, particularly in mouse models, provides a way to address these questions. In the past decade, the advances in mouse fMRI have demonstrated the translatability of functional networks between mouse and human [15–18]. Importantly, fine-detail mapping of the connectome and genome of the entire mouse brain has been made. By mapping the data onto a common brain atlas space, such as the Allen mouse brain reference atlas (ABA; www.brain-map.org), in vivo functional dynamics can be associated with ex vivo axonal, cellular, and molecular organization. These analyses have revealed how network organization and function may depend on interregional axonal projection or regional gene expression or transcription [19–24]. Yet, the causality of the cellular and molecular changes on the brain network cannot be determined by these correlational analyses [25].

Conventionally, the causal role can be tested by lesion, electrical stimulation, or drug treatment to silence or activate a targeted brain area or receptor. However, these methods suffer from nonspecific effects or a lack of temporal control. The development of optogenetic or chemogenetic tools allows cell-specific manipulations to determine the causality. Optogenetics expresses light-sensitive proteins, such as Channelrhodopsin-2 (ChR2), on specific cells to allow excitation or inhibition of neural activity with high temporal precision [26, 27]. In conjunction with functional brain imaging, it enables mapping the brain-wide and local neural circuit responses of specific cells [28]. Delivered with a particular promotor or into transgenic animals, the opsin can be expressed on a specific cell type, such as dopaminergic neurons, for understanding the impact of cell-specific neural pathways on brain network function and behavior [29, 30]. Furthermore, by modulating neural activity at a different rate, frequency-dependent network and cell-type activities, as well as brain state transition, have been identified [31–34]. These studies highlighted the importance of using whole-brain functional imaging to understand how network dynamics is mediated and controlled. Similar to optogenetics, chemogenetics (or pharmacogenetics) [35], can achieve cell-type specific modulation to understand how sustained activity change impacts the brain-wide activity and behavior [36–38]. Although it does not have precise temporal control, it can modulate neural activity chronically without the need for a light-emitting device or a surgical implant of optic fibers. By manipulating a specific cell type, axonal projection, or circuit, their causal roles in brain network dynamics and their relationship with underlying cellular and molecular substrates can be verified. In the following, the methods for conducting optogenetic fMRI in mice, analyzing functional readouts, and comparing with brain-wide connectomic, cellular, and genetic data are described.

2 Materials

2.1 Animal and Virus

To conduct optogenetic fMRI in a specific region, cell type, or connection, careful selections of the viral vector, promotor, and mouse strain are needed. Adeno-Associated Viral (AAV) vectors are the most common choice for targeting a specific cell type in mice. AAV with certain promotors can be used to target a couple of cell types, such as all neurons (using human Synapsin [hSyn] promotor), excitatory neurons (using Calmodulin-dependent kinase II [CaMKII] promotor), adrenergic neurons (using Dopamine β hydroxylase [DBH] promotor), and astrocytes (using glial fibrillary acidic protein [GFAP] promotor). For these types of cells, C57Bl/6 mice (8–10 weeks, mixed sex) can be used. To target other cell types, transgenic mice with knock in *Cre* or *Flp* recombinases can be used. For instance, injecting viruses carrying double-floxed inverse orientation (DIO) element into parvalbumin (PV)-Cre mice can selectively express transgenes in the PV-expressing interneurons. It should be noted that there are several subtypes of AAV. Each subtype has different transfection efficiency in different brain regions, cell types, and anterograde/retrograde capability (see [39] for a comprehensive review).

Besides the ChR2, many types of opsins can be used for optogenetics depending on the need for excitation, inhibition, and stimulating frequency [40]. The ChR2 is the most common choice for excitation due to its good sensitivity. However, it has a moderate kinetics, making it not an ideal choice when high frequency (>20 Hz) stimulation is needed [41]. In such cases, a fast opsin, such as the Chronos, would be preferable [42].

These AAV vectors or plasmids can be purchased from Addgene (https://www.addgene.org/). Transgenic mice could be purchased from breeders, such as The Jackson Laboratory (https://www.jax.org/).

2.2 Optogenetic fMRI

Stereotaxic surgery will be conducted to deliver AAV-opsins at least 3 weeks before the imaging experiment. For a detailed list of the viral surgery tools and materials needed, please refer to [43, 44].

For optogenetics, the following tools and equipment are needed:

1. Ceramic fiber-optic cannula: 1.25 mm ferrules, 200 μm core, 0.37NA (RWD). A smaller 100 μm fiber core could be used to target a smaller area and to reduce susceptibility artifact in fMRI. Besides the conventional silica fiber, a flexible polymethyl methacrylate (PMMA) fiber could be used to fit into an enclosed RF coil, such as the cryoprobe, while minimizing susceptibility artifact [30]. Furthermore, choosing a black cannula would help to reduce light leakage.
2. Optic fiber patch cable: two sets of fiber cables would be needed, one for optogenetics and the other for visual stimulation. The length would be 6–10 m, depending on the distance from the light source to the center of the MRI bore. All the metal parts on the cable should be removed before use.
3. Cannula mating sleeves: for connecting a patch cable to a fiber cannula. Black paint could be applied to the sleeve to reduce light leakage.
4. Fiber cleaving and polishing tools: to cut the fiber cannula to the length that matches the depth of the target area.
5. Light source: 470 nm light for ChR2. Either laser or LED light source can be used, with the latter being a more economical choice (e.g., Model M470F3, Thorlabs). The light source will need computer control (via TTL trigger) to synchronize with the fMRI scan.
6. Light power meter: to calibrate light power at the tip of the patch cable (e.g., Model S151C, Thorlabs).
7. Metal-free dental cement: C&B Metabond (Parkell).

For fMRI:

1. MRI hardware: 7 Tesla or higher. A high-performance gradient with high slew rate and at least 40 G/cm strength is needed for good image quality. A volume transmit RF coil and a 10-mm single loop surface receive coil are needed for good fMRI sensitivity and access for optic fibers.

2. MRI pulse sequence: Fast spin-echo (RARE) sequence will be used for acquiring structural MRI. A single-shot gradient-echo echo-planar imaging (EPI) sequence will be used for BOLD fMRI. To detect fast hemodynamic transient responses, simultaneous multislice (SMS) EPI could be used [45].
3. Laptop PC and USB digital/analog data acquisition card with driver compatible with MATLAB for receiving/sending TTL triggers or stimulus waveforms.
4. Software for synchronizing fMRI scans and controlling stimuli and fiber output. Typical human fMRI stimulus control software, such as Psychtoolbox or E-Prime, can be used.
5. 50-ohm coaxial cables: one for receiving trigger output from the MRI scanner and the other for sending trigger or waveform to the light source.
6. MRI-compatible physiological monitors: pulse oximeter (SpO_2), respiration sensor, and temperature probe.
7. Kwik-Cast (World Precision Instruments) or toothpaste: applying Kwik-Cast over the head can effectively minimize EPI susceptibility artifacts near the cortical surface. Fomblin Y (Sigma-Aldrich) could be used to reduce the susceptibility artifact near the ear canal [46].
8. Anesthesia and saline (or phosphate-buffered saline [PBS]): a mixture of (dex-) medetomidine (MED) sedative and low level of isoflurane (ISO) anesthesia is a preferred choice for good fMRI sensitivity [15, 45]. Other anesthetics, such as just using isoflurane alone, could be considered when targeting the noradrenergic neural pathways that MED acts on.
9. Syringe pump (KDS 100, KD Scientific): for constant infusion of MED.
10. Eye gel: to avoid dry eyes under anesthesia.
11. A long PE10 catheter, 30 G needle, and 1 mL syringe.
12. Visual stimulation. Before the optogenetic stimulation, a sensory (somatosensory or visual) evoked fMRI should be conducted to test whether a suitable physiological condition that preserves neurovascular coupling is achieved [45]. Typically, visual stimulation is preferred as the same light source can be used to deliver the light stimulation via another fiber cable.

2.3 fMRI Data Processing

Data Format Conversion

- bru2nii (https://github.com/neurolabusc/Bru2Nii). There are two useful features: rescale the intensity based on the receiver gain setting, and magnify the resolution by 10×, which makes the data more compatible with software designed for the human brain, such as SPM and FSL.

- BrkRaw (https://github.com/BrkRaw/brkraw). It can import accurate geometry into the NIFTI header and convert to BIDS format.
- MRIcroGL (https://github.com/rordenlab/MRIcroGL). It is useful for converting the NRRD format used by the Allen Brain to NIFTI.

Data Processing

- MATLAB (MathWorks Inc.) and the following tools:
  - Aedes (https://github.com/mjnissi/aedes) for online fMRI analysis;
  - PCNN3D (https://sites.google.com/site/chuanglab/software/3d-pcnn) for rodent brain extraction [47];
  - NIFTI toolbox (https://www.mathworks.com/matlabcentral/fileexchange/8797-tools-for-nifti-and-analyze-image);
  - Network-Based Statistic (https://www.nitrc.org/projects/nbs/) for group-wise comparison [48]; and
  - Brain Connectivity Toolbox (https://sites.google.com/site/bctnet/) for graph theoretic analysis.
- FSL (https://www.fmrib.ox.ac.uk/fsl).
- AFNI (https://afni.nimh.nih.gov/).
- ANTs (https://stnava.github.io/ANTs/).
- Rodent whole brain processing pipeline (https://github.com/GT-EmoryMINDlab/rodent-whole-brain-preprocessing-recipe) [49].

Data Visualization

- MRIcroGL
- FSLeyes (part of FSL)

2.4 Brain Atlas

Structural Atlas

Most of the cellular and molecular atlases are in the ABA Common Coordinate Framework (CCF) version 3. The ABA atlas is constructed by population average specimens in 3D 10 μm isotropic resolution, and parcellated into ~300 gray matter structures, cortical layers, and ~80 fiber tracts. The volumetric data for the atlas template in various resolutions, and labeled atlas (i.e., annotation) in 25 μm resolution can be downloaded from https://download.alleninstitute.org/informatics-archive/current-release/mouse_ccf/. The full list of anatomical labels can be found in https://download.alleninstitute.org/informatics-archive/current-release/mouse_ccf/annotation/ccf_2022/voxel_count_and_differences.csv. For matching the structural MRI, the 100 μm template could be used. For fMRI, the template would need to be downsampled to a resolution, such as 200 μm, that is closer to that of fMRI data to avoid excessive interpolation error. Since the NRRD data format does not have correct orientation information, it can be reoriented and converted to NIFTI format using MRIcroGL.

As the ABA template is created by histological sections instead of ex vivo or in situ MRI, the different image contrast and brain geometry would make it difficult to coregister with in vivo MRI data. A workaround is to use the ex vivo MRI-based AMBMC template (https://imaging.org.au/AMBMC/) as an intermediate. Download the symmetric version of the template and use MRIcroGL to correct the orientation. The original template is in 15 μm, thus downsampling is needed. The individual MRI data can be registered to the AMBMC space first and then warped to the ABA space.

Connectivity Atlas

The Allen Brain Connectivity atlas uses anterograde tract tracer to define monosynaptic axonal projection from a brain structure, cortical layer, or cell type. It provides several unique pieces of information on the structural connectivity (SC):

a. A clear definition of the direction with the use of anterograde tracers.
b. Differentiate direct vs indirect connections and derive the hierarchy of connectome organization.
c. Quantify the connection strength (or density), which is typically defined as the summed axonal projection volumes, normalized by the tracer injection volume in each nonoverlapping brain region. It should be noted that it is not the number/density of axon/synapse.
d. Pinpoint cell-type or layer-specific connection using *Cre*-transgenic mice, such as cholinergic neurons (using *ChAT-Cre* mouse line) or cortical layer 4 neurons (using *Nr5a1-Cre* mouse line). However, there are only a few regions injected with cell-type targeting tracers.

The volumetric data can be obtained using the commands described in https://community.brain-map.org/t/api-allen-brain-connectivity/2988. To extract regional projection density in tabular format, here are example codes https://neuroinformatics.nl/HBP/ABA_mouse/. The axonal projection density from major brain structures or cortical layers in tabular format can be downloaded from the supplementary data of the original papers [50, 51].

Genetic Atlas

There are two kinds of genetic atlases. The first is the Allen Brain Anatomic Gene Expression atlas, which is based on the in situ hybridization (ISH) gene expression images [52]. The volumetric data of a specific gene in 200 μm resolution can be downloaded following the instructions on https://community.brain-map.org/t/downloading-3-d-expression-grid-data/2859. The tabulated data can be retrieved using the codes from https://github.com/benfulcher/AllenSDK [53].

The second is the molecular atlas of the mouse brain [54]. It is based on the spatial transcriptome of messenger RNA [55]. The image and tabulated data can be downloaded from https://www.molecularatlas.org/download-data.

Cell-Type Atlas

There are at least three kinds of cell-type atlases. The first is created by cell-type-specific knock in mice and targeted fluorescent probes that quantify the distribution of seven inhibitory neuron types, including three major types of inhibitory neurons that express parvalbumin (PV), somatostatin (SST), or vasoactive intestinal peptide (VIP) [56]. The SST+ and VIP+ neurons are further divided into subtypes that coexpress calretinin, nitric oxide synthase, or cholecystokinin. The tabulated cell count, density, and sex difference data can be downloaded from the supplementary information of the paper [56].

The second atlas uses genetic markers from the Allen Brain to distinguish excitatory and inhibitory neurons, astrocytes, oligodendrocytes, and microglia in each of the brain regions defined in the ABA space [57]. The tabulated data can be visualized and downloaded from https://bbp.epfl.ch/nexus/cell-atlas/.

The third atlas is the Allen Brain Cell Type Atlas (https://portal.brain-map.org/atlases-and-data/bkp/abc-atlas), which provides the most detailed cell-type classification. It combines whole-brain single-cell RNA-sequencing and spatial-resolved in situ hybridization to distinguish cell types based on the transcriptomic profiles of ~10 million cells [58]. A total of 34 main classes and 5322 subtypes have been identified. The data can be downloaded using the commands described in https://alleninstitute.github.io/abc_atlas_access/intro.html.

3 Methods

3.1 Surgery for Optogenetics

1. Determine the coordinates for injection. Typically, the injection coordinates are determined from the Paxinos and Franklin mouse brain atlas [59]. As the actual brain size could be different depending on the age, sex, and strain, pilot studies would be needed, particularly for targeting small subcortical nuclei or a specific cortical layer.
2. Conduct stereotaxic intracerebral injection and cannula implantation (refer to [60] for detailed procedures).
3. As blood and air can lead to severe susceptibility artifacts in EPI, care should be taken to ensure no more bleeding before inserting the implant and no air bubbles when applying dental cement.

4. After the animal recovered from the surgery, confirm the coordinates by inspecting the optic fiber location in T2-weighted structural MRI. Refine the injection coordinate if needed.
5. To ensure enough expression of opsins, a pilot histological examination is needed to inspect the virus expression at 3–4 weeks after the injection. If the expression level is low, use a higher titer of virus, more freshly prepared virus, and/or a larger volume.

3.2 Preparation for Imaging

Stimulation Design

For visual stimulation, flashing blue light (5 Hz, 10 ms pulse) with a block design consisting of 39 s off, 21 s on, repeated three times, and ending with a final off period of 60 s could be used.

For optogenetic stimulation, an event-related design is preferable to the conventional block design as the latter could induce nonlinear effects in the neural and vascular responses. A stimulation on duration of 1–2 s followed by an off duration of 10–15 s could be used to avoid overlapping of the hemodynamic responses from repeated stimuli. Light pulses (10–20 ms) with various intensities (e.g., from 5 to 60 mW/mm^2 peak power) and frequencies (e.g., from 1 to 20 Hz) could be presented in randomized order within a stimulation run. For such event-related design, a 7–10 min paradigm can be used. Avoid using a very long scan as it would be too intense for the animal.

Equipment Preparation

1. Prepare a 100× dilution from the MED stock solution (Domitor® 1 mg/mL injectable vital) using PBS and prefill the PE10 catheter and syringe. Note that normal saline is less preferable as it is acidic. Test and ensure the syringe pump is functioning.
2. Calibrate the light output at the fiber tip of the patch cable using a power meter. Note that the relationship between light power and the voltage/current setting of the light source may not be linear. Thus, each stimulation intensity in the task design should be calibrated.
3. Ensure the physiological monitor and sensors are functioning and the battery is fully charged.
4. Test the MRI scanner and its synchronization with stimulation delivery.

Animal Preparation

1. Anesthetize the animal with 2%–3% ISO on the bench. Insert an intraperitoneal catheter (PE10) for the infusion of a sedative.
2. Kwik-Cast can be applied on top of the head to minimize susceptibility artifacts in EPI. Shave the animal's hair and apply Kwik-Cast.

3. Switch the ISO to the nose cone on the MRI holder. Carefully move the animal and the catheter over. Secure the head with ear bars and tooth bar.
4. Connect the rectal temperature probe, the respiration sensor, and SpO_2 sensor.
5. Monitor the temperature. Cover the animal with a thick paper towel to keep it warm.
6. Reduce ISO to 2%. Inject the initial bolus of MED (0.05 mg/kg or 0.025 mg/kg for dexmedetomidine) and then start to gradually reduce the ISO level toward 0.5% over 10 min.
7. Tape both ears down to make them close to the head, as the protruding ears could be aliased into the brain MRI.
8. Put the receiver RF coil. When using a small surface coil to maximize the sensitivity, carefully adjust the coil position to properly cover the entire cerebrum and cerebellum.
9. Connect the patch cable to the fiber-optic cannula using a ceramic mating sleeve. Black tape can be used to cover the sleeve to minimize leakage of light from the interface.
10. Set up visual stimulation. To avoid shining light directly into the animal's eyes, a semicircular reflective surface (e.g., a cardboard) can serve as a screen and be placed in front of the animal's head. The fiber tip of a patch cable is then placed facing toward the screen.
11. Start constant infusion of MED (0.1 mg/kg/h or 0.05 mg/kg/h for dexmedetomidine) at 10 min after the initial bolus.

3.3 MRI Acquisition

1. After system calibration and acquiring a localizer scan, use an ellipsoid shim volume that matches the mouse brain to conduct high-order shim (e.g., MAPshim on Bruker MRI).
2. Acquire a structural MRI using 2D T_2-weighted fast spin-echo with repetition time (TR)/echo time (TE) = 5500/40 ms, echo train length = 8, field of view (FOV) = 25.6×25.6 mm^2, matrix = 256×256, 56 axial slices, slice thickness/gap = 0.3/0 mm, and 2–4 averages.
3. Setup and calibrate fMRI scan using single-shot 2D gradient-echo EPI with TR/TE = 1000/15 ms, flip angle = 52°, FOV = 19.2×19.2 mm^2, matrix = 64×64, 20 axial slices, thickness/gap=0.5/0.1 mm, bandwidth = 300 kHz. Make sure the readout is in the left-right direction, which helps to reduce geometric distortion. Dummy scans lasting about 10 s can be included to ensure the signal reaches steady state. Enable trigger output per volume. To avoid aliasing artifact from the tissue signal below the brain, a saturation slab can be applied. To distinguish cortical layer-dependent response, in-plane resolution could be increased to 150 μm with FOV = 19.2×9.6 mm^2 and matrix = 128×64. If SMS-EPI

is available, TR can be reduced to 300 ms to increase the temporal resolution and statistical power [45]. It should be noted that the optimal TE and flip angle should be based on the T2* and T1 of the rodent brain at the particular MRI field strength. The parameters described above are based on 9.4 T. For other field strengths, please refer to [61].

4. Acquire reverse phase-encoding EPI using the same parameters but with four averages for distortion correction.
5. As animal physiology generally stabilizes around 45 min after the initial MED bolus [62], conducting fMRI before that could lead to a weak and variable response. Once the structural and fMRI calibration scans are completed, simply wait for the physiology to stabilize. Do not start fMRI before the physiology is stable.
6. Acquired sensory-evoked fMRI.
7. Use Aedes to analyze the sensory fMRI data on the MRI scanner. If activation is not detected in the expected sensory cortex, the physiological condition (e.g., the ventilation and SpO_2) and the anesthesia dose (e.g., the isoflurane level) could be adjusted. Repeat the sensory fMRI scan and physiological adjustment until an expected response can be detected.
8. Acquire resting-state fMRI. For reliable FC analysis, at least a 10-min resting-state scan should be collected. 30-min scan is recommended.
9. Acquire optogenetic fMRI. The paradigm can be repeated multiple times for data averaging, with a 3–4 min break in between runs.
10. Use Aedes to analyze the optogenetic fMRI data to make sure evoked responses can be detected.
11. After the experiment is completed, the animal should be perfused and fixed for histological assessment of the expression and localization of opsins.

3.4 fMRI Data Processing

Format Conversion and Reorientation

1. Convert the MRI vendor data format to NIFTI.
2. As several tools, such as FSL and SPM, have internal parameters optimized for the size of the human brain, it is preferable to magnify the voxel size by ten-fold (using FSL *fslchpixdim* command) to make the rodent brain have a comparable size to that of the human.
3. Some converters do not translate the orientation of the MRI scanner accurately. It can be corrected by AFNI *3drotate* and *3dLRflip* commands.
4. Check orientation using FSLeyes.

Create Study-Specific Template

As the ABA template is based on histological images instead of in vivo MRI, creating a study-specific brain template using the structural MRI can improve the coregistration accuracy.

1. Use automatic rodent brain extraction methods, such as PCNN3D, to create a brain mask for the T2-weighted structural MRI. As coregistration requires an accurate brain mask, manual editing of the automatically generated brain mask is typically required.
2. Correct the coil inhomogeneity using the ANTs *N4BiasFieldCorrection* command together with the brain mask to improve the accuracy.
3. Once all the structural MRIs from the study cohort are preprocessed, use the ANTs *antsMultivariateTemplateConstruction.sh* command to create the study-specific template with the ABA (or preferably AMBMC) 100-μm template as the reference.
4. If using the AMBMC template as an intermediate, nonlinearly coregister the AMBMC template to ABA using ANTs *antsRegistrationSyNQuick.sh*. Then, apply the transformation to the study-specific template using ANTs *antsApplyTransforms* with the interpolation option *-n HammingWindowedSinc*.
5. As coregistering low-resolution fMRI to a high-resolution template could introduce interpolation error, and some of the genetic and connectivity atlases are in 200 μm resolution, a lower resolution template will be created. Downsample the high-resolution template using the FSL *flirt* command with the *-applyisoxfm* option.
6. Nonlinearly coregister the structural MRI of each animal to the 200-μm template. If the default "deformable syn" transform is too aggressive, try the "deformable b-spline syn" transform (i.e., option *-t b*).
7. Create an atlas label in 200-μm resolution for ROI analysis. First, linearly register the high-resolution ABA template to the low-resolution version using *antsRegistrationSyNQuick.sh* with the rigid-body transform option *-t r*. Then, apply the transformation to the atlas label image using *antsApplyTransforms* with the interpolation option *-n MultiLabel*.
8. As some of the small nuclei are not detectable at 200-μm, calculate the voxel number for each anatomical label and remove the label with a voxel number less than 2–3 corresponding voxels in fMRI. For example, if the fMRI resolution is 0.3×0.3×0.6 mm^3, anatomical labels with <14–20 voxels in the 200-μm template should be removed.

fMRI Preprocessing

The fMRI data is preprocessed following an optimized data processing pipeline [63] that includes the following steps:

1. Slice timing correction using FSL *slicetimer*, particularly for data with event-related design.
2. Motion correction by FSL *mcflirt* or AFNI *3dvolreg*. Although head motion is minimal with a head-fixed anesthetized animal, motion correction and the motion regressors derived can help to reduce certain signal drift or vibration during the scan.
3. EPI distortion correction by FSL *TOPUP* [64]. A configuration file used for human EPI (e.g., b02b0.cnf, which can be found in the FSL installation folder) can be used and then fine-tuned for the mouse data if needed. Another parameter text file that indicates the direction of phase encoding and echo train length should be created according to the user guide (https://fsl.fmrib.ox.ac.uk/fsl/docs/#/diffusion/topup/users_guide/index).
4. Extract an EPI brain mask automatically from the averaged EPI using PCNN3D, followed by manual editing.
5. Normalize the fMRI signal intensity of each voxel by the mean intensity of the initial baseline period for the task fMRI or the entire temporal mean for the resting fMRI data. This corrects the coil profile inhomogeneity and converts the BOLD signal into a percent signal change, which is useful for data averaging across runs. The temporal mean of the baseline duration can be calculated using the AFNI *3dTstat* command and normalized using FSL *fslmaths* command.
6. Removing nonneural or systemic confounds is essential for fMRI analysis. Nuisance regression is an effective method to remove specific confounds. Typical regressors include quadratic drift, six motion parameters and their first-order derivatives, and ten principal components from tissues outside the brain [63]. The temporal derivatives can be calculated by the AFNI *1d_tool.py* command, and principal components can be extracted using the AFNI *3dpc* command. Alternatively, independent component analysis with many components can be used to identify artifactual components [65].
7. Temporal filters are used to further reduce noise. For stimulation fMRI, a highpass filter with the cutoff frequency set at half of the stimulation block/event repetition frequency. For example, if the stimulation on and off duration together is 15 s, then, the filter cutoff can be $1/15/2 = 0.033$ Hz. For the resting fMRI data, a typical band-pass filter cutoff is between 0.01 and 0.2 Hz, which is slightly broader than that used in humans.

8. The nuisance regressors and filtering can be combined and removed using AFNI *3dTproject*.
9. Coregister the fMRI data to the study-specific template. As residual EPI distortion and signal dropout near the ear canal can affect the coregistration accuracy in these regions, the EPI data should not be directly coregistered to the template. Instead, the registration should combine a linear affine registration to the animal's structural MRI (using the ANTs *antsRegistrationSyNQuick.sh* with *-t a* option), and the nonlinear transformation from the structural MRI to the template using ANTs *antsApplyTransforms* with a spline interpolation option *-n BSpline*.
10. Apply spatial smoothing of a 0.6-mm (or 6-mm if the resolution is magnified by 10×) Gaussian kernel using the FSL *fslmaths* command with *-s sigma* option. As the 0.6-mm kernel refers to the full-width-at-half-maximum (FWHM) instead of the standard deviation, *sigma*, of the Gaussian function, it can be converted using *sigma*=FWHM / 2.3548.
11. Transform the EPI brain mask to the template space and create a group EPI mask with 80%–90% overlap across all animals. Use this EPI mask to remove anatomical labels not covered by the fMRI scan.

Optogenetic fMRI Analysis: Spatial Mapping

The optogenetic fMRI data will be analyzed using a general linear model (GLM).

1. Create a design matrix with each intensity/frequency as an explanatory variable (EV).
2. Model each EV as a boxcar waveform convolved by an HRF. Typical BOLD activation under MED-ISO anesthesia has minimal poststimulus undershoot; thus, the HRF could be modeled by a single gamma variate function. An empirical setting for event-related fMRI is to use a "gamma" function with standard deviation = 2 s and mean lag = 4 s in FSL FEAT.
3. Create a contrast file with one row for each EV.
4. Use the FSL *fsl_glm* command to conduct the first-level GLM analysis. It will generate a beta contrast map for each stimulation condition.
5. Average the beta maps of repeated runs in each subject using the FSL *fslmaths* command.
6. Conduct second-level random effect analysis on the averaged beta maps using a one-sample t-test design in *fsl_glm* or AFNI's *3dttest++*.

7. The statistical map can be thresholded at voxel-level $p<0.05$, with multiple comparisons corrected by false discovery rate (FDR) using the FSL *fdr* or AFNI *3dfdr* command.

Optogenetic fMRI Analysis: Temporal Characteristics

As the regional activation could be highly variable depending on the axonal connection, brain hierarchy, regional cell type, and gene expression, further region-of-interest (ROI) analyses will be useful.

1. Extract the nonzero mean BOLD signal in each brain region (or cortical layer) based on the ABA atlas label using the FSL *fslstats* command with preoption-*K* *<label image>* and option-*M*. This will output a text file including the signals from all the ROI in the *<label image>*. Noted: for regions not fully covered by fMRI, such as those affected by susceptibility artifacts, should be removed.
2. Import the ROI time-series into MATLAB using the *load* command.
3. Average the time blocks of the same stimulus intensity or frequency in each animal to derive the evoked response.
4. Wavelet denoise can be applied to reduce high-frequency noise using the MATLAB *wdenoise* command [66].
5. Estimate temporal characteristics, such as:

   a) Stimulus intensity responsiveness: the light power-dependent response can be modeled as a sigmoid function by fitting the peak BOLD amplitude, S, to the function:

$$S(i) = \frac{a}{1 + e^{-b(i-c)}}, \tag{1}$$

   where i denotes different stimuli, and a, b, and c are the fitted parameters.

   b) Hemodynamic properties: the regional BOLD signal can be fitted to a double gamma-variate function that convolved with the stimulation paradigm to the evoked BOLD signal. The hemodynamic response is defined as:

$$HR(t) = a_1 t^{b_1} e^{-t/c_1} - a_2 t^{b_2} e^{-t/c_2}, \tag{2}$$

   where t is the time, and a_1, b_1, c_1, a_2, b_2 and c_2 are the fitted parameters. Before fitting, the signal can be normalized by its peak amplitude. As certain brain regions may present biphasic responses, the hemodynamic response in these regions could also be modeled as triple gamma-variate functions.

c) The time-to-rise and time-to-peak: The hemodynamic response timing can be estimated from the fitted $HR(t)$, sampled at 0.1 s resolution. The time-to-rise can be defined as the signal rising above 10%–30% of the peak amplitude.

Functional Connectivity Analysis: ROI-Based

FC can be estimated under resting state or optogenetic stimulation.

1. Create the interregional FC matrix and seed-based FC maps by Pearson's correlation coefficients, *r*, between the ROI time-courses using AFNI *3dNetCorr* command.
2. FC under optogenetic stimulation can be either calculated from the time-series signals with or without regressing out the evoked response (e.g., the HRF-modelled response). By removing the evoked response, the former could be used to evaluate the effects of stimulation on the resting-state FC. The latter could be used to reveal the dependency between regional evoked responses. Since the evoked response in each brain region could be different from the HRF model, care should be taken to check whether there are residual stimuli-dependent changes.
3. Significant connectivity is determined by a one-sample t-test of Fisher's *z*-transformed *r* using region-/cluster-wise threshold with FDR-correction, or network-based statistic.
4. To account for interregional time delay, cross-correlation can also be used to estimate time-lagged FC using the MATLAB *xcorr* command.
5. Graph theoretic analysis can be applied to the FC matrix to estimate global or regional features of the network graph.

Functional Connectivity Analysis: Voxel-Wise

As the anatomically defined brain regions may not align with functional and molecular parcels, FC could also be calculated at the voxel level to take advantage of the detailed cellular and molecular maps. As this approach will create a much larger connectivity matrix, the following summary measures could be used:

1. Weighted or unweighted global FC: this calculates the averaged FC strength or the number of connections, respectively, for each voxel [67]. They can be estimated using AFNI *3dTcorrMap* with the *-mean* or *-threshold* option, respectively.
2. FC density mapping [68]: This only considers the connections within local clusters. It can be calculated using the AFNI *3dLFCD* command.

3.5 Multimodal Data Analysis and Integration

The fMRI and the axonal/cellular/molecular data could be compared at (1) voxel; (2) regional, (3) interregional, or (4) network levels [69]. The voxel-level comparison can be used to identify the underlying cellular and molecular substrates that constitute the observed functional maps at the finest scale without the assumption of anatomical borders. The regional comparison can be used to elucidate the structure–function relationship between clearly defined brain structures, connections, and hierarchy. The interregional analysis focuses on the correlated changes between structural/functional connectivity with respect to regional cell-type and genetic profiles. The network-level analysis intends to define the structure–function topological features (based on graph theoretic analysis).

Structural Connectivity Analysis

1. Load the tabulated SC data in MATLAB using the *readtable* command.
2. The original SC matrix is organized as sources x targets, where the sources are the tracer injection sites and the targets are the axonal projected areas that are in the ipsilateral (left-hand side of the matrix) and contralateral (right-hand side of the matrix) hemispheres.
3. As the sources are unilateral, the matrix is rectangular instead of a square. A workaround is to swap the target hemispheres and concatenate to create a square matrix.
4. The ROIs used in fMRI analysis may not have the same brain coverage. Select and reorder the SC ROIs to form an SC matrix that matches those used in fMRI.
5. Make sure the hemisphere for optogenetic stimulation is the same as the tracer injection side, which is the left-hand side of the matrix.
6. As the SC strength has a log-normal distribution, use the *log10* function to make it become a normal distribution, which makes it more suitable for parametric statistical tests.
7. To compare SC and FC in regions with direct connection, select the regions with nonzero SC strength and compare using linear regression (MATLAB *fitlm* command) or Spearman correlation (MATLAB *corr* command with *"Type" "Spearman"* option) if the trend is not linear.
8. Regions with indirect connections are those with zero SC strength.

Structural Connectivity Hierarchy

The brain connectome is organized as a hierarchy in which sensory inputs are received and processed at the low-order areas, integrated at high-order areas (association cortices), and then passed down to low-order motor areas for output. This wiring hierarchy forms the backbone of neural information flow. Once the hierarchy of brain regions is defined, the regional fMRI activation characteristics can be sorted and compared within the hierarchy. The relationship between FC and SC can be compared across hierarchy and between feedforward and feedback connections. The connectomic hierarchy can be constructed from the ABA connectivity matrix.

1. Unlike FC, the SC strength can differ by several orders. To eliminate weak connections, a certain threshold can be applied to select the strong connections (e.g., top 20%) for constructing the hierarchy.
2. The first-order projection areas can be defined as brain regions that receive feedforward projections from the optogenetic stimulation site. Thus, locate the optogenetic site from the source ROI. Assign it as the source node.
3. Assign ROIs with nonzero strength in the entire row as the target nodes.
4. Assign the SC strength as the edge weights.
5. Combine them into a network graph using the MATLAB *digraph* command, such as $G =$ *digraph(source_node, target_node, edge_weight)*.
6. The second-order projection areas are brain regions receiving either feedforward or feedback projections from the first-order areas. Thus, go through the first-order areas one by one to identify their projection targets and add them into the network graph using the *digraph* command.
7. Iterate the same procedures to identify third-order or fourth-order projection areas.
8. If needed, remove self or repeated connections in the graph using the MATLAB *simplify* command.
9. Finally, convert the graph to a connectivity matrix using MATLAB's *adjacency(G, 'weighted')* command.

Interregional Covariance

The cellular and genetic maps only depict their spatial distribution. To explore the interrelationship between regional cellular distribution or gene expression, the interregional covariance of the cell-type or genetic profile can be calculated. This will elucidate the cell-type or genetic similarity between regions.

1. Load the tabulated data in MATLAB using the *readtable* command.
2. Select and reorder the table into a numerical array based on the ordering of ROIs used in fMRI analysis.
3. As the density (or expression level) of a cell type (or gene) can vary a lot, normalization is needed to avoid bias toward a high-density cell (or gene). For cell-type covariance, the density of each cell type across regions is first normalized into a z-value.

4. For the genetic covariance, the expression levels of each gene across regions are normalized as a z-value, and then transformed and rescaled between 0 and 1 by a sigmoid function, $S(z) = \frac{1}{1+e^{-z}}$ [20]. The normalized values in each region formed a genetic profile.
5. Construct a covariance matrix using Pearson's correlation coefficient between pairs of regional profiles. This can be calculated using the MATLAB *corr* command.

Region-Wise Comparison

1. Derive the regional excitatory-inhibitory (E/I) ratio by dividing the excitatory neuronal density by the sum of the densities of all inhibitory neuron subtypes.
2. To identify the overall relationship between structure and function, the activation area/amplitude and hemodynamic characteristics could be correlated with E/I ratio, each cell type or gene over the ROIs using linear regression (MATLAB *fitlm* command) or Spearman correlation (MATLAB *corr* command with *'Type' 'Spearman'* option) if the trend is not linear.
3. Stepwise regression analysis (MATLAB *stepwiselm* command) could be used to estimate the joint effects of several cell types or genes on the functional readouts.
4. Between-group comparisons can be used to distinguish which regional structural and functional characteristics differ by their hierarchy (e.g., low-order versus high-order regions) or their structural division (e.g., isocortex versus thalamus).

Interregional Comparison

1. The interregional connections in the FC, SC, and the covariances of cell type and gene expression can be directly compared using correlation analysis or linear regression.
2. Between-group comparisons can be conducted by dividing the interregional connections based on the connection type (e.g., feedforward versus feedback, or direct versus indirect).

Network-Level Comparison

1. The topological features of the FC, SC, cell-type, and genetic covariance matrices can be characterized using functions in the Brain Connectivity Toolbox. Useful measures include centrality, rich club, efficiency, community, and modularity. Note that different functions are used for weighted/unweighted and directed/undirected networks.
2. The nodal (regional) graph features, such as degree centrality or local efficiency, can be correlated with regional structural and functional characteristics to identify the biological substrates that underlie the topology.

3. Between-group comparisons can be used to distinguish regional characteristics based on their topology (e.g., hub versus nonhub) or community (e.g., within versus between network modules).

4 Notes

4.1 The Animal is Difficult to Sedate with MED

A major challenge of using MED-ISO for mouse fMRI is that the animal cannot be sedated when it is anxious. A trick is to habituate the animal to the experimenter by handling the animal for 5 min over 3–5 days before the fMRI experiment. In addition, transport the animal in its home cage or put the bedding from the home cage in the transportation box to reduce anxiety.

Younger mice or some disease models may be difficult to sedate. A higher initial MED bolus dose up to 0.15 mg/kg (or 0.075 mg/kg for dexmedetomidine) would help in such cases. The maintenance dose should remain unchanged, as a higher dose could suppress functional connectivity [70]. Furthermore, as the pharmacokinetics of single bolus MED injection changes drastically [71], constant infusion should be used to maintain consistent sedation of the animal throughout the scanning session.

It should also be noted that once the vial is opened, MED will gradually lose its potency. Therefore, the diluted MED solution should be freshly prepared, and the original vial should be replaced every few months.

4.2 No Activation in Mouse fMRI

There are several factors that could affect BOLD response in mice. The first is a prolonged preparation duration under high ISO, which usually happens due to a more complicated optogenetic fMRI setup. Although the anesthesia depth of ISO can be adjusted quickly, it has long-lasting effects on brain function. A study in rats showed that behavioral performance of a choice task could take an hour to recover [72]. The longer the animal is exposed under high ISO, the longer it may take for the brain to return to a functional state. The preparation time should be less than 20 min (from the start of anesthetizing the animal to the time that ISO is reduced below 0.5%).

The second factor is low body temperature during the preparation on the MRI scanner. As anesthetized mice lose temperature very quickly, the acute hypothermia could have a profound impact on the neural and vascular functions. It is critical to keep animals warm during the preparation using additional MRI-compatible heating pads or a warmer. Empirically, when rectal temperature reduces more than 2°C even briefly, it could take more than an hour to observe the BOLD response.

The third factor is to maintain proper physiological conditions throughout the entire scanning session. For instance, rectal temperature should be maintained within 36.5–37.5°C. SpO_2 should be maintained above 98%. The heart rates for normal mice under MED-ISO are in the range of 200–320 beats per minute, and the respiratory rates are 90–140 breaths per minute (bpm). The physiology should be monitored and adjusted constantly to ensure a stable physiological response. If SpO_2 is low, the airway may not be smooth, the tubing may be clogged, or the O_2 in the gas mixture may be too low. If the respiration rate is low (e.g., below 60 bpm), it may indicate that the effective ISO dose is high which could suppress neural activity and respiration. If the heart rate keeps increasing, it indicates that the sedative effect is wearing off and the animal would tend to wake up. To ensure physiological consistency, fMRI data acquired outside these ranges may need to be discarded.

4.3 SpO_2 is Difficult to Detect in Mice

The SpO_2 sensor provides two essential readouts, blood oxygenation and heart rate, which are critical for BOLD fMRI. Suitable contact with the tail artery, but not compressing the artery, is a key to obtaining a robust measure of SpO_2 from the small blood vessel in mice. A typical SpO_2 sensor is made with a clip that is either too loose or too tight. Designing a half-circular holder for the SpO_2 sensor with a small hole just enough for the mouse tail to pass through will allow easy setup and reliable detection.

4.4 Susceptibility Artifacts in EPI

EPI of the mouse brain on ultrahigh field MRI suffers from severe susceptibility artifacts, including geometric distortion and signal dropout, which are exacerbated by optic fiber implants [44, 73]. Whereas part of the geometric distortion could be compensated by postprocessing [64], applying susceptibility-matching materials could greatly reduce the artifacts. For the distortion over the cortex and around the fiber implant, Kwik-Cast or toothpaste can effectively reduce the magnetic field inhomogeneity. If a surface coil is used, the Kwik-Cast can be molded by the coil to make the coil closer to the brain. If using an array coil, make sure Kwik-Cast is not too thick.

To further reduce susceptibility artifacts in mouse fMRI, multi-shot gradient-echo EPI or single-shot spin-echo EPI could be considered. However, the latter will have lower BOLD sensitivity. Alternatively, zero-TE sequences, such as the SWeep Imaging with Fourier Transformation (SWIFT), On-the-Ramp Detection of INduction-decay with Oversampling (SORDINO), and ZTE, could provide artifact-free functional imaging, though with a more moderate temporal resolution [74, 81].

4.5 Off-Target Effects

Although optogenetic stimulation has high specificity. Two potential confounds may lead to off-target effects. The first is the use of high light power for activating larger brain areas and inducing stronger neural activation. High power would cause tissue heating, which results in pseudoactivation due to altered T2/T2* and vasodilation [75]. To ensure that the light power is adequate, pilot studies could be conducted in naïve animals without opsins.

The second is that light leakage from the fiber interface would elicit additional visual activation. Additionally, the light scattered from the fiber tip inside the brain may activate the retina. This confound could be evaluated in naïve animals without delivering opsins.

4.6 Motion and Drift Caused by Temperature Control

Changing the heating temperature, particularly the water bath, could lead to a large deformation of the holder itself and result in apparent "head movement." As the movement could be easily more than half of the voxel size, the signal change induced would not be well-corrected by motion correction. The use of an air heater could reduce but not eliminate the holder "movement." Therefore, it is better not to change the heating temperature during an fMRI scan. Also, once changing the heater temperature, wait for a few minutes for the temperature change to stabilize.

4.7 Global Signal Regression

Global signal regression is a popular method for reducing nonspecific signal variations and motion artifacts in human fMRI. However, there are several factors that need to be considered before applying it to rodent studies: First of all, the global signal has been linked to brain state change, such as arousal. Second, as the majority of the rodent brain is the gray matter, the global signal would mostly come from neural activity. Third, optogenetic stimulation could elicit strong brain-wide responses. Finally, it has been shown to change the distribution of FC and create strong negative FC. Therefore, its use should be carefully evaluated to ensure the findings are not dramatically changed by the regression.

4.8 Optimal HRF

As the HRF is species, region, and anesthesia (and dosage) dependent, using the default HRF designed for humans will hamper the detection sensitivity. There have been several studies investigating the HRF for humans and rats, but few in mice. A recent study estimated the mouse HRF using mechanical stimulation and found a more delayed HRF than that of the rats [76], which may be due to the long stimulation duration used. As optogenetic stimulation could induce very different neural activation in downstream areas, there may not be an optimal HRF that suits all the experimental conditions and brain regions. Empirically, the HRF could be adjusted by matching to the evoked response in the first-order projection areas.

4.9 Anesthesia Effect

Anesthesia has strong effects on neural and vascular responses, therefore impacting the interpretation of findings. Comparing the fMRI readout with the electrophysiological recording could help to address part of the neurovascular confound. Although awake optogenetic fMRI in mice has been demonstrated in a few labs, there is still no consensus on the optimal design and the impact of chronic restraint on stress and brain activity [77]. As this is an active area of research and development, future studies will benefit from more improved and standardized protocols.

4.10 Other Techniques for Whole-Brain Functional Imaging

A limitation of mouse fMRI is the low sensitivity and low resolution. Functional ultrasound imaging (fUSI) is another modality that has high sensitivity and better spatial and temporal resolution. However, due to the position of the ultrasound probe, combining fUSI with optogenetics is still restrictive [78]. A major limitation of current whole-brain functional imaging is that they all rely on the hemodynamic signal. The combination of optical fluorescent imaging of neuronal activity, either by wide-field imaging [79] or fiber photometry [80] would address some of the issues.

Acknowledgments

Dr. Chuang is supported by Australian Research Council grants DP240101321.

Conflict of Interest The author has no conflicts of interest to declare that are relevant to the content of this chapter.

References

1. Raj A et al (2018) Models of network spread and network degeneration in brain disorders. Biol Psychiatry Cogn Neurosci Neuroimaging 3:788–797
2. van den Heuvel MP et al (2019) A cross-disorder connectome landscape of brain dysconnectivity. Nat Rev Neurosci 20:435–446
3. Park H-J et al (2013) Structural and functional brain networks: from connections to cognition. Science 342:1238411
4. Sporns O (2014) Contributions and challenges for network models in cognitive neuroscience. Nat Neurosci 17:652–660
5. Breakspear M (2017) Dynamic models of large-scale brain activity. Nat Neurosci 20: 340–352
6. Fornito A et al (2015) The connectomics of brain disorders. Nat Rev Neurosci 16:159–172
7. Cole MW et al (2016) Activity flow over resting-state networks shapes cognitive task activations. Nat Neurosci 58:7250–7257
8. Soleimani G et al (2023) Closing the loop between brain and electrical stimulation: towards precision neuromodulation treatments. Transl Psychiatry 13:279
9. Matthews PM et al (2006) Applications of fMRI in translational medicine and clinical practice. Nat Rev Neurosci 7:732–744
10. Straathof M et al (2019) A systematic review on the quantitative relationship between structural and functional network connectivity strength in mammalian brains. J Cereb Blood Flow Metab 39:189–209
11. Suárez LE et al (2020) Linking structure and function in macroscale brain networks. Trends Cogn Sci 24:302–315
12. Logothetis NK et al (2004) Interpreting the BOLD signal. Annu Rev Physiol 66:735–769
13. Craddock RC et al (2013) Imaging human connectomes at the macroscale. Nat Methods 10:524–539
14. Lynn CW et al (2019) The physics of brain network structure, function and control. Nat Rev Phys 1:318–332
15. Chuang K-H et al (2017) Functional networks and network perturbations in rodents. Neuroimage 163:419–436
16. Mandino F et al (2022) A triple-network organization for the mouse brain. Mol Psychiatry 27:865–872

17. Sforazzini F et al (2014) Distributed BOLD and CBV-weighted resting-state networks in the mouse brain. Neuroimage 87:403–415
18. Pagani M et al (2021) mTOR-related synaptic pathology causes autism spectrum disorder-associated functional hyperconnectivity. Nat Commun 12:6084
19. Melozzi F et al (2019) Individual structural features constrain the mouse functional connectome. Proc Natl Acad Sci USA 116:26961–26969
20. Fulcher BD et al (2016) A transcriptional signature of hub connectivity in the mouse connectome. Proc Natl Acad Sci USA 113:1435–1440
21. Stafford JM et al (2014) Large-scale topology and the default mode network in the mouse connectome. Proc Natl Acad Sci 111:18745–18750
22. Grandjean J et al (2017) Structural basis of large-scale functional connectivity in the mouse. J Neurosci 37:8092–8101
23. Chang W-T et al (2018) Connectomic imaging reveals Huntington-related pathological and pharmaceutical effects in a mouse model. NMR Biomed 31:e4007
24. Chuang K-H et al (2023) Hemodynamic transient and functional connectivity follow structural connectivity and cell type over the brain hierarchy. Proc Natl Acad Sci 120:e2202435120
25. Reid AT et al (2019) Advancing functional connectivity research from association to causation. Nat Neurosci 22:1751–1760
26. Zhang F et al (2007) Circuit-breakers: optical technologies for probing neural signals and systems. Nat Rev Neurosci 8:577–581
27. Wiegert JS et al (2017) Silencing neurons: tools, applications, and experimental constraints. Neuron 95:504–529
28. Lee JH et al (2010) Global and local fMRI signals driven by neurons defined optogenetically by type and wiring. Nature 465:788–792
29. Bernal-Casas D et al (2017) Studying brain circuit function with dynamic causal modeling for optogenetic fMRI. Neuron 93:522–532.e5
30. Zou Y et al (2024) Cell-type-specific optogenetic fMRI on basal forebrain reveals functional network basis of behavioral preference. Neuron 112:1342–1357.e6
31. Salvan P et al (2021) Frequency modulation of entorhinal cortex neuronal activity drives distinct frequency-dependent states of brain-wide dynamics. Cell Rep 37:109954
32. Chan RW et al (2017) Low-frequency hippocampal-cortical activity drives brain-wide resting-state functional MRI connectivity. Proc Natl Acad Sci USA 114:E6972–E6981
33. Liu J et al (2015) Frequency-selective control of cortical and subcortical networks by central thalamus. eLife 4:1–27
34. Leong ATL et al (2016) Long-range projections coordinate distributed brain-wide neural activity with a specific spatiotemporal profile. Proc Natl Acad Sci USA 113:E8306–E8315
35. Roth BL (2016) DREADDs for neuroscientists. Neuron 89:683–694
36. Giorgi A et al (2017) Brain-wide mapping of endogenous serotonergic transmission via chemogenetic fMRI. Cell Rep 21:910–918
37. Tu W et al (2021) Brain network reorganization after targeted attack at a hub region. Neuroimage 237:118219
38. Zerbi V et al (2019) Rapid reconfiguration of the functional connectome after chemogenetic locus coeruleus activation. Neuron 103:702–718.e5
39. Haery L et al (2019) Adeno-associated virus technologies and methods for targeted neuronal manipulation. Front Neuroanat 13:93
40. Yizhar O et al (2011) Optogenetics in neural systems. Neuron 71:9–34
41. Li Y et al (2024) Optogenetic fMRI reveals therapeutic circuits of subthalamic nucleus deep brain stimulation. Brain Stimul 17: 947–957
42. Klapoetke NC et al (2014) Independent optical excitation of distinct neural populations. Nat Methods 11:338–346
43. Lin P et al (2016) Optogenetic functional MRI. J Vis Exp 110:e53346. https://doi.org/10.3791/53346-v
44. Schlegel F et al (2018) Fiber-optic implant for simultaneous fluorescence-based calcium recordings and BOLD fMRI in mice. Nat Protoc 13:840–855
45. Lee H-L et al (2019) Ultrafast fMRI of the rodent brain using simultaneous multi-slice EPI. Neuroimage 195:48–58
46. Li R et al (2015) Restoring susceptibility induced MRI signal loss in rat brain at 9.4 T: a step towards whole brain functional connectivity imaging. PLoS One 10:e0119450
47. Chou N et al (2011) Robust automatic rodent brain extraction using 3-D pulse-coupled neural networks (PCNN). IEEE Trans Image Process 20:2554–2564
48. Zalesky A et al (2010) Network-based statistic: identifying differences in brain networks. Neuroimage 53:1197–1207
49. Xu N et al (2023) Rodent whole-brain fMRI data processing toolbox. Aperture neuro 3: |3–5. http://www.ncbi.nlm.nih.gov/pubmed/37654427
50. Harris JA et al (2019) Hierarchical organization of cortical and thalamic connectivity. Nature 575:195–202
51. Oh SW et al (2014) A mesoscale connectome of the mouse brain. Nature 508:207–214
52. Lein ES et al (2007) Genome-wide atlas of gene expression in the adult mouse brain. Nature 445:168–176

53. Fulcher BD et al (2019) Multimodal gradients across mouse cortex. Proc Natl Acad Sci USA 116:4689–4695
54. Ortiz C et al (2020) Molecular atlas of the adult mouse brain. Sci Adv 6:eabb3446
55. Ståhl PL et al (2016) Visualization and analysis of gene expression in tissue sections by spatial transcriptomics. Science 353:78–82
56. Kim Y et al (2017) Brain-wide maps reveal stereotyped cell-type-based cortical architecture and subcortical sexual dimorphism. Cell 171:456–469. e22
57. Erö C et al (2018) A cell atlas for the mouse brain. Front Neuroinform 12:1–16
58. Zhang M et al (2023) Molecularly defined and spatially resolved cell atlas of the whole mouse brain. Nature 624:343–354
59. Paxinos G et al (2019) Paxinos and Franklin's the Mouse Brain in Stereotaxic Coordinates, 5th edn. Academic Press
60. Grimm C et al (2022) An optimized protocol for assessing changes in mouse whole-brain activity using opto-fMRI. STAR Protoc 3:101761
61. Grandjean J et al (2023) A consensus protocol for functional connectivity analysis in the rat brain. Nat Neurosci 26:673–681
62. Nasrallah FA et al (2014) Detection of functional connectivity in the resting mouse brain. Neuroimage 86:417–424
63. Chuang K-H et al (2019) Evaluation of nuisance removal for functional MRI of rodent brain. Neuroimage 188:694–709
64. Hong X et al (2015) Evaluation of EPI distortion correction methods for quantitative MRI of the brain at high magnetic field. Magn Reson Imaging 33:1098–1105
65. Zerbi V et al (2015) Mapping the mouse brain with rs-fMRI: an optimized pipeline for functional network identification. Neuroimage 123:11–21
66. Gil R et al (2021) Neuroplasticity-driven timing modulations revealed by ultrafast functional magnetic resonance imaging. Neuroimage 225:117446
67. Cole MW et al (2010) Identifying the brain's most globally connected regions. Neuroimage 49:3132–3148
68. Tomasi D et al (2010) Functional connectivity density mapping. Proc Natl Acad Sci 107:9885–9890
69. Fornito A et al (2019) Bridging the gap between connectome and transcriptome. Trends Cogn Sci 23:34–50
70. Nasrallah FA et al (2012) Pharmacological modulation of functional connectivity: α2-adrenergic receptor agonist alters synchrony but not neural activation. Neuroimage 60:436–446
71. Kint LT et al (2020) The pharmacokinetics of medetomidine administered subcutaneously during isoflurane anaesthesia in Sprague-Dawley rats. Anim 10:1050
72. Vincent KF et al (2021) Return of the righting reflex does not portend recovery of cognitive function in anesthetized rats. Front Syst Neurosci 15:762096
73. Desai M et al (2011) Mapping brain networks in awake mice using combined optical neural control and fMRI. J Neurophysiol 105:1393–1405
74. Paasonen J et al (2020) Multi-band SWIFT enables quiet and artefact-free EEG-fMRI and awake fMRI studies in rat. Neuroimage 206:116338
75. Christie IN et al (2013) fMRI response to blue light delivery in the naïve brain: implications for combined optogenetic fMRI studies. Neuroimage 66:634–641
76. Chen H-F et al (2023) Generation of a whole-brain hemodynamic response function and sex-specific differences in cerebral processing of mechano-sensation in mice detected by BOLD fMRI. Front Neurosci 17:1187328
77. Mandino F et al (2024) Where do we stand on fMRI in awake mice? Cereb Cortex 34:bhad478
78. Sans-Dublanc A et al (2021) Optogenetic fUSI for brain-wide mapping of neural activity mediating collicular-dependent behaviors. Neuron 109:1888–1905.e10
79. Lake EMR et al (2020) Simultaneous cortex-wide fluorescence Ca2+ imaging and whole-brain fMRI. Nat Methods 17:1262–1271
80. Oyarzabal EA et al (2022) Chemogenetic stimulation of tonic locus coeruleus activity strengthens the default mode network. Sci Adv 8:eabm9898
81. Mackinnon MJ et al (2025) SORDINO for silent, sensitive, specific, and artifact-resisting fMRI in awake behaving mice. bioRxiv. 2025.03.10.642406 https://doi.org/10.1101/2025.03.10.642406

Chapter 15

Characterization of Neural Networks in the Brain and Spinal Cord of Non-Human Primates by Integrating Optogenetics, fMRI, Electrophysiology, and Histology

Li Min Chen, Pai-Feng Yang, Feng Wang, Zhangyan Yang, Ning Zheng and John C. Gore

Abstract

The neocortex consists of excitatory and inhibitory neurons, each playing distinct roles in brain function, and the loss of their integrity contributes to various brain disorders. Optogenetic neuromodulation combined with functional magnetic resonance imaging (opto-fMRI), intracranial electrophysiology, and tract-tracing histology enables comprehensive monitoring of neural activity across the whole brain and spinal cord. These techniques can be used to derive causal connections and quantify both functional and anatomical connectivity within and between networks. To date, while most opto-fMRI studies have been conducted in rodent models, the number of optogenetic studies in non-human primates (NHPs) has grown rapidly. Given the greater complexity of brain and spinal cord circuits and neuronal types in NHPs, which more closely resemble those in the human brain and spinal cord, opto-fMRI in NHPs presents unique technical challenges. Nonetheless, it represents a crucial step toward potential human applications, especially as the technology continues to advance. This approach has proven especially valuable for dissecting the neural circuits underlying complex behavior and for providing new insights into the brain's functional architecture—from specific neurons in particular brain regions to large-scale functional and anatomical connections. In this chapter, we focus on the technical considerations involved in conducting optogenetic fMRI, electrophysiology, and anatomical tracing studies in an NHP model system.

Keywords Circuit dissection, Somatosensory network, Excitatory neuron, Optogenetics, fMRI, Primate, LFP, Spiking

1 Introduction

fMRI has become the primary tool for mapping distinct functional circuits and understanding human brain functions. The proper interpretation of human fMRI findings relies on the assumption that fMRI signals reflect the underlying activity of neurons in each brain region. Given that the neocortex is primarily composed of excitatory and inhibitory neurons, each playing distinct roles in cortical information integration and cognition, it is crucial to determine to what extent

Kevin Chuen Wing Chan and Seong-Gi Kim (eds.), *Optogenetic Functional Neuroimaging*, Neuromethods, vol. 232,
https://doi.org/10.1007/978-1-0716-5178-0_15,

activity from specific neuron subtypes influences neuron–hemodynamic relationships. Understanding these relationships is particularly significant, as imbalances between excitatory and inhibitory activity have been implicated in numerous neurological and psychiatric disorders, including autism, schizophrenia, depression, and anxiety [1–3], as well as chronic pain [4].

Optogenetic photostimulation of neural populations in the brain in vivo is a cutting-edge technique that uses light-sensitive proteins to achieve precise spatial and temporal control of specific cell types, enabling the investigation of cellular functions and neural circuits, and their causal relationships with behaviors [5–8]. FMRI provides whole-brain coverage, and researchers can non-invasively map activity related to particular brain functions within local and global networks. By combining fMRI with optogenetics (termed opto-fMRI), researchers can causally manipulate specific brain circuits and observe the resulting changes in brain activity across regions [9]. This approach helps clarify the roles of different brain circuits in behavior and disease by linking activities of specific neuron types (such as excitatory or inhibitory neurons) to the functions of large-scale brain circuits.

To date, most of the opto-fMRI studies have been conducted in rodent models [9–12]. In recent years, optogenetic studies in non-human primates (NHPs) have grown rapidly [13] (for a review, see Galvan et al. [14]). Given the greater complexity of brain circuits and neuronal types in NHP, which more closely resemble those in the human brain, establishing opto-fMRI in NHP is a crucial step toward using this technology to understand human brain organization.

We focused our study on the secondary somatosensory cortex (S2) of NHPs, because it is a critical region involved in somatic sensory and nociceptive information processing and integration. Both animal and human fMRI studies suggest that the S2 cortex and its corresponding parietal operculum region in the human brain serve as key hubs for gating sensory information to higher cognitive and emotional circuits [15], making it a potential target for neuromodulation. For instance, our previous research in the same NHP model used in this study identified brain regions and networks involved in processing touch and thermal painful information, including the S2 cortex and posterior insula [16–18], through stimulus-driven and resting-state fMRI. Gaining a better understanding of the causal, functional, and anatomical connections of the S2 cortex is essential for providing novel insights into the circuit-level pathology in chronic pain disorders and related neurological conditions [15, 19–21], as well as for developing new therapies.

In this chapter, we mainly review the procedures and results of our multimodal studies, discuss the role of the S2 cortex in the context of tactile and pain processing, the interactions between the brain and spinal cord, and highlight some future applications of such an approach in dissecting circuits and selecting targets for other neuromodulation procedures. Six scientific goals are considered: (1) to demonstrate the ability of optogenetics to selectively activate excitatory neurons and map functional connections in the NHP brain; (2) to understand how excitatory neuronal activity is related to the BOLD signal; (3) to quantify the relationships between opto-fMRI BOLD signal changes and corresponding electrophysiological local field potentials (LFP) and spiking activities of cortical excitatory neurons; (4) to evaluate whether the functional connectivity patterns derived from fMRI correspond to verified anatomical connections in the NHP brain; (5) to illustrate how cortical optogenetic stimulation may offer a novel tool to probe brain–spinal cord interactions; and (6) to demonstrate how MRI-based thermometry can be used to measure potential thermal effects associated with optogenetic stimulation. The compositions and arrangements of excitatory and inhibitory neurons vary across brain regions—contributing to the specialized function of each area, such as the S2 cortex—so our current study allowed us to elucidate the specific circuits through that the innocuous tactile and nociceptive sensory information are processed and integrated. These studies provide novel insights into the interpretation of opto-fMRI data and help delineate the causal, functional, and anatomical circuits of the primate S2 cortex.

2 Materials

2.1 Animal Preparations

Adult female and male squirrel monkeys (*Saimiri sciureus*) were studied. Each animal underwent repeated MRI and fMRI scans before invasive procedures. Before adeno-associated viral vector (AAV) injection, the hand representation within the S2 cortex was first defined by receptive field mapping using microelectrode electrophysiology on one or both hemispheres, followed by AAV injections into these electrophysiologically defined hand regions. Weeks after AAV injection, an optogenetic electrophysiology evaluation was conducted to ensure the successful expression of light-sensitive ChR2 (channelrhodopsin-2) inside S2 neurons. After confirmation, optical probes were implanted for the combined optogenetic and fMRI experiments of the brain. After the completion of the fMRI and electrophysiological data acquisitions, we harvested the brain and conducted immunohistology to confirm successful opsin expression and evaluate the anatomical connections to the S2 cortex, where the AAV was injected. All procedures adhered to National Institutes of Health and ARRIVE guidelines and were approved by the Institutional Animal Care and Use Committee of Vanderbilt University.

2.2 Choice AAV

We chose to inject AAV 5 and AAV9 (pAAV-CaMKIIa-hChR2(H134R)-mCherry) (www.addgene.org), which have demonstrated transfection efficacy in NHPs [22, 23]. Monkeys received either AAV5/AAV9 or AAV9 injections in the S2 hand regions of each hemisphere. So far, AAV expression has been successful in all animals, regardless of whether the AAV was injected into both hemispheres on the same day or on separate days. The AAV expression appears to be robust in our squirrel monkey models.

3 Methods

3.1 Surgical Procedures

The maintenance of a stable physiological condition during surgery and experimentation for each animal is critical to ensure a consistent signal-to-noise ratio and robust preparation for data collection. Animals were initially sedated with ketamine hydrochloride (10 mg/kg, i.m.) and given atropine sulfate (0.05 mg/kg, i.m.), followed by anesthesia with isoflurane (in a 1.5%–2% range for surgery and 0.8%–1.2% range for fMRI). Our experience is that an animal's physiology and vital signs are much more stable when isoflurane is delivered with a 70:30 N_2O/O_2 mixture, which allows us to maintain each animal at a relatively low isoflurane concentration. Ringer's solution was infused intravenously (3 mL/kg/h) to prevent dehydration and facilitate animal recovery from anesthesia after a long procedure. The fluid supplement can be administered via subcutaneous injections. The animals were artificially ventilated. Rectal temperature was maintained between 37.5°C and 38.5°C using a circulating water blanket. Heart rate, peripheral capillary oxygen saturation (SpO_2), respiration pattern, and end-tidal CO_2 (22–26 mm Hg) were continuously monitored and maintained within appropriate ranges throughout the procedure.

3.2 Viral Injection After Electrophysiology Mapping of the Secondary Somatosensory Cortex, Based on fMRI Activation Maps

Prior to virus injection, the hand-digit region in the S2 cortex of each hemisphere was identified electrophysiologically by mapping the somatotopic organization and characterizing the neuronal responses of receptive fields and preferred peripheral stimuli using an intracranial microelectrode. Local field potentials (LFP) and spiking activity were recorded with a linear array, followed by the virus injection. This step is to ensure that the successful expression of ChR2 and light-elicited electrophysiological activity in S2 neurons can be directly compared with the activity evoked by natural tactile stimulation. The comparison allows us to examine and interpret the functions of light-activated excitatory (or inhibitory) neurons within the S2 cortex.

For S2 mapping, a craniotomy was made at the juncture of the lateral and central sulci, where somatosensory area 3b and S2 cortices are located. Single epoxylite-coated tungsten microelectrodes (FHC Inst., Bowdoinham, ME) with a sharp tip (<3 μm) and ~1 MΩ impedance were inserted into the upper bank of the lateral sulcus, where the S2 digit region is located. At each electrode penetration

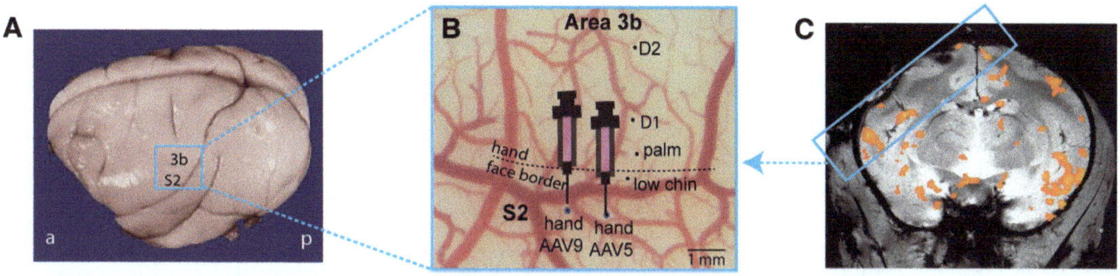

Fig. 1 AAV injection into the touch and heat clusters within the hand regions in the S2 cortex. (**A**) Post-Mortem brain image showing the location of the electrophysiology mapping of S2 digit representation (**B**) which was used to guide the injection of viral vectors into the S2 cortex. (**C**) A coronal fMRI activation map shows the tissue volume corresponding to the electrophysiology mapping

site, the preferred stimulus type (i.e., gentle stroking with a wooden probe, squeeze, light touch) and response strength (scored on a six-level scale) were characterized along the cortical depth. Receptive fields of neurons were identified by palpating and squeezing the contralateral hand, while listening to an audio amplifier for spike activity and viewing action potential traces on a display. Based on these characteristics—receptive field properties, preferred stimuli, and the somatotopic organization of the digits, as established in this species [24–26]—we identified the S2 hand/digit region. We specifically mapped and localized clusters of both low-threshold mechanoreceptive neurons and heat-sensitive neurons. Detailed characterization of these two regions was previously described in Ye et al. [18].

AAV was injected into two locations within the S2 hand representation map: AAV 5 into the hand tactile neuron cluster and AAV 9 into a posterior heat-sensitive neuron cluster (approximately 1 mm posterior), both within the S2 digit region, at two cortical depths: the middle granular layer (3500–4500 μm from the brain surface and the infragranular layer (500 μm above middle layer). Figure 1 shows one case. A total of 2–3 injection tracks were made in the hemisphere of the dominant hand, with 2 μL of virus solution injected per track. We purposely constrained the injection volume within the digit tactile and heat neuron cluster to map its anatomical connections later using immunohistology.

3.3 Optogenetics and Electrophysiology

Six to eight weeks after AAV injection, a 32-channel NeuroNexus (A32-Poly2, optical fiber core: 125 μm with 0.22 NA) or Cambridge (H7b, optical fiber core: 100 μm with 0.37 NA) optoelectrode was inserted into the AAV injection site to record laser-evoked spiking and LFP activity (CerePlex Direct, BlackRock Neurotech). During recording, the exposed cortex was covered and stabilized with 4% agar. The Fieldtrip data analysis toolkit was used to process LFP signals. Raw LFP data were filtered with 60 Hz and 120 Hz band-stop

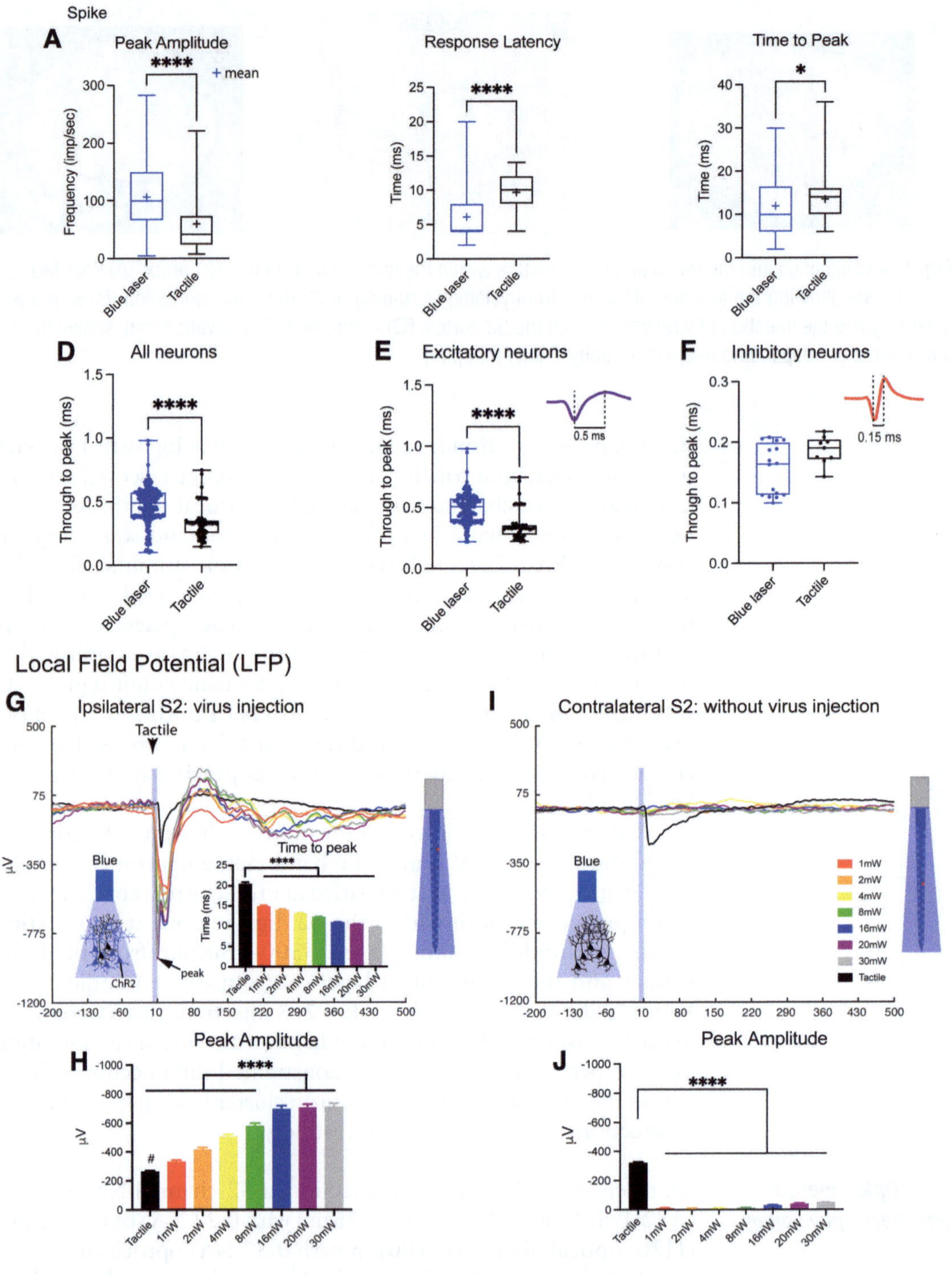

Fig. 2 Comparison between optogenetically and tactile stimulus-evoked spiking and LFP activity in the S2 cortex. (**A–C**) Box plots of peak amplitude (**A**), response latency (**B**), and time to peak (**C**) of spiking activity as a function of blue light stimulation and tactile stimulation. (**D–F**) Group-level comparison of spike widths (through to peak) between neurons responsive to blue laser stimulation and those responsive to tactile stimulation, across three groups: all neurons (**D**), excitatory neurons (**E**), and inhibitory neurons (**F**). Unpaired *t*-test:

Chebyshev filters to remove power line frequency interference, then band-pass filtered between 1 Hz and 150 Hz for quantification. Stimulus-evoked LFP signals were averaged across 30–50 trials within each run to increase the signal-to-noise ratio. Spike activity was initially sorted offline using Plexon's Offline Sorter software (Plexon Inc.), and the results were presented as scatter plots and peristimulus time histograms (PSTH, 2 ms bin width, Neuroexplorer). The 99% confidence intervals for the spike count were computed to determine whether increases in spiking rate during stimulation were statistically significant, as well as to detect the response latency at which the spike count met or exceeded the upper confidence interval.

One of the goals of electrophysiology is to confirm that light-responsive neurons are expressed in the intended neuron population, the excitatory neurons in our case. We classified and sorted single units into putative excitatory or inhibitory neurons by measuring the trough-to-peak duration of spike waveforms. Excitatory neurons typically exhibit longer trough-to-peak durations compared to inhibitory neurons, with commonly used cutoffs around 0.2–0.22 milliseconds (ms) [27–29]. In our study, we employed a cutoff of 0.22 ms: waveforms wider than 0.22 ms were categorized as putative excitatory neurons, while those narrower were considered putative inhibitory neurons.

We compared the results of optogenetic stimulation with those produced by vibrotactile stimulation of the finger pads (Fig. 2). Unlike natural peripheral stimuli, such as tactile stimulation of body parts or electrical stimulation of the cortex, which may activate both excitatory and inhibitory neurons and axons, optogenetic fMRI allows for the selective activation of primarily one type of neuron—in this case, excitatory neurons, which are responsible for interareal communication. Compared to peripheral tactile stimulation, blue laser-elicited spiking activity was significantly stronger in amplitude, exhibited much shorter response latency (mean ± SD: 6.1 ± 3.9 ms for laser versus 9.7 ± 2.6 ms for tactile), and peaked faster (mean ± SD: 12.0 ± 6.6 ms for laser vs. 13.5 ± 4.5 ms for tactile, Fig. 2A–F). These results indicate that blue light directly activates S2 excitatory neurons, eliciting robust and faster responses compared to natural tactile stimulation, where signals are transmitted from the peripheral skin and afferents to the S2 cortex via ascending pathways. Taken together, these observations led us to conclude that opto-fMRI signals reflect underlying neuronal activity at the population level, as measured

Fig. 2 (continued) *$p < 0.05$; ****$p < 0.0001$. (**G**) Average LFP signal time courses in response to blue light stimulation at varying powers and 8 Hz tactile stimulation. Inset: bar plots of the time-to-peak measures of the LFP response. (**H**) Bar plots of LFP peak amplitudes (mean ± standard error) at different blue laser powers. (**I**) Average LFPs to green laser stimulation at varying powers. Tactile-evoked LFP signals are shown as a control. (**J**) Bar plots of the LFP response peak amplitudes (mean ± standard error) at different green laser powers in the virus-free S2 cortex (without AAV injections). ****$p < 0.0001$, One-way ANOVA with Tukey post hoc

by LFP. The relationship between opto-fMRI signal amplitude and spiking rates suggests that changes in the spiking activity of individual neurons within a population are not the primary contributors to BOLD signal increases.

3.4 Optogenetic Excitation and BOLD fMRI of the Brain and the Spinal Cord

We conducted separate MR imaging sessions of the brain and spinal cord while optogenetically stimulating the S2 cortex in anesthetized squirrel monkeys at 9.4 Tesla (Bruker BioSpec) using protocols we have previously described [30, 31]. In our standard protocols, high-resolution T2*-weighted anatomic images are acquired to guide invasive electrophysiological mapping and for the placement of optical fibers. Functional MRI data are obtained using multi-shot gradient-echo echo-planar imaging (GE-EPI) with a volume radiofrequency coil. Experimental details of our animal preparation and imaging protocols have been published in previous reports [18, 32–36].

Blue light stimuli in three intensities (1, 2, and 4 mW laser output, equivalent to approximately 8, 16, and 32 mW/mm^2) were applied as trains of 10-ms pulses at 8 Hz for 6 s, followed by an interstimulus interval of 24 s.

During each MRI scan, the monkeys were anesthetized with isoflurane (0.8%–1.2%) and mechanically ventilated, with head and body stabilized in a MR-compatible frame. Vital signals (cardiac and respiratory cycles, core body temperature, end-tidal CO_2, peripheral capillary oxygen saturation SpO_2 via pulse oximetry) were monitored and maintained within appropriate ranges throughout each imaging session. The respiration pattern during each fMRI scan was recorded and later used as a regressor in data preprocessing.

For the brain, an 85 mm ID volume coil was used to collect both high-resolution anatomic images and BOLD-sensitive fMRI images. T_2*-weighted anatomic images were obtained using a gradient echo sequence (TR = 480 ms, TE = 10 ms, FOV = 55 × 55 mm^2, matrix size = 512 × 512, in-plane resolution = 0.107 × 0.107 mm^2, thickness = 1 mm, 24 slices). BOLD fMRI images were obtained using a four-shot T_2*-weighted 2D GE-EPI sequence (TR = 750 ms, TE = 10.8 ms, FOV = 55 × 55 mm^2, matrix size = 128 × 128, in-plan resolution = 0.430 × 0.430 mm^2, thickness = 1 mm, 24 slices, 3 s per volume). The stimulus protocol included a 30-s baseline, followed by interleaved 6-s ON and 24-s OFF periods of blue light stimuli, delivered to the left AAV-treated hemisphere. Each fMRI run comprised 12 epochs (six for each condition) and yielded a total of 250 image volumes.

For the spine, a custom-built saddle-shaped transmit–receive surface coil (2.5 × 3 cm^2 in size) was positioned over each animal's neck. Five contiguous axial image slices (3 mm thickness) covering C3–C7 cervical segments, with the third slice positioned over the C5 segment, were acquired in each imaging session. High-resolution structural gradient echo images (TR = 220 ms, TE = 3.24 ms,

FOV = 32×32 mm^2, matrix size = 128×128, in-plane resolution = 0.25×0.25 mm^2, thickness = 3 mm, five slices) with magnetization transfer contrast (MTC) were acquired following a Gaussian saturation pulse (flip angle = 820°, pulse width = 12 ms, RF offset = 5000 Hz). Multiple runs of stimulus-evoked fMRI data were acquired using the same block-design stimulation protocol as that applied for brain fMRI. BOLD fMRI data were acquired using a fast gradient echo sequence (flip angle = ~18°, TR = 46.88 ms, TE = 6.5 ms, in-plane resolution = 0.5×0.5 mm^2, 3 s per volume, 250 volumes) covering the same geometry as anatomic imaging.

fMRI data underwent standard preprocessing steps using SPM12 [37] in MATLAB (The MathWorks Inc., USA), following procedures similar to those used in our previous study [38]. Five EPI image volumes were discarded at the beginning of each epoch. For brain data, slice timing correction was applied first, followed by 2D motion correction. Nuisance signals from white matter (WM) and muscle were regressed for each run. Physiological noise was further corrected using RETROICOR [39]. A 0.01–0.25 Hz band-pass filter was then applied to the BOLD time series. The general linear model (GLM) analysis was performed using FEAT (FMRI Expert Analysis Tool), part of the FSL toolbox [40], to generate fMRI activation maps based on a canonical hemodynamic response function (HRF). For each fMRI EPI run, activated voxels were defined as those showing stimulus-related signal changes at a statistically significant level ($t > 1.2$, $p < 0.05$, FDR corrected) [41]. Second-level group analyses using a fixed-effects model were also performed using FSL [42].

For spinal cord data, similar preprocessing steps were followed. The difference is that 3D motion parameters and time-varying signals extracted from voxels in the muscle and CSF regions—those capturing at least 70% of the cumulative variance via principal component analysis—were regressed out as nuisance variables. The median fMRI image from each run was registered to the corresponding anatomical MTC image, and each subject's anatomical image was then registered to a custom spinal cord template using FSL. To localize the stimulus-driven activation, we quantified percentage BOLD signal changes by averaging the signal over a six-second window prior to the onset of each stimulus epoch and used it as the baseline. This baseline was then used to calculate the average percentage signal changes during and after the light stimuli. BOLD time courses were extracted from 7 ROIs in gray matter (GM) and 8 in WM ROIs, which were defined by independent component analysis (ICA) of resting state fMRI signals within the spinal cord [31, 35, 43], as shown in Fig. 3. For group analysis (second level), the mean signal (three EPI volumes) prior to the stimulus onset in each block was considered as the baseline for the calculation of percent signal change (PSC) at each time point after stimulus onset. The block-averaged mean time courses were fitted with a double gamma variate function to characterize the overall

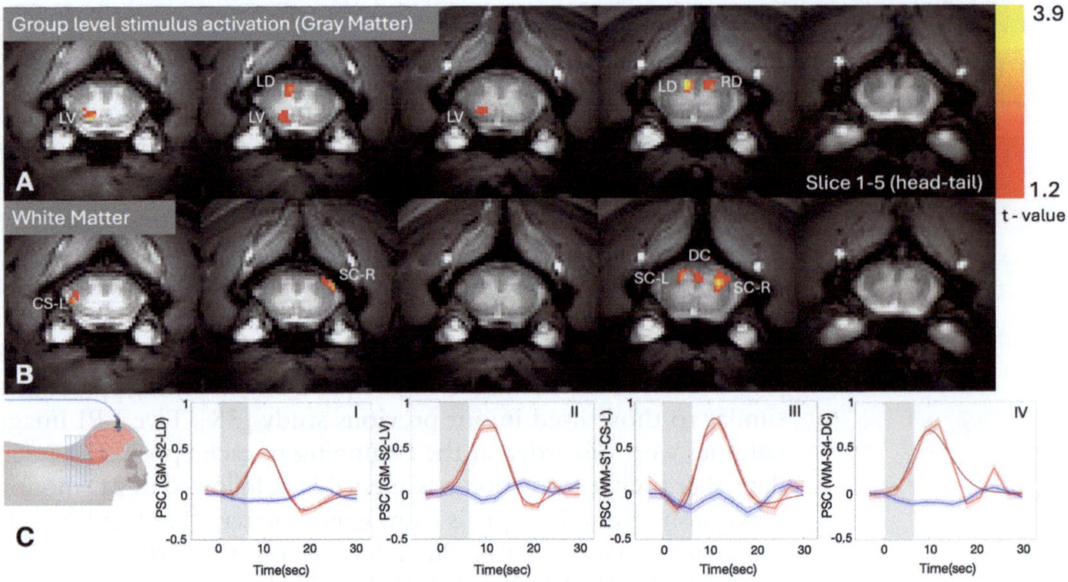

Fig. 3 fMRI BOLD signal detected in the cervical spinal cord in response to S2 light stimulation. (**A**) Group level ($n = 2$ animals and $n = 34$ fMRI runs) light stimulation evoked activation maps (thresholded at $t > 1.2$, $p < 0.05$, FDR corrected), showing within the gray matter. (**B**) Group-level activation in white matter tracts. (**C**) Left: Schematic of the experimental setup. Right: Plots of stimulus cycle averaged percentage BOLD time courses (blue lines, mean ± std error) and two-gamma fitted curves shown on two gray matter regions and white matter tracts (GM ROI: I–II: GM Slice 2-LD and LV, WM ROI: III–IV: WM Slice 1 CS-L and WM Slice 4-DC). CS cerebral spinal tract, DC dorsal column tract. Blue curve on each panel refers to the main signal variation of voxels unresponsive to the stimulus within the same ROI

stimulus response, such as delay and time to reach the peak signal change. The mean and peak PSC were calculated for comparison.

3.5 Using fMRI to Monitor Potential Thermal Confounds Associated with Optogenetic Experiments

The light delivery during optogenetic experiments may induce tissue heating, which in turn could produce pseudo-BOLD responses or affect neural activity in the brain [44–46]. Many neural processes are temperature-sensitive, so the delivery of light at high intensities is a potential concern [47–49]. To avoid such confounding effects, light energy deposition during optogenetic stimulation needs to be carefully controlled. MRI imaging may itself be adapted to measure temperature changes produced by lasers within tissues. We performed simulations of heat generation within our experiments and also developed MR thermometry based on measurements of proton resonance frequency (PRF) changes produced by heating [46]. Heat generation simulations were performed to predict potential temperature increases during laser stimulation, and actual temperature changes were calculated using T2*-weighted images collected during optogenetic irradiations. Both simulation and thermometry measurements indicated temperature increases were smaller than 0.02°C around the probe tip at 1–4 mW power levels.

3.6 Simulation and Measurement of Temperature Effects

Simulations of possible temperature increases during the laser stimulations were performed in MATLAB, assuming specific brain tissue optical properties. We used a Monte Carlo simulation model for light intensity spread and Pennes' bioheat transfer equation for temperature diffusion. Optical properties (absorption coefficient: m_a, scattering coefficient: m_s, and anisotropy parameter: g) of 561 nm green light and 473 nm blue light were calculated from Johansson's in-vivo data [50] green: $m_a = 0.2565\ \text{mm}^{-1}$, $m_s = 9.9470\ \text{mm}^{-1}$, $g = 0.8900$; blue: $m_a = 0.3796\ \text{mm}^{-1}$, $m_s = 11.0910\ \text{mm}^{-1}$, $g = 0.8900$). The light intensity spread was first predicted by a Monte Carlo simulation using an absorption coefficient, a scattering coefficient, and an anisotropy parameter calculated from in-vivo data [50]. Then, the diffusion of heat within the brain was calculated using a biophysically realistic heat diffusion model [51] utilizing the Pennes' bioheat transfer equation [51], taking into account perfusion by blood vessels, metabolic heat production, and heat diffusion. Calculations were based on our optical fiber parameters (numerical aperture: 0.39, diameter: 200 μm, implant depth: 4200–4500 μm) and laser delivery paradigms (6 s ON and 24 s OFF, laser power of 1, or 2, or 4 mW, 10 ms pulses with a frequency of 8 Hz, imaging resolution: $0.43 \times 0.43 \times 1\ \text{mm}^3$).

MR thermometry has been widely used to monitor temperature in vivo during thermal ablation and hyperthermia [52]. Maps of temperature changes can be measured using a gradient-recalled echo (GRE) sequence by measuring the phase difference between images before and during heating, which is proportional to the temperature-dependent PRF change and the echo time (TE). Several PRF temperature mapping techniques that are robust to motion and dynamic background magnetic field changes unrelated to heating have been developed for monitoring temperature during thermal therapies, such as multibaseline subtraction [53], referenceless thermometry [54], and a combination of the two [55].

We recently demonstrated how T_2*-weighted GE-EPI collected in fMRI can be used to simultaneously image brain function while monitoring small temperature changes on the order of tenths of a degree Celsius caused by the stimulating light delivery, but a motion-robust and respiration-robust temperature map processing is required [46, 52]. The simulation process is shown in Fig. 4; and one example is shown in Fig. 5. The raw fMRI EPI image data were exported and reconstructed to complex-valued images offline in MATLAB. Temperature change maps for each time point were then calculated using a hybrid referenceless [55, 56] method based on shifts in the PRF of fMRI data. The phase shift θ can be converted to the temperature ΔT in units of $\Delta^\circ C$ using the following equation:

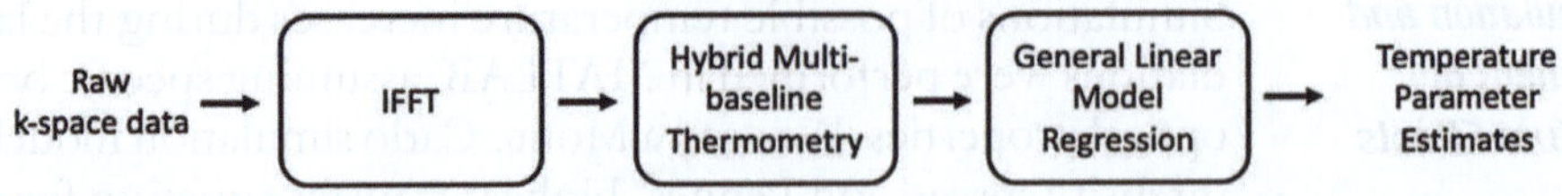

Fig. 4 Temperature map calculation and processing pipeline. Raw images were reconstructed to a complex-valued time series, which was converted to temperature maps using the hybrid multibaseline thermometry method

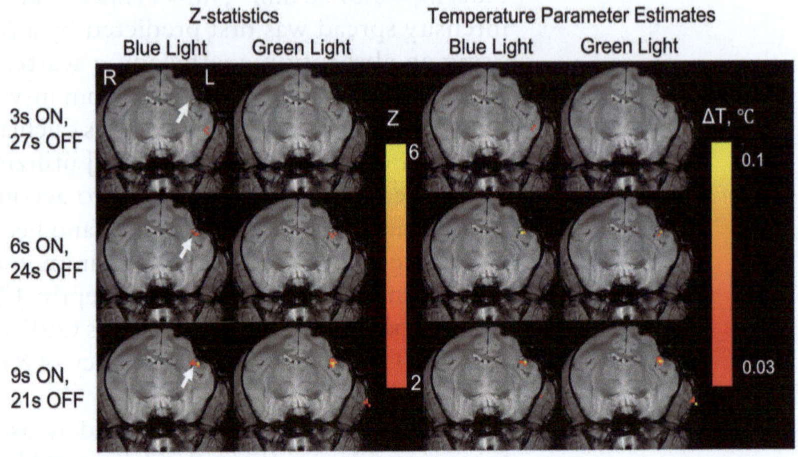

Fig. 5 Maps of temperature change *z*-statistics and quantitative parametric temperature change maps of single runs obtained from general linear modeling. The white arrows point to the positions of the laser probes. Detectable temperature rises were seen for 6-s and higher laser stimulations (30 mW) with both blue and green lasers. Modified from Fig. 5 in Luo et al. [46].

$$\Delta T = -\frac{\theta}{\gamma \alpha B_0 TE}$$

where γ is the gyromagnetic radio, $\alpha = -0.01$ ppm/°C is the PRF change coefficient, B_0 is the magnetic field strength, and TE is the echo time of the GE-EPI sequence. Once the temperature maps were calculated in the first monkey, they were statistically analyzed to generate z-statistic maps identifying voxels with significant temperature changes and parametric temperature maps using general linear modeling with FEAT.

temperature signals were then extracted from voxels around the tip of the optical probe in the S2 region that exhibited significant temperature changes (FDR <0.05) and were then averaged to obtain their time courses of temperature. Figure 5 shows the maps of temperature change z-statistics (left panel) and quantitative temperature parameter estimates (right panel) when high intensity (30 mW) blue and green laser power was delivered.

Figure 6A, B show the calculated spread of light intensity in three dimensions produced by green and blue light separately at the intensities used in our experiments. The intensity scale in the figure is logarithmic i.e., a value −4 is equivalent to a 99.99% light intensity reduction. In our case, blue light intensity has a 99.99% loss at a depth of 3.25 mm, and green light has a 99.99% loss at a depth of 4.08 mm. Temperature changes from the laser fiber across time under 4 mW green and blue light are shown in Fig. 6C, D, respectively. The estimated maximum temperature rises (blue light/green light) were 0.0276/0.0208°C.

Figure 7 illustrates the temperature changes measured using (a) Monte Carlo simulation and (b) MR thermometry under optogenetic stimulation with blue and green light at power settings of 1, 2,

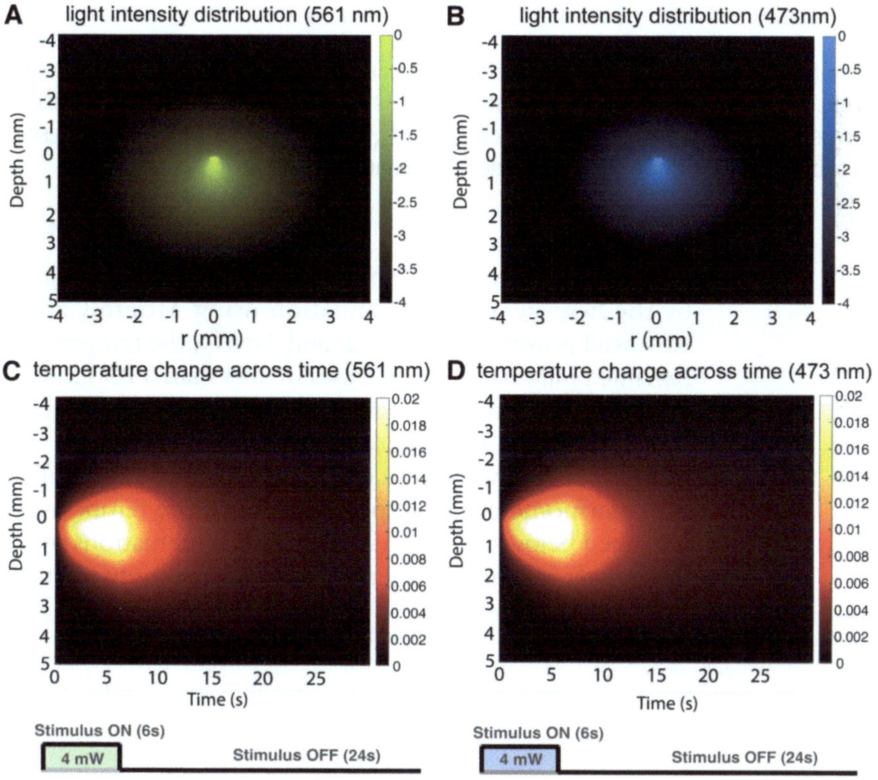

Fig. 6 Light intensity distribution and temperature change across time. Light intensity distribution predicted from Monte Carlo simulation with an optical fiber (NA: 0.39, diameter: 200 μm, implant depth: 4200–4500 μm) for 561 nm green light. (**B**) Light intensity distribution predicted from a Monte Carlo simulation with the same optical fiber parameters for 473 nm blue light. In (**A**) and (**B**), x-axis: radial distance (mm) from the tip. y-axis (left): depth (mm), the tip is defined at depth = 0 mm. y-axis (right): log reduction values: for example, 0 represents no signal loss and −4 represents 99.99% signal loss. (**C, D**) Maps of temperature changes under green and blue light, separately simulated from the bioheat model, are plotted as a function of time and depth. x-axis: time course (s). y-axis (left): depth (mm). y-axis (right): temperature changes. Stimulus patterns were shown under the map (power: 4 mW, stimulus ON duration: 6 s, stimulus OFF duration: 24 s)

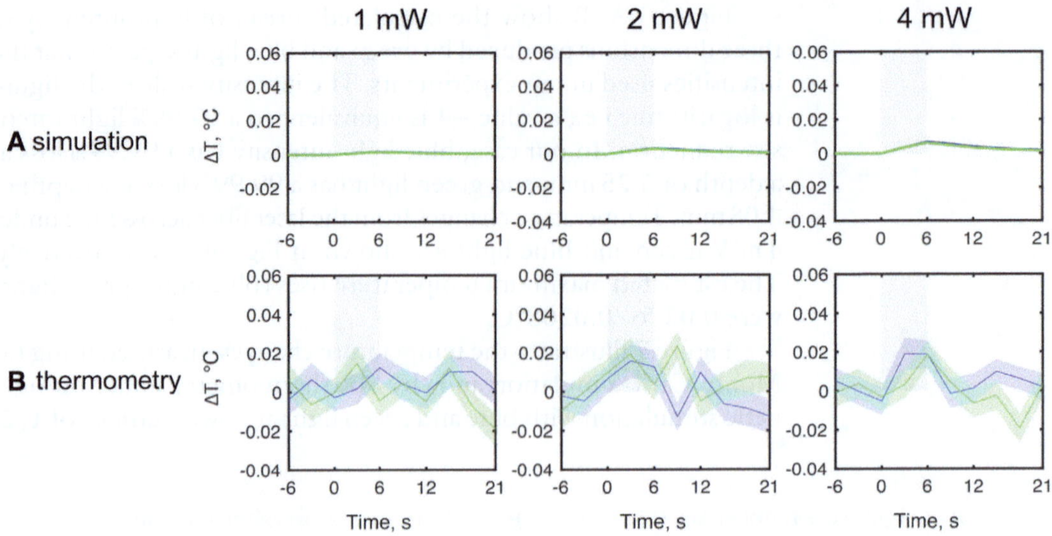

Fig. 7 Temperature changes of blue and green light under the power settings of 1, 2, and 4 mW (**A**) Temperature changes using a Monte Carlo simulation. (**B**) The calculated temperature changes using MR thermometry. The green line represents temperature changes of green light, and the blue line represents temperature changes of blue light. The shaded area represents standard deviations

and 4 mW. The top row shows the temperature changes predicted by the heat generation simulation using Monte Carlo methods. Across all power settings (1, 2, and 4 mW), the temperature changes remained minimal, with increases smaller than 0.02°C. Both blue and green light resulted in similarly low temperature changes, indicating minimal heating effects predicted by the simulations. The bottom row presents temperature changes measured experimentally using MR thermometry. At all power settings, the temperature changes remained below 0.03°C, with noticeable fluctuations over time. The low-power setting was then used for a later study to ensure limited thermal effects.

4 Notes

4.1 Optogenetic Stimulation with Increasing Powers Elicits Stronger and Faster Spiking and Local Field Potential Activity than Tactile Stimulus

Optogenetic stimulation-evoked LFP signal changes reflect predominantly the integrated synaptic activity of a population of excitatory neurons. In the same monkey model, as the blue laser power was increased from 1 to 16 mW, the LFP signals exhibited progressively larger negative deflections (Fig. 2G, H). However, further increases in laser power beyond 16 mW (up to 30 mW) did not result in any further significant increase in the evoked response amplitude. In comparison, the LFP amplitude evoked by natural tactile stimulation of the monkey's fingers was significantly weaker than that evoked by the laser.

To compare the speeds of information processing under blue laser versus tactile stimulation, we also measured the times to the peak of LFP responses. We found that the latencies of LFP peaks, regardless of laser intensity, were consistently shorter than those evoked by tactile stimulation (Fig. 2G inset). When green laser light stimuli were delivered at the same power levels, no LFP signal changes were observed (data not shown) [57]. Additionally, we replicated the experiment in the AAV-naïve S2 cortex, which lacks ChR2 opsin expression. No apparent LFP signal changes were detected during blue laser or green laser illumination (Fig. 2I, J). Tactile stimulation evoked distinctive LFP responses from the S2 cortex, confirming that the recording site was functioning properly. The LFP signal changes associated with excitatory neuron activation differed from those elicited by natural tactile stimulation, confirming that the BOLD signals detected in S2 and across the brain are associated with the selective activation of excitatory neurons in this region [58, 59].

4.2 Spike Waveform Analysis is Reliable in Classifying Excitatory Versus Inhibitory Neuron Types

Using spike widths (trough to peak duration) to classify neurons as putative excitatory or inhibitory is a commonly accepted method for analyzing extracellular recording data. Using 0.22 ms as the cutoff [27–29, 60], we determined that approximately 93% of blue-light-responsive neurons isolated in the S2 cortex were putative excitatory neurons, a finding consistent with previous reports [61]. By contrast, the proportions of excitatory and inhibitory neurons sampled in the S2 cortex during tactile information were 86% and 14%, respectively. Using a multichannel depth array for recording, we observed that the blue light-responsive neurons were scattered across various cortical depths, with some channels exhibiting increased firing rates during and after blue laser illumination, while other channels showed lower firing rates (Fig. 2 in Yang et al. [57]). This finding supports the notion that excitatory neurons are distributed differently across cortical layers [62, 63]. The excitatory neurons in S2 primarily function as conduits for relaying, processing, and integrating sensory information, while inhibitory neurons play a critical role in modulating and shaping sensory signals. The proportion of excitatory (86%) and inhibitory (14%) neurons sampled in the S2 cortex during tactile information processing closely mirrors that observed in the rat's medial prefrontal cortex [64] and generally aligns with typical cortical neuron populations [65–67].

4.3 Disassociated Dose-Response Relationships of Opto-fMRI, LFP, and Spiking Activity

We observed that blue light, delivered in 6-s duration bursts via an implanted optical fiber, elicited robust, laser power-dependent BOLD signal increases at low intensities (Fig. 8A–C). The time courses of BOLD signal changes extracted from the voxels surrounding the optical probe exhibited a standard hemodynamic response function

(HRF), with peak signal changes occurring between 9 and 12 s from the stimulation onset (Fig. 8D). Plots of normalized percentage BOLD signal changes as a function of laser output power (1 mW: 0.30%, 2 mW: 0.33%, and 4 mW: 0.40%) showed a monotonically increasing trend (Fig. 8E, dotted blue line, R-squared = 0.99). In addition to the BOLD activation detected at the stimulated S2 location, fMRI activation across the brain was measurable in multiple locations (Yang et al., Figs. 5B and 6A). In addition, LFPs and electrical spiking rates were recorded from electrodes implanted in the S2 cortex. To help interpret the opto-fMRI results, we characterized the dose (laser power) versus response curves for opto-fMRI, LFP, and spiking signals in the S2 cortex, and examined the correlation strengths between changes in these three signals. We found that changes in evoked BOLD signals strongly correlated with LFP signals, whereas corresponding changes in spiking activity were not detected at low laser power levels (1–4 mW) (Fig. 8E).

Our finding of the disassociated dose-response relationships of opto-fMRI, LFP, and spiking activity differs from those observed in the normal cortex [34] but is similar to those seen in the deafferented cortex, such as after sensory input deprivation caused by traumatic spinal cord injury [68]. Our findings support the notion that the relationship between fMRI signals and underlying intracranial electrophysiology activity is significantly altered when only specific neuron types are selectively manipulated by optogenetics or when afferent inputs to the cortex are disrupted.

4.4 Neuron-Type Specific Structural and Functional Relationships of the S2 Cortex and Beyond

We and others have previously demonstrated that the S2 cortex functions as a hub, connecting somatosensory cortices to other brain networks during the processing of both tactile and painful information [15]. Selectively activating S2 excitatory neurons leads to brain-wide fMRI responses, indicating that the S2 cortex is causally and functionally connected to numerous brain regions with distinct roles (Fig. 9A). For example, S2 cortex activation was directly associated with fMRI responses in the midcingulate cortex (MCC), anterior cingulate cortex (ACC), and subcortical regions, such as the putamen. Other activated regions include somatosensory area 3b, primary motor cortex (M1), putamen, thalamic ventral posterolateral nucleus (VPL), insula cortex, and amygdala. The BOLD time course sampled at the S2 and other selected regions exhibits a standard HRF (Fig. 9B, C). Significant differences were observed in the BOLD response amplitudes between the stimulated S2 cortex and off-target sensorimotor cortical regions, such as area 3b and M1 ($p<0.05$, Brown-Forsythe and Welch ANOVA test), as well as cortical and subcortical regions involved in the processing of touch, affective, and emotional aspects of pain, such as the ACC, MCC, amygdala, and putamen (Fig. 8F). Notably, the BOLD signal was stronger in regions such as the insula and ACC compared to sensorimotor areas such as the thalamic VPL

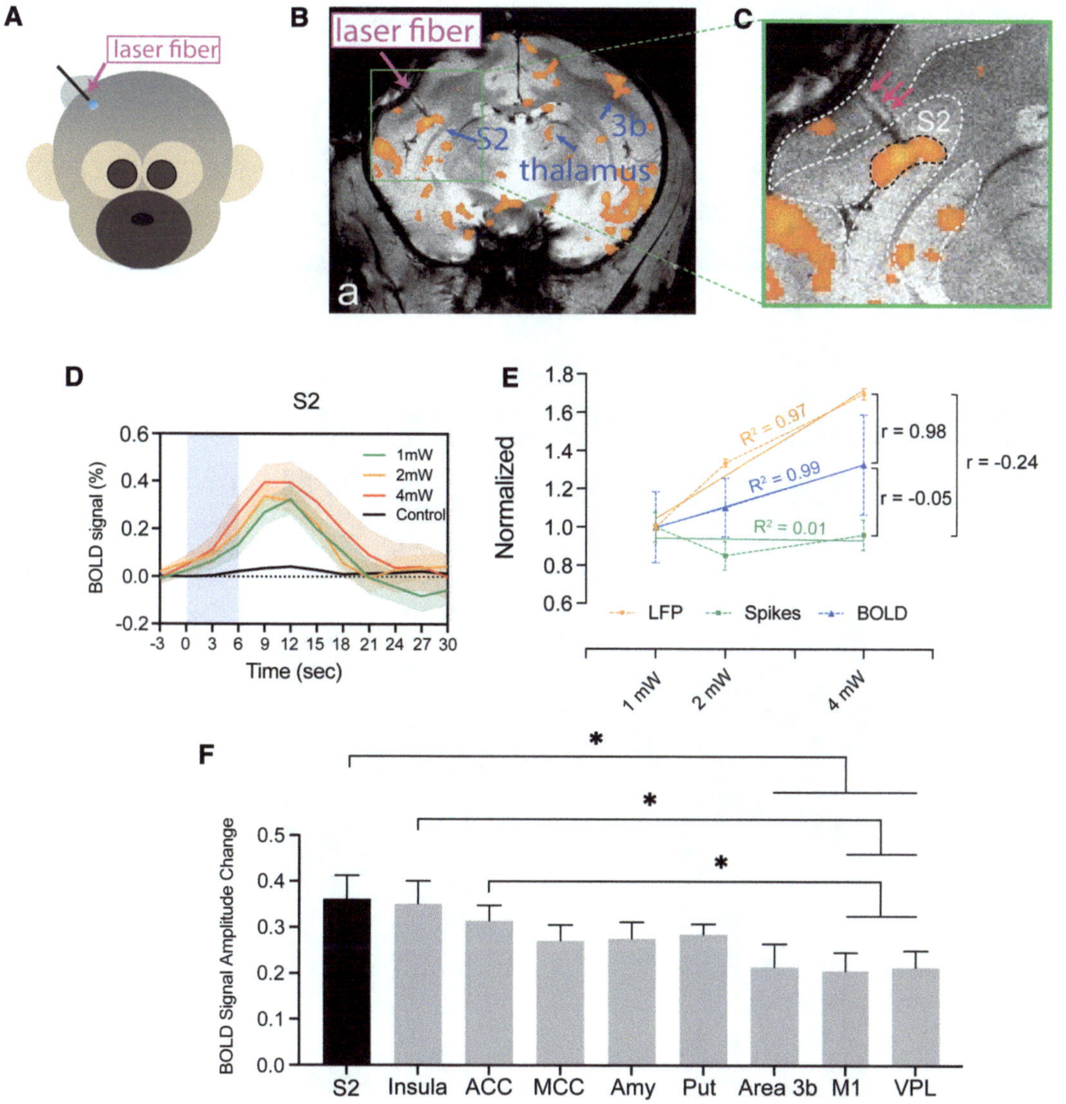

Fig. 8 Optogenetic BOLD fMRI signal change at the S2 cortex and its relationship with spiking and LFP activity. (**A**) Schematic illustration of the Setup for optogenetic fMRI and electrophysiological data acquisition. (**B–C**) Opto-fMRI activation maps are shown on one coronal brain slice and a zoomed-in view of the S2 region with the implanted optical fiber (indicated by arrows). 3b: primary somatosensory area 3b. (**D**) Averaged BOLD time courses extracted from the S2 cortex with AAV injection during blue light stimulation at three laser power levels (1 mW, green; 2 mW, orange; 4 mW, red) compared to those from the AAV-free S2 cortex (control: black line) during the same stimulation. (**E**) Plots of normalized signal changes (mean ± SEM) of BOLD, LFP, and spikes (dotted color lines) at three blue laser powers, with linear fitting (solid color lines). r: correlation coefficient between the three curves of signal changes: BOLD and LFP, BOLD and spikes, and LFP and spikes. (**F**) Comparison of fMRI BOLD signal changes in off-target brain regions during optogenetic stimulation of S2 excitatory neurons. Bars represent the mean peak percentage change in BOLD signal, with error bars indicating the standard error of the mean.

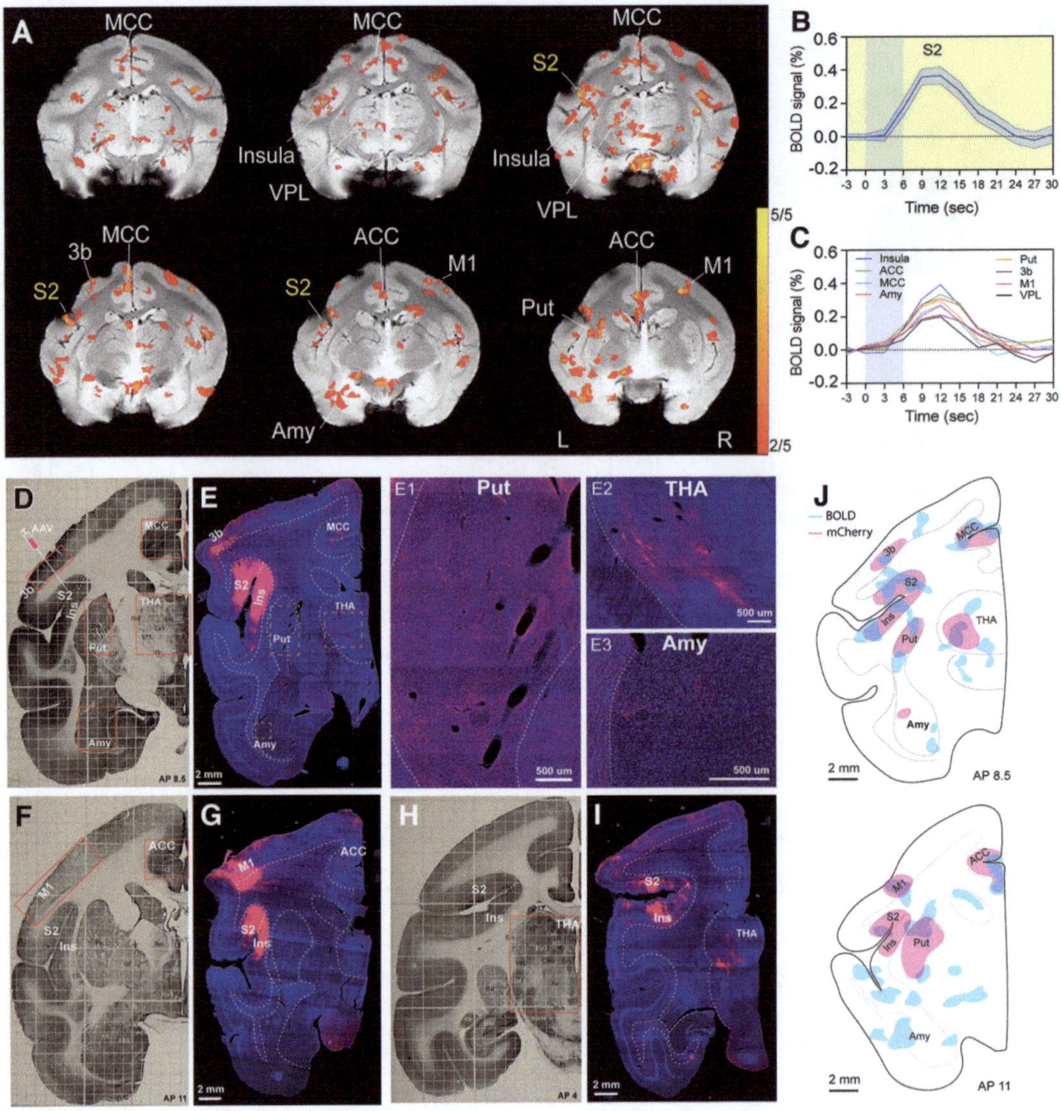

Fig. 9 Spatial correspondence between optogenetically evoked fMRI BOLD activation maps and global anatomical connections of AAV tracer injected into the S2 cortex. (**A**) Multislice BOLD activation probability maps during 30 mW blue laser stimulation of the S2 cortex, which received the AAV injection. (**B, C**) Averaged BOLD time courses extracted from voxels surrounding the implanted optical fiber in the S2 (**B**, shadow: SEM [standard error]) and eight brain regions: Insula: insular cortex, ACC anterior cingulate cortex, MCC medial cingulate cortex, Amy amygdala, Put putamen, 3b somatosensory area 3b, M1 the primary motor cortex, VPL thalamic ventroposterior lateral nucleus. (**D–I**) Histological evaluation of AAV expression in one subject. Cortical and subcortical regions showing mCherry expression from the AAV injection (**E, G, and I**) are annotated on corresponding brain slices from a squirrel monkey atlas (**D, F, H**). The small letters in the bottom right corners indicate the stereotaxic AP coordinates of each sectional image. E1–E3: High magnification views of mCherry immunoreactive fibers in three deeper brain regions: the putamen (Put), thalamus (THA), and amygdala (Amy). The AAV injection site in the S2 digit region is depicted in A. (**J**) Overlays of blue laser-evoked fMRI activations (blue patches) and mCherry expression (pink patches). Brain structure outlines were modified from images in A Stereotaxic Atlas of the Squirrel Monkey's brain, Public Health Service Publication No 933, 1962

nucleus and M1. These fMRI results suggest that excitatory neuron activity elicited by optogenetic stimulation of S2 propagates robustly to remote cortical and subcortical regions, inducing stronger BOLD responses in perception-related and emotion-related regions than in sensorimotor regions. Previous studies in macaque monkeys have provided evidence of S2 connections with the cingulate cortex [69, 70] and the putamen [71].

Our experimental setting enabled us to directly link the brain regions exhibiting opto-fMRI signals with the functional roles of the S2 cortex, as AAV vectors were injected into a functionally defined hand region (Fig. 1). Our previous studies have shown that this hand region contains clusters of low-threshold tactile and high-threshold thermal nociceptive neurons, so the brain regions showing opto-fMRI signals are involved in the processing and integration of touch and pain [18].

The postmortem immunohistological staining of the brain tissue not only validated the successful expression of AAV vectors but also delineated anatomical connections. The uptake of the AAV and expression of the mCherry fluorescent tag were very strong at the injection site in the S2 hand region. Additionally, coronal histological sections of the brain revealed widespread fluorophore-positive neurons and axons in many brain regions, as shown in Fig. 9D-I. Importantly, the cortical and subcortical regions exhibiting robust opto-fMRI signals also showed fluorophore uptake, indicating strong anatomical tracer connections with the S2 cortex, where the AAV tracer was injected. Zoomed-in images highlight axons with fluorescence expression in key regions, including the thalamus, putamen, and amygdala (Fig. 9E1–E3). There was strong spatial correspondence between the S2 opto-fMRI activation map and the S2 anatomical tracer connectivity map (Fig. 9J). Moreover, AAV tracer uptake was detected in axons within these brain regions, showing strong opto-fMRI signals, confirming their extensive connections to the S2 cortex. Many of these regions are not classical somatosensory regions but are involved in higher-order brain functions. Our data provide the first optogenetic evidence supporting the direct connection of the S2 cortex to networks engaged in functions such as perception and cognition. At the circuit level, the tracer uptake pattern also indicates opsin expression in neurons and axons connected to the S2 cortex, including in S1, M1, the insula, and the thalamic VPL nucleus, which is largely consistent with known anatomical connections between S2 and area 3b [72], area 1 [73], area 3a [74], M1 [75], the insula, and the thalamus [76]. For reviews, see [77].

It is worth noting that AAV uptake may have included nearby regions, such as the posterior insular cortex, which has known connections with S2 in macaque monkeys, and projections to cognitive and emotional processing regions, including the amygdala, as part of a "corticolimbic pathway for touch" [78]. Our results are consistent with regions affected in patients with impaired touch perception [79], and support a model of ventral stream somatosensory processing [80]. This strong correspondence between opto-fMRI and tracer-based anatomical networks of the S2 cortex is in line with our previous studies of functional pathways of the S2 cortex [16].

4.5 fMRI Detection of Spinal Cord White and Gray Matter Activity During S2 Optogenetic Cortical Stimulation

Execution of coordinated skilled hand use in primates, including humans, engages both the brain and cervical spinal cord interactively. Understanding this interaction is fundamental to interpreting changes seen in diseased conditions, such as following spinal cord or brain injuries. The spinal cord serves as the first processing and relay station of ascending peripheral inputs to the brain, as well as descending outputs. Current concepts of spinal cord function posit that neural signals from peripheral stimuli are processed by the dorsal horn of GM and are transmitted to the brain via specific WM tracts, while information from the cortex is transmitted down the WM of the spinal cord and produces ventral horn GM responses. In previous work [30, 31, 81], we identified the presence of synchronized hubs of resting state BOLD signals in a set of GM regions in each section of the spinal cord, which together form intraspinal networks, as well as in a set of regions that colocate with major WM tracts. These previous observations underscore the complexity of spinal cord functional organization and indicate that the interactions among spinal cord ascending and descending pathways are more extensive than we have previously recognized.

We have also previously reported robust detection of peripheral stimulation-evoked neural activity using BOLD fMRI within both the GM and WM of the cervical spinal cord of nonhuman primates [30, 31, 43]. Of more novel significance, we also detected robust and focal BOLD signal changes in spinal GM and WM in response to optogenetic stimulation of the S2 cortex hand region. Figure 3 shows the locations and time courses of the epoch-averaged percentage signal changes extracted from GM (left dorsal and ventral horns), which exhibited a classical BOLD signal hemodynamic response to 6-s laser stimulation. The peak signal amplitudes varied from approximately 0.5%–0.9%. Time courses extracted from some WM ROIs also demonstrated a robust response of similar form (Fig. 2). In particular, the left corticospinal, and left and right spinocerebellar tracts and dorsal column showed transient increases in BOLD signal that were comparable ($\approx$0.5%–0.8%) to those seen in GM and had similar time courses. Note that, we also interrogated the signal changes in regions of interest (ROIs) defined from a previous study, in which 8 WM and 7 GM regions were identified using independent component analysis as showing synchronous BOLD signal fluctuations in a resting state. Not all voxels within each of those ROIs showed a strong response.

The spinal cord serves as the first processing and relay station of ascending peripheral inputs to the brain, as well as descending outputs. The functional integrity of the spinal cord may be disrupted in several clinical disorders and after traumatic injury, so understanding the functional organization of the normal spinal cord is of high clinical significance. BOLD signals are robustly detectable in WM, but to date, they have been largely ignored by others and are often treated as nuisance regressors in fMRI analyses. BOLD signal increases in GM are usually interpreted as physiological responses to increased demands for nutrients that are required for increased activity, but the nature and driving force for such responses in WM are unclear. The ratio of glial cells to neurons is much higher in WM than in GM, so it is possible that BOLD effects in WM may partially reflect the energy demands of different cellular processes than in GM.

There is evidence in rodents and NHP models suggesting direct termination of somatosensory cortical neurons in spinal cord GM [82–87]. Another possibility is that activating excitatory neurons in the S2 cortex can drive activations of spinal cord WM tracts and GM horns beyond those involved in somatic sensory processing. Future work is needed to address these questions. It is possible that the S2 cortex is a higher-order cortical area that processes distinct features of somatosensory stimuli. For example, more recent work [88] has suggested both S1 (e.g., area 3b) and S2 may operate in parallel. In our study, laser stimulation of the left S2 cortex elicited brain-wide activation in cortical and subcortical regions (Fig. 9), including area 3b/1 of S1, M1 cortex, and MCC cortex [57]. In line with the activation of ascending and descending WM tracts, both ventral and dorsal horns, along with the middle zone, were all activated in response to S2 stimulation, supporting the functional relevance of the activated GW and WM regions. The successful detection of evoked BOLD signals in both central GM regions of the spinal cord, as well as selected WM tracts, captured information transfer between the brain and spine, suggesting that BOLD effects may be produced by reverse signal transduction. The ability to induce BOLD signals antidromically from the cortex potentially provides a new way to interrogate functional connections more closely. The neuronal mechanisms underlying the antidromically evoked BOLD signals within the spinal cord in GM and WM warrant future investigation.

4.6 MR Thermometry

The precision of PRF shift thermometry is usually defined in terms of the standard deviation of temperature measurements, which was measured to be 0.05°C in this study. This is much smaller than the typical ~1°C standard deviations reported for temperature mapping during ablation [89]; this improvement resulted from a combination of approximately 2×-longer TE than is typically used during ablation due to phase wrapping and scan time concerns, as well as the GLM processing, which effectively averages the signal across multiple heating repetitions. The standard deviation of temperature measurements [90] is inversely proportional to the imaging SNR, which increases with voxel size, matrix size, and the sampling time. The precision can be improved by further improving imaging SNR. There are also some limitations when employing PRF shift thermometry to detect laser-associated heating. For optogenetic stimulation, the inserted optical fiber will cause local tissue damage and lost signals around its tip. The temperature changes at the immediate tip of the optical fiber could not be reliably measured because of the signal loss.

4.7 Optogenetic fMRI in NHPs and Technical Considerations

To date, most of the opto-fMRI studies have been conducted in rodent models [9–12]. In recent years, optogenetic studies in non-human primates have grown rapidly [13]. For a review, see [14]. Given the greater complexity of brain circuits and neuronal types in NHP, which more closely resemble those in the human brain, establishing opto-fMRI in NHP is a crucial step toward potential human applications, especially as technology continues to advance. The growing use of optogenetics in NHPs has significantly advanced our understanding of brain function and holds promise for developing therapeutic interventions for human disorders [91, 92]. This approach has been especially valuable for dissecting the neural circuits underlying complex behavior and providing new insights into the brain's functional architecture—from specific neurons in particular brain regions to large-scale functional and anatomical connections. In this study, we successfully combined opto-fMRI with intracranial electrophysiology, providing a method to causally manipulate defined excitatory neurons in the S2 cortex to observe the resulting changes in brain-wide BOLD signals, and demonstrating the reliability of our approach and its potential for future preclinical and clinical studies of primate sensorimotor behavior and brain circuits.

We attribute our success to the advanced AAV viral vectors and data acquisition at high-MRI fields (9.4 T in our case). The high signal-to-noise and contrast-to-noise ratios offered at 9.4 T enabled us to push the spatial resolution, which is critical for separating fine-scale structures within the brain and the spinal cord.

4.8 Significance of Using Optogenetic fMRI to Advance Neuroscience Knowledge and Establish Foundations for Clinical Applications

fMRI has become the primary tool for mapping distinct functional circuits and understanding human brain functions. The proper interpretation of human fMRI findings relies on the assumption that fMRI signals reflect the underlying activity of neurons in each brain region. Given that the neocortex is primarily composed of excitatory and inhibitory neurons, each playing distinct roles in cortical information integration and cognition, it is crucial to determine to what extent activity from specific neuron subtypes influences the neurovascular coupling. Understanding this relationship is particularly significant, as imbalances between excitatory and inhibitory activity have been implicated in numerous neurological and psychiatric disorders, including autism, schizophrenia, depression, and anxiety [1–3], as well as chronic pain [4].

Optogenetic photostimulation of neural populations in the brain in vivo is a cutting-edge technique that uses light-sensitive proteins to achieve precise spatial and temporal control of specific cell types, enabling the investigation of cellular functions and neural circuits, and their causal relationships with behaviors [5–8]. While fMRI provides whole-brain coverage, researchers can noninvasively map activity related to particular brain functions at local and global networks. By combining fMRI with optogenetics, researchers can causally manipulate specific brain circuits and observe the resulting changes in brain activity across regions [9]. This causal approach helps clarify the roles of different brain circuits in behavior and disease by linking activities of specific neuron types (such as excitatory or inhibitory neurons) to the functions of large-scale brain circuits.

We focused our study on the secondary somatosensory cortex (S2) of NHPs, because it is a critical region involved in somatic sensory and nociceptive information processing and integration. Both animal and human fMRI studies suggest that the S2 cortex and its corresponding parietal operculum region in the human brain serve as key hubs for gating sensory information to high cognitive and emotional circuits [15], making it a potential target for neuromodulation. For instance, our previous research in the same NHP model used in this study identified brain regions and networks involved in processing touch and thermal painful information, including the S2 cortex and posterior insula [16–18], through stimulus-driven and resting-state fMRI. Therefore, gaining a better understanding of the causal, functional, and anatomical connections of the S2 cortex is essential for providing novel insights into the circuit-level pathology in chronic pain disorders and related neurological conditions [15, 19–21], as well as for developing new therapies.

4.9 Conclusion

Our studies represent the first successful application of a combined fMRI, electrophysiology, and optogenetic approach to selectively stimulate excitatory neurons within the somatosensory S2 cortex of monkeys. Our results reveal clear differential relationships between increased optical power and changes in BOLD, LFP, and spiking activity, underscoring the unique underlying properties of excitatory neuron-driven opto-fMRI signals. Additionally, we established a strong spatial correspondence between brain-wide opto-fMRI activity and anatomical tracer-based connections in the S2 hand region. These findings offer novel insights into the electrophysiological underpinnings of excitatory neuron-dominated opto-fMRI signals and the structure–function relationship of S2 circuits.

The study makes three key contributions: First, it establishes a strong correlative relationship between opto-fMRI and LFP signals in populations of a specific type of neuron (excitatory neurons) in the NHP cortex. Second, we achieved precise, neuron-type-specific activation of excitatory neurons through optogenetic stimulation, enabling the dissection of causal connections in the primate S2 circuits. This advancement opens new avenues for future research into neuron-type-specific microcircuitry and large-scale cortical networks using opto-fMRI. Third, our findings provide both electrophysiological and fMRI evidence supporting the pivotal role of the S2 cortex in connecting functionally distinct brain networks for the integration and processing of touch and pain-related information. The translational relevance of this work lies in its potential to inform the development of targeted neuromodulation therapies for neurological and psychiatric disorders in humans. The success of this technique in NHPs, combined with advancements in noninvasive AAV delivery tools, highlights promising future clinical applications for treating brain-related disorders.

Acknowledgments

These studies were funded by NIH NINDS grant 2R01NS078680. We thank Ms. Chaohui Tang for her assistance with animal care and data acquisition, and we acknowledge the valuable contributions of Huixin Qi and Laura Trace to the immunohistology work.

Author Contributions Performed the experiments: P-F. Yang, F. Wang; analyzed the data: P-F. Yang, Z. Yang, N. Zheng; writing-original draft: P-F. Yang, L. M. Chen; writing-reviewing & editing: J. C. Gore, L. M. Chen.

Conflict of Interest All authors claim no competing interests in publishing the observations. The author has no conflicts of interest to declare that are relevant to the content of this chapter.

References

1. Culotta L, Penzes P (2020) Exploring the mechanisms underlying excitation/inhibition imbalance in human iPSC-derived models of ASD. Mol Autism 11(1):32. https://doi.org/10.1186/s13229-020-00339-0
2. Yizhar O, Fenno LE, Prigge M, Schneider F, Davidson TJ, O'Shea DJ, Sohal VS, Goshen I, Finkelstein J, Paz JT, Stehfest K, Fudim R, Ramakrishnan C, Huguenard JR, Hegemann P, Deisseroth K (2011) Neocortical excitation/inhibition balance in information processing and social dysfunction. Nature 477(7363):171–178. https://doi.org/10.1038/nature10360
3. Kaul D, Schwab SG, Mechawar N, Ooi L, Matosin N (2022) Alterations in astrocytic regulation of excitation and inhibition by stress exposure and in severe psychopathology. J Neurosci 42(36):6823–6834. https://doi.org/10.1523/JNEUROSCI.2410-21.2022
4. Schweinhardt P, Seminowicz DA, Jaeger E, Duncan GH, Bushnell MC (2009) The anatomy of the mesolimbic reward system: a link between personality and the placebo analgesic response. J Neurosci 29(15):4882–4887. https://doi.org/10.1523/JNEUROSCI.5634-08.2009
5. Kim CK, Adhikari A, Deisseroth K (2017) Integration of optogenetics with complementary methodologies in systems neuroscience. Nat Rev Neurosci 18(4):222–235. https://doi.org/10.1038/nrn.2017.15
6. Boyden ES, Zhang F, Bamberg E, Nagel G, Deisseroth K (2005) Millisecond-timescale, genetically targeted optical control of neural activity. Nat Neurosci 8(9):1263–1268. https://doi.org/10.1038/nn1525
7. Gerits A, Farivar R, Rosen BR, Wald LL, Boyden ES, Vanduffel W (2012) Optogenetically induced behavioral and functional network changes

in primates. Curr Biol 22(18):1722–1726. https://doi.org/10.1016/j.cub.2012.07.023
8. Cavanaugh J, Monosov IE, McAlonan K, Berman R, Smith MK, Cao V, Wang KH, Boyden ES, Wurtz RH (2012) Optogenetic inactivation modifies monkey visuomotor behavior. Neuron 76(5):901–907. https://doi.org/10.1016/j.neuron.2012.10.016
9. Chuang KH, Li Z, Huang HH, Khorasani Gerdekoohi S, Athwal D (2023) Hemodynamic transient and functional connectivity follow structural connectivity and cell type over the brain hierarchy. Proc Natl Acad Sci USA 120(5):e2202435120. https://doi.org/10.1073/pnas.2202435120
10. Lee JH, Durand R, Gradinaru V, Zhang F, Goshen I, Kim DS, Fenno LE, Ramakrishnan C, Deisseroth K (2010) Global and local fMRI signals driven by neurons defined optogenetically by type and wiring. Nature 465(7299):788–792. https://doi.org/10.1038/nature09108
11. Desai M, Kahn I, Knoblich U, Bernstein J, Atallah H, Yang A, Kopell N, Buckner RL, Graybiel AM, Moore CI, Boyden ES (2011) Mapping brain networks in awake mice using combined optical neural control and fMRI. J Neurophysiol 105(3):1393–1405. https://doi.org/10.1152/jn.00828.2010
12. Kim S, Moon HS, Vo TT, Kim CH, Im GH, Lee S, Choi M, Kim SG (2023) Whole-brain mapping of effective connectivity by fMRI with cortex-wide patterned optogenetics. Neuron 111(11):1732–1747. e1736. https://doi.org/10.1016/j.neuron.2023.03.002
13. Klink PC, Aubry JF, Ferrera VP, Fox AS, Froudist-Walsh S, Jarraya B, Konofagou EE, Krauzlis RJ, Messinger A, Mitchell AS, Ortiz-Rios M, Oya H, Roberts AC, Roe AW, Rushworth MFS, Sallet J, Schmid MC, Schroeder CE, Tasserie J, Tsao DY, Uhrig L, Vanduffel W, Wilke M, Kagan I, Petkov CI (2021) Combining brain perturbation and neuroimaging in non-human primates. Neuroimage 235:118017. https://doi.org/10.1016/j.neuroimage.2021.118017
14. Galvan A, Stauffer WR, Acker L, El-Shamayleh Y, Inoue KI, Ohayon S, Schmid MC (2017) Nonhuman primate optogenetics: recent advances and future directions. J Neurosci 37(45):10894–10903. https://doi.org/10.1523/JNEUROSCI.1839-17.2017
15. Taub DG, Jiang Q, Pietrafesa F, Su J, Carroll A, Greene C, Blanchard MR, Jain A, El-Rifai M, Callen A, Yager K, Chung C, He Z, Chen C, Woolf CJ (2024) The secondary somatosensory cortex gates mechanical and heat sensitivity. Nat Commun 15(1):1289. https://doi.org/10.1038/s41467-024-45729-7
16. Wu R, Wang F, Yang PF, Chen LM (2017) High-resolution functional MRI identified distinct global intrinsic functional networks of nociceptive posterior insula and S2 regions in squirrel monkey brain. Neuroimage 155:147–158. https://doi.org/10.1016/j.neuroimage.2017.04.067
17. Wu R, Wang F, Yang PF, Gore JC, Chen LM (2022) Graph theory analysis identified two hubs that connect sensorimotor and cognitive and cortical and subcortical nociceptive networks in the non-human primate. Neuroimage 257:119244. https://doi.org/10.1016/j.neuroimage.2022.119244
18. Ye X, Yang PF, Liu Q, Dillenburger BD, Friedman RM, Chen LM (2021) A thermal nociceptive patch in the S2 cortex of nonhuman primates: a combined functional magnetic resonance imaging and electrophysiology study. Pain 162(11):2705–2716. https://doi.org/10.1097/j.pain.0000000000002247
19. Maihofner C, Herzner B, Otto Handwerker H (2006) Secondary somatosensory cortex is important for the sensory-discriminative dimension of pain: a functional MRI study. Eur J Neurosci 23(5):1377–1383. https://doi.org/10.1111/j.1460-9568.2006.04632.x
20. Condylis C, Lowet E, Ni J, Bistrong K, Ouellette T, Josephs N, Chen JL (2020) Context-dependent sensory processing across primary and secondary somatosensory cortex. Neuron 106(3):515–525. e515. https://doi.org/10.1016/j.neuron.2020.02.004
21. Ishida H, Grandi LC, Fornia L (2024) Secondary somatosensory and posterior insular cortices: a somatomotor hub for object prehension and manipulation movements. Front Integr Neurosci 18:1346968. https://doi.org/10.3389/fnint.2024.1346968
22. Chernov MM, Friedman RM, Chen G, Stoner GR, Roe AW (2018) Functionally specific optogenetic modulation in primate visual cortex. Proc Natl Acad Sci USA 115(41):10505–10510. https://doi.org/10.1073/pnas.1802018115
23. O'Shea DJ, Kalanithi P, Ferenczi EA, Hsueh B, Chandrasekaran C, Goo W, Diester I, Ramakrishnan C, Kaufman MT, Ryu SI, Yeom KW, Deisseroth K, Shenoy KV (2018) Development of an optogenetic toolkit for neural circuit dissection in squirrel monkeys. Sci Rep Apr 30;8(1):6775-6795. https://doi.org/10.1038/s41598-018-24362-7
24. Kaas JH, Nelson RJ, Sur M, Dykes RW, Merzenich MM (1984) The somatotopic organization of the ventroposterior thalamus of the squirrel monkey, *Saimiri sciureus*. J Comp Neurol 226(1):111–140. https://doi.org/10.1002/cne.902260109
25. Wang Z, Qi HX, Kaas JH, Roe AW, Chen LM (2013) Functional signature of recovering cortex: dissociation of local field potentials and spiking activity in somatosensory cortices of spinal cord injured monkeys. Exp Neurol 249:132–143. https://doi.org/10.1016/j.expneurol.2013.08.013

26. Jain N, Qi HX, Collins CE, Kaas JH (2008) Large-scale reorganization in the somatosensory cortex and thalamus after sensory loss in macaque monkeys. J Neurosci 28(43):11042–11060. https://doi.org/10.1523/JNEUROSCI.2334-08.2008
27. Hussar CR, Pasternak T (2009) Flexibility of sensory representations in prefrontal cortex depends on cell type. Neuron 64(5):730–743. https://doi.org/10.1016/j.neuron.2009.11.018
28. Jacob SN, Ott T, Nieder A (2013) Dopamine regulates two classes of primate prefrontal neurons that represent sensory signals. J Neurosci 33(34):13724–13734. https://doi.org/10.1523/JNEUROSCI.0210-13.2013
29. Thiele A, Brandt C, Dasilva M, Gotthardt S, Chicharro D, Panzeri S, Distler C (2016) Attention induced gain stabilization in broad and narrow-spiking cells in the frontal eye-field of Macaque monkeys. J Neurosci 36(29):7601–7612. https://doi.org/10.1523/JNEUROSCI.0872-16.2016
30. Chen LM, Mishra A, Yang PF, Wang F, Gore JC (2015) Injury alters intrinsic functional connectivity within the primate spinal cord. Proc Natl Acad Sci USA 112(19):5991–5996. https://doi.org/10.1073/pnas.1424106112
31. Sengupta A, Mishra A, Wang F, Chen LM, Gore JC (2024) Characteristic BOLD signals are detectable in white matter of the spinal cord at rest and after a stimulus. Proc Natl Acad Sci USA 121(22):e2316117121. https://doi.org/10.1073/pnas.2316117121
32. Wu TL, Mishra A, Wang F, Yang PF, Gore JC, Chen LM (2016) Effects of isoflurane anesthesia on resting-state fMRI signals and functional connectivity within primary somatosensory cortex of monkeys. Brain Behav 6(12):e00591. https://doi.org/10.1002/brb3.591
33. Wu TL, Yang PF, Wang F, Shi Z, Mishra A, Wu R, Chen LM, Gore JC (2019) Intrinsic functional architecture of the non-human primate spinal cord derived from fMRI and electrophysiology. Nat Commun 10(1):1416. https://doi.org/10.1038/s41467-019-09485-3
34. Wang Z, Chen LM, Negyessy L, Friedman RM, Mishra A, Gore JC, Roe AW (2013) The relationship of anatomical and functional connectivity to resting-state connectivity in primate somatosensory cortex. Neuron 78(6):1116–1126. https://doi.org/10.1016/j.neuron.2013.04.023
35. Sengupta A, Wang F, Mishra A, Reed JL, Chen LM, Gore JC (2023) Detection and characterization of resting state functional networks in squirrel monkey brain. Cereb Cortex Commun 4(3):tgad018. https://doi.org/10.1093/texcom/tgad018
36. Sengupta A, Yang PF, Reed JL, Mishra A, Wang F, Manzanera Esteve IV, Yang Z, Chen LM, Gore JC (2025) Correspondence between thalamic injury-induced changes in resting-state fMRI of monkeys and their sensorimotor behaviors and neural activities. Neuroimage Clin 45:103753. https://doi.org/10.1016/j.nicl.2025.103753
37. Friston K, Ashburner J, Kiebel S, Nichols T, Penny W (2007) Statisitical parametric mapping: the analysis of functional brain images. Academic Press, London and San Diego. https://doi.org/10.1016/B978-0-12-372560-8.X5000-1
38. Mishra A, Majumdar S, Wang F, Wilson GH 3rd, Gore JC, Chen LM (2019) Functional connectivity with cortical depth assessed by resting state fMRI of subregions of S1 in squirrel monkeys. Hum Brain Mapp 40(1):329–339. https://doi.org/10.1002/hbm.24375
39. Glover GH, Li TQ, Ress D (2000) Image-based method for retrospective correction of physiological motion effects in fMRI: RETROICOR. Magn Reson Med 44(1):162–167. 10.1002/1522-2594(200007)44:1<162::aid-mrm23>3.0.co;2-e
40. Smith SM, Jenkinson M, Woolrich MW, Beckmann CF, Behrens TE, Johansen-Berg H, Bannister PR, De Luca M, Drobnjak I, Flitney DE, Niazy RK, Saunders J, Vickers J, Zhang Y, De Stefano N, Brady JM, Matthews PM (2004) Advances in functional and structural MR image analysis and implementation as FSL. Neuroimage 23(Suppl 1):S208–S219. https://doi.org/10.1016/j.neuroimage.2004.07.051
41. Benjamini Y, Hochberg Y (1995) Controlling the false discovery rate: a practical and powerful approach to multiple testing. J R Stat Soc B (Methodol) 57(1):289–300. https://doi.org/10.1111/j.2517-6161.1995.tb02031.x
42. Woolrich MW, Behrens TE, Beckmann CF, Jenkinson M, Smith SM (2004) Multilevel linear modelling for fMRI group analysis using Bayesian inference. Neuroimage 21(4):1732–1747. https://doi.org/10.1016/j.neuroimage.2003.12.023
43. Sengupta A, Mishra A, Wang F, Li M, Yang PF, Chen LM, Gore JC (2021) Functional networks in non-human primate spinal cord and the effects of injury. Neuroimage 240:118391. https://doi.org/10.1016/j.neuroimage.2021.118391
44. Shin Y, Yoo M, Kim HS, Nam SK, Kim HI, Lee SK, Kim S, Kwon HS (2016) Characterization of fiber-optic light delivery and light-induced temperature changes in a rodent brain for precise optogenetic neuromodulation. Biomed Opt Express 7(11):4450–4471. https://doi.org/10.1364/BOE.7.004450
45. Albers F, Wachsmuth L, Schache D, Lambers H, Faber C (2019) Functional MRI readouts from BOLD and diffusion measurements differentially respond to optogenetic activation

and tissue heating. Front Neurosci 13:1104. https://doi.org/10.3389/fnins.2019.01104
46. Luo H, Yang Z, Yang PF, Wang F, Reed JL, Gore JC, Grissom WA, Chen LM (2023) Detection of laser-associated heating in the brain during simultaneous fMRI and optogenetic stimulation. Magn Reson Med 89(2):729–737. https://doi.org/10.1002/mrm.29464
47. Cardozo Pinto DF, Lammel S (2019) Hot topic in optogenetics: new implications of in vivo tissue heating. Nat Neurosci 22(7):1039–1041. https://doi.org/10.1038/s41593-019-0426-z
48. Owen SF, Liu MH, Kreitzer AC (2019) Thermal constraints on in vivo optogenetic manipulations. Nat Neurosci 22(7):1061–1065. https://doi.org/10.1038/s41593-019-0422-3
49. Arias-Gil G, Ohl FW, Takagaki K, Lippert MT (2016) Measurement, modeling, and prediction of temperature rise due to optogenetic brain stimulation. Neurophotonics 3(4):045007. https://doi.org/10.1117/1.NPh.3.4.045007
50. Johansson JD (2010) Spectroscopic method for determination of the absorption coefficient in brain tissue. J Biomed Opt 15(5):057005. https://doi.org/10.1117/1.3495719
51. Stujenske JM, Spellman T, Gordon JA (2015) Modeling the spatiotemporal dynamics of light and heat propagation for in vivo optogenetics. Cell Rep 12(3):525–534. https://doi.org/10.1016/j.celrep.2015.06.036
52. Rieke V, Butts Pauly K (2008) MR thermometry. J Magn Reson Imaging 27(2):376–390. https://doi.org/10.1002/jmri.21265
53. Vigen KK, Daniel BL, Pauly JM, Butts K (2003) Triggered, navigated, multi-baseline method for proton resonance frequency temperature mapping with respiratory motion. Magn Reson Med 50(5):1003–1010. https://doi.org/10.1002/mrm.10608
54. Rieke V, Vigen KK, Sommer G, Daniel BL, Pauly JM, Butts K (2004) Referenceless PRF shift thermometry. Magn Reson Med 51(6):1223–1231. https://doi.org/10.1002/mrm.20090
55. Grissom WA, Rieke V, Holbrook AB, Medan Y, Lustig M, Santos J, McConnell MV, Pauly KB (2010) Hybrid referenceless and multibaseline subtraction MR thermometry for monitoring thermal therapies in moving organs. Med Phys 37(9):5014–5026. https://doi.org/10.1118/1.3475943
56. Rieke V, Instrella R, Rosenberg J, Grissom W, Werner B, Martin E, Pauly KB (2013) Comparison of temperature processing methods for monitoring focused ultrasound ablation in the brain. J Magn Reson Imaging 38(6):1462–1471. https://doi.org/10.1002/jmri.24117
57. Yang PF, Reed J, Yang Z, Wang F, Zheng N, Gore JC, Chen LM (2025) Multimodal correspondence between optogenetic fMRI, electrophysiology, and anatomical maps of the secondary somatosensory cortex in non-human primates. J Neurosci. https://doi.org/10.1523/JNEUROSCI.2375-24.2025
58. Fenno L, Yizhar O, Deisseroth K (2011) The development and application of optogenetics. Annu Rev Neurosci 34:389–412. https://doi.org/10.1146/annurev-neuro-061010-113817
59. Deisseroth K, Etkin A, Malenka RC (2015) Optogenetics and the circuit dynamics of psychiatric disease. JAMA 313(20):2019–2020. https://doi.org/10.1001/jama.2015.2544
60. Mitchell JF, Sundberg KA, Reynolds JH (2007) Differential attention-dependent response modulation across cell classes in macaque visual area V4. Neuron 55(1):131–141. https://doi.org/10.1016/j.neuron.2007.06.018
61. Scheyltjens I, Laramee ME, Van den Haute C, Gijsbers R, Debyser Z, Baekelandt V, Vreysen S, Arckens L (2015) Evaluation of the expression pattern of rAAV2/1, 2/5, 2/7, 2/8, and 2/9 serotypes with different promoters in the mouse visual cortex. J Comp Neurol 523(14):2019–2042. https://doi.org/10.1002/cne.23819
62. Radnikow G, Feldmeyer D (2018) Layer- and cell type-specific modulation of excitatory neuronal activity in the neocortex. Front Neuroanat 12:1. https://doi.org/10.3389/fnana.2018.00001
63. Lubke J, Feldmeyer D (2007) Excitatory signal flow and connectivity in a cortical column: focus on barrel cortex. Brain Struct Funct 212(1):3–17. https://doi.org/10.1007/s00429-007-0144-2
64. Gabbott PL, Dickie BG, Vaid RR, Headlam AJ, Bacon SJ (1997) Local-circuit neurones in the medial prefrontal cortex (areas 25, 32 and 24b) in the rat: morphology and quantitative distribution. J Comp Neurol 377(4):465–499. 10.1002/(sici)1096-9861(19970127)377:4<465::aid-cne1>3.0.co;2-0
65. Rudy B, Fishell G, Lee S, Hjerling-Leffler J (2011) Three groups of interneurons account for nearly 100% of neocortical GABAergic neurons. Dev Neurobiol 71(1):45–61. https://doi.org/10.1002/dneu.20853
66. Hayashi A, Yoshida T, Ohki K (2018) Cell type specific representation of vibro-tactile stimuli in the mouse primary somatosensory cortex. Front Neural Circuits 12:109. https://doi.org/10.3389/fncir.2018.00109
67. Markram H, Toledo-Rodriguez M, Wang Y, Gupta A, Silberberg G, Wu C (2004) Interneurons of the neocortical inhibitory system. Nat Rev Neurosci 5(10):793–807. https://doi.org/10.1038/nrn1519

68. Wu R, Yang PF, Wang F, Liu Q, Gore JC, Chen LM (2022) Differential recovery of submodality touch neurons and interareal communication in sensory input-deprived area 3b and S2 cortices. J Neurosci 42(50):9330–9342. https://doi.org/10.1523/jneurosci.0034-22.2022
69. Morecraft RJ, Stilwell-Morecraft KS, Cipolloni PB, Ge J, McNeal DW, Pandya DN (2012) Cytoarchitecture and cortical connections of the anterior cingulate and adjacent somatomotor fields in the rhesus monkey. Brain Res Bull 87(4–5):457–497. https://doi.org/10.1016/j.brainresbull.2011.12.005
70. Morecraft RJ, Cipolloni PB, Stilwell-Morecraft KS, Gedney MT, Pandya DN (2004) Cytoarchitecture and cortical connections of the posterior cingulate and adjacent somatosensory fields in the rhesus monkey. J Comp Neurol 469(1):37–69. https://doi.org/10.1002/cne.10980
71. Yeterian EH, Pandya DN (1993) Striatal connections of the parietal association cortices in rhesus monkeys. J Comp Neurol 332(2):175–197. https://doi.org/10.1002/cne.903320204
72. Liao CC, Gharbawie OA, Qi H, Kaas JH (2013) Cortical connections to single digit representations in area 3b of somatosensory cortex in squirrel monkeys and prosimian galagos. J Comp Neurol 521(16):3768–3790. https://doi.org/10.1002/cne.23377
73. Cerkevich CM, Kaas JH (2019) Corticocortical projections to area 1 in squirrel monkeys (*Saimiri sciureus*). Eur J Neurosci 49(8):1024–1040. https://doi.org/10.1111/ejn.13884
74. Guldin WO, Akbarian S, Grusser OJ (1992) Cortico-cortical connections and cytoarchitectonics of the primate vestibular cortex: a study in squirrel monkeys (*Saimiri sciureus*). J Comp Neurol 326(3):375–401. https://doi.org/10.1002/cne.903260306
75. Gao Y, Choe AS, Stepniewska I, Li X, Avison MJ, Anderson AW (2013) Validation of DTI tractography-based measures of primary motor area connectivity in the squirrel monkey brain. PLoS One 8(10):e75065. https://doi.org/10.1371/journal.pone.0075065
76. Stevens RT, London SM, Apkarian AV (1993) Spinothalamocortical projections to the secondary somatosensory cortex (SII) in squirrel monkey. Brain Res 631(2):241–246. https://doi.org/10.1016/0006-8993(93)91541-y
77. Delhaye BP, Long KH, Bensmaia SJ (2018) Neural basis of touch and proprioception in primate cortex. Compr Physiol 8(4):1575–1602. https://doi.org/10.1002/cphy.c170033
78. Friedman DP, Murray EA, O'Neill JB, Mishkin M (1986) Cortical connections of the somatosensory fields of the lateral sulcus of macaques: evidence for a corticolimbic pathway for touch. J Comp Neurol 252(3):323–347. https://doi.org/10.1002/cne.902520304
79. Preusser S, Thiel SD, Rook C, Roggenhofer E, Kosatschek A, Draganski B, Blankenburg F, Driver J, Villringer A, Pleger B (2015) The perception of touch and the ventral somatosensory pathway. Brain 138(3):540–548. https://doi.org/10.1093/brain/awu370
80. Dijkerman HC, de Haan EH (2007) Somatosensory processes subserving perception and action. Behav Brain Sci 30(2):189–201; discussion 201–139. https://doi.org/10.1017/S0140525X07001392
81. Sengupta A, Mishra A, Wang F, Chen L, Gore J (2024) Characteristic BOLD signals are detectable in white matter of the spinal cord at rest and after a stimulus. Proc Natl Acad Sci U S A May 28;121(22):e231677121 PMID: 38776372. https://doi.org/10.1073/pnas.2316117121
82. Macias M, Lopez-Virgen V, Olivares-Moreno R, Rojas-Piloni G (2022) Corticospinal neurons from motor and somatosensory cortices exhibit different temporal activity dynamics during motor learning. Front Hum Neurosci 16:1043501. https://doi.org/10.3389/fnhum.2022.1043501
83. Suter BA, Shepherd GM (2015) Reciprocal interareal connections to corticospinal neurons in mouse M1 and S2. J Neurosci 35(7):2959–2974. https://doi.org/10.1523/JNEUROSCI.4287-14.2015
84. Karadimas SK, Satkunendrarajah K, Laliberte AM, Ringuette D, Weisspapir I, Li L, Gosgnach S, Fehlings MG (2020) Sensory cortical control of movement. Nat Neurosci 23(1):75–84. https://doi.org/10.1038/s41593-019-0536-7
85. Frezel N, Platonova E, Voigt FF, Mateos JM, Kastli R, Ziegler U, Karayannis T, Helmchen F, Wildner H, Zeilhofer HU (2020) In-depth characterization of layer 5 output neurons of the primary somatosensory cortex innervating the mouse dorsal spinal cord. Cereb Cortex Commun 1(1):tgaa052. https://doi.org/10.1093/texcom/tgaa052
86. Galea MP, Darian-Smith I (1994) Multiple corticospinal neuron populations in the macaque monkey are specified by their unique cortical origins, spinal terminations, and connections. Cereb Cortex 4(2):166–194. https://doi.org/10.1093/cercor/4.2.166
87. Kameda H, Murabe N, Odagaki K, Mizukami H, Ozawa K, Sakurai M (2019) Differential innervation within a transverse plane of spinal gray matter by sensorimotor cortices, with special reference to the somatosensory cortices. J Comp Neurol 527(8):1401–1415. https://doi.org/10.1002/cne.24626
88. Orban GA (2016) Functional definitions of parietal areas in human and non-human primates.

Proc Biol Sci Apr 13;283(1828-1837). https://doi.org/10.1098/rspb.2016.0118

89. Weidensteiner C, Quesson B, Caire-Gana B, Kerioui N, Rullier A, Trillaud H, Moonen CT (2003) Real-time MR temperature mapping of rabbit liver in vivo during thermal ablation. Magn Reson Med 50(2):322–330. https://doi.org/10.1002/mrm.10521
90. Odeen H, Parker DL (2019) Improved MR thermometry for laser interstitial thermotherapy. Lasers Surg Med 51(3):286–300. https://doi.org/10.1002/lsm.23049
91. Tremblay S, Acker L, Afraz A, Albaugh DL, Amita H, Andrei AR, Angelucci A, Aschner A, Balan PF, Basso MA, Benvenuti G, Bohlen MO, Caiola MJ, Calcedo R, Cavanaugh J, Chen Y, Chen S, Chernov MM, Clark AM, Dai J, Debes SR, Deisseroth K, Desimone R, Dragoi V, Egger SW, Eldridge MAG, El-Nahal HG, Fabbrini F, Federer F, Fetsch CR, Fortuna MG, Friedman RM, Fujii N, Gail A, Galvan A, Ghosh S, Gieselmann MA, Gulli RA, Hikosaka O, Hosseini EA, Hu X, Huer J, Inoue KI, Janz R, Jazayeri M, Jiang R, Ju N, Kar K, Klein C, Kohn A, Komatsu M, Maeda K, Martinez-Trujillo JC, Matsumoto M, Maunsell JHR, Mendoza-Halliday D, Monosov IE, Muers RS, Nurminen L, Ortiz-Rios M, O'Shea DJ, Palfi S, Petkov CI, Pojoga S, Rajalingham R, Ramakrishnan C, Remington ED, Revsine C, Roe AW, Sabes PN, Saunders RC, Scherberger H, Schmid MC, Schultz W, Seidemann E, Senova YS, Shadlen MN, Sheinberg DL, Siu C, Smith Y, Solomon SS, Sommer MA, Spudich JL, Stauffer WR, Takada M, Tang S, Thiele A, Treue S, Vanduffel W, Vogels R, Whitmire MP, Wichmann T, Wurtz RH, Xu H, Yazdan-Shahmorad A, Shenoy KV, DiCarlo JJ, Platt ML (2020) An open resource for non-human primate optogenetics. Neuron 108(6):1075–1090. e1076. https://doi.org/10.1016/j.neuron.2020.09.027
92. O'Shea DJ, Kalanithi P, Ferenczi EA, Hsueh B, Chandrasekaran C, Goo W, Diester I, Ramakrishnan C, Kaufman MT, Ryu SI, Yeom KW, Deisseroth K, Shenoy KV (2018) Development of an optogenetic toolkit for neural circuit dissection in squirrel monkeys. Sci Rep 8(1):6775. https://doi.org/10.1038/s41598-018-24362-7

Check for updates

Chapter 16

Leveraging Optogenetic fMRI for Understanding Disease Processes in Neurological Disorders

Michael A. Kelberman, Nmachi Anumba, Lauren Daley, Alexia Marriott, Wen-Ju Pan, David Weinshenker and Shella Keilholz

Abstract

Optogenetics and functional magnetic resonance imaging (fMRI) are two tools that have become indispensable for neuroscientists, though they have traditionally been used to ask distinct questions regarding neurophysiology. Namely, optogenetics is used to manipulate neuronal activity by activating or inhibiting specific brain regions and/or neuronal subtypes, while fMRI detects and quantifies brain-wide activity and dynamics under different conditions. However, when combined, optogenetics and fMRI (opto-fMRI) can be a powerful technique for understanding the brain-wide effects of stimulating or inhibiting a specific brain region/neuronal subtype. Opto-fMRI has mainly been applied under "baseline" conditions but holds promise for understanding pathological processes in neurological disorders. Here, we describe the general setup for opto-fMRI experiments and their potential for translational relevance, using our own experiments as a proof-of-concept. We stress the importance of carefully considered control groups and proper confirmation of optogenetic function in employing opto-fMRI. Moreover, beyond traditional measures of static functional connectivity, additional analyses of dynamic functional connectivity could shed light on currently unexplored disease processes and potential therapeutic avenues. Finally, although we highlight the application of opto-fMRI using the locus coeruleus–norepinephrine system in an Alzheimer's disease model rat as an example, this technique is broadly applicable for the study of other neurological and neuropsychiatric diseases.

Keywords Optogenetics, Functional MRI (fMRI), Locus coeruleus, Neurological disorders, Alzheimer's disease, Dynamic functional connectivity

1 Introduction: Combining Optogenetics and Functional Magnetic Resonance Imaging for Studying Neurological Disorders

Optogenetics and functional magnetic resonance imaging (fMRI) are two seemingly disparate tools that have become staples in neuroscience research. Optogenetics is perhaps the most widely used technique for interrogating causal relationships between the activity of specific neuronal populations and behavior/physiology. It involves the use of opsins, which are light-gated ion channels or pumps that

Kevin Chuen Wing Chan and Seong-Gi Kim (eds.), *Optogenetic Functional Neuroimaging*, Neuromethods, vol. 232, https://doi.org/10.1007/978-1-0716-5178-0_16,

open or close in response to specific wavelengths of light. These opsins are expressed in neuronal populations of interest via viral vector or transgene, along with an optic ferrule implant to deliver light for modulating activity. Opsins come in many different varieties that enable specific electrochemical responses based on their subclass. The most common opsins are excitatory channelrhodopsins, which depolarize neurons by allowing the influx of cations when stimulated. There are also channelrhodopsins, halorhodopsins, and archaerhodopsins that inhibit neuronal activity when stimulated by hyperpolarizing neurons. Other variants, such as red-shifted and step-function options, afford different kinetics, deeper light penetration, and the ability for bidirectional control of neural activity. Such modulation can be performed at either the level of the cell bodies, affecting efferent connections of all activated neurons, or along axon segments and nerve terminals to target specific neuronal pathways. The major benefit of optogenetics compared to other neural modulation techniques like chemogenetics is the ability to precisely control the timing and pattern of stimulation to mimic physiological conditions.

On the other hand, MRI is used to assess the structure and function of biological systems. Clinical applications involve primarily structural examinations like volumetric analysis, diffusion tensor imaging, and contrast. MRI is minimally invasive, widely available, and offers superior spatial resolution compared to other techniques like positron emission tomography. fMRI occupies a distinct niche within MRI, using changes in blood oxygenation as a proxy for brain activity. The basic premise behind fMRI is that neurons require oxygen when they become more metabolically active, which is delivered via the blood supply. The influx of oxygenated blood causes differences in the magnetic resonance signal, which can be read by MRI machinery. Changes in activity can now be detected on sub-second timescales within submillimeter voxel sizes, and these resolution scales are improving constantly. Importantly, fMRI is a valid technique to probe whole-brain networks simultaneously, unlike approaches that are optimized for a few to a few hundred cells in one brain region, such as electrophysiology and optical techniques, and can be performed at rest or during tasks.

fMRI and optogenetics have each been employed in numerous organisms and models of neurological disorders, allowing unprecedented insights into the neural circuits underlying behavior and its dysfunction in disease states [1–11]. In human studies and clinical settings, fMRI can be used to help diagnose neurological disorders. When used in combination with interventions, such as pharmacology, fMRI can be leveraged to understand pathological processes and their response to treatment. A major advantage of fMRI is that it can be deployed in animal models while maintaining translational relevance to humans and disease states [12, 13]. In such animal studies, interventions can be even more fine-tuned and controlled, especially with optogenetics. The combination of optogenetics and fMRI (opto-fMRI), therefore, provides a unique opportunity for researchers to probe brain-wide circuits in healthy and disease states through the selective modulation of a singular brain region with cell-type specificity [8–11, 14–20]. Opto-fMRI has been used to study the mechanisms underlying deep brain stimulation for Parkinson's disease and neuropsychiatric disorders [8, 9], the involvement of the anterior cingulate cortex in chronic pain [21], stimulation of the motor cortex to promote poststroke recovery [11], and modulating the progression of α-synuclein pathology [10]. However, studies using opto-fMRI have yet to achieve their full potential for exploring disease processes.

The opto-fMRI approach could be particularly useful for studying subcortical nuclei, which have widespread efferent neuromodulatory projection patterns and exhibit vulnerability and extensive dysfunction in multiple neurological disorders [22–24]. Such nuclei are difficult to access using fMRI alone, given their small size and location deep within the brain [22, 25–27], but are routinely manipulated using optogenetics [8, 9, 14–16, 28]. Our labs are particularly interested in dysfunction of the locus coeruleus–norepinephrine (LC–NE) system across a variety of disorders. Our opto-fMRI studies focus on Alzheimer's disease (AD), employing the TgF344-AD rat model that contains a transgene expressing mutant human amyloid precursor protein (APPswe) and presenilin-1 (PS1ΔE9) mutations [29]. Importantly for our purposes, like humans, these rats develop hyperphosphorylated tau in the LC prior to exhibiting amyloid or tau pathology in other brain regions [30]. This tau pathology results in early and progressive deficits in electrophysiological and molecular properties of LC neurons, and behavioral deficits reminiscent of AD [30–33]. We have shown loss of LC fibers/terminals and LC neuron hypoactivity in 15-month TgF344-AD rats, which is associated with spatial reversal learning deficits [30–32]. Behavioral rescue can be achieved by chemogenetically stimulating the LC in these aged rats [30]. Through the use of opto-fMRI, we can selectively modulate LC activity and probe impacts on brain-wide dynamics, which are altered in TgF344-AD rats [34–40]. In the following chapter, we will broadly cover the implementation of opto-fMRI for studying neurological disorders, and then use our lab's work to delve deeper into the applications and potential of this technique for studying nervous system disorders.

2 General Applications of Opto-fMRI

The first component of designing experiments using opto-fMRI is to assess whether such methodology is appropriate to answer the specific question being asked. There are several broad categories of investigation that are well-suited opto-fMRI, some of which include:

- *What are the brain-wide effects of activity within a single brain region?* Opto-fMRI takes advantage of the divergent scales of study using these two techniques, specifically the targeted stimulation of optogenetics paired with monitoring brain-wide activity with fMRI. This can be used to investigate how stimulation of distinct brain regions impacts global brain states [41–46]. These studies are especially powerful when employed in a subregion (i.e., dorsal vs. ventral hippocampus), cell-type (i.e., dopaminergic and non-dopaminergic ventral tegmental area neurons), or circuit-specific manner (i.e., stimulating axonal segments/terminals in a particular output region) [11, 17, 41, 43, 47, 48].
- *How do specific rates and patterns of neural activity impact functional connectivity?* One of the major advantages of optogenetics is the ability to precisely control the rate, timing, and pattern of stimulation to mimic physiological conditions. With opto-fMRI, studies could differentiate how different types of neural activity within the same region may be responsible for different behavioral/physiological responses. LC neurons fire in tonic and phasic (burst-like) modes, and one study showed that these distinct patterns of LC stimulation caused disparate changes in functional connectivity [15]. Specifically, tonic stimulation targeted associative regions, while burst-like stimulation recruited sensory regions. Another study examined changes in brain network dynamics as a function of different stimulation frequencies of the dorsal and intermediate hippocampus [49]. Recruitment of cortical and subcortical networks mainly occurred by stimulation of the intermediate hippocampus and was frequency dependent.
- *What are the neural circuits underlying the processing of sensory information?* Some of the most well-described experiments utilizing opto-fMRI have dissected the circuits involved in sensory processing [21, 44, 50–53]. Such studies have shown enhancement of noise-evoked fMRI signals in the inferior colliculus due to visual cortex pyramidal neuron stimulation, refined the role of the anterior cingulate cortex in gating sensorimotor information related to pain, revealed the recruitment of downstream brain regions in response to stimulation of primary sensory areas, and, conversely, the response of somatosensory areas to stimulation of inhibitory neuronal populations.
- *How does modulating a specific brain region impact disease processes?* Deep brain stimulation has been a revolutionary intervention for Parkinson's disease and neuropsychiatric disorders, but the mechanisms underlying its beneficial effects are debated. Using opto-fMRI, one study showed that stimulation of the subthalamic nucleus improved circling behavior in a rat model of Parkinson's disease [9]. Stimulation also led

to blood-oxygenation level-dependent (BOLD) signal changes in the substantia nigra, globus pallidus, and caudate/putamen, which could be responsible for the behavioral improvements. In a separate study, stimulation of the nucleus accumbens revealed only local increases in cerebral blood volume, suggesting that beneficial effects may be due to antidromic stimulation [8]. Optogenetics has also been used non-concurrently with fMRI to improve functional recovery after stroke and modulate α-synuclein pathology [10, 11]. Specifically, following treatment with optogenetic stimulation, opto-fMRI was used to show positive BOLD signal changes in brain regions previously affected by stroke and those with reduced α-synuclein pathology.

The above are just some of the many questions that can be addressed using opto-fMRI. Our studies fit in the last category, asking how stimulation of a specific brain region (i.e., the LC) could improve behavioral phenotypes by altering or normalizing whole brain networks. Protocols for implementing opto-fMRI have been covered elsewhere [54, 55]. Here, we will provide expanded instructions for the implementation of opto-fMRI for specific use in studying neurological disorders, using our own study as an example, and suggest ways to expand opto-fMRI beyond its current use.

3 Materials

3.1 Animals

Once a concrete hypothesis is made, the next step is to determine what type of model to employ that will best answer the question. Both mice and rats are routinely used for opto-fMRI studies. However, with respect to neurological disorders, rodent models can vary dramatically. For example, rat models of AD, unlike mice, can develop endogenous tau pathology even in the absence of a tauopathy transgene [29]. Furthermore, mouse tau does not contain the full complement of isoforms found in humans and is aggregation-resistant [56, 57]. On the other hand, the suite of genetic tools available for mice is much more extensive than for their rat counterparts. Therefore, models should be chosen with the utmost care, paying careful attention to construct validity with respect to disease-specific characteristics and patterns of temporal progression. Finally, optogenetics, fMRI, and disease models have been or are being developed for use in nonhuman primates, with the potential for increased translational relevance to the human condition once all three can be deployed simultaneously [26, 58–65].

For the purposes of our study investigating the LC–NE system in AD, we utilized TgF344-AD rats on a Fischer background and their wild-type littermates, which were bred in-house at Emory University. We used both sexes, and animals were housed in cages of two to three until being singly housed following optic ferrule implantation. All animals were maintained on a 12-h light/dark cycle (lights on at 7:00 a.m.) and provided ad libitum access to food and water. All procedures were approved by and performed in accordance with protocols from the Institutional Animal Care and Use Committee (IACUC) at Emory University. As previously stated, this model was ideal for our purposes because it displays temporal progression of tau pathology akin to the human condition, heavily impacting the LC–NE system prior to other brain regions [30–32].

3.2 Optogenetics

Equipment for optogenetics will generally include: a virus or transgene to express a specific optogenetic actuator (ideally with a way to achieve cell-type specificity), a laser source, and fiber optic cable, ferrules, and connecting sleeves that are all MRI-compatible. Optogenetic equipment that is fMRI-compatible is commercially available and inherently easier to implement than other techniques that manipulate neural activity, such as electrical stimulation. One of the other major advantages of opto-fMRI is precise control of the rate, pattern, and timing of stimulation to the onset of scanning, which can be achieved through the use of a pulse generator.

We used lentiviruses containing the noradrenergic-specific promoter, PRSx8, to selectively express ChR2-mCherry or mCherry alone in the LC [66–68]. Our optogenetics setup consisted of a DPSS laser (Opto Engine LLC) and MRI-compatible optogenetic equipment (fiber optic cables, fiber optic ferrule, and ceramic split mating sleeves, all from Thorlabs). A Master-8-cp programmable pulse stimulator (AMPI) was used to program and trigger stimulation patterns based on the start of scanning. BNC cables were used to connect the DPSS laser module, the Master-8-cp, and the scanner.

3.3 Complementary Tools and Methods to Verify Optogenetic Stimulation

It is paramount to ensure proper viral targeting, opsin expression, and optic ferrule placement when employing optogenetics. This can be achieved in a variety of ways, but the most common is postmortem histology. Histology can first be used to determine optic fiber placement above or adjacent to the cell population of interest with a basic counterstain. Neuronal activation can also be confirmed with histology via c-Fos staining, the time course of which will need to be optimized for each brain region. These methods will require a way to slice harvested brains, necessary reagents to stain for markers of interest, and a way to image stained brain slices. Pharmacology has also been deployed to confirm optogenetic efficacy through pretreatment with an antagonist to block the effects of stimulation [47, 69, 70]. Behavior is another low-cost way to validate optogenetic stimulation using well-established paradigms, such as observing anxiety-like behaviors following high-tonic LC stimulation or acquisition of self-stimulation during ventral tegmental area stimulation [47, 69]. Finally, stimulation can also be confirmed in real-time with calcium imaging and electrophysiology to measure cell activity or fiber photometry with genetically encoded sensors to monitor changes in neurotransmitter release during stimulation. These may be the most reliable indicators of optogenetic efficacy, especially when determining how stimulation rate and pattern impact cell activity and transmitter release. However, such methods require dedicated setups for data acquisition and processing, in addition to additional viruses needed to express calcium/neurotransmitter indicators, making them more time, labor, and cost-intensive.

In our study, we complemented a conventional postmortem brain tissue analysis approach with a unique feature of the LC–NE system to confirm fidelity of optogenetic stimulation, namely that LC stimulation results in pupil dilation [15, 69, 71, 72]. We employed a low-cost pupillometry setup [72], which consisted of a Raspberry Pi NoIR Camera Module V2 night vision camera, an infrared light source, a Raspberry Pi 4 Model B (CanaKit), and a basic desktop computer.

For immunohistochemistry, we used 0.1 M kPBS and 4% paraformaldehyde for perfusions and stored full brains in 30% sucrose and brain slices in cryoprotectant. When staining, we used 1× PBS, blocking buffer (5% normal goat serum, 3% bovine serum albumin in 0.1% PBS-Triton), and the following antibodies: chicken anti-tyrosine hydroxylase, rabbit anti-c-Fos, rabbit anti-DsRed, goat anti-chicken 633, goat anti-chicken 488, goat anti-rabbit 568, and goat anti-rabbit 488. A Leica DM6000B epifluorescent upright microscope was used to image sections, which were used to visually confirm viral expression and imported into ImageJ for quantification of c-Fos.

3.4 fMRI Data Acquisition and Analysis

MRI scanners, radio frequency coils, and acquisition and analysis parameters and software are variable. Scanners up to 21.1 T have been used successfully in rodents [8, 10, 11, 14, 21, 73, 74]. Many types of coils are available, but transmit–receiver surface coils are also easy to fabricate. Custom-built coils can also be optimized for sensitivity to the brain while minimizing outside sources of noise. Generally, custom-built coils have an open center (unlike many commercial versions), which can better accommodate fiber optic ferrules used for optogenetics. Physiological monitoring is important to ensure the collection of high-quality data and can include instruments to measure real-time heart rate, blood oxygen levels, and temperature during scanning. Rodent cradles are a crucial part of animal setup and physiological monitoring. They can be purchased or custom-designed, but should contain components to head fix the animal (with ear and tooth bars), a nose cone to induce anesthesia, and tubing to circulate water to maintain animal temperature. While a setup for intubation is not strictly necessary, synchronizing breathing frequency with image acquisition can help minimize potential motion artifacts [75].

In our case, all MRI scanning took place using a 9.4 T Bruker Biospec MRI scanner with a 20 cm horizontal bore running on ParaVision 360 software. Animal setup included the use of an intubation rig, MRI cradle, and transmit–receive surface coil, which were all custom-made in-house. During scanning, a pulse oximeter was used to monitor heart rate and blood oxygen levels, while a circulating water bath was used to control animal body temperature. For fMRI, we used isoflurane as an anesthetic in a mixture of O_2 and medical-grade air. The paralytic pancuronium dibromide (Selleck) was administered continuously after intubation, and breathing was maintained at a constant rate using a ventilator. We use open-source software developed in-house in MATLAB for preprocessing, analysis, and visualization of fMRI data [76, 77].

4 Methods

4.1 Surgical Procedures

Unless transgenic lines that express opsins in the regions/cell types of interest are available, animals will receive a viral injection to express the chosen optogenetic actuator in the selected location. Then, either during the same surgery or at a later time, animals will be implanted with an optic fiber targeting the brain region of interest, either at the same location as the viral infusion to manipulate cell body activity, or at a projection site to manipulate axon/terminal activity. The main factor influencing whether viral injections and implants are performed within the same surgery is the amount of time between surgery and testing, with each route having its own considerations. Expression of viral constructs, especially with high expression levels over extended periods, can change brain structure and function [13, 78, 79]. However, some studies may necessitate older animals (1+ years old) to study age-related disorders. The benefits of a two-step surgery in this case are implant stability and the ability to confirm optic fiber placement in real-time (see multimodal optogenetic validation; Fig. 1A). Regardless of the route chosen, careful care must be taken to control for any potential confounding factors.

In our case, all animals received stereotaxic surgery at approximately 2 months of age to infuse PRSx8-ChR2-mCherry or PRSx8-mCherry lentivirus. Rats were anesthetized with 5% isoflurane and transferred to a bite bar, after which isoflurane was lowered to 2% for the remainder of the surgery. Ketoprofen (5 mg/kg, s.c.) was provided directly after incision. A 15° head tilt was employed to avoid contact with the sagittal sinus, a common feature of surgeries targeting the LC in rats. Craniotomies were drilled over the approximate location of the LC, and rats received bilateral injections of lentivirus to target the LC (1.3 μL/hemisphere; AP: −3.8 mm from lambda, ML: ±1.2 mm, DV: −7.0 mm from skull surface). Infusion needles

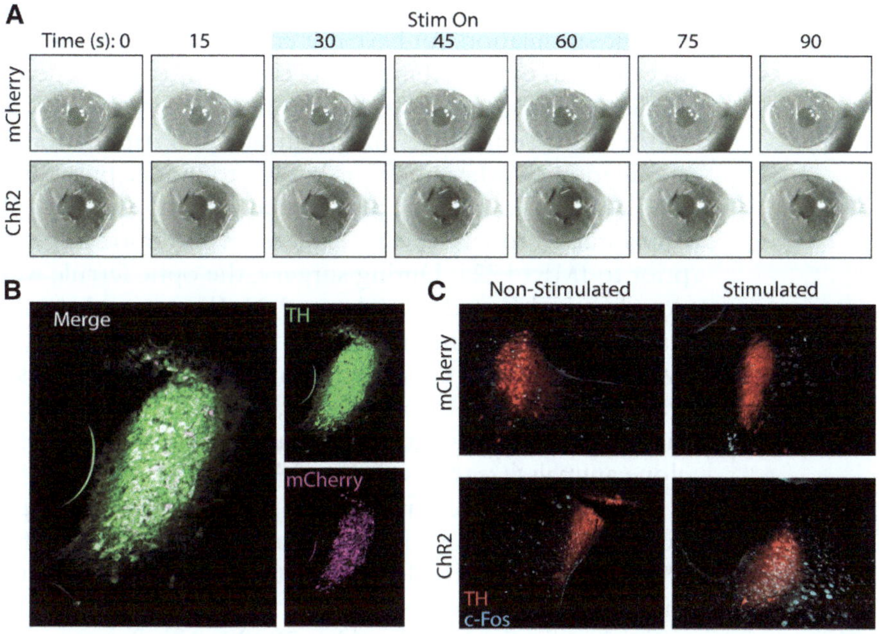

Fig. 1 Real-time and postmortem validation of optogenetic stimulation. (**A**) Pupillometry time series during optogenetic stimulation in animals expressing either mCherry or ChR2 in the locus coeruleus (LC). Robust pupil dilation is only observed during optogenetic stimulation in the ChR2-expressing animal. (**B**) Immunofluorescence displaying viral expression of mCherry in LC neurons. TH - tyrosine hydroxylase. (**C**) c-Fos expression in the stimulated and nonstimulated hemispheres of animals expressing mCherry or ChR2 in the LC. High levels of c-Fos are only observed in the stimulated hemisphere of the ChR2-expressing animal

remained in place for 5 min before being moved dorsally 1 mm and left for another 2 min to allow for diffusion of the virus at the injection site. Rats were then allowed to age until ~5.5 and ~14.5 months prior to optic fiber implant to limit the possibility of the implants falling out during the aging process and to carry out real-time confirmation of optogenetic stimulation.

Approximately 2–4 weeks prior to undergoing MRI scanning, rats underwent a second surgery to implant an optic ferrule above the LC. The implanted hemisphere was randomized and counterbalanced across groups. Rats received ketoprofen (5 mg/kg, s.c.) directly following incision. Then, the skull was implanted with three MRI-compatible screws (*see* **Note 1**). Metabond was applied to the skull to ensure adherence of the headcap, which was composed of dental cement. An optic ferrule was implanted dorsal to the LC (AP: −3.8 mm from lambda, ML: ± 1.2 mm, DV: 6.5–6.8 mm from skull surface). Following proper targeting (see Sect. 4.2), dental cement was used to secure the implant to the skull. Rats were then single-housed until fMRI scanning.

4.2 Multimodal Optogenetic Validation

Since we have mentioned several potential ways for confirming optogenetic stimulation but have not employed them all ourselves, here we will focus on describing the specific methods we undertook in our studies. Our approach was multipronged, leveraging unique features of the LC system (Fig. 1). Specifically, LC stimulation, as well as stimulation of other nuclei, is known to induce pupil dilation [15, 69, 71, 72]. We utilized a pupillometry toolbox during optic ferrule implantation to ensure viral expression and optic ferrule targeting prior to fMRI [72]. During surgery, the optic ferrule was lowered dorsal to the LC, and a Raspberry Pi NoIR Camera Module V2 night vision camera and infrared light source were set up to capture the eye ipsilateral to the implanted hemisphere. Optogenetic stimulation was delivered in a 30 s off, 30 s on, 30 s off protocol to observe the induction of pupil dilation (ChR2 animals) or lack thereof (mCherry alone animals) (*see* **Note 2**, Fig. 1A).

Following the completion of fMRI scanning (see procedures below), rats were immediately perfused with 0.1 M kPBS followed by 4% paraformaldehyde. Perfusions took place approximately 90 min following optogenetic stimulation to allow for optimal detection and quantification of c-Fos [14, 28, 80, 81]. Brains were stored in 4% paraformaldehyde overnight and then moved to 30% sucrose for a minimum of 3 days. Brains were sectioned at 30 μm at the level of the LC. For staining, sections were washed three times for 5 min in 1× PBS followed by 1 h incubation in blocking buffer. Primary antibodies were added for 24–48 h to detect tyrosine hydroxylase (1:1000) and DsRed/mCherry (1:1000) or tyrosine hydroxylase (1:1000) and c-Fos (1:3000). On the second day, sections were again washed three times for 5 min in 1× PBS before the appropriate secondary antibody was added to blocking buffer: goat anti-chicken 488 (1:500) or 633 (1:500) and goat anti-rabbit 568 (1:500). After 2 h, sections were washed in 1× PBS mounted on Colorfrost Plus slides and coverslipped with DAPI Fluoromount-G® (Southern Biotech).

Sections stained for mCherry were checked visually for viral expression (Fig. 1B). c-Fos images were acquired at 10× at the level of the LC using uniform exposure parameters (Fig. 1C) and imported into ImageJ for analysis as previously described [14, 28, 80, 81]. Tyrosine hydroxylase (TH) was used as an LC marker to manually outline the region of interest. A common Otsu threshold was set for all images, followed by despeckling. The number of c-Fos+ cells per region of interest was counted using standard criteria (pixel size: 100-infinity, circularity: 0.7–1.0).

4.3 Opto-fMRI Data Acquisition

Because specific acquisition parameters will depend on MRI hardware and software, in addition to the specific questions being asked, we will cover procedures specific to our study while highlighting areas that we believe are important for standardization across labs to improve translatability. Rats were prepared for MRI scanning by anesthetizing them with 5% isoflurane for 5 min in a combination of

2:3 O_2 and medical-grade air. Isoflurane was then lowered to 2%–3% for intubation. Immediately after intubation, rats were administered the paralytic pancuronium dibromide (1.5 mg/kg/h, s.c.). Animals were placed in the MRI cradle and equipped with a transmit–receive surface coil, maintaining a breathing rate of 1.6 Hz using the ventilator. The optic fiber patch cord was connected to the optic ferrule, and animals were head-fixed to limit motion artifacts (*see* **Note 3**). Eyes were covered with eye ointment prior to scanning, and isoflurane was lowered to 1.3% for the duration of scanning (*see* **Note 4**). Most rodent fMRI studies utilize a combination of low-dose isoflurane and dexmedetomidine for anesthesia, but the latter is an adrenergic agonist known to inhibit LC activity and the effects of LC stimulation on functional connectivity [69]. Therefore, low-dose isoflurane was the best anesthetic regimen for our study. A pulse oximeter was attached to the hindpaw and secured with tape, while a rectal thermometer was used to monitor and maintain the animal's body temperature at 37 ± 0.5°C using the circulating heated water bath. An image of our setup is shown in Fig. 2.

fMRI scans were 10 min gradient-echo echo-planar imaging (EPI) with partial Fourier encoding with a factor of 1.4 (*see* **Note 5**). Images were acquired with the following parameters: isotropic voxel size = 0.5 mm^3, matrix size = 70 × 70, slice number = 24, TE = 15 ms, TR = 1250 ms, flip angle = 90°, and bandwidth = 216.45 kHz. Scans were phase-locked such that each acquisition occurred at every other breath, corresponding to a repetition time of 1250 ms (0.8 Hz), effectively minimizing motion-induced artifacts [82]. Top-up correction

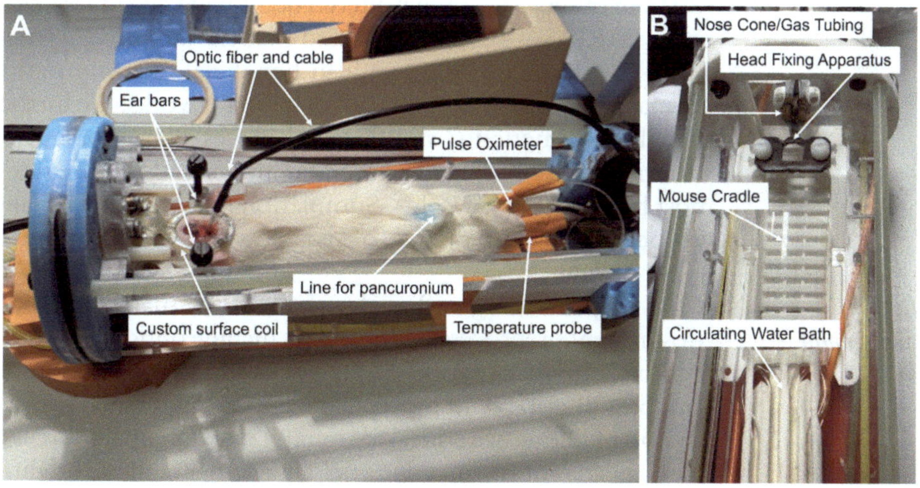

Fig. 2 Opto-fMRI rodent setup. Images depicting our custom rodent opto-fMRI setup for use in (**A**) rats and (**B**) mice. Before scanning, the red head cap would be "blacked" out with a permanent marker, and tape would be used to cover the optic fiber to block escaping light. The pulse generator, syringe pump for continuous pancuronium delivery, water bath for temperature maintenance, ventilator, and physiological monitoring systems are not depicted

was performed using a single-volume reversed blip EPI image with the same acquisition parameters [83, 84]. Each scan was preceded by ten dummy scans for system calibration, and saturation bands were used to minimize non-brain signals.

Our study employed four levels of optogenetic stimulation: baseline (no stimulation), 2 Hz tonic, 5 Hz tonic, and 15 Hz phasic (three pulses of light given at 15 Hz every 10 s). These parameters will vary between studies and brain regions, but were chosen here to mimic the different firing patterns of LC neurons: low tonic, high tonic, and phasic burst firing [23, 85–91]. Each stimulation level was repeated during two separate scans. Furthermore, for the 2 Hz and 5 Hz stimulation scans, stimulation was given either continuously or in a block design (10 s on, 20 s off). Continuous stimulation may be useful when determining the effects of longer-lasting changes to neural activity, such as elevated LC activity in response to stress or novelty, but could exhaust neurons and deplete neurotransmitter levels [92]. Meanwhile, block stimulation allows for quicker manipulation of neural activity and recovery of neurotransmitters. Block designs could also shorten scanning time due to the inclusion of "rest" periods, eliminating the need for baseline scans as long as proper controls are included. Stimulation was applied in a pseudorandomized order with 3 min between scans (*see* **Note 6**). We underscore the choice of stimulation patterns as crucial for the translational relevance of neurological disease. For example, TgF344-AD rat LC neurons display tonic hypoactivity at 6 and 15 months. Evoked phasic activity diverges with age, with hyperactivity and hypoactivity being observed at 6 and 15 months, respectively [31]. Because chemogenetic LC stimulation in 15-month TgF344-AD rats rescues reversal learning [30] and others have shown altered brain-wide dynamics in this strain using fMRI [34–40], we predicted that elevating LC activity in old TgF344-AD rats would result in beneficial reconfiguration (i.e., normalization toward wild-type) of brain networks. We further know that chemogenetic LC stimulation induces a high tonic (~3–4 Hz) pattern of firing [68], indicating that inducing this firing rate may be most beneficial (e.g., 5 Hz). Stimulation for fMRI and pupillometry was performed with 60 mW of laser power and 10-ms-long pulses.

4.4 Opto-fMRI Preprocessing and Analysis

fMRI preprocessing and analysis can follow many different paths. Most rodent preprocessing toolboxes include options for various corrections, nuisance regression, spatial normalization and smoothing, bandpass filtering, and more. Corrections, nuisance regression, and bandpass filtering are critical for minimizing potential sources of noise due to head motion, magnetic field inhomogeneities, and differences in the timing of image collection. Normalization and smoothing permit group-level analyses by warping individual volume scans into a common reference space. Similarly, once preprocessed, analysis can take many forms based on the specific questions being asked (Fig. 3). Two popular output metrics for fMRI are static and dynamic

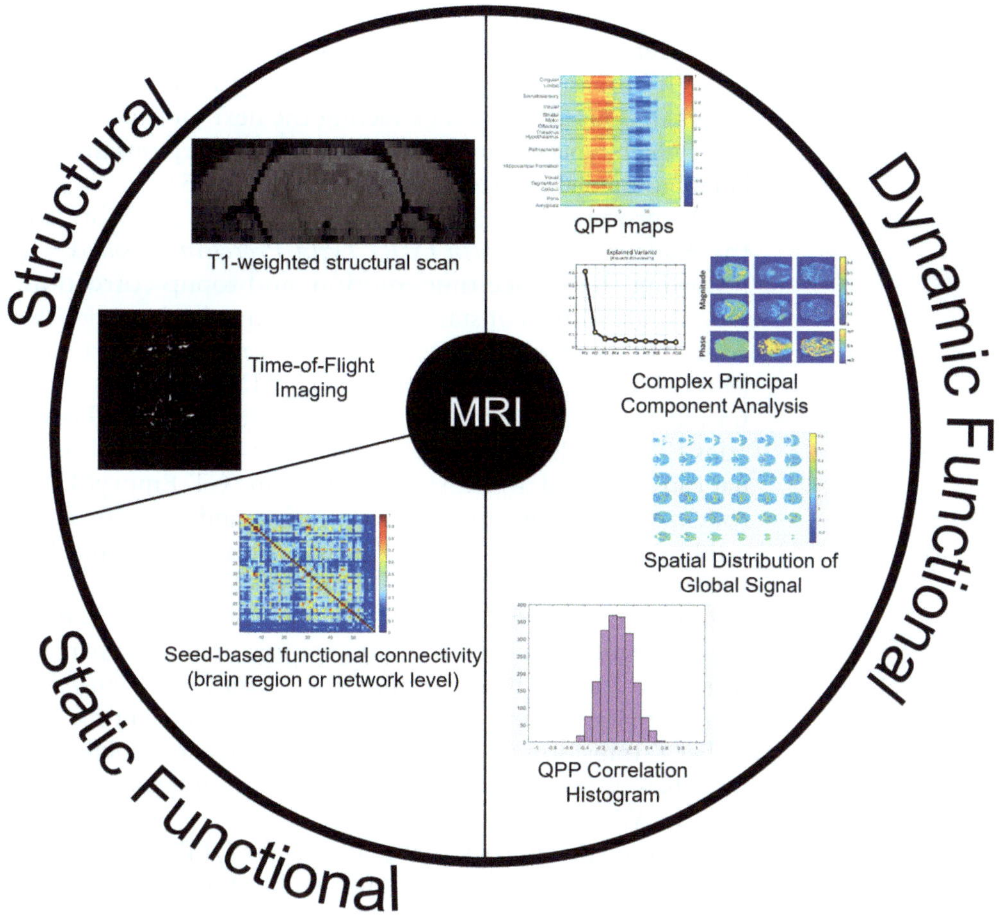

Fig. 3 Overview of analysis techniques possible with MRI. An overview of the potential analysis techniques, both structural and functional, which can be performed on MRI data. In particular, we highlight the wide range of dynamic functional connectivity analyses that can be used for opto-fMRI studies

functional connectivity [93, 94]. Traditional static functional connectivity consists of analyzing average measures of connectivity over an entire scan and is the most common metric reported in opto-fMRI studies. On the other hand, dynamic functional connectivity considers how spatiotemporal patterns evolve over time. Within the domain of dynamic functional connectivity, we focus mainly on quasi-periodic patterns (QPPs). QPPs are spatially structured signals within fMRI that are composed of three primary components [95–98]. The first and most prominent is QPP1, which captures the BOLD signal and is composed of the global signal. The global signal captures the average brain activity of the entire course of a scan and is often used as a nuisance regressor in hopes of removing noise in resting-state fMRI. This practice has been questioned recently, as global signal patterns may reflect relevant neural information [99–102]. For this reason, we preprocessed fMRI data both with and without global signal regression. Meanwhile, QPP2 is mainly composed of cyclical anticorrelation between the default mode and task-positive networks in both humans and rodents [96, 99]. QPP3 can be assessed by regressing out QPP2 and extracting the next most prominent pattern, which typically involves bilateral cortical regions that exhibit a lateral to medial propagation in rodents [75, 95].

Preprocessing of fMRI data in our study was performed using the *Rodent Whole-Brain fMRI Data Processing Toolbox* with the following steps: slice time, motion, and topup correction, nuisance regression (constant, linear, and quadratic trends with six additional motion regressors), normalization, bandpass filtering (0.01–0.1 Hz), spatial smoothing (FWHM = 0.5 mm), and registration to and seed extraction from the functional SIGMA-Wistar atlas [77, 103]. The complete software toolbox and user manual can be found at https://github.com/GT-EmoryMINDlab/rodent-whole-brain-preprocessing-recipe, and a step-by-step video guide can be found at https://www.youtube.com/playlist?list=PLzl6lxEF9yCb3i0Coc5noXWzINKqWXTOd.

For static functional connectivity, we performed Pearson's linear correlation between the time courses of all atlas parcellations using the *Rodent Whole-Brain fMRI Data Preprocessing Toolbox.* Connectivity matrices were made in MATLAB using the averaged Fisher Z-transformed data per group. We further simplified our data by collating regions of interest into canonical networks using the SIGMA-Wistar functional atlas [77, 103].

We used the global signal as a proxy for QPP1, which was calculated by extracting all time courses within the brain and averaging them into a single time course. These global signals were z-scored and concatenated for each group. Pearson's linear correlation was applied between the time course of each voxel within the brain and the global signal. The output for such an analysis is a spatial map of voxel-wise correlations to the global signal. For specific regions of interest, correlation values for different parcellations can be averaged within a group and compared across groups.

QPP2 detection was based on an original report [96] that was later modified [97]. An algorithm was used to define a template comprised of several volumes manually set by the experimenter within a defined window length. A sliding window approach was used to correlate this template to each timepoint of an acquired scan. A threshold is set for the correlation time course, and all points where the correlation between the template and scan exceeds this threshold, the volumes are taken and averaged to create a new template. When the template does not change after two iterations, this process is complete (cc > 0.9999).

A variety of analyses can be done with extracted QPP data, including frequency of QPP occurrence, changes in regional contribution to QPPs, and the strength of contribution of individual brain regions to QPPs as a function of disease, age, and stimulation parameters. The recent release of our software package, QPPLab, will promote standardization for the detection, analysis, and visualization of QPPs for comparison across studies [76]. The entire software package and user manual can be found at https://github.com/imnanxu/QPPLab.

Interestingly, TgF344-AD rats show deficits in both static and dynamic functional connectivity [34–40]. Under normal conditions, activity of subcortical nuclei, including the LC, contributes to both static and dynamic functional connectivity [15, 69, 104–109]. Thus, it will be interesting to determine whether normalizing the activity of the LC in TgF344-AD rats using optogenetics has beneficial effects on brain-wide dynamics.

5 Considerations and Recommendations

With the potential for significant translational relevance for numerous neurological disorders, there are many aspects that need to be considered when performing opto-fMRI studies. As previously stated, viruses can induce changes in brain structure and function [78, 79], but optogenetics itself can induce heat-related artifacts, damage neural tissue, and produce nonspecific effects [13, 110–115]. Therefore, when designing opto-fMRI experiments, it is important to include the proper controls, such as fluorophore-only viruses, as employed here, and/or apply a wavelength of light that does not activate the opsin. Confirmation of optogenetic stimulation should also include a multi-method approach. A new histological marker of neuronal inhibition may assist with confirmation of neuronal inhibition [116], in addition to other techniques like electrophysiology, calcium imaging, and fiber photometry. It may be tempting to interrogate specific pathways in the brain by targeting stimulation to terminals rather than cell bodies. However, this practice needs more investigation, as a few studies have produced mixed results [11, 17], suggesting that the effects of optogenetic stimulation may be region/circuit specific.

For rodent fMRI, there are many variables to consider during setup, data acquisition, preprocessing, and analysis. Standard protocols for animal imaging have been suggested and will assist with comparisons between studies [117, 118], improving translational value. Our suite of software for rodent MRI preprocessing and analysis should assist with such standardization [76, 77]. However, some methods might be incompatible with specific studies. For example, some studies indicate that the use or addition of dexmedetomidine to anesthetic protocols maintains connectivity patterns better than other regimens [119, 120]. However, dexmedetomidine is an α 2-adrenergic receptor agonist that inhibits LC activity and can block the effects of LC stimulation on whole brain functional connectivity

[69]. Therefore, such an anesthetic regimen could not be used in our study. Instead, we chose to use a low dose of isoflurane as our anesthetic, combined with a paralytic to limit motion-induced artifacts. Scanning in awake rodents should be considered, since some studies report more robust optogenetic effects in awake animals [42, 44]. This must be weighed against potential drawbacks and confounds, such as the need for a high amount of animal training [117, 118, 120, 121]. Thus, it is important to consider many of these factors on a study-by-study basis.

While MRI hardware and software will largely depend on what is available at a particular study site, tools to implement optogenetics, especially the choice of opsin variant, can vary widely [122, 123]. The first question to pose is whether the desired effect is neuronal excitation or inhibition. The most common variant for excitation is channelrhodopsin, which exhibits millisecond on-kinetics and can produce reliable action potentials in response to light stimulation [122, 123]. Ultrafast opsins have also been developed, which have been used to improve kinetics, drive activity of highly active neuronal populations, reduce extra spikes caused by single laser pulses, and abolish plateau potentials [122–124]. Step-function opsins can be considered, if prolonged periods of excitation or inhibition are desired due to their slower off-kinetics [123]. Meanwhile, red-shifted opsins have been developed for better light penetration, which could lead to less invasive optogenetic procedures and may be used in combination with other opsins for bidirectional activity modulation [122, 123, 125, 126]. Separately, halorhodopsins and archaerhodopsins can be used for neuronal inhibition, though there are still some concerns over the fidelity of inhibition, post-inhibitory rebound, cellular localization and trafficking, and toxicity [122, 123, 127]. For all these opsin variants and subclasses, it is important to consider the underlying neurochemistry allowing specific neuronal control. For example, halorhodopsins hyperpolarize through an inward chloride pump versus archaerhodopsins that hyperpolarize through an outward proton pump.

For certain studies, it may also be worth considering whether chemogenetics or pharmacology are viable alternatives to optogenetics. These methods are less invasive, have been reported in combination with fMRI [13, 69, 109], and, at least in the case of pharmacology, may be more translationally relevant. The limitations of chemogenetics and pharmacology compared to optogenetics are the lack of precise control of stimulation patterns and timing. It has already been shown that differences in stimulation rate and pattern have diverging effects on fMRI measures, which are important when designing clinical interventions [9, 14, 15, 128–130].

Overall, results from opto-fMRI studies have the potential for significant clinical relevance. In our studies, we aim to uncover molecular and circuit mechanisms of the benefits of LC stimulation and increasing NE levels in a rat model of AD. We have shown cognitive benefits of LC stimulation in TgF344-AD rats, and follow-up studies show promising benefits of increasing NE levels for both AD and Parkinson's patients [30, 128–130]. We propose that opto-fMRI methods be expanded beyond current use in three key categories:

1. *Expansion beyond healthy states and current disease models.* Many of the current studies have focused on using opto-fMRI to understand the brain-wide impact of brain region or cell-type-specific stimulation. Some have moved toward using such methods to understand disease processes. We focus on the positive effects of LC stimulation in a transgenic rat model of AD, while others have shed light on the potential mechanisms underlying deep brain stimulation and beneficial effects of optogenetic stimulation on poststroke recovery, alleviating α-synuclein pathology, and modulating pain [8–11, 21]. This could be further extended to studying other neurological disorders, such as sensorimotor disorders, substance use disorder, neurodegenerative disorders, developmental syndromes, and psychiatric conditions. Opto-fMRI should be deployed in various model systems, including nontraditional and non-rodent species. For example, monogamous prairie voles are emerging as a potential model of prolonged grief disorder, and opto-fMRI may help identify circuits involved in these pathological processes [131–134]. As another example, the use of opto-fMRI in songbirds [135–137], though currently difficult, could be used to dissect systems involved in motor learning and developmental deficits. Meanwhile, nonhuman primate models of disease are in development alongside optogenetic and chemogenetic techniques [26, 58–65], which may enhance translatability to human conditions. This could be particularly useful for understanding neurodegenerative disorders, as nonhuman primates recapitulate processes observed in human aging, such as accumulating neuromelanin in catecholaminergic nuclei and developing protein aggregates (i.e., amyloid, tau) [63, 138–142]. Combining optogenetics with the widespread use of fMRI in nonhuman primates, including monitoring event- and task-related activity [65, 143, 144], could therefore be essential for bridging the translational gap between preclinical animal models and humans.
2. *Deployment of varied optogenetic actuators.* The specific question being posed will likely guide the selection of opsin subclass, but excitatory channelrhodopsins have been the main choice for opto-fMRI thus far. However, the use of other opsin subclasses is warranted for studying neurological disorders using opto-fMRI. For example, pathologies associated with neurodegenerative disorders can cause hyperactivity prior to substantial cell loss [24, 31, 145]. Therefore, the use of inhibitory opsins during fMRI could be useful for understanding

the benefits of mitigating neuronal hyperactivity. However, applications of inhibitory opsins in opto-fMRI are limited in favor of studying the effects of stimulating inhibitory neuronal populations on brain-wide dynamics [21, 53]. Ultrafast opsin variants are another interesting consideration for studies that want to probe the dysfunction of neuronal populations with high spike rates [124, 146–148], such as interneurons in epilepsy and seizures. These explain just a few of the potential applications for opsin variants for studying neurological disorders where different types of neural control are desired.

3. *In-depth analysis of fMRI data beyond static functional connectivity.* BOLD changes during fMRI in response to optogenetic stimulation have been the primary outcome measure of opto-fMRI studies. However, there is a wealth of information to be obtained through fMRI, looking at temporal and spatial features of BOLD functional connectivity. These include analysis of QPPs, spatial distribution of the global signal, and complex principal component, among others [93, 94]. Of note, changes in dynamic functional connectivity have been reported in various neurological disorders [149–152]. Importantly, most of these methods do not require the collection of extra scans or special fMRI sequences and instead can be analyzed with already collected data.

6 Conclusions

Opto-fMRI has been cleverly deployed to study the brain-wide engagement of circuits following stimulation of a single brain region or cell type. It has already shown promise for uncovering the mechanisms and identifying therapeutic targets for various neurological disorders. However, to take full advantage of this complex technique, it should be deployed in more preclinical disease models using different subclasses of opsins while examining measures beyond traditional static functional connectivity. Such expansion should come with equal caution in confirming the quality of collected fMRI data (see [117]), proper optogenetic controls, and validation of stimulation. If all these suggestions are taken into account, we foresee opto-fMRI becoming a staple tool for researching neurological disorders.

7 Notes

1. It is important to be careful when drilling and implanting screws, as these can cause significant damage to the cortical blood supply and brain tissue, potentially inducing changes in

functional and structural signals. Furthermore, any blood that dries around the surface of the skull or implant can produce strong disruptions during imaging and should therefore be cleaned during surgery.

2. If one hemisphere did not yield pupil dilation in ChR2 animals, we probed pupil dilation in response to optogenetic stimulation in the other hemisphere. Animals that did not show pupil dilation following stimulation to either hemisphere were moved to the control groups. This beneficial practice not only helps ensure optogenetic stimulation but also increases efficiency and uses fewer animals. Such practices can also be employed with fiber photometry, electrophysiology, and calcium imaging.
3. Every effort should be made to limit the potential of light-induced artifacts in brain signals. We used a black permanent marker to "blackout" the transparent headcaps, and tape was placed over the connection between the optic fiber and optic ferrule implant to block escaping light. Control groups can also serve as a comparison for the effects of any visual stimulation caused by optogenetic stimulation.
4. We recommend practicing animal preparation such that setup takes between 30 min and 45 min, which will reduce exposure to prolonged anesthesia and result in more stable scans.
5. Our total time from setup start to perfusion took approximately 2–2.5 h per animal and consisted of eight 10-min functional scans. Although longer scan times would afford better resolution, we carefully considered this against the potential deleterious effects of prolonged anesthesia on functional connectivity measures.
6. A comprehensive study investigating the persistent effects of optogenetic stimulation on functional connectivity has not yet been performed, but it will be important for considering the effects of stimulation on ensuing scans. Notably, we did not observe carryover effects of stimulation on subsequent scans following a 3 min recovery period [14].

Acknowledgments

This work was supported by funding from the National Institutes of Aging (AG062581 to D.W. and S.K., AG069502 to M.A.K.), the National Institute of Neurological Disorders and Stroke (NS96050 to M.A.K., NS078095 to S.K.), and the National Institute of Biomedical Imaging and Bioengineering (EB029857 to S.K.).

Conflict of Interest The author has no conflicts of interest to declare that are relevant to the content of this chapter.

References

1. Matthews PM, Honey GD, Bullmore ET (2006) Applications of fMRI in translational medicine and clinical practice. Nat Rev Neurosci 7:732–744. https://doi.org/10.1038/nrn1929
2. Moningka H, Lichenstein S, Worhunsky PD et al (2019) Can neuroimaging help combat the opioid epidemic? A systematic review of clinical and pharmacological challenge fMRI studies with recommendations for future research. Neuropsychopharmacol 44:259–273. https://doi.org/10.1038/s41386-018-0232-4
3. Dennis EL, Thompson PM (2014) Functional brain connectivity using fMRI in aging and Alzheimer's disease. Neuropsychol Rev 24:49–62. https://doi.org/10.1007/s11065-014-9249-6
4. Tessitore A, Cirillo M, De Micco R (2019) Functional connectivity signatures of Parkinson's disease. J Parkinsons Dis 9:637–652. https://doi.org/10.3233/JPD-191592
5. Bohanna I, Georgiou-Karistianis N, Hannan AJ, Egan GF (2008) Magnetic resonance imaging as an approach towards identifying neuropathological biomarkers for Huntington's disease. Brain Res Rev 58:209–225. https://doi.org/10.1016/j.brainresrev.2008.04.001
6. D'Esposito M, Deouell LY, Gazzaley A (2003) Alterations in the BOLD fMRI signal with ageing and disease: a challenge for neuroimaging. Nat Rev Neurosci 4:863–872. https://doi.org/10.1038/nrn1246
7. Lee MH, Smyser CD, Shimony JS (2013) Resting-state fMRI: a review of methods and clinical applications. AJNR Am J Neuroradiol 34:1866–1872. https://doi.org/10.3174/ajnr.A3263
8. Albaugh DL, Salzwedel A, Van Den Berge N et al (2016) Functional magnetic resonance imaging of electrical and optogenetic deep brain stimulation at the rat nucleus accumbens. Sci Rep 6:31613. https://doi.org/10.1038/srep31613
9. Li Y, Lee S-H, Yu C et al (2024) Optogenetic fMRI reveals therapeutic circuits of subthalamic nucleus deep brain stimulation. Brain Stimulation 17:947–957. https://doi.org/10.1016/j.brs.2024.07.022
10. Dadgar-Kiani E, Bieri G, Melki R et al (2024) Neuromodulation modifies α-synuclein spreading dynamics in vivo and the pattern is predicted by changes in whole-brain function. Brain Stimul 17:938–946. https://doi.org/10.1016/j.brs.2024.07.021
11. Vahdat S, Pendharkar AV, Chiang T et al (2021) Brain-wide neural dynamics of poststroke recovery induced by optogenetic stimulation. Sci Adv 7:eabd9465. https://doi.org/10.1126/sciadv.abd9465
12. Xu N, LaGrow TJ, Anumba N et al (2022) Functional connectivity of the brain across rodents and humans. Front Neurosci 16. https://doi.org/10.3389/fnins.2022.816331
13. Mandino F, Cerri DH, Garin CM et al (2020) Animal functional magnetic resonance imaging: trends and path toward standardization. Front Neuroinform 13. https://doi.org/10.3389/fninf.2019.00078
14. Anumba N, Kelberman MA, Pan W et al (2024) The effects of locus coeruleus optogenetic stimulation on global spatiotemporal patterns in rats. Imaging Neurosci. https://doi.org/10.1162/imag_a_00314
15. Grimm C, Duss SN, Privitera M et al (2024) Tonic and burst-like locus coeruleus stimulation distinctly shift network activity across the cortical hierarchy. Nat Neurosci 27:2167–2177. https://doi.org/10.1038/s41593-024-01755-8
16. Hamada HT, Abe Y, Takata N et al (2024) Optogenetic activation of dorsal raphe serotonin neurons induces brain-wide activation. Nat Commun 15:4152. https://doi.org/10.1038/s41467-024-48489-6
17. Hsu L-M, Cerri DH, Carelli RM, Shih Y-YI (2025) Optogenetic stimulation of cell bodies versus axonal terminals generate comparable activity and functional connectivity patterns in the brain. Brain Stimul. https://doi.org/10.1016/j.brs.2025.03.006
18. Grandjean J, Corcoba A, Kahn MC et al (2019) A brain-wide functional map of the serotonergic responses to acute stress and fluoxetine. Nat Commun 10:350. https://doi.org/10.1038/s41467-018-08256-w
19. Asaad M, Lee JH (2018) A guide to using functional magnetic resonance imaging to study Alzheimer's disease in animal models. Dis Model Mech 11:dmm031724. https://doi.org/10.1242/dmm.031724
20. Lee JH, Liu Q, Dadgar-Kiani E (2022) Solving brain circuit function and dysfunction with computational modeling and optogenetic fMRI. Science 378:493–499. https://doi.org/10.1126/science.abq3868
21. Lee J-Y, You T, Lee C-H et al (2022) Role of anterior cingulate cortex inputs to periaqueductal gray for pain avoidance. Curr Biol 32:2834–2847.e5. https://doi.org/10.1016/j.cub.2022.04.090
22. Ehrenberg AJ, Kelberman MA, Liu KY et al (2023) Priorities for research on

neuromodulatory subcortical systems in Alzheimer's disease: position paper from the NSS PIA of ISTAART. Alzheimers Dement 19:2182–2196. https://doi.org/10.1002/alz.12937
23. Poe GR, Foote S, Eschenko O et al (2020) Locus coeruleus: a new look at the blue spot. Nat Rev Neurosci 21:644–659. https://doi.org/10.1038/s41583-020-0360-9
24. Weinshenker D (2018) Long road to ruin: noradrenergic dysfunction in neurodegenerative disease. Trends Neurosci 41:211–223. https://doi.org/10.1016/j.tins.2018.01.010
25. Engels-Domínguez N, Koops EA, Prokopiou PC et al (2023) State-of-the-art imaging of neuromodulatory subcortical systems in aging and Alzheimer's disease: challenges and opportunities. Neurosci Biobehav Rev 144:104998. https://doi.org/10.1016/j.neubiorev.2022.104998
26. Kelberman M, Keilholz S, Weinshenker D (2020) What's that (blue) spot on my MRI? Multimodal neuroimaging of the locus coeruleus in neurodegenerative disease. Front Neurosci 14:583421. https://doi.org/10.3389/fnins.2020.583421
27. Betts MJ, Kirilina E, Otaduy MCG et al (2019) Locus coeruleus imaging as a biomarker for noradrenergic dysfunction in neurodegenerative diseases. Brain 142:2558–2571. https://doi.org/10.1093/brain/awz193
28. Tillage RP, Foster SL, Lustberg D et al (2021) Co-released norepinephrine and galanin act on different timescales to promote stress-induced anxiety-like behavior. Neuropsychopharmacology 46:1535–1543. https://doi.org/10.1038/s41386-021-01011-8
29. Cohen RM, Rezai-Zadeh K, Weitz TM et al (2013) A transgenic Alzheimer rat with plaques, tau pathology, behavioral impairment, oligomeric aβ, and frank neuronal loss. J Neurosci 33:6245–6256. https://doi.org/10.1523/JNEUROSCI.3672-12.2013
30. Rorabaugh JM, Chalermpalanupap T, Botz-Zapp CAet al (2017) Chemogenetic locus coeruleus activation restores reversal learning in a rat model of Alzheimer's disease. Brain 140:3023–3038. https://doi.org/10.1093/brain/awx232
31. Kelberman MA, Rorabaugh JM, Anderson CR et al (2023) Age-dependent dysregulation of locus coeruleus firing in a transgenic rat model of Alzheimer's disease. Neurobiol Aging 125:98–108. https://doi.org/10.1016/j.neurobiolaging.2023.01.016
32. Kelberman MA, Anderson CR, Chlan E et al (2022) Consequences of hyperphosphorylated tau in the locus coeruleus on behavior and cognition in a rat model of Alzheimer's disease. J Alzheimers Dis 86:1037–1059. https://doi.org/10.3233/JAD-215546
33. Kelberman M, McCann K, Hurst CD et al (2023) Transcriptomic and proteomic characterization of the locus coeruleus in a rat model of Alzheimer's disease. Alzheimers Dement 19:e075403. https://doi.org/10.1002/alz.075403
34. Rezaei A, van den Berg M, Mirlohi H et al (2024) Recurrence quantification analysis of rs-fMRI data: a method to detect subtle changes in the TgF344-AD rat model. Comput Methods Programs Biomed 257:108378. https://doi.org/10.1016/j.cmpb.2024.108378
35. van den Berg M, Adhikari MH, Verschuuren M et al (2022) Altered basal forebrain function during whole-brain network activity at pre- and early-plaque stages of Alzheimer's disease in TgF344-AD rats. Alzheimer's Res Ther 14:148. https://doi.org/10.1186/s13195-022-01089-2
36. Anckaerts C, Blockx I, Summer P et al (2019) Early functional connectivity deficits and progressive microstructural alterations in the TgF344-AD rat model of Alzheimer's disease: a longitudinal MRI study. Neurobiol Dis 124:93–107. https://doi.org/10.1016/j.nbd.2018.11.010
37. Tudela R, Muñoz-Moreno E, Sala-Llonch R et al (2019) Resting state networks in the TgF344-AD rat model of Alzheimer's disease are altered from early stages. Front Aging Neurosci 11:213.
38. De Waegenaere S, van den Berg M, Keliris GA et al (2024) Early altered directionality of resting brain network state transitions in the TgF344-AD rat model of Alzheimer's disease. Front Hum Neurosci 18:1379923. https://doi.org/10.3389/fnhum.2024.1379923
39. Amiri S, van den Berg M, Nazem-Zadeh M-R et al (2024) Nodal degree centrality in the default mode-like network of the TgF344-AD Alzheimer's disease rat model as a measure of early network alterations. NPJ Aging 10:1–11. https://doi.org/10.1038/s41514-024-00151-7
40. Muñoz-Moreno E, Tudela R, López-Gil X, Soria G (2018) Early brain connectivity alterations and cognitive impairment in a rat model of Alzheimer's disease. Alzheimer's Res Ther 10:16. https://doi.org/10.1186/s13195-018-0346-2
41. Takata N, Yoshida K, Komaki Y et al (2015) Optogenetic activation of CA1 pyramidal neurons at the dorsal and ventral hippocampus evokes distinct brain-wide responses revealed

by mouse fMRI. PLoS One 10:e0121417. https://doi.org/10.1371/journal.pone.0121417
42. Liang Z, Watson GDR, Alloway KD et al (2015) Mapping the functional network of medial prefrontal cortex by combining optogenetics and fMRI in awake rats. Neuroimage 117:114–123. https://doi.org/10.1016/j.neuroimage.2015.05.036
43. Abe Y, Sekino M, Terazono Y et al (2012) Opto-fMRI analysis for exploring the neuronal connectivity of the hippocampal formation in rats. Neurosci Res 74:248–255. https://doi.org/10.1016/j.neures.2012.08.007
44. Desai M, Kahn I, Knoblich U et al (2011) Mapping brain networks in awake mice using combined optical neural control and fMRI. J Neurophysiol 105:1393–1405. https://doi.org/10.1152/jn.00828.2010
45. Menon V, Cerri D, Lee B et al (2023) Optogenetic stimulation of anterior insular cortex neurons in male rats reveals causal mechanisms underlying suppression of the default mode network by the salience network. Nat Commun 14:866. https://doi.org/10.1038/s41467-023-36616-8
46. Kim S, Moon HS, Vo TT et al (2023) Whole-brain mapping of effective connectivity by fMRI with cortex-wide patterned optogenetics. Neuron 111:1732–1747.e6. https://doi.org/10.1016/j.neuron.2023.03.002
47. Brocka M, Helbing C, Vincenz D et al (2018) Contributions of dopaminergic and non-dopaminergic neurons to VTA-stimulation induced neurovascular responses in brain reward circuits. Neuroimage 177:88–97. https://doi.org/10.1016/j.neuroimage.2018.04.059
48. Lohani S, Poplawsky AJ, Kim S-G, Moghaddam B (2017) Unexpected global impact of VTA dopamine neuron activation as measured by opto-fMRI. Mol Psychiatry 22:585–594. https://doi.org/10.1038/mp.2016.102
49. Weitz AJ, Fang Z, Lee HJ et al (2015) Optogenetic fMRI reveals distinct, frequency-dependent networks recruited by dorsal and intermediate hippocampus stimulations. Neuroimage 107:229–241. https://doi.org/10.1016/j.neuroimage.2014.10.039
50. Lee J-Y, You T, Woo C-W, Kim S-G (2022) Optogenetic fMRI for brain-wide circuit analysis of sensory processing. Int J Mol Sci 23:12268. https://doi.org/10.3390/ijms232012268
51. Leong ATL, Dong CM, Gao PP et al (2018) Optogenetic auditory fMRI reveals the effects of visual cortical inputs on auditory midbrain response. Sci Rep 8:8736. https://doi.org/10.1038/s41598-018-26568-1
52. Lee JH, Durand R, Gradinaru V et al (2010) Global and local fMRI signals driven by neurons defined optogenetically by type and wiring. Nature 465:788–792. https://doi.org/10.1038/nature09108
53. Jung WB, Jiang H, Lee S, Kim S-G (2022) Dissection of brain-wide resting-state and functional somatosensory circuits by fMRI with optogenetic silencing. Proc Natl Acad Sci 119:e2113313119. https://doi.org/10.1073/pnas.2113313119
54. Lin P, Fang Z, Liu J, Lee JH (2016) Optogenetic functional MRI. J Vis Exp 53346. https://doi.org/10.3791/53346
55. Shim H-J, Im GH, Jung WB et al (2022) Protocol for mouse optogenetic fMRI at ultrahigh magnetic fields. STAR Protocols 3:101846. https://doi.org/10.1016/j.xpro.2022.101846
56. Drummond E, Wisniewski T (2017) Alzheimer's disease: experimental models and reality. Acta Neuropathol 133:155–175. https://doi.org/10.1007/s00401-016-1662-x
57. Jankowsky JL, Zheng H (2017) Practical considerations for choosing a mouse model of Alzheimer's disease. Mol Neurodegener 12:89. https://doi.org/10.1186/s13024-017-0231-7
58. Galvan A, Caiola MJ, Albaugh DL (2018) Advances in optogenetic and chemogenetic methods to study brain circuits in non-human primates. J Neural Transm (Vienna) 125:547–563. https://doi.org/10.1007/s00702-017-1697-8
59. Tremblay S, Acker L, Afraz A et al (2020) An open resource for non-human primate optogenetics. Neuron 108:1075–1090.e6. https://doi.org/10.1016/j.neuron.2020.09.027
60. Galvan A, Stauffer WR, Acker L et al (2017) Nonhuman primate optogenetics: recent advances and future directions. J Neurosci 37:10894–10903. https://doi.org/10.1523/JNEUROSCI.1839-17.2017
61. Emborg ME (2017) Nonhuman primate models of neurodegenerative disorders. ILAR J 58:190–201. https://doi.org/10.1093/ilar/ilx021
62. Marmion DJ, Kordower JH (2018) α-Synuclein nonhuman primate models of Parkinson's disease. J Neural Transm (Vienna) 125:385–400. https://doi.org/10.1007/s00702-017-1720-0
63. Arnsten AFT, Datta D, Leslie S et al (2019) Alzheimer's-like pathology in aging rhesus macaques: unique opportunity to study the etiology and treatment of Alzheimer's disease. Proc Natl Acad Sci 116:26230–26238. https://doi.org/10.1073/pnas.1903671116

64. Seita Y, Morimura T, Watanabe N et al (2020) Generation of transgenic cynomolgus monkeys overexpressing the gene for amyloid-β precursor protein. J Alzheimers Dis 75:45–60. https://doi.org/10.3233/JAD-191081
65. Chen Y, Menegas W, Zhang Q, Feng G (2025) Common marmoset: an emerging non-human primate model for translational applications in brain disorders. Curr Opin Neurobiol 92:102998. https://doi.org/10.1016/j.conb.2025.102998
66. Abbott SBG, Stornetta RL, Fortuna MG et al (2009) Photostimulation of retrotrapezoid nucleus phox2b-expressing neurons in vivo produces long-lasting activation of breathing in rats. J Neurosci 29:5806–5819. https://doi.org/10.1523/JNEUROSCI.1106-09.2009
67. Hwang DY, Carlezon WA, Isacson O, Kim KS (2001) A high-efficiency synthetic promoter that drives transgene expression selectively in noradrenergic neurons. Hum Gene Ther 12:1731–1740. https://doi.org/10.1089/104303401750476230
68. Vazey EM, Aston-Jones G (2014) Designer receptor manipulations reveal a role of the locus coeruleus noradrenergic system in isoflurane general anesthesia. Proc Natl Acad Sci USA 111:3859–3864. https://doi.org/10.1073/pnas.1310025111
69. Zerbi V, Floriou-Servou A, Markicevic M et al (2019) Rapid reconfiguration of the functional connectome after chemogenetic locus coeruleus activation. Neuron 103:702–718.e5. https://doi.org/10.1016/j.neuron.2019.05.034
70. Scott NA, Murphy TH (2012) Hemodynamic responses evoked by neuronal stimulation via channelrhodopsin-2 can be independent of intracortical glutamatergic synaptic transmission. PLoS One 7:e29859. https://doi.org/10.1371/journal.pone.0029859
71. Liu Y, Rodenkirch C, Moskowitz N et al (2017) Dynamic lateralization of pupil dilation evoked by locus coeruleus activation results from sympathetic, not parasympathetic, contributions. Cell Rep 20:3099–3112. https://doi.org/10.1016/j.celrep.2017.08.094
72. Privitera M, Ferrari KD, von Ziegler LM et al (2020) A complete pupillometry toolbox for real-time monitoring of locus coeruleus activity in rodents. Nat Protoc 15:2301–2320. https://doi.org/10.1038/s41596-020-0324-6
73. Jung WB, Im GH, Jiang H, Kim S-G (2021) Early fMRI responses to somatosensory and optogenetic stimulation reflect neural information flow. Proc Natl Acad Sci 118:e2023265118. https://doi.org/10.1073/pnas.2023265118
74. Roussel T, Rosenberg JT, Grant SC, Frydman L (2018) Brain investigations of rodent disease models by chemical exchange saturation transfer at 21.1T. NMR Biomed 31:e3995. https://doi.org/10.1002/nbm.3995
75. Meyer-Baese L, Anumba N, Bolt T et al (2024) Variation in the distribution of large-scale spatiotemporal patterns of activity across brain states. Front Syst Neurosci 18. https://doi.org/10.3389/fnsys.2024.1425491
76. Xu N, Yousefi B, Anumba N et al (2025) QPPLab: a generally applicable software package for detecting, analyzing, and visualizing large-scale quasiperiodic spatiotemporal patterns (QPPs) of brain activity. SoftwareX 29:102067. https://doi.org/10.1016/j.softx.2025.102067
77. Xu N, Zhang L, Larson S et al (2023) Rodent whole-brain fMRI data preprocessing toolbox. Aperture Neuro 3:1–3. https://doi.org/10.52294/001c.85075
78. Miyashita T, Shao YR, Chung J et al (2013) Long-term channelrhodopsin-2 (ChR2) expression can induce abnormal axonal morphology and targeting in cerebral cortex. Front Neural Circuits 7. https://doi.org/10.3389/fncir.2013.00008
79. Kabanova A, Yang M, Logothetis NK, Eschenko O (2024) Partial chemogenetic inhibition of the locus coeruleus due to heterogeneous transduction of noradrenergic neurons preserved auditory salience processing in wild-type rats. Eur J Neurosci 60:6237–6253. https://doi.org/10.1111/ejn.16550
80. Lustberg D, Tillage RP, Bai Y et al (2020) Noradrenergic circuits in the forebrain control affective responses to novelty. Psychopharmacology 237:3337–3355. https://doi.org/10.1007/s00213-020-05615-8
81. Tillage RP, Wilson GE, Liles LC et al (2020) Chronic environmental or genetic elevation of galanin in noradrenergic neurons confers stress resilience in mice. J Neurosci 40:7464–7474. https://doi.org/10.1523/JNEUROSCI.0973-20.2020
82. Pan W-J, Khalizad Sharghi V, Zhang X, Keilholz S (2020) Brain mechanism of anesthesia and sedation: fMRI functional connectivity study with minimized impact of physiological background noise in rats. ISMRM. https://doi.org/https://archive.ismrm.org/2020/3958.html
83. Smith SM, Jenkinson M, Woolrich MW et al (2004) Advances in functional and structural MR image analysis and implementation as FSL. Neuroimage 23:S208–S219. https://doi.org/10.1016/j.neuroimage.2004.07.051

84. Andersson JLR, Skare S, Ashburner J (2003) How to correct susceptibility distortions in spin-echo echo-planar images: application to diffusion tensor imaging. Neuroimage 20:870–888. https://doi.org/10.1016/S1053-8119(03)00336-7
85. Aston-Jones G, Bloom FE (1981) Activity of norepinephrine-containing locus coeruleus neurons in behaving rats anticipates fluctuations in the sleep-waking cycle. J Neurosci 1:876–886. https://doi.org/10.1523/JNEUROSCI.01-08-00876.1981
86. Aston-Jones G, Cohen JD (2005) An integrative theory of locus coeruleus-norepinephrine function: adaptive gain and optimal performance. Annu Rev Neurosci 28:403–450. https://doi.org/10.1146/annurev.neuro.28.061604.135709
87. Berridge CW, Waterhouse BD (2003) The locus coeruleus–noradrenergic system: modulation of behavioral state and state-dependent cognitive processes. Brain Res Rev 42:33–84. https://doi.org/10.1016/S0165-0173(03)00143-7
88. Prokopiou PC, Engels-Domínguez N, Papp KV et al (2022) Lower novelty-related locus coeruleus function is associated with Aβ-related cognitive decline in clinically healthy individuals. Nat Commun 13:1571. https://doi.org/10.1038/s41467-022-28986-2
89. Valentino RJ, Foote SL (1988) Corticotropin-releasing hormone increases tonic but not sensory-evoked activity of noradrenergic locus coeruleus neurons in unanesthetized rats. J Neurosci 8:1016–1025. https://doi.org/10.1523/JNEUROSCI.08-03-01016.1988
90. McCall JG, Al-Hasani R, Siuda ER et al (2015) CRH engagement of the locus coeruleus noradrenergic system mediates stress-induced anxiety. Neuron 87:605–620. https://doi.org/10.1016/j.neuron.2015.07.002
91. Carter ME, Yizhar O, Chikahisa S et al (2010) Tuning arousal with optogenetic modulation of locus coeruleus neurons. Nat Neurosci 13:1526–1533. https://doi.org/10.1038/nn.2682
92. Li L, Rana AN, Li EM et al (2023) Activity-dependent constraints on catecholamine signaling. Cell Rep 42. https://doi.org/10.1016/j.celrep.2023.113566
93. Hutchison RM, Womelsdorf T, Allen EA et al (2013) Dynamic functional connectivity: promise, issues, and interpretations. Neuroimage 80:360–378. https://doi.org/10.1016/j.neuroimage.2013.05.079
94. Lurie DJ, Kessler D, Bassett DS et al (2020) Questions and controversies in the study of time-varying functional connectivity in resting fMRI. Network Neuroscience 4:30–69. https://doi.org/10.1162/netn_a_00116
95. Bolt T, Nomi JS, Bzdok D et al (2022) A parsimonious description of global functional brain organization in three spatiotemporal patterns. Nat Neurosci 25:1093–1103. https://doi.org/10.1038/s41593-022-01118-1
96. Majeed W, Magnuson M, Hasenkamp W et al (2011) Spatiotemporal dynamics of low frequency BOLD fluctuations in rats and humans. Neuroimage 54:1140–1150. https://doi.org/10.1016/j.neuroimage.2010.08.030
97. Yousefi B, Shin J, Schumacher EH, Keilholz SD (2018) Quasi-periodic patterns of intrinsic brain activity in individuals and their relationship to global signal. Neuroimage 167:297–308. https://doi.org/10.1016/j.neuroimage.2017.11.043
98. Yousefi B, Keilholz S (2021) Propagating patterns of intrinsic activity along macroscale gradients coordinate functional connections across the whole brain. Neuroimage 231:117827. https://doi.org/10.1016/j.neuroimage.2021.117827
99. Belloy ME, Naeyaert M, Abbas A et al (2018) Dynamic resting state fMRI analysis in mice reveals a set of quasi-periodic patterns and illustrates their relationship with the global signal. Neuroimage 180:463–484. https://doi.org/10.1016/j.neuroimage.2018.01.075
100. Liu TT, Nalci A, Falahpour M (2017) The global signal in fMRI: nuisance or information. Neuroimage 150:213–229. https://doi.org/10.1016/j.neuroimage.2017.02.036
101. Murphy K, Fox MD (2017) Towards a consensus regarding global signal regression for resting state functional connectivity MRI. Neuroimage 154:169–173. https://doi.org/10.1016/j.neuroimage.2016.11.052
102. Anumba N, Maltbie E, Pan W-J et al (2023) Spatial and spectral components of the BOLD global signal in rat resting-state functional MRI. Magn Reson Med 90:2486–2499. https://doi.org/10.1002/mrm.29824
103. Barrière DA, Magalhães R, Novais A et al (2019) The SIGMA rat brain templates and atlases for multimodal MRI data analysis and visualization. Nat Commun 10:5699. https://doi.org/10.1038/s41467-019-13575-7
104. Shine JM (2019) Neuromodulatory influences on integration and segregation in the brain. Trends Cogn Sci 23:572–583. https://doi.org/10.1016/j.tics.2019.04.002
105. Abbas A, Belloy M, Kashyap A et al (2019) Quasi-periodic patterns contribute to functional connectivity in the brain. Neuroimage 191:193–204. https://doi.org/10.1016/j.neuroimage.2019.01.076

106. Turchi J, Chang C, Ye FQ et al (2018) The basal forebrain regulates global resting-state fMRI fluctuations. Neuron 97:940–952.e4. https://doi.org/10.1016/j.neuron.2018.01.032
107. van den Brink RL, Pfeffer T, Donner TH (2019) Brainstem modulation of large-scale intrinsic cortical activity correlations. Front Hum Neurosci 13.
108. Bekar LK, Wei HS, Nedergaard M (2012) The locus coeruleus-norepinephrine network optimizes coupling of cerebral blood volume with oxygen demand. J Cereb Blood Flow Metab 32:2135–2145. https://doi.org/10.1038/jcbfm.2012.115
109. Oyarzabal EA, Hsu L-M, Das M et al (2021) Chemogenetic stimulation of tonic locus coeruleus activity strengthens the default mode network.
110. Acharya AR, Vandekerckhove B, Larsen LE et al (2021) In vivo blue light illumination for optogenetic inhibition: effect on local temperature and excitability of the rat hippocampus. J Neural Eng 18:066038. https://doi.org/10.1088/1741-2552/ac3ef4
111. Otchy TM, Wolff SBE, Rhee JY et al (2015) Acute off-target effects of neural circuit manipulations. Nature 528:358–363. https://doi.org/10.1038/nature16442
112. Christie IN, Wells JA, Southern P et al (2013) fMRI response to blue light delivery in the naïve brain: implications for combined optogenetic fMRI studies. Neuroimage 66:634–641. https://doi.org/10.1016/j.neuroimage.2012.10.074
113. Stujenske JM, Spellman T, Gordon JA (2015) Modeling the spatiotemporal dynamics of light and heat propagation for in vivo optogenetics. Cell Rep 12:525–534. https://doi.org/10.1016/j.celrep.2015.06.036
114. Schmid F, Wachsmuth L, Schwalm M et al (2016) Assessing sensory versus optogenetic network activation by combining (o)fMRI with optical Ca2+ recordings. J Cereb Blood Flow Metab 36:1885–1900. https://doi.org/10.1177/0271678X15619428
115. Rungta RL, Osmanski B-F, Boido D et al (2017) Light controls cerebral blood flow in naive animals. Nat Commun 8:14191. https://doi.org/10.1038/ncomms14191
116. Yang D, Wang Y, Qi T et al (2024) Phosphorylation of pyruvate dehydrogenase inversely associates with neuronal activity. Neuron 112:959–971.e8. https://doi.org/10.1016/j.neuron.2023.12.015
117. Grandjean J, Desrosiers-Gregoire G, Anckaerts C et al (2023) A consensus protocol for functional connectivity analysis in the rat brain. Nat Neurosci 26:673–681. https://doi.org/10.1038/s41593-023-01286-8
118. Gozzi A, Stuefer A, Alvino FG et al (2025) Charting the path in rodent functional neuroimaging. Imaging Neurosci 3:IMAG.a.12. https://doi.org/10.1162/IMAG.a.12
119. Grandjean J, Schroeter A, Batata I, Rudin M (2014) Optimization of anesthesia protocol for resting-state fMRI in mice based on differential effects of anesthetics on functional connectivity patterns. Neuroimage 102:838–847. https://doi.org/10.1016/j.neuroimage.2014.08.043
120. Pan W-J, Billings JCW, Grooms JK et al (2015) Considerations for resting state functional MRI and functional connectivity studies in rodents. Front Neurosci 9. https://doi.org/10.3389/fnins.2015.00269
121. Vrooman RM, van den Berg M, Desrosiers-Gregoire G et al (2025) fMRI data acquisition and analysis for task-free, anesthetized rats. Nat Protoc 1–20. https://doi.org/10.1038/s41596-024-01110-y
122. Joshi J, Rubart M, Zhu W (2020) Optogenetics: background, methodological advances and potential applications for cardiovascular research and medicine. Front Bioeng Biotechnol 7. https://doi.org/10.3389/fbioe.2019.00466
123. Guru A, Post RJ, Ho Y-Y, Warden MR (2015) Making sense of optogenetics. Int J Neuropsychopharmacology 18:pyv079. https://doi.org/10.1093/ijnp/pyv079
124. Gunaydin LA, Yizhar O, Berndt A et al (2010) Ultrafast optogenetic control. Nat Neurosci 13:387–392. https://doi.org/10.1038/nn.2495
125. Lin JY, Knutsen PM, Muller A et al (2013) ReaChR: a red-shifted variant of channelrhodopsin enables deep transcranial optogenetic excitation. Nat Neurosci 16:1499–1508. https://doi.org/10.1038/nn.3502
126. Mermet-Joret N, Moreno A, Zbela A et al (2025) Dual-color optical activation and suppression of neurons with high temporal precision. eLife 12. https://doi.org/10.7554/eLife.90327.2
127. Li N, Chen S, Guo ZV et al (2019) Spatiotemporal constraints on optogenetic inactivation in cortical circuits. eLife 8:e48622. https://doi.org/10.7554/eLife.48622
128. Levey AI, Qiu D, Zhao L et al (2022) A phase II study repurposing atomoxetine for neuroprotection in mild cognitive impairment. Brain 145:1924–1938. https://doi.org/10.1093/brain/awab452
129. Dammer EB, Shantaraman A, Ping L et al (2024) Proteomic analysis of Alzheimer's disease cerebrospinal fluid reveals alterations associated with APOE ε4 and atomoxetine treatment. Sci Transl Med 16:eadn3504.

https://doi.org/10.1126/scitranslmed.adn3504

130. O'Callaghan C, Hezemans FH, Ye R et al (2021) Locus coeruleus integrity and the effect of atomoxetine on response inhibition in Parkinson's disease. Brain 144:2513–2526. https://doi.org/10.1093/brain/awab142
131. Sadino JM, Donaldson ZR (2024) Prairie voles as a model for adaptive reward remodeling following loss of a bonded partner. Ann NY Acad Sci 1535:20–30. https://doi.org/10.1111/nyas.15134
132. Donaldson ZR, Shear MK (2024) Neurobiology and treatment advances for prolonged grief disorder. Neuropsychopharmacology 49:309–310. https://doi.org/10.1038/s41386-023-01663-8
133. Yee JR, Kenkel WM, Kulkarni P et al (2016) BOLD fMRI in awake prairie voles: a platform for translational social and affective neuroscience. Neuroimage 138:221–232. https://doi.org/10.1016/j.neuroimage.2016.05.046
134. Ortiz JJ, Portillo W, Paredes RG et al (2018) Resting state brain networks in the prairie vole. Sci Rep 8:1231. https://doi.org/10.1038/s41598-017-17610-9
135. Van Ruijssevelt L, Van der Kant A, De Groof G, Van der Linden A (2013) Current state-of-the-art of auditory functional MRI (fMRI) on zebra finches: technique and scientific achievements. J Physiol Paris 107:156–169. https://doi.org/10.1016/j.jphysparis.2012.08.005
136. Van Ruijssevelt L, De Groof G, Van der Kant A et al (2013) Functional magnetic resonance imaging (fMRI) with auditory stimulation in songbirds. J Vis Exp 4369. https://doi.org/10.3791/4369
137. Ruijssevelt LV, Chen Y, Eugen K et al (2018) fMRI reveals a novel region for evaluating acoustic information for mate choice in a female songbird. Curr Biol 28:711–721.e6. https://doi.org/10.1016/j.cub.2018.01.048
138. Li H-W, Zhang L, Qin C (2019) Current state of research on non-human primate models of Alzheimer's disease. Anim Models Exp Med 2:227–238. https://doi.org/10.1002/ame2.12092
139. Adler A (1942) Melanin pigment in the brain of the gorilla. J Comp Neurol 76:501–507. https://doi.org/10.1002/cne.900760309
140. Herrero MT, Hirsch EC, Kastner A et al (1993) Does neuromelanin contribute to the vulnerability of catecholaminergic neurons in monkeys intoxicated with MPTP? Neuroscience 56:499–511. https://doi.org/10.1016/0306-4522(93)90349-K
141. Scherer HJ (1939) Melanin pigmentation of the substantia nigra in primates. J Comp Neurol 71:91–98. https://doi.org/10.1002/cne.900710106
142. McCormack AL, Di Monte DA, Delfani K et al (2004) Aging of the nigrostriatal system in the squirrel monkey. J Comp Neurol 471:387–395. https://doi.org/10.1002/cne.20036
143. Pelekanos V, Mok RM, Joly O et al (2020) Rapid event-related, BOLD fMRI, non-human primates (NHP): choose two out of three. Sci Rep 10:7485. https://doi.org/10.1038/s41598-020-64376-8
144. Schaeffer DJ, Gilbert KM, Hori Y et al (2019) Task-based fMRI of a free-viewing visuo-saccadic network in the marmoset monkey. Neuroimage 202:116147. https://doi.org/10.1016/j.neuroimage.2019.116147
145. Targa Dias Anastacio H, Matosin N, Ooi L (2022) Neuronal hyperexcitability in Alzheimer's disease: what are the drivers behind this aberrant phenotype? Transl Psychiatry 12:257. https://doi.org/10.1038/s41398-022-02024-7
146. Marín O (2012) Interneuron dysfunction in psychiatric disorders. Nat Rev Neurosci 13:107–120. https://doi.org/10.1038/nrn3155
147. Leitch B (2024) Parvalbumin interneuron dysfunction in neurological disorders: focus on epilepsy and Alzheimer's disease. Int J Mol Sci 25:5549. https://doi.org/10.3390/ijms25105549
148. Magloire V, Mercier MS, Kullmann DM, Pavlov I (2019) GABAergic interneurons in seizures: investigating causality with optogenetics. Neuroscientist 25:344–358. https://doi.org/10.1177/1073858418805002
149. Fiorenzato E, Strafella AP, Kim J et al (2019) Dynamic functional connectivity changes associated with dementia in Parkinson's disease. Brain 142:2860–2872. https://doi.org/10.1093/brain/awz192
150. Demirtaş M, Tornador C, Falcón C et al (2016) Dynamic functional connectivity reveals altered variability in functional connectivity among patients with major depressive disorder. Hum Brain Mapp 37:2918–2930. https://doi.org/10.1002/hbm.23215
151. Abdallah M, Zahr NM, Saranathan M et al (2021) Altered cerebro-cerebellar dynamic functional connectivity in alcohol use disorder: a resting-state fMRI study. Cerebellum 20:823–835. https://doi.org/10.1007/s12311-021-01241-y
152. Abbas A, Bassil Y, Keilholz S (2019) Quasi-periodic patterns of brain activity in individuals with attention-deficit/hyperactivity disorder. Neuroimage Clin 21:101653. https://doi.org/10.1016/j.nicl.2019.101653

Subject Index

A

B

C

D

E

F

Kevin Chuen Wing Chan and Seong-Gi Kim (eds.), *Optogenetic Functional Neuroimaging*, Neuromethods, vol. 232, https://doi.org/10.1007/978-1-0716-5178-0,

MIX
Papier aus verantwortungsvollen Quellen
Paper from responsible sources
FSC® C105338

If you have any concerns about our products,
you can contact us on
ProductSafety@springernature.com

In case Publisher is established outside the EU,
the EU authorized representative is:
Springer Nature Customer Service Center GmbH
Europaplatz 3, 69115 Heidelberg, Germany

Printed by Libri Plureos GmbH
in Hamburg, Germany